Handbook of
Preformulation

Handbook of Preformulation

Chemical, Biological, and Botanical Drugs

Sarfaraz K. Niazi
*Pharmaceutical Scientist Inc.
Deerfield, Illinois, U.S.A.*

informa
healthcare

New York London

First published in 2007 by Informa Healthcare, Telephone House, 69-77 Paul Street, London EC2A 4LQ, UK.

Simultaneously published in the USA by Informa Healthcare, 52 Vanderbilt Avenue, 7th Floor, New York, NY 10017, USA.

Informa Healthcare is a trading division of Informa UK Ltd. Registered Office: 37–41 Mortimer Street, London W1T 3JH, UK. Registered in England and Wales number 1072954.

©2007 Informa Healthcare, except as otherwise indicated

No claim to original U.S. Government works

Reprinted material is quoted with permission. Although every effort has been made to ensure that all owners of copyright material have been acknowledged in this publication, we would be glad to acknowledge in subsequent reprints or editions any omissions brought to our attention.

All rights reserved. No part of this publication may be reproduced, stored in a retrieval system, or transmitted, in any form or by any means, electronic, mechanical, photocopying, recording, or otherwise, unless with the prior written permission of the publisher or in accordance with the provisions of the Copyright, Designs and Patents Act 1988 or under the terms of any licence permitting limited copying issued by the Copyright Licensing Agency, 90 Tottenham Court Road, London W1P 0LP, UK, or the Copyright Clearance Center, Inc., 222 Rosewood Drive, Danvers, MA 01923, USA (http://www.copyright.com/ or telephone 978-750-8400).

Product or corporate names may be trademarks or registered trademarks, and are used only for identification and explanation without intent to infringe.

This book contains information from reputable sources and although reasonable efforts have been made to publish accurate information, the publisher makes no warranties (either express or implied) as to the accuracy or fitness for a particular purpose of the information or advice contained herein. The publisher wishes to make it clear that any views or opinions expressed in this book by individual authors or contributors are their personal views and opinions and do not necessarily reflect the views/opinions of the publisher. Any information or guidance contained in this book is intended for use solely by medical professionals strictly as a supplement to the medical professional's own judgement, knowledge of the patient's medical history, relevant manufacturer's instructions and the appropriate best practice guidelines. Because of the rapid advances in medical science, any information or advice on dosages, procedures, or diagnoses should be independently verified. This book does not indicate whether a particular treatment is appropriate or suitable for a particular individual. Ultimately it is the sole responsibility of the medical professional to make his or her own professional judgements, so as appropriately to advise and treat patients. Save for death or personal injury caused by the publisher's negligence and to the fullest extent otherwise permitted by law, neither the publisher nor any person engaged or employed by the publisher shall be responsible or liable for any loss, injury or damage caused to any person or property arising in any way from the use of this book.

A CIP record for this book is available from the British Library.

Library of Congress Cataloging-in-Publication Data available on application

ISBN-13: 9780849371936

Orders may be sent to: Informa Healthcare, Sheepen Place, Colchester, Essex CO3 3LP, UK
Telephone: +44 (0)20 7017 5540
Email: CSDhealthcarebooks@informa.com
Website: http://informahealthcarebooks.com/

For corporate sales please contact: CorporateBooksIHC@informa.com
For foreign rights please contact: RightsIHC@informa.com
For reprint permissions please contact: PermissionsIHC@informa.com

Printed and bound in India by Replika Press Pvt. Ltd.

To Professor Atta-ur-Rahman, FRS: a scholar, a visionary, and a man with inexhaustible energy.

Preface

Preformulation studies constitute the delicate connection between the two major groups of scientists: those at the drug discovery end and those at the drug delivery end. Whereas scientific camaraderie, or perhaps stubbornness, at the two ends of new drug development has historic roots, it is the preformulation group of comrades that brings peace to the table. It is often humbling for the drug discovery group to bring out a novel molecule with remarkable potential only to be shot down by the formulation group as a worthless exercise in taking it to a deliverable form. The preformulation group works with both ends and helps reduce the overall cost and shrink the timeline of drug development. Whereas in some companies the dividing line between preformulation and formulation is often a gray zone, those who have understood the significance of keeping the two groups separate have reaped great rewards. This book is a practical manual for those involved in the preformulation stages of drug development, yet it would be a good read for the drug discovery and drug development group as well. The traditional scope of textbooks and manuals on this topic is expanded here to include biological drugs, particularly therapeutic proteins and also botanical drugs or phytomedicines. The latter category is particularly significant as it is fast becoming evident that such regulatory authorities as the U.S. Food and Drug Administration and European Medicines Evaluation Agency will start approving them in the same system of approval as is ordinarily reserved for small- and large-molecule, well-characterized drugs.

The goals of preformulation studies are to choose the correct form of the drug substance, evaluate its physical and chemical properties, and generate a thorough understanding of the material's stability under the conditions that will lead to the development of a practical drug delivery system. Preformulation is a science that serves as a big umbrella for the fingerprinting of a drug substance or product both at the early and latter stage of development in pharmaceutical manufacturing. The preformulation phase is a critical learning time about candidate drugs. Typically, it begins during the lead optimization phase and continues through prenomination and into the early phases of development. Decisions made on the information generated during this phase can have a profound effect on the subsequent development of those compounds. Therefore, it is imperative that preformulation should be performed as carefully as possible to enable rational decisions to be made. The quantity and quality of the drugs available at this stage can affect the quality and quantity of data generated—so can the equipment available and the expertise of the personnel conducting the investigations. In some companies, there are specialized preformulation teams, but in others the information is generated by a number of scattered teams. Whichever way a company chooses to organize its preformulation information gathering, one of the most important practices to adopt is to keep close communication among the various groups involved.

The classic definitions and the management systems to conduct preformulation studies are discussed in the chapters; it would suffice for now to claim that there is a need to apply the most current knowledge and analytical sophistication available to deliver these goals. Over the past quarter of a century, the science of analysis and characterization has taken a giant leap, and so have the options that are now available to scientists regarding the basic substances as we enter the era of nano-particles and intelligent delivery designs. A recent study authorized by the National Security Agency and conducted by Rand Corporation listed nano and material science to be the leading sciences in the year 2015; this is an important indication of what is to come and the science of drug discovery will be greatly affected by it. With new possibilities of materials and systems such as nano pumps, the evaluation of new drug entities will have to take a different perspective.

Crystalline structure studies form the core of preformulation studies because molecules make crystals, crystals make particles, and particles make dosage forms. Novel research in crystallography of new entities involves studies of amorphous forms to learn how local properties contribute to the chemical reactivity of these short-interacting forms. Solid solutions are better understood today, and studies on the effect of different solvents on the formation of solid solutions remain a challenging opportunity made easy by computer simulation models. Predicting accurate behavior of amorphous forms will allow great opportunities in drug delivery. By understanding the effects of various cations and anions on crystal structure and properties, one can move towards the rational selection of counter-ions for salt formation and the design of stable salt forms. Neutral pharmaceuticals and excipients can and do form stable co-crystals. These co-crystals may possess superior physical properties, such as solubility, melting point, and compaction behavior. These studies seek to understand the forces that promote the formation of neutral co-crystals, in particular intermolecular hydrogen bonding. There is an increasingly strong driving force to predict the properties of a drug in its solid form and this, in turn, drives the need to predict three-dimensional structure. It is important to compute all reasonable low-energy conformers from a known chemical structure and place these conformers into ranked three-dimensional crystal structures. Studies of how particle shape influences packing properties and how this affects the translational stress into strain in the tableting process are useful. By studying such material constants as friction coefficients, Young's modulus, and viscoelasticity, one can obtain properties of the bulk powder. This then should be followed to study polydispersity (in particle size) and to validate the model with X-ray microtomography of glass powder. The controlled crystallization of organic molecules on surfaces controls nucleation and deposition of drug crystals of chosen structure as formed by manipulation of the substrate surface. This provides the groundwork for the development of self assembled three-dimensional and chemical cage formulations. The idea is that by controlling the nature of the surface of the substrate, particular polymorphs may be encouraged to grow. Additionally, the directing surface may prove to be a way of stabilizing more favorable, but higher energy, polymorphs such that solid form selection may be made on the basis of polymorph properties and not be limited to the thermodynamically most stable form. Self-assembled functional dosage forms could provide the next generation of pharmaceutical products, obviating the need for extreme conditions encountered during tableting. Areas of research include the manipulation of surfaces to allow coating and filling and the controlled crystallization and controlled production of porous polymeric structures using poly methyl methacrylate as a "proof of concept" polymer.

The availability of new scientific tools is not limited to the development of small molecules; large-molecule drugs, particularly therapeutic proteins, are now studied with greater accuracy than was possible when they were first approved a quarter of a century ago. With almost certainty that biogeneric (or biosimilar or follow-on, depending on which side of the Atlantic you live on) products will be approved by the U.S. Food and Drug Administration, as the European Medicines Evaluation Agency is already approving them, there will be a great rush to develop tools to study the equivalence of protein products. Given the complexities involved in predicting the side effects related to three-dimensional and even four-dimensional structures, a lot of science is yet to be developed in executing the comparability protocols. The *Handbook of Preformulation of Pharmaceutical and Biopharmaceutical Products* provides a broad discussion of testing of biological products at the preformulation level, and it is anticipated that many new techniques will become available in the near future. The reader is strongly advised to review the *Handbook of Pharmaceutical Manufacturing Formulations* and the *Handbook of Biogeneric Therapeutic Proteins*, publications of CRC Press (see Bibliography for further reference).

Botanical drugs had long been set aside by the busy regulatory authorities, partly because of lack of resources to monitor them, and partly because they were so poorly understood that it was better to leave them alone. This is changing fast as all regulatory authorities worldwide have announced the guidelines on the development and submission of marketing authorization applications for phytomedicines. These are likely to be treated somewhere between the small- and large-molecule categories; a detailed discussion of the new requirements for Chemistry, Manufacturing, and Control package submission are presented in this book, with particular emphasis on the preformulation work that needs to be performed in the development of these drugs. There is a need to devote an entire volume to these studies.

Given the fast-changing backdrop in the studies that constitute preformulation, it is heartening to know that there are several excellent resources that can be tapped by scientists on a routine basis. The American Association of Pharmaceutical Scientists (www.aaps.org) offers, among many other useful sources of information, the Preformulation Focus Group, which was established in 1993 to bring together all American Association of Pharmaceutical Scientists members with a common interest in the broad area of preformulation research. The ultimate goal of the group is to provide a forum for the exchange of ideas and development of strategies, thereby collectively improving our capabilities in this important area of research. The goals of the group are:

- To provide a forum for information exchange of issues relating to solid- and liquid-state characterization of chemical entities, active pharmaceutical ingredients, excipients, and early-phase development of pharmaceutical dosage forms and delivery systems
- To promote and disseminate novel scientific advances in the areas of discovery and development of drug substances and products through themed meetings and workshops in pharmaceutical technologies or pharmaceutics and drug delivery sections
- To encourage and promote joint programs with other focus groups, academia, and regulatory agencies, for example, Food and Drug Administration, United States Pharmacopeia and other national regulatory agencies worldwide

- To strengthen the membership through such different portals as the focus group newsletters, student chapters, conferences, volunteer activities, and American Association of Pharmaceutical Scientists list serve
- To support nominations of preformulation scientists for American Association of Pharmaceutical Scientists awards

The reader is highly encouraged to join the group and benefit from a ready knowledge base, particularly as the harmonization of international standards of the technical packages of new drug applications is achieved. Besides these discussion groups, several journals publish relevant articles on preformulation, including:

- American Association of Pharmaceutical Scientists Pharmaceutica: www.aapspharmaceutica.com/index.asp, Pharmaceutical Research.
- www.ADMET.net is the newer web portal, which aims to keep researchers and business leaders up to date on what's happening in the dynamic fields of absorption, distribution, metabolism and excretion, and toxicology. Updated on a daily basis, the unique mix of exclusive interviews, specialist supplier listings, events, news, and new products makes this a one-stop shop for essential information. For reference there are links to research centers, journals, books, reviews, and market reports.
- www.HTScreening.net is the main web information portal that aims to keep researchers and business leaders up to date on what's happening in the dynamic field of biomolecular screening. Updated on a daily basis, the unique mix of exclusive interviews, specialist supplier listings, events, news, and new products makes this a one-stop shop for essential information. For reference there are links to research centers, journals, books, reviews, and market reports.
- www.CombiChem.net is the main web information portal dedicated to combinatorial and medicinal chemistry. Updated on a daily basis, the unique mix of exclusive interviews, specialist supplier listings, events, news and new products makes this a one-stop shop for essential information. For reference there are links to research centers, journals, books, reviews, and market reports.
- The Drug Delivery Insight (www.espicom.com/ddi) offers a detailed review and takes the hard work out of staying in touch with the companies, products, alliances, and research activities shaping the industry.

The work of scientists has been substantially eased through the availability of search engines on the Internet. A search for "preformulation" yielded 40,000 possible hits in 0.28 second; for "pharmaceutical preformulation" the number was 29,000. The problems, therefore, arise regarding how to parse the data; there is a lot of redundant and superfluous information available. It is for this reason that, despite the wide availability of information, knowledge must be gained from such condensed sources as this book and other books written on the subject.

Over the past two decades a large number of highly sophisticated laboratories have emerged that offer excellent opportunities to outsource the preformulation work. Some of these include:

- ABC Laboratories: www.abclabs.com
- Agami: www.aagami.net/
- Aptuit: www.aptuit.com

- Azopharma: www.azopharma.com
- Baxter Biopharma Solutions: www.baxter.com
- Bilcare, Inc.: www.bilcare.com
- Cardinal Health: www.cardinal.com
- Chemic Laboratories, Inc.: www.chemiclabs.com
- Chemir Analytical Services: www.chemir.com
- Dow Pharmaceutical Sciences: www.dowpharmsci.com
- DPT: www.dptlabs.com
- Emerson Resources, Inc.: www.emersonresources.com
- Formatech: www.formatech.com
- Fortitech: www.fortitech.com
- Fresenius Product Partnering: www.fresenius-pp.com
- Frontage Laboratories, Inc.: www.frontagelab.com
- Fulcrum Pharma Developments, Inc.: www.fulcrumpharma.com
- Glatt Pharmaceutical Services: www.glattpharmaceuticals.com
- Groupe Parima, Inc.: www.groupeparima.com
- KP Pharmaceutical Technology, Inc.: www.KPPT.com
- MedPharm Ltd.: www.medpharm.co.uk
- Metrics, Inc.: www.metricsinc.com
- Micron Technologies, Inc.: www.microntech.com
- Mikart, Inc.: www.mikart.com
- Murty Pharmaceuticals, Inc.: www.mpirx.com
- Neurogen: www.neurogen.com/
- Newport Scientific, Inc.: www.newport.com.
- OctoPlus B.V.: www.octoplus.nl
- Orbus Pharma, Inc.: www.orbus.com
- Patheon, Inc.: www.newportsci.com
- Penn Pharmaceutical Services, Ltd.: www.pennpharm.co.uk
- Pharmaceutical Development & Manufacturing Service: www.pdms-almac.com
- PharmaFab: www.PharmaFab.com
- Pharmatek Laboratories, Inc.: www.pharmatek.com
- Pharmaterials: www.pharmaterials.co.uk/
- PharmPro (Division of Fluid Air): www.pharmproservices.com
- Pharmquest Corp.: www.pharmquest.com
- PII: www.pharmproservices.com
- Quality Assistance SA: www.quality_assistance.com
- Quintiles: www.quintiles.com
- Ricerca Biosciences, LLC: www.ricerca.com
- SBS Pharma: www.sbspharma.com
- Stason Pharmaceuticals, Inc.: www.stason.com
- Temmler Pharma: www.temmler.de
- TetraGenX: www.tetragenx.com
- UPM Pharmaceuticals, Inc.: www.upm-inc.com

Many of these fine companies have extended their courtesy to me in providing technical information, for which I am very grateful.

The organization of this book was at first difficult as I kept sifting away those materials that would classically constitute formulation work or be more closely related to drug discovery. However, preformulation studies, slotted in between these two, must, by necessity, overlap these disciplines to some degree. Starting with an overview of the drug discovery process in the first chapter, the book takes the reader to more specific topics, including the regulatory environment and the intellectual property requirements for both understanding how it is developed to avoiding a drug infringement. Classical studies of basic property evaluation of new drug substances is presented for the three types of products—chemical, biological, and botanicals—is presented along with examples of the newest trends in the use of newer techniques.

In writing this book I have benefited greatly from those who have ventured this road before me and produced great published works. I have tried to acknowledge them in the bibliographies; however, it is impossible for me to fully credit these authors and quote their written works for they may be embedded in the very language of preformulation sciences. However, despite much care, it is inevitable that errors remain; these are all mine and I would appreciate it if the reader would bring them to my attention at niazi@niazi.com so that I may correct them in the future editions of the book.

The support of editors and publishers was exemplary as it has been in the previous books that I wrote for CRC Press (Informa Healthcare). The continuous support and encouragement of Stephen Zollo can never be fully acknowledged. Judith Spiegel is my new editor at CRC press in Florida and I look forward to working with her. My new editor at Informa Healthcare, Yvonne Honigsberger, has been more than encouraging as we have gone ahead and signed another author agreement. I would like to acknowledge the assistance of Sherri Niziolek, Pat Roberson, and Tara Kuboski, and many others at Informa Healthcare, without whose attention this work could not have been completed.

Sarfaraz K. Niazi

BIBLIOGRAPHY

Anton P, Silberglitt R, Schneider R. The Global Technology Revolution: Bio/Nano/Materials Trends and their Synergies with Information Technology by 2015, Rand Corporation 2001. Santa Monica, California (http://www.rand.org/pubs/monograph_reports/MR1307/index.html).
Blumenthal M, Daniel E, Farnsworth N, Riggins C. Botanical Medicine: Efficacy, Quality Assurance and Regulation. Mary Ann Liebert, 2003.
Carpenter JF, Manning MC, eds. Rational Design of Stable Protein Formulations: Theory and Practice (Pharmaceutical Biotechnology). New York: Plenum Press, 2003.
Carstensen JT. Pharmaceutical Preformulation. Boca Raton, FL: CRC Press, 1998.
Florence AT, ed. Materials Used in Pharmaceutical Formulation (Critical Reports on Applied Chemistry, Vol 6). Blackwell Science, 1984.
Gibson, Mark. Pharmaceutical Preformulation and Formulation: A Practical Guide from Candidate Drug Selection to Commercial Dosage Form. Boca Raton, FL: CRC Press, 2001.
Hovgaard SF Jr. Pharmaceutical Formulation Development of Peptides and Proteins. Boca Raton, FL: CRC Press, 1999.
Niazi SK. Handbook of Biogeneric Therapeutic Proteins: Manufacturing, Regulatory, Testing and Patent Issues. Boca Raton, FL: CRC Press, 2005.
Niazi SK. Handbook of Pharmaceutical Manufacturing Formulations, Vol. 1: Compressed Solids. Boca Raton, FL: CRC Press, 2004.
Niazi SK. Handbook of Pharmaceutical Manufacturing Formulations, Vol. 2: Uncompressed Solids. Boca Raton, FL: CRC Press, 2004.

Niazi SK. Handbook of Pharmaceutical Manufacturing Formulations, Vol. 3: Liquid Products. Boca Raton, FL: CRC Press, 2004.

Niazi SK. Handbook of Pharmaceutical Manufacturing Formulations, Vol. 4: Semisolid Products. Boca Raton, FL: CRC Press, 2004.

Niazi SK. Handbook of Pharmaceutical Manufacturing Formulations, Vol. 5: Over the Counter Products. Boca Raton, FL: CRC Press, 2004.

Niazi SK. Handbook of Pharmaceutical Manufacturing Formulations, Vol. 6: Sterile Products. Boca Raton, FL: CRC Press, 2004.

Wells JI. Pharmaceutical Preformulation: The Physicochemical Properties of Drug Substances. West Sussex, UK: Ellis Horwood Ltd., 1988.

List of Appendices, Figures, Schemes, and Tables

APPENDICES
Appendix 4.1 Some representative studies on pK_a and log P determination reported in the literature.
Appendix 5.1 Recent studies reported in literature.

FIGURES
Figure 1.1 Typical pharmaceutical product life cycle: income potential against time.
Figure 1.2 Drugs approved by the U.S. Food and Drug Administration.

Figure 3.1 The NanoCrystal technology.

Figure 4.1 Typical Bronsted acids and their conjugate bases.
Figure 4.2 pH of common fluids.
Figure 4.3 pH probe design.
Figure 4.4 Relationship between pH and electrode potential as a function of temperature.
Figure 4.5 Percent ionization of an acid with pK_a of 8.0.
Figure 4.6 Percent ionization of a base with pKa of 8.0.
Figure 4.7 Example of a graph for a linear free-energy relationship.
Figure 4.8 Log D profile of an acid with $pK_a = 8$.
Figure 4.9 Log D profile of a base with $pK_a = 8$.
Figure 4.10 Log D profile of a zwitter ion (base) with $pK_a = 5.6$ and 7.0 for acid and base, respectively.
Figure 4.11 Optimal log P values for absorption from different parts of the gastrointestinal tract.
Figure 4.12 Filter plate method for solubility testing.

Figure 5.1 Diffusion layer model of dissolution.
Figure 5.2 Calculation of solubility as a function of pH.
Figure 5.3 Assay complexity versus correlation with human absorption.
Figure 5.4 The 96-well permeability testing method.
Figure 5.5 MultiScreen Caco-2 assay system.

Figure 6.1 Crystal lattice.
Figure 6.2 Scalars of lattice structure.
Figure 6.3 Bravais lattice system.
Figure 6.4 Common crystal habits.
Figure 6.5 Monotropic system as a function of temperature (x-axis).
Figure 6.6 Enantiotropic system as a function of temperature (x-axis).
Figure 6.7 Enantiotropic system with metastable phases as a function of temperature (x-axis).
Figure 6.8 Braggs diffraction.

Figure 6.9   Design plan of a typical X-ray powder diffractometer.
Figure 7.1   The dynamic vapor sorption chart for microcrystalline cellulose.
Figure 7.2   The dynamic vapor sorption chart for lactose.
Figure 7.3   Inverse gas chromatography principles.
Figure 7.4   Techniques of particle size detection and their limits.
Figure 7.5   The six types of International Union for Physical and Applied Chemistry isotherms.
Figure 7.6   Static volumetric gas adsorption system.
Figure 7.7   Emulsion particle size dependence on the method of manufacture.
Figure 9.1   Structure of polyethylene glycol.
Figure 9.2   Improved drug performance using advanced PEGylation.
Figure 9.3   *m*-PEGylation using *N*-hydroxysuccinimide ester.

SCHEMES
Scheme 3.1   Salt selection tree.
Scheme 6.1   Some acceptance criteria for polymorphism in drug substances and drug products.
Scheme 6.2   Decision flow diagram to select the best salt form.
Scheme 7.1   Selection criteria for dosage form selection.
Scheme 7.2   Setting acceptance criteria for drug product dissolution, that is, what types of drug release acceptance criteria are acceptable?
Scheme 7.3   Setting acceptance criteria for drug product dissolution, that is, what specific test conditions and acceptance criteria are needed for the immediate release of dosage forms?
Scheme 7.4   Setting acceptance criteria for drug product dissolution, that is, the appropriate acceptance ranges for extended release dosage forms.
Scheme 8.1   Steps that lead to the development of specifications for new lead compounds.
Scheme 8.2   Decision tree for chiral compound resolution.
Scheme 8.3   Decision tree for disposition of degradation products.
Scheme 8.4   Decision tree for disposition of impurities in the drug substances.
Scheme 8.5   Decision tree for the disposition of particle size variation.
Scheme 8.6   Decision tree for the disposition of polymorphism.
Scheme 8.7   Decision tree for the disposition of microbiological attributes.
Scheme 8.8   Decision tree for impurities in drug substances.

TABLES
Table 1.1   Drugs approved by the U.S. Food and Drug Administration during 1995 to 2005.
Table 1.2   Most active categories for new drug development.
Table 1.3   List of monographs of phytomedicines available from European Scientific Cooperative on Phytotherapy.
Table 4.1 ρ values of various substituents.
Table 4.2 σ values at meta and para substitution positions.
Table 4.3   Hammett σ constants.
Table 4.4   The United States Pharmacopeia solubility classification.
Table 4.5   Advantages and disadvantages of pK_a and log P measurement techniques.

Table 4.6	Equipments available for sirius for pK_a, log P and solubility measurement, and data handling.
Table 5.1	The biopharmaceutics drug classification system as defined by the Food and Drug Administration and modified by recent findings.
Table 6.1	Seven crystal systems.
Table 6.2	Common crystal habits.
Table 6.3	Thermodynamic rules for polymorphic transitions.
Table 6.4	Drug substance hydrate forms as reported in the pharmacopeia.
Table 6.5	The pK_a of common weak acids used in salt formation.
Table 6.6	Standards for thermal analysis in the order of increasing melting point.
Table 8.1	Study factors for various prospective dosage forms.
Table 8.2	Common functional groups of drugs.
Table 8.3	Typical oxidation numbers of carbon.
Table 8.4	General case of study, storage condition, and time covered.
Table 8.5	Storage and minimum time covered for drug substances intended for storage in a refrigerator.
Table 8.6	Storage conditions and minimum time for drug substances intended for storage in a freezer.
Table 8.7	Retest study, storage condition, and minimum time period.
Table 8.8	Thresholds of impurities.
Table 8.9	Application of good manufacturing practice to active pharmaceutical ingredient manufacturing.
Table 9.1	Parts of Title 21 of the Code of Federal Regulations relevant to biological drugs.
Table 9.2	Approved products transfer from center for biologics evaluation and research to center for drug evaluation and research.
Table 9.3	Composition of approved biological products.
Table 9.4	Common excipients in biological products.
Table 9.5	Preventive actions against proteolysis.
Table 9.6	Preventive actions against deamidation.
Table 9.7	Preventive actions against oxidation.
Table 9.8	Preventive actions against β-elimination.
Table 9.9	Preventive actions against racemization.
Table 9.10	Preventive actions against cysteinyl residue loss.
Table 9.11	Preventive actions against hydrolysis.
Table 9.12	Preventive actions against hydrolysis.
Table 9.13	Preventive actions against aggregation.
Table 9.14	Preventive actions against precipitation.
Table 10.1	Herbal drug approval rules in different countries.
Table 10.2	Chemistry–Manufacturing–control considerations for National Center for Complementary and Alternative Medicine clinical trials.
Table 10.3	Example of component combinations for a botanical drug.

Contents

Preface .... *v*
List of Appendices, Figures, Schemes, and Tables .... *xiii*

1. Drug Discovery Trends   *1*
    Introduction *1*
    Development Phases *2*
    Product Life Cycle *6*
    New Pathways to Discovery *11*
    References *23*
    Bibliography *23*

2. Intellectual Property Considerations   *33*
    Introduction *33*
    Patenting Strategies *33*
    Patenting Systems *35*
    What Is a Patent? *35*
    Patent Myths *36*
    What Is Not Patentable? *38*
    Patent Search *38*
    Components of a Patent Application *44*
    Understanding Claims *49*
    Food and Drug Administration *55*
    References *56*
    Bibliography *56*

3. The Scope of Preformulation Studies   *57*
    Introduction *57*
    Preformulation Testing Criteria *57*
    Regulatory Requirements *58*
    Testing Systems *65*
    Solid-State Characterization *69*
    Transport Across Biological Membranes *77*
    References *79*
    Bibliography *79*

4. Dissociation, Partitioning, and Solubility   *87*
    Introduction *87*
    The Ionization Principle *87*
    Quantitative Structure–Activity Relationships *98*
    Partitioning *105*
    Measurement Strategies *113*
    References *122*
    Bibliography *122*
    Appendix 1 .... *131*

5. Release, Dissolution, and Permeation 141
Introduction 141
Release 141
Assay Systems 147
The Biopharmaceutics Drug Classification Systems 156
References 159
Bibliography 159
Appendix 1 168

6. Solid-State Properties 197
Introduction 197
Crystal Morphology 197
Polymorphism 203
High-Throughput Crystal Screening 207
Solvates 210
Hydrates 210
Amorphous Forms 212
Hygroscopicity 212
Solubility 213
Study Methods 218
References 231
Bibliography 231

7. Dosage Form Considerations in Preformulation 241
Introduction 241
Solid Dosage Form Considerations 241
Solution Formulations 262
Emulsion Formulations 263
Freeze-Dried Formulations 271
Suspensions 272
Topical 273
Pulmonary Delivery 274
General Compatibility 275
References 275
Bibliography 276

8. Chemical Drug Substance Characterization 287
Introduction 287
Scheme of Characterization 288
Impurities 317
Good Manufacturing Practice 322
References 326
Bibliography 326

9. Characterization of Biopharmaceutical Drugs 329
Introduction 329
Preformulation Studies 333
Packaging and Materials 348
Physio-Chemical Characterization Tests 385
Design of Preformulation Studies 387
References 388
Bibliography 388

10. Characterization of Phytomedicines 391
Introduction 391
Regulatory Status 392

Characteristics of Phytomedicines *392*
Efficacy and Safety *399*
Regulatory Filing Procedure *400*
Starting Material *403*
Stability Testing *406*
Bibliography *410*

Glossary of Terms *419*
Index *431*

1 Drug Discovery Trends

INTRODUCTION

Drug discovery precedes preformulation studies but often a preformulation feedback helps faster drug discovery. Much has changed in the discovery of drugs over the past 25 years. Microprocessor-driven instrumentation has revolutionized data handling systems. Robotic systems have eased large sample processing, and integration of various physical and chemical sciences has resulted in the emergence of newer techniques. For example, high-throughput screening (HTS) is now an integral component of the drug discovery process practiced by most pharmaceutical and large biotechnology companies. It has evolved over the past decade from crude automation to use of sophisticated computer-driven array systems using robotic devices. Improved physicochemical data on prospective new active entities (NAEs) provide a great stimulus to new drug development as well as offering insights for the preformulation scientists to project the characteristics of the NAEs that would prove useful in their downstream processing. The downstream processes include hit-to-lead (HTL), lead optimization (LO), and in vitro absorption, distribution, metabolism, elimination, and toxicology (ADMET) studies, which are all driven by the peculiar characteristics of the NAE. What used to be the application of preformulation studies in the formulation steps has now shifted somewhat backwards into evaluation of lead compounds. In many instances, HTL and preformulation studies (theoretical or experimental) are combined to maximize the probability of clinical success.

There is always an additional burden on scientists to assure that in their pursuit of novelty in creating a value for their company, they must review the vast databases of patents involved. It is for this reason that in describing the overall drug development scene, an in-depth discussion of what constitutes intellectual property, how to evaluate it, and search for new ideas is presented in a separate chapter.

The drugs developed today fall into three main categories: small molecules, large molecules, and botanical extracts. The preformulation work required for each category is quite different from the others. The classic small molecule chemistry involves the measurement of solubility, pK_a, crystallinity, polymorphism, and so on. For large molecules, particularly proteins and peptides, often the three-dimensional (3D) structure or even interaction with formulation components requiring a 4D evaluation of drugs in solution must be studied. Botanical drugs are the newest category for the preformulation scientist, as the regulatory authorities like the United States Food and Drug Administration (U.S. FDA) and European Medicines Evaluation Agency (EMEA) have established guidelines for their approval. For botanical drugs (often called phytomedicines in Europe), the fingerprinting methods using thin layer chromatography (TLC), infrared (IR), or mass spectra (MS) are more relevant.

DEVELOPMENT PHASES
The development of NAEs undergoes a lengthy and expensive cycle that varies greatly depending on the type of drug developed. What follows is a typical development cycle.

Stage 1: Lead Finding or Establishing Directions (One to Two Years)
These are studies to understand how alteration of a biological function or mechanism would create a therapeutically useful entity or process. The strategic research of a particular company is usually guided by factors such as its inherent research competence and expertise, therapeutic areas of unmet medical need, and market potential. Market potential is the strongest motivation and as a result companies invest heavily in specific therapy areas. It is not difficult to label companies as cardiovascular, anticancer, and endocrine and other such focussed companies. It takes years, often decades, to pool the scientific and clinical expertise that would provide the combination of elements necessary for efficient new discoveries. However, in recent years, many large companies have begun to look outside for new drug leads; smaller research companies have often proven more efficient. Outsourcing of research, even at the level of research lead compounds, is now a common place, and companies like Abbott Laboratories have developed an expertise in securing new drugs from outside vendors. The in-house expertise development is not limited only to scientific expertise to provide a specific direction, but also at times to a specific type of product, such as biological drugs versus the small molecules, innovative drug delivery systems versus the traditional systems, and so on.

It is not uncommon for large companies to suddenly establish a direction through serendipitous discoveries. Alexander Fleming discovered penicillin as a result of "serendipity." Although many of the drugs that are in use today have been discovered in this way, it is a difficult route for the pharmaceutical industry to follow and breaks are few and far between. More recently, the discovery that the phosphodiesterase type 5 inhibitor (originally tested as antihypertensive agents) can be a good candidate for male erectile dysfunction has changed the focus of research at Pfizer, and as a result, at several other companies.

Stage 2: Candidate Drug Screening (One to Ten Years)
In the previous step (stage 1), a particular biological mechanism is identified or targeted; the phase that follows it involves identifying chemicals or modalities that would interfere or interact with it and thereby identify what the industry calls, "leads." This stage requires extensive screening for biological activity, a process that used to be tedious and slow; however, over the last decade, the use of techniques as "combinatorial chemistry" and automated HTS make it possible to identify a large number of synthetic or biosynthetic molecules.

Organizations may differ in library acquisition methods (e.g., synthesis versus purchase), compound inclusion criteria, synthesis methodology, or how they mine their collections to uncover leads. The rationale, however, remains constant: to find new, patentable structures as efficiently as possible. Automated synthesis and HTS, pioneered during the 1990s, enabled companies to synthesize, test, and maintain compound libraries populated with hundreds of thousands—even millions—of unique compounds. Automation opened the door to new chemistry and a very large number of compounds, with a typical large pharmaceutical company possessing between 1 and 10 million compounds in its various compound collections. The reliance on large libraries has now become an effort with diminishing returns. Now outsourcing the lead compounds appears more plausible. This allows greater molecular diversity, expanding what may end up as a myopic view of a company. Companies like Abbott Laboratories have done well in emphasizing chemical diversity. As combinatorial chemistry became more automated and specialized, companies created separate groups to handle synthesis, library acquisition, and compound management. Aventis, for example, maintains a 60-person combinatorial chemistry group in Tucson, Arizona that feeds libraries of various sizes to the rest of the company. Other firms take a more traditional view, preferring not to segregate discovery-related chemical competencies. Like most large discovery organizations, Roche constructs libraries to target gene families, protein targets, or to expand the chemical diversity of the company's compound collection. Roche's lead-generation strategy includes acquiring libraries from specialist compound suppliers, contracting with external partners for library synthesis, and developing libraries in-house. Roche's view on library size is in line with current thinking and Roche has gradually moved away from very large compound libraries. Roche uses all modern synthetic tools, including parallel synthesis, combinatorial chemistry, and solid-phase methods, principally (although not limited to) solid-phase reagents and scavengers in the latter. Resin-bound synthesis is also used, but only in situations where the technique can offer an advantage such as when very large numbers of related structures are desired. In today's discovery paradigm, the prime focus is more toward activity than on numbers of compounds. Today, the pendulum has swung back toward rational design. Large libraries, despite their waning appeal, are however still in demand from specialty synthesis vendors. The hit rate can be improved when refined compound libraries are created from an existing hit through designing libraries for a given chemotype or pharmacophore, which amounts to second-generation combinatorial methods coupled with computational chemistry.

Many large companies do not purify individual compounds in 100,000-entry libraries, instead preferring to bias reaction pathways toward products that are reasonably pure. Eventually, when discovery chemists settle on focused panels of between 200 and 1000 compounds, they rely on a number of tools to help clean up these smaller libraries. One method is solid-phase scavenger reagents, which can soak up substantial amounts of impurities and side products, and even drive reactions toward completion by shifting equilibrium. Simple anion or cation exchange resins are popular for removing anionic or cationic species, respectively. Scavengers work well to remove major impurities from individual reactions, but top-tier libraries in the 200–1000 compound range probably require more careful cleaning up. Newer techniques such as solution-phase mixture synthesis obviate the problems associated with solid-phase synthesis since the kinetics are easier to study in homogeneous systems.

Genomics is emerging as a useful technique in the drug selection process and has begun to play a significant role in the identification of lead compounds.

The distinct phases of stage 2 include:

- *Research planning*: This may involve classical structure–activity correlation studies, rational drug design modeling, HTS of libraries obtained by combinatorial chemistry or leads from natural sources, and so on.

- *Obtaining test compounds or samples*: Laboratory-scale preparation, preparation of compounds or sample libraries, determination of in vitro or animal models to test activity, and setting up of HTS.
- *Screening*: Basic pharmacological and biochemical screening. Selection of "hits" and identification of active compounds. It is at this stage that the patent application for the compound will normally be filed. However, strategies on the timing of patent filing will be discussed later.
- Preformulation studies consist of all that it takes to characterize a drug substance to enable its formulation into a practical drug delivery system. The preformulation studies begin immediately after a drug molecule (at this stage a lead) has been recognized. The process of lead optimization integrates preformulation studies to assure that all optimal forms of the drug (e.g., salt or crystalline forms) have been identified.

Stage 3: Candidate Drug Selection (One to Two Years)

With a solid chemical lead on-hand, companies can create compounds with more optimal characteristics, pharmacological and formulation compatible forms that lead to nomination candidates. Ideally, the nominated candidate would have the following features:

- a simple structure
- no chiral centers
- fewer steps in synthesis
- passes carcinogenicity, mutagenicity, and lethal dose 50 (LD50) level testing
- nonhygroscopic
- crystalline with sufficient solid-state stability
- possible to be administered orally with sufficient bioavailability
- no strong colors or odors
- compatible with standard excipients

Obviously, not all conditions are met, but by every compound we keep this direction in mind, it is easier to sift through many possibilities offered. It requires testing a range of selected compounds in in vitro and in vivo animal studies and thus, preformulation work gets combined with biopharmaceutic studies to identify product design issues.

One of the major factors that often slows down preformulation studies is the availability of a sufficient quantity of a compound at this stage, especially if biopharmaceutic studies are conducted; as a result, methods need to be devised that would utilize the smallest quantity of the substance; this is amply emphasized throughout the testing phases in this book. Another factor that often slows down work at this stage is often misplaced importance on validated test methods and documentation; while this is desired, the perennial shortage of manpower requires that some things take secondary importance.

Stage 4: Preclinical Studies (One to Two Years)

Preclinical trials (four to six years):

- Pre-investigational new drug (IND) meetings with the FDA.
- *Preclinical trials stage I*: Acute toxicity, detailed pharmacological studies (main effect, side effect, and duration of effect), analytical methods for active substance, and stability studies.

- *Preclinical trials stage II*: Pharmacokinetics (absorption, distribution, metabolism, and excretion), subchronic toxicity, teratogenicity, mutagenicity, scale-up of synthesis, development of final dosage form, and production of clinical samples [Chemistry, Manufacturing, and Control (CMC) section for FDA].

Stage 5: Phase I Clinical Studies (One to Two Years)

This is the stage of proof of concept or Phase I testing to understand how the candidate drug is absorbed and metabolized in healthy human volunteers before testing it in patients. Often small-scale studies are done in patients when the cost factors are high to make sure there is sufficient indication for the drug's utility. There are significant differences in regulatory filings at this stage; in the United States, a fully approved IND is required, whereas in Europe these filings are not required. Also, once an IND has been filed with the U.S. FDA, any study conducted overseas comes under the purview of this IND. Some companies may therefore decide to conduct their Phase I studies prior to filing INDs in the United States to circumvent the issue of informing FDA of their work.

Stage 6: Phase II and Phase III Studies and Launch (Four to Six Years)

The longer-term safety and clinical studies come under Phases II and III wherein patients suffering from the disease are studied. In Phase II studies, the goal is to perform studies on different dose ranges in perhaps a few hundred patients to evaluate the effectiveness of the drug and its common side effects. It is at this stage that commercial formulation is developed and its scale-up is optimized. This is necessary because whatever formulation goes into Phase III trials becomes the final formulation. In Phase III studies, there may be several thousand patients involved. Most regulatory authorities require sufficient data to demonstrate the safety and efficacy data and almost always, it is a mutually agreed upon limit to which the drug companies are required to conduct the studies. The EMEA has started a program of technical consultation wherein for a sizeable fee of around $150,000, companies can sit down with the review staff and fine-tune their protocols; the investment is worth the effort. Once the regulatory authorities declare an application to be approvable, a marketing authorization request is made and trucks are loaded pending issuance of the letter of approval that may have cost companies hundreds of million dollars.

- Sequence of clinical trials (four to six years):
  - *Phase I*: Tolerance in healthy volunteers, pharmacokinetics in man, and supplementary animal pharmacology.
  - *Phase II*: First controlled trials on efficacy in patients, dose range studies, chronic toxicity, and carcinogenicity studies in animals.
  - *Phase III*: Large-scale trial at several centers for final establishment of therapeutic profile (indications, dosages and types of administration, contraindications, and side effects), proof of efficacy and safety in long-term administration, demonstration of therapeutic advantages in comparison with known drugs, and clarification of interactions with other medication.
- Registration, launch, and sales (two to three years):
  - *Registration with Health Authorities*: Documentation of all relevant data, expert opinions on clinical trials and toxicology, preparation for launch, information for doctors, wholesalers, and pharmacists, training of sales staff, preparation of packaging and package inserts, and dispatch of samples.
  - *Launch and sales*: Production and packaging of final form and quality control.

Stage 7: Postmarket Surveillance (Three to Five Years)
Phase IV Studies

Whereas the intent of larger Phase III trials is to bring out any statistically driven side effects, it is nevertheless impossible to predict them even with several thousand patients. Phase IV studies continue to collect clinical efficacy and toxicity data; the pharmacovigilance program required by EMEA is one such example. It is important to recognize that it is during this phase of studies that drugs are recalled. The most recent example of the recall of the cyclo-oxygenase (COX)-2 selective inhibitor medication Vioxx (generic drug name is rofecoxib) in September 2004 shows the importance of this phase of study. Whereas studies of Phases I–III are expensive to conduct, a failed Phase IV study can bankrupt companies.

PRODUCT LIFE CYCLE

Patent expiry periods of newly approved NAEs determine how companies manage the product life cycle (Fig. 1).

The majority of the patent term may be over by the time a new drug reaches the market, particularly in the non-U.S. markets. The U.S. patent laws are quite different from the laws of the rest of the world; in United States, companies may request a patent extension for the duration of time the regulatory applications are pending with the FDA (see Drug Price Competition and Patent Term Restoration Act). Once marketed, it takes some time for the medicine to build up sales and achieve an optimal market share. This phase is followed by the maturity phase where most profits are made. Finally, as the medicine loses patent protection, copies (generic products) hit the market at a lower cost and the sales of the original medicine start to fall. To retain market share once the medicine loses patent protection, a number of things can be done. Marketing strategies which might reduce the impact include improving the product and relaunching the "new improved" version that includes only cosmetic changes. What does pay off is the identity of brand developed over years, including any physical dosage form attributes. An effective way of branding is to make the product look unique. This could be achieved by making the tablet or capsule of a special shape. Also it can be colored or produced with special graphics printed on it. A distinctive use of unusual shapes and colors makes it more difficult for competitors to produce a

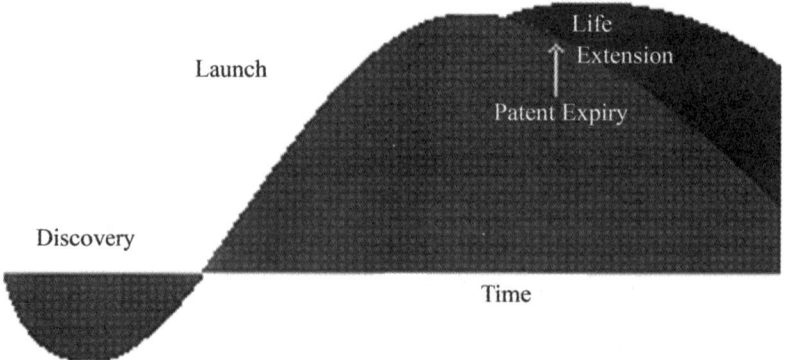

Figure 1 Typical pharmaceutical product life cycle: income potential against time.

drug that the consumer might confuse with the original. Names, logos, and packaging are all protected by registering them as trademarks. The blue diamond-shaped tablet of Viagra and purple Nexium tablets are just two examples of this approach.

Long before patents expire, companies begin to develop additional intellectual property into their products; for example, patented delivery systems can extend the life of a drug long after the chemical patent is expired, as the formulations of the drug must be the same so as to get qualified for the generic drug status. Companies like Elan have developed highly sophisticated drug delivery systems that allow companies to extend their product life (1). Companies also resort to changes in specifications as one route to keep generic competitors out; Abbott recently did this when its patent on calcitriol expired. The company reduced the level of oxygen in its specification and modified other specifications, leaving it to the generic manufacturers to claim the exact formulation, even though it made no difference in the efficacy, stability, or safety of the product. Biological drugs have just begun to come off patent and one of the most widely used techniques is to launch a pegylated form of drug that results in a modified dosing regimen; this will be considered as a new product, which is then used to establish the proprietary market as the patents expire for the drug entity. The leader in pegylation technology is Nektar (2) which has been involved in almost all of today's marketed pegylated therapeutic proteins. These are some of the reasons why all scientists involved in the development of drugs are familiar with patent laws.

Given the statistical reality that only one out of 5000 lead compounds reaches the market, the amortized cost over the entire development program is fast approaching the billion dollar mark for each new approval, though for the molecule in question it may range from $100 to $300 million depending upon the complexity of the testing involved. The requirement for such huge investment is one of the reasons for the mega-mergers of the pharmaceutical companies to cut down on research cost and also absorb the gyrations of profitability. For example, as the blockbuster drugs come off patent, companies lose billions of dollars of sales as did Glaxo-Wellcome when Zantac® came off patent, when the sales dropped by $1.6 billion. Much bigger swings in the sales are expected as even bigger drugs like Erythropoietin come off patent within the next couple of years. Generally, the price drops by 30–50% with the introduction of generic version of patented molecules.

According to pharmaceutical research–based manufacturers (PhRMa), the representative body of the research-based pharmaceutical companies (3), in 2003, PhRMa member companies invested an estimated $33.2 billion in research to develop new treatments for diseases—an estimated 17.7% of domestic sales on R&D—a higher R&D to sales ratio than any other U.S. industry, which averages around 4% of sales (3).

Approval Trends

If the number of drugs approved by the U.S. FDA can be taken as a measure of success, the results have been very dismal for the large pharmaceutical companies in the past five years (Fig. 2, Table 1). If we examine the data over the past decade, we find that two-thirds of the drugs approved during this period were approved in the 1990s and numbers approved every year have been declining since then. The top three categories of drugs include endocrinology, immunology/infectious diseases, and oncology.

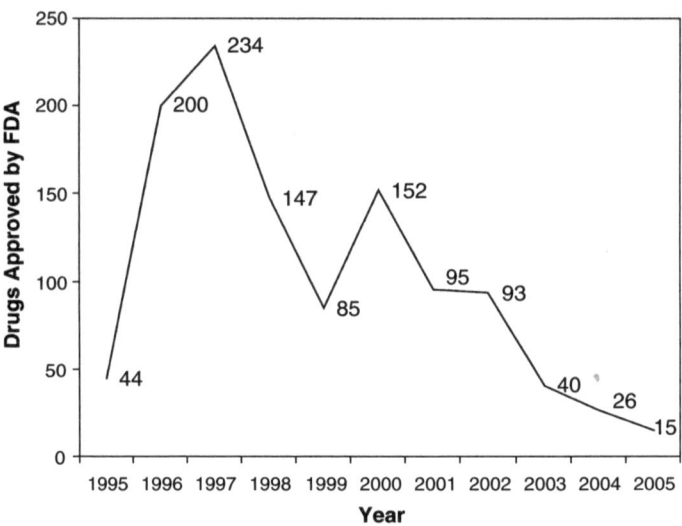

Figure 2 Drugs approved by the U.S. Food and Drug Administration (FDA). *Source*: From Ref. 14.

The new drugs under development (Table 2) provide an insight on the type of molecules that will soon reach the markets.

The poor performance of the drug industry in recent years was predictable to scientists, as the low rate of approval observed today is the result of decade-old research that did not go well. Hence, the pharmaceutical industry is pressed to develop newer methods of drug lead identification. As a result, the past decade saw development in areas such as combinatorial chemistry, combinatorial biosynthesis, metabolic pathway engineering, gene shuffling, and directed evolution of proteins. What was historically the greatest minefield of new molecules, the natural products, was placed on a back burner because of the complexity of characterizing naturally derived product leads. But, now this is about to change. The science of bioinformatics is revolutionizing exploitable biology, making a return to emphasis on detecting novel organisms, novel chemical structures, and novel biocatalytic activities from natural products. This program shift has been driven by a convergence of complementary technologies, exemplified by DNA sequencing and amplification, genome sequencing and annotation, proteome analysis, and phenotypic inventorying, all resulting in the establishment of huge databases that can be mined to generate useful knowledge, such as the identity and characterization of organisms and the identity of biotechnology targets. Concurrently there have been major advances in understanding the extent of microbial diversity, how uncultured organisms might be grown, and how expression of the metabolic potential of microorganisms can be maximized. The integration of information from complementary databases presents a significant challenge. Such integration should facilitate answers to complex questions involving sequential, biochemical, physiological, taxonomic, and ecological information of the sort required in exploitable biology.

Table 1 Drugs Approved by the U.S. Food and Drug Administration During 1995 to 2005

Category/year	2005	2004	2003	2002	2001	2000	1999	1998	1997	1996	1995	Total
Cardiology/vascular diseases	1	1	2	6	4	9	3	12	23	11	3	74
Dental/maxillofacial surgery	0	0	0	0	2	0	1	2	1	1	0	7
Dermatology/plastic surgery	0	0	2	1	5	7	2	8	15	5	1	46
Endocrinology	3	0	3	9	13	15	13	23	45	35	12	169
Gastroenterology	0	2	3	11	6	5	1	4	8	10	0	50
Hematology	1	1	4	1	1	3	5	6	7	7	0	35
Immunology/infectious diseases	2	3	5	10	9	14	10	9	19	17	9	106
Musculoskeletal	2	0	4	7	3	3	2	5	8	6	3	41
Nephrology/urology	1	4	3	2	4	8	3	6	5	10	1	46
Neurology	2	3	3	6	8	8	8	9	17	13	2	78
Obstetrics/gynecology	0	1	0	5	7	6	13	12	15	10	6	75
Oncology	2	6	7	8	7	4	8	14	12	19	4	89
Ophthalmology	0	1	0	1	3	8	2	6	3	6	0	30
Otolaryngology	0	0	0	0	0	23	1	2	2	1	1	30
Pediatrics/neonatology	0	0	1	8	5	4	2	10	17	9	0	56
Pharmacology/toxicology	0	1	0	2	0	6	1	0	7	3	0	20
Psychiatry/psychology	0	1	0	8	6	9	3	6	8	13	0	54
Pulmonary/respiratory diseases	0	2	3	4	8	5	6	10	17	19	2	76
Rheumatology	1	0	0	4	4	6	1	3	5	3	0	26
Trauma/emergency medicine	0	0	0	0	0	9	0	0	0	2	0	11
Total	15	26	40	93	95	152	85	147	234	200	44	1119

Note: This table does not include diagnostic agents, generics, over-the-counter products, medical devices, biologic compounds not approved by CDER, and most vaccines; also excluded are new dosages, new indications, and new administrations of previously approved compounds.

Category definitions:

Cardiology/vascular diseases: Diseases concerned with the structure and function of the heart and blood vessels. Studies in these areas include: heart failure, coronary artery disease, high cholesterol, blood clots, circulation disorders, and others.

Dental/maxillofacial surgery: Surgery relating to the teeth, jaw, face, and its structures. Studies in these areas include: acute and chronic dental pain, oral cavity cancer, oral facial pain, oral medicine, and saliva and salivary gland dysfunction.

Dermatology/plastic surgery: Fields concerned with skin disorders and the reconstruction or replacement of deformed, damaged, or lost parts of the body. Also concerns cosmetic surgery. Studies in these areas include: acne, congenital skin diseases, genital herpes, genital warts, liposuction, psoriasis, skin wounds, athlete's foot, venous leg ulcers, and others.

Endocrinology: Field relating to hormone-manufacturing glands, such as the pituitary, thyroid, parathyroid, and adrenal glands, as well as the ovary and testis, the placenta, and the pancreas. Studies in this area include: diabetes and diabetes-related disorders, diet and nutrition, hormone-replacement therapy, menopause, obesity, and others.

(*Continued*)

Table 1 Drugs Approved by the U.S. Food and Drug Administration During 1995 to 2005 (*Continued*)

Gastroenterology: The study of the gastrointestinal organs and diseases that relate to them. This includes any part of the digestive tract from mouth to anus, liver, biliary tract, and the pancreas. Studies in this area include: constipation, Crohn's disease, diarrhea, gall bladder disease, heartburn, hemorrhoids, irritable bowel syndrome (IBS), ulcers, liver disease, stomach cancer, and others.

Hematology: Field regarding blood, blood-forming tissues, and the diseases associated with them. Studies in this area include: anemia, blood clots, bone marrow transplant, leukemia, platelet disorders, red-cell disorders, T-cell lymphoma, vitamin deficiencies, white-cell disorders, and others.

Immunology/infectious diseases: Diseases affecting the defense mechanisms of the body. Studies in this area include: AIDS, auto-immune diseases, bacterial infections, chronic fatigue syndrome, common cold, genital herpes, genital warts, hepatitis, HIV infections, immunosuppressive, influenza, lyme disease, meningitis, parasite and protozoan infections, strep throat, vaccines, viral infections, and others.

Musculoskeletal: Field concerned with the muscles and the bones of the body. Studies in this area include: aging, bone density, bone fractures, chronic back pain, hip replacement, osteoarthritis, osteoporosis, rheumatoid arthritis, spinal cord injuries, and others.

Nephrology/urology: The studies and the treatment of diseases of the kidney and the urinary tract. Studies in these areas include: bladder cancer, impotence, kidney disease, kidney stones, nocturia, renal cell carcinoma, urinary tract infections, and others.

Neurology: Field concerning the nervous system, especially the brain, peripheral nerves, and spinal cord. Studies in this field include: Alzheimer's disease, attention deficit hyperactivity disorder (ADHD), carpal tunnel syndrome, Huntington's disease, dementia, memory loss, migraine headaches, multiple sclerosis, muscular dystrophy, Parkinson's disease, strokes, Tourette's syndrome, and others.

Obstetrics/gynecology: Research pertaining to the care of women during pregnancy and childbirth, as well as to the study of the women's reproductive system in general. Studies in these areas include: contraception, hormone-replacement therapy, menopause, menstrual disorders, ovarian cysts, postmenopausal syndrome, pregnancy/labor/delivery, yeast infections, and others.

Oncology: The medical, surgical, and radiation treatment of tumors (cancerous, especially). Studies in this area include most types of cancer and treatments thereof.

Ophthalmology: Field concerning the eye and eye diseases. Studies in this area include: cataracts, eye infections, glaucoma, macular degeneration, near-sighted corrective surgery, and others.

Otolaryngology: Also known as ENT (ears, nose, and throat), this is the study of diseases involving the ears and the larynx (organ that helps produce vocal sounds and serves as an air passageway, located in the neck/throat). Studies in this area include: allergy, ear infections, pneumonia, rhinitis, sinus infections, strep throat, and others.

Pediatrics/neonatology: The medical treatment and study of children and infants, respectively. Studies in these areas include: anorexia, asthma, ADHD, birth defects, cancers in children, child depression, growth deficiencies, juvenile diabetes, obesity, strep throat, vaccines, and others.

Pharmacology/toxicology: The science of drugs and poisonous materials (respectively) and their effects on the body. Studies in these areas include: diet and nutrition, overdoses, and vitamin deficiencies.

Psychiatry/psychology: Fields relating to mental disorders and their treatment and prevention. Also involves the study of human behavior. Studies in these areas include: addictions, anxiety, dementia, depression, bipolar disorders, manic disorders, mood disorders, post-traumatic stress disorders, schizophrenia, social phobia, substance abuse, and others.

Pulmonary/respiratory diseases: Diseases concerned with the lungs and/or breathing. Studies in these areas include: acute respiratory distress syndrome (ARDS), allergy, asthma, bronchitis, chronic obstructive pulmonary disease (COPD), cystic fibrosis, emphysema, lung disease, pneumonia, sinus infections, smoking cessation, and others.

Rheumatology: The field that relates to joints, tendons, muscles, ligaments, and associated structures. Studies in this area include: arthritis, osteoarthritis, rheumatic fever, rheumatoid arthritis, and others.

Trauma/emergency medicine: The medical specialties involving physical wounds or injuries. Studies in these areas include: athletic injuries, burns, traumatic brain injuries, and others.

Source: From Refs. 6 and 7.

Table 2 Most Active Categories for New Drug Development

Alzheimer's disease
Asthma and allergy
Attention deficit/hyperactivity disorders (ADHD)
Breast cancer
Chronic obstructive pulmonary disease (COPD)
Crohn's disease
Cystic fibrosis
Depression
Diabetes mellitus—Types 1 and 2
Eye disorders
Head and neck cancer
Hepatitis C
Hepatitis—all types
HIV/AIDS
Hyperlipidemia
Irritable bowel syndrome (IBS)
Kidney cancer
Leukemia
Lung cancer
Melanoma
Migraine
Obesity
Osteoporosis
Ovarian cancer
Parkinson's disease
Rheumatoid arthritis
Sexual dysfunction

Abbreviations: ADHD, attention deficit/hyperactivity disorders; COPD, chronic obstructive pulmonary disease; IBS, irritable bowel syndrome.

NEW PATHWAYS TO DISCOVERY

Whereas pharmaceutical companies take pride in their long-term research programs, the recent changes at the National Institutes of Health (NIH) must be looked at as remarkable opportunities for new drug development. The "New Pathways to Discovery" is the NIH Roadmap theme that sets out to advance our understanding of biological systems and to build a better "toolbox" for medical research in the twenty-first century (4), a theme that can truly revolutionize medicine and improve human health. We need a more detailed understanding of the vast networks of molecules that make up our cells and tissues, their interactions, and their regulation. We also must have a more precise knowledge of the combinations of molecular events that lead to diseases.

To capitalize on the completion of the human genome sequence and recent exciting discoveries in molecular and cell biology, the research community needs wide access to technologies, databases, and other scientific resources that are more sensitive, more robust, and more easily adaptable to researchers' individual needs. Resources are made available through five components of the New Pathways to Discovery theme: Building Blocks, Biological Pathways, and Networks; Molecular Libraries and Molecular Imaging; Structural Biology; Bioinformatics and Computational Biology; and Nanomedicine.

1. *Building blocks, biological pathways, and networks*: Complex elements—from individual genes to entire organs—work together in a feat of biological teamwork to promote normal development and sustain health. These systems

work because of intricate and interconnected pathways that enable communication among genes, molecules, and cells. Scientists are still working to discover all of these pathways and to determine how disturbances in them may lead to disease. A central component of an organism's biological pathways and networks is the set of proteins encoded by its genome, commonly referred to as the proteome. This NIH Roadmap initiative promotes the development of new proteomic technologies to enable researchers to expand the identification of biological pathways, with the ultimate goal being to understand diseases involving such pathways and, ideally, to develop potential treatments. Another critical focus is providing researchers with novel analytical tools to better understand the metabolic components and networks within the cell, commonly referred to as the metabolome. Ultimately, scientists hope to be able to completely map an organism's protein and metabolic networks, and create models to help predict the human body's response to disease, injury, or infection.

- *National technology centers for networks and pathways*: This initiative will encourage the development of highly sensitive tools to quantitatively measure the activity, translocation, and interactions of intracellular protein molecules. The National Technology Centers for Networks and Pathways will cooperate in a networked, national effort to develop highly novel technologies in the field of proteomics. These technologies will be directed at gathering information at the level needed to characterize subcellular processes. Beyond the cataloging of proteins and their interactions within cells, these methods will be aimed at defining the dynamics of complex intracellular systems. Ultimately, additional research grants supported by individual NIH Institutes will apply the advanced technologies in the centers to address a broad range of challenging biomedical research problems.
- *Metabolomics technology development*: The emerging field of metabolomics seeks to understand all the small molecules found within cells and tissues. To date, there is no single technology that can effectively measure, with sufficient sensitivity and precision, the diverse range of metabolites and their dynamic fluctuations within cells. This initiative will encourage the development of highly innovative and sensitive tools to identify and quantify cellular metabolites. The technologies developed under this initiative will make a major contribution to research on the chemical and molecular pathways in cells involved in development, normal function, aging, and disease.

2. *Molecular Libraries and molecular imaging*: The Molecular Libraries' initiative provides public sector biomedical researchers access to small organic molecules that can be used as chemical probes to study the functions of genes, cells, and biochemical pathways in health and disease. This component is also expected to facilitate the development of new drugs by providing early stage chemical compounds to researchers so they can find successful matches between a chemical and its target and thus help validate new targets with potential for therapeutic intervention. Unlike anatomical imaging, molecular imaging is an emerging research area that aims to display the biochemical and physiological abnormalities that underlie disease, rather than simply show the consequences of these abnormalities. The molecular imaging initiative will enhance the discovery and availability of technologies and reagents for imaging molecules

or molecular events within single cells and whole organisms. The ultimate goal is to enable a detailed molecular understanding of cell and tissue function in normal and disease states, which may lead to greater power to diagnose and treat disease.

- *Molecular Libraries Small Molecule Repository*: The Small Molecule Repository acquires, maintains, and distributes a collection of up to 500,000 compounds, obtained from both commercial and academic sources, with diverse chemical structures, and known or unknown biological activities. The repository will provide these compounds to the Molecular Libraries Screening Centers Network (MLSCN) for use in HTS of a diverse set of biological assays submitted by the research community and implemented within the centers. It is anticipated that "hits" identified by screening will be further developed into optimized chemical analogs that can be used by the scientific community as bioactive probes to study molecular targets and cellular pathways, and potentially as starting points for therapeutics development outside the MLSCN. The probes will be available to researchers via the repository, and the chemical structures of the compounds in the repository, along with the associated screening data obtained from the MLSCN, will be shared with the public through PubChem (5).
- *Molecular Libraries Screening Centers Network*: The MLSCN will be a national resource to empower scientists to explore biology using small molecules. This initiative will provide researchers in the public and private sectors with small molecules and will be linked to a larger database of biological information on small organic molecules (PubChem). These compounds will be useful as biological probes for the study of molecular and cellular pathways and phenotypes, and their functions in health and disease.
- *Molecular Libraries Screening Instrumentation*: The Molecular Libraries Screening Instrumentation (MLSI) initiative is designed to develop innovative instrumentation that can be integrated with HTS systems to identify small molecules to explore the biological mechanisms within living cells. HTS allows automated testing of vast numbers of these molecules at an extremely rapid rate. This initiative seeks to develop HTS instrumentation that will be faster, more efficient, and more accurate than currently available systems. Such new technologies will be necessary to achieve the ambitious goals of the Molecular Libraries Initiative to discover probes for biological processes on an unprecedented scale. These investigations will ultimately lead to the identification of novel targets for the treatment of diseases.
- *High-throughput molecular screening assay development*: HTS has great potential to provide insights into the mechanisms of cellular biology and disease, but to date, use of the technology has been limited because of inadequate access to screening facilities and compound libraries. This initiative will provide unprecedented access to these resources and allow a much broader research application of HTS. Its goal is to create a continuous stream of biological assays that can be used for automated screening at the NIH Molecular Libraries Screening Centers. The initiative will emphasize assays that provide insight into new cellular or molecular targets that have not been the focus of current HTS approaches.
- *Innovation in molecular imaging probes*: Molecular imaging is an emerging research area aimed at imaging specific molecular pathways in living tissues and cells, particularly those that are key targets in disease processes.

Unlike anatomical imaging, molecular imaging displays the biochemical and physiological abnormalities that underlie the disease, rather than simply the consequences of these abnormalities. Thus far, however, these methods have been limited by the poor sensitivity and specificity of currently used molecular probes. This initiative will encourage the development of new probes that will achieve an improvement of one to two orders of magnitude in the ability to detect and image specific molecular events in vivo. The new probes will also have potential for clinical applications.

- *Development of high-resolution probes for cellular imaging*: The goal of this initiative is to create molecular probes and imaging systems that are sensitive enough to detect and image individual molecules within living cells. The information gleaned will allow significant new insights into cellular processes, events, and changes over time. The improved technology that will result from these efforts will increase the resolution of the images within living cells by 10- to 100-fold, and will constitute an important step forward in understanding cell biology.

3. *Structural biology*: Proteins are indispensable molecules in our bodies, and each has a unique 3D shape that is well suited for its particular function. Some proteins build our cells and others work like miniature machines to allow us to think, smell, eat, and breathe. If the shape of even one protein goes awry, there can be major consequences for human health, such as cystic fibrosis, Alzheimer's disease, and countless other diseases. The NIH Roadmap Structural Biology initiative is an effort to create a "picture gallery" of the molecular shapes of proteins in the body, and is designed to advance our understanding of how proteins and their component parts function in the body. As an initial step, this will require the development of rapid, efficient, and dependable methods to produce protein samples that scientists can use to determine the structure of proteins and clarify the role of protein shape in normal and abnormal protein function. The data and tools will be shared with researchers across the nation. In later years, the initiative will focus on finding ways to discern structures of protein biomolecular "machines"—sets of proteins that act together to carry out essential cellular functions.

4. *Bioinformatics and computational biology*: The NIH Roadmap is paving a future, "information superhighway," dedicated to the advancement of medical research. Scientists today are using computers and robots to separate molecules in solution, read genetic information, reveal the 3D shapes of proteins, and take pictures of the brain in action. These techniques generate huge amounts of data, and biology is changing fast into a science of information management. Researchers need software programs and other tools to analyze, integrate, visualize, and model these data. Through the Bioinformatics and Computational Biology initiative, researchers will be able to seamlessly share data gathered from large experiments, such as the role of heredity in individuals' different responses to medicines.

5. *Nanomedicine*: A long-term goal of this NIH Roadmap is to create materials and devices at the level of molecules and atoms to cure diseases or repair damaged tissues, such as bones, muscles, or nerves. A nanometer is one-billionth of a meter, too small to be seen with a conventional lab microscope. And it is at this scale that biological molecules and structures inside living cells operate. Researchers have set their sights on replacing broken parts of a cell with miniature biological devices and searching out and destroying infectious

agents before they do harm. The opportunities for drug delivery systems based on small engines opens a new era of development for the pharmaceutical companies.

Research Trends

Several new disciplines of research have developed into drug discovery modules including genomics, bioinformatics, proteomics, microarray plans, and others to give the drug discovery groups newer tools to secure lead compounds. What follows are some of the current topics of research in drug discovery. Generally, a good assessment of these trends would be found in journals like *Nature Reviews Drug Discovery* (6).

Computer-assisted drug design (CADD), also called computer-assisted molecular design (CAMD), represents more recent applications of computers as tools in the drug design process, though it does not replace the human element in analyzing the data.

In considering this topic, it is important to emphasize that computers cannot substitute for a clear understanding of the system being studied. That is, a computer is only an additional tool to gain better insight into the chemistry and biology of the problem at hand.

Once potential drug designs have been identified by the methods described earlier, other molecular modeling techniques may then be applied. For example, geometric optimization may be used to "relax" the structures and to identify low-energy orientations of drugs in receptor sites. Molecular dynamics may assist in exploring the energy landscape, and free energy simulations can be used to compute the relative binding of free energies of a series of putative drugs. Many of these tools are available at the National Institutes of Health website (7).

In vitro screening for drugs that inhibit cytochrome P450 enzymes is well established as a means of predicting potential metabolism-mediated drug interactions in vivo. Given that these predictions are based on enzyme kinetic parameters observed from in vitro experiments, the miscalculation of the inhibitory potency of a compound can lead to an inaccurate prediction of an in vivo drug interaction, potentially preventing a safe drug from advancing in development or allowing a potent inhibitor to "slip" into the patient population. The processes that underlie the generation of in vitro drug metabolism data, and commonly encountered uncertainties and sources of bias and error that can affect extrapolation of drug–drug interaction information to the clinical setting, offer remarkable drug discovery opportunities.

Carbohydrates present both potential and problems—their biological relevance has been recognized, but problems in procuring sugars rendered them a difficult class of compounds to handle in drug discovery efforts. The development of the first automated solid-phase oligosaccharide synthesizer and other methods to rapidly assemble defined oligosaccharides have fundamentally altered this situation. There are now opportunities for the development of carbohydrate-based vaccines, defined heparin oligosaccharides, and aminoglycosides that have recently begun to influence drug discovery.

New perspectives on the complexity of G-protein-coupled receptor (GPCR) signaling and the increased resolution of existing tools for studying GPCR behavior has led to the conception of new hypotheses that affect the discovery of drugs acting at GPCRs. The novel concepts of collateral efficacy and permissive antagonism in the search for synthetic agonists and antagonists, respectively, will be essential in the search for drugs with unique therapeutic profiles. These concepts have been applied to design drugs against HIV as an example of how such concepts might be taken into consideration for GPCR-targeted drugs in general.

The phenomenon of multidrug efflux, whereby a single transporter is capable of recognizing and transporting multiple drugs with no apparent common structural similarity, was first described in higher eukaryotes where P-glycoprotein was found to provide resistance to anticancer chemotherapeutic agents via an adenosine triphosphate (ATP)-driven efflux process. In the late 1980s and early 1990s, it became apparent that multidrug efflux systems were also present in microorganisms, with the identification of bacterial multidrug transporters such as Bmr from *Bacillus subtilis*, QacA from *Staphylococcus aureus*, and EmrB from *Escherichia coli*. Since that time, the number of characterized multidrug efflux transporters has expanded dramatically, and it appears from genomic analyses that multidrug efflux systems are probably essentially ubiquitous.

Several disciplines of research have emerged to create remarkable new opportunities for drug development. Some important ones are described in the following sections.

Bioinformatics

In bioinformatics, the search strategy is based upon data collection and storage and the mining (retrieval and integration) of the databases in order to generate knowledge, that is, generation of knowledge (understanding what is important about a situation) from information or data (the sum of everything we know about that situation). Outcomes from this approach will include the identification of new drug targets via functional genomics, and it is driven by a number of significant changes in how we discover new lead compounds. Bioinformatics databases include DNA (genomes), RNA, protein sequences, proteomes, macromolecular structures, chemical diversity, biotransformations, metabolic pathways (metabolomes), biodiversity, and systematics. Thus, innovative "experiments" can be made in silico rather than in vivo or in vitro, so that only essential experiments need be undertaken.

Genomics

Whereas the chemical libraries of "drug-like" compounds can be readily screened to yield chemically novel compounds, the road to taking these leads to useful drugs remains a challenge. The off-patent and old drugs should not be shelved, as they offer a remarkable opportunity for newer uses; for example, apomorphine, now used to treat nausea, finds a role in treating erectile dysfunction. These older drugs can be useful in generating genomic databases. Genomics involves sequencing of genomes and leads to the derivation of theoretical information from the analysis of such sequences with computational tools. In contrast, functional genomics defines the transcriptome and proteome status of a cell, tissue, or organism under a prescribed set of conditions. The term transcriptome describes the transcription (mRNA) profile, whereas proteome describes the translation (protein) complement derived from a genome, including post-translational modifications (PTM) of proteins. The proteome information on the distribution of proteins within a cell or organism in time, space, and response to the environment. Together, genomics and functional genomics provide a precise molecular blueprint of a cell or organism, and in this and the following section we examine how they can reveal novel targets for search-and-discovery developments.

The Human Genome Project has a major impact on the identification of potential drug targets which influence the design of specific screens for therapeutic drugs. Potential therapeutic targets, such as Alzheimer's disease, angiogenesis, asthma, stroke, and cystic fibrosis, which are human genome-specific, multifactorial, and often involve complex signal cascades, may continue to dominate technology development. In addition, specific and sensitive molecular screens are now available using the same molecular biology technologies that drive the genome programs and use the sequence data from those studies to give high-throughput robotic screening. Initial success in the rational design for targets, such as HIV-1 protease, leads to strategies for rational design involving gene identification, metabolic pathway analysis, or determination of protein–protein interactions using affinity methods, such as the yeast two-hybrid system, phage display, or fluorescent-protein biosensors, structure prediction (8), and modeling.

Rational drug design strategies have not been as successful as anticipated, but other strategies that involve semirational design and HTS of massive libraries have worked better. Whereas combinatorial chemical libraries have served well, the current trend seems to shift towards biological libraries, such as those based on peptides and antibodies that control human disease. Many human diseases of interest to the pharmaceutical industry involve multiple gene pathways, environmental interactions, and genetic predisposition rather than simply direct causal effects. These factors also mediate adverse drug reactions and dictate the effectiveness of drug treatments. These considerations result in extensive comparative genome studies of ethnic populations and human disease states and expectations of personal genetic profiles.

Genomics has contributed to this rational search for drug targets by providing a large set of almost complete catalogues of genes, across a wide range of organisms, which can be compared at many levels. Conservation of genes across a wide range of organisms may prove to be a good indication of an essential function, and a minimal set of essential genes for life can be identified. Transposon mutagenesis and PCR can be used to directly screen for essential genes, and signature-tagged mutagenesis can be used to analyze multiple pools of mutants for loss of function. Identification of probable targets in silico allows these experimental molecular techniques to be used to search for a smaller set of target genes, making them more directed.

These search strategies can be applied to characterized or uncharacterized genes, and the chance of identifying a novel target may well be higher for uncharacterized genes. Uncharacterized gene targets may be identified in databases, such as COG (the database of Clusters of Orthologous Groups of proteins, which represents an attempt at a phylogenetic classification of the proteins encoded in complete genomes), and PROSITE [a database of protein families and domains. It consists of biologically significant sites, patterns, and profiles that help to reliably identify to which known protein family (if any) a new sequence belongs (9)] as uncharacterized gene targets that are conserved across groups, such as microbial pathogens. Characterized gene targets can be sought by using strategies to identify taxon-specific genes by employing subtractive techniques, most directly between a specific pathogen and the human genome. Another strategy that can be used to define subsets of genes is to search using subtractive techniques. In concordance analysis, the sequences present in one set of genomes and absent from others are determined, for example, bacterial genomes compared with eukaryotic genomes.

Searching DNA sequence databases for targets homologous to known drug targets in other organisms has revealed an aspartic protease, cyclophilin, and calcineurin, explaining the antimalarial activity of cyclosporin A. The full genome can be expected to provide many more potential targets.

Combinatorial pharmacogenetics seeks to characterize genetic variations that affect reactions to potentially toxic agents within the complex metabolic networks of the human body. Polymorphic drug-metabolizing enzymes are likely to represent some of the most common inheritable risk factors associated with common "disease" phenotypes, such as adverse drug reactions. The relatively high concordance between polymorphisms in drug-metabolizing enzymes and clinical phenotypes indicates that research into this class of polymorphisms could benefit patients in the near future. Characterization of other genes that affect drug disposition (absorption, distribution, metabolism, and elimination) will further enhance this process. As with most questions concerning biological systems, the complexity arises out of the combinatorial magnitude of all the possible interactions and pathways. The high-dimensionality of the resulting analysis problem will often overwhelm traditional analysis methods. Novel analysis techniques, such as multifactor dimensionality reduction, offer viable options for the evaluation of such data.

Pharmacoproteomics

Proteome analysis is composed of three sequential steps: sample preparation, protein separation and mapping, and protein characterization. Sample preparation may entail cell fractionation and preliminary removal of more abundant proteins in order to detect those present in low concentration. Analysis is always dependent upon effective protein separation, and 2D gel electrophoresis (most usually immobilized pH gradient followed by molecular weight separation) is the present method of choice. Between 2500 and 10,000 proteins are claimed to be resolvable on such gels. In addition to determining protein inventories, the analyses can be made quantitative with respect to individual proteins (detection is possible at 1 ng with silver staining and at less than 1 pg with fluorescent dyes). It is important to note that PTMs will significantly increase the number of separate proteins expressed from a genome and will not be revealed by genome annotation; the estimates are 1.2–1.3-fold for bacteria and threefold for eukaryotic microorganisms like *Saccharomyces cerevisiae*. Protein characterization is achieved by mass spectrometric amino acid sequencing and identity of PTMs, followed by interrogation of protein databases. In turn, this reverse genetics enables the identification of genes that are responsible for producing a particular protein expression profile.

At present, proteomics is being applied most actively in pharmaceutical research and development in two principal areas: drug discovery and target selection (e.g., via proteome difference analysis of pathogenic versus nonpathogenic organisms, normal versus dysfunctional states, and disruption of stress-induced protein synthesis) as well as drug mode of action, toxicological screening, and the monitoring of disease progression during clinical trials. The latter group of clinical interests, which is directed toward gaining a complete understanding of pharmacological mechanisms of drugs, drives the new field of pharmacoproteomics. On the one hand, natural product discovery and combinatorial synthesis can generate an enormous repertoire of candidate drugs; on the other hand, the demonstration of their mode of action, efficacy, and safety is hugely demanding in resources and time. The advent of pharmacoproteomics is set to transform these aspects of pharmaceutical development.

The development and application of proteomics constitute a very recent field of technology. Present limitations and areas in need of improvement include the resolution and characterization of hydrophobic proteins, which include major targets for pharmaceutical intervention (membrane enzymes and receptors), quality of protein separation, ability to detect very low copy number of proteins, and improved throughput and automation.

Phytomedicines

The search for and exploitation of natural products and properties has been the mainstay of the biotechnology industries. Natural product search and discovery, however, is not synonymous with drug discovery. But, if we examine how novel natural product chemotypes along with interesting structures and biological activities continue to be reported, this becomes the mainstay of drug discovery as it has been in the therapeutic areas, such as neurodegenerative disease, cardiovascular disease, most solid tumors, and immune-inflammatory diseases. Today, well over half the drugs are either directly derived from biological sources or have been produced as a result of biodiversity evaluation. Antibiotics remain the largest sellers of all drugs. Significantly, however, the reported discovery of microbial metabolites with nonantibiotic activities has increased progressively over the past 30 years, and now exceeds that of antibiotic compounds.

One prerequisite of natural product discovery that remains paramount is the range and novelty of molecular diversity. This diversity surpasses that of combinatorial chemical libraries and consequently provides unique lead compounds for drug and other developments. Newly discovered bioactive products do not usually become drugs per se, but may enter a chemical transformation program in which the bioactivity and pharmacodynamic properties are modified to suit particular therapeutic needs. In recent years, most regulatory authorities worldwide have carved out a path to approve natural products without having to identify the exact active ingredients of their specifications. The use of markers is suggested and greater emphasis is placed on characterization of extracts. This means greater involvement of the preformulation group in drug discovery for natural or botanical products. Recent efforts to harmonize standards are evident from the monographs developed by the American Herbal Pharmacopoeia (10). Another major effort is the European Scientific Cooperative on Phytotherapy (11). ESCOP was founded in June 1989 as an umbrella organization representing national phytotherapy associations across Europe, especially in their discussions with European medicine regulators. Since 1996, it has been a company in the United Kingdom. Its goals are to advance the scientific status of phytomedicines and to assist with the harmonization of their regulatory status at the European level, to develop a coordinated scientific framework to assess phytomedicines, to promote the acceptance of phytomedicines especially within general medical practice, to support and initiate clinical and experimental research in phytotherapy, to improve and extend the international accumulation of scientific and practical knowledge in the field of phytotherapy, to support all appropriate measures that will secure optimum protection for those who use phytomedicines, and to produce reference monographs on the therapeutic use of plant drugs. ESCOP has published a large number of monographs and established uniform standards for several botanical drugs (Table 3).

Table 3 List of Monographs of Phytomedicines Available from European Scientific Cooperative on Phytotherapy

Absinthii herba	Alexandrian Senna Pods
Aloe capensis	Aniseed
Anisi fructus	Arnica Flower
Arnicae flos	Bearberry Leaf
Blackcurrant Leaf	Cape Aloes
Caraway	*Carvi fructus*
Cascara	*Centaurii herba*
Centaury	*Crataegi folium* cum *flore*
Dandelion Leaf	Dandelion Root
Devil's Claw	*Echinaceae pallidae radix*
Echinaceae purpureae radix	*Eucalypti aetheroleum*
Eucalyptus Oil	Fennel
Feverfew	Frangula Bark
Frangulae cortex	Garlic Bulb
Gentan Root	*Gentianae radix*
Ginger	Golden Rod
Hamamelidis folium	Hamamelis Leaf
Harpagophyti radix	Hawthorn Leaf and Flower
Hippocastani semen	Hop Strobiles
Horse-Chestnut Seed	*Hyperici herba*
Iceland Moss	Ispaghula
Ispaghula Husk	Java Tea
Juniper Berries	*Juniperi fructus*
Lichen islandicus	*Lini semen*
Linseed	*Lupuli flos*
Matricaria Flower	*Matricariae flos*
Meliloti herba	Melilotus
Melissa Leaf	*Melissae folium*
Menthae piperitae aetheroleum	*Menthae piperitae folium*
Myrrh	*Myrrha*
Nettle Leaf and Herb	Nettle Root
Allii sativi bulbus	*Ononidis radix*
Orthosiphonis folium	Pale Coneflower Root
Passiflora	*Passiflorae herba*
Peppermint Leaf	Peppermint Oil
Plantaginis ovatae semen	*Plantaginis ovatae testa*
Polygalae radix	Primula Root
Primulae radix	*Psyllii semen*
Psyllium Seed	Purple Coneflower Root
Restharrow Root	*Rhamni purshiani cortex*
Rhei radix	Rhubarb Root
Ribis nigri folium	Rosemary
Rosmarini folium cum *flore*	Sage Leaf

(Continued)

Table 3 List of Monographs of Phytomedicines Available from European Scientific Cooperative on Phytotherapy (*Continued*)

Salicis cortex	*Salviae folium*
Senega Root	Senna Leaf
Sennae folium	*Sennae fructus acutifoliae*
Sennae fructus angustifoliae	*Solidaginis virgaureae herba*
St. John's Wort	*Tanaceti parthenii herba/folium*
Taraxaci folium	*Taraxaci radix*
Thyme	*Thymi herba*
Tinnevelly Senna Pods	*Urticae folium/herba*
Urticae radix	*Uvae ursi folium*
Valerian Root	*Valerianae radix*
Willow Bark	Wormwood
Zingiberis rhizoma	

ESCOP publishes a professional newsletter, *The European Phytojournal*, available online (12). Also available from ESCOP is the old *British Herbal Pharmacopoeia* and other important books.

The search for drugs through botanical sources has its roots in the realization that it is unlikely that some of the most successful drugs could have been discovered by any process of rational or semirational design. For example, the mode of action of the immunosuppressants, cyclosporin A and rapamycin, is that both bind to *cis-trans* prolyl isomerase, but only cyclosporine A is involved in further steps of signal transduction cascades through calcineurin; this would be too complex to design on an a priori basis. Similarly, two of the most successful antimalarial drugs, quinine and chloroquinine, exert their effect by inhibiting host-encoded functions rather than activities encoded by *Plasmodium falciparum* itself. Chloroquine resistance in *P. falciparum* resides in a 36-kDa nucleotide sequence, which contains genes of unknown function, along with 40% of the *P. falciparum* genome. In the search for new classes of antibiotics over the last 25 years, traditional approaches have also failed to deliver new drugs fast enough to keep up with the loss of effectiveness of existing drugs against increasingly resistant pathogens [95% of *Staphylococcus aureus* are penicillin resistant and 60% are methicillin resistant, and there are cases in China, Japan, Europe, and the United States of vancomycin resistance (13)]. The development of resistance may be followed by compensatory mechanisms to adjust for reduced fitness, which may then lock the resistance mechanism. The search for antibiotics through random screening seems to have been abandoned in view of the poor performance in finding leads in favor of rational, target-based approaches. Molecular biology, robotics, miniaturization, massively parallel preparation and detection systems, and automatic data analysis now dominate the search for drug discovery leads.

Natural product extracts and bacterial culture collections do not necessarily work well together on drug discovery platforms. The separation, identification, characterization, scale-up, and purification of natural products for large-scale libraries suitable for these HTS are daunting, and rational arguments for the selection of organisms and/or natural product molecules are often absent, especially given the poor taxonomic characterization of strains in natural product bacterial strain collections.

Many of these screening systems are not sufficiently robust to handle complex mixtures of natural products from ill-defined biological systems and may be inhibited by interactions with uncontrolled physicochemical conditions, simple toxic chemicals, and known bioactive compounds. This has led to significant efforts in rational drug design, combinatorial chemistry, peptide libraries, antibody libraries, and combinatorial biosynthesis and other synthetic and semisynthetic methods to provide clean inputs to screens. However, natural products are still unsurpassed in their ability to provide novelty and complexity. In chemical screening of natural products, complex mixtures of metabolites from growth and fermentation are separated, purified, and identified using high-pressure liquid chromatography, diode array ultraviolet (UV)/visible spectra, and mass spectrometry. Novel chemical structures are passed on for screening, now uncontaminated with background interference from the original complex mixture, and built up into high-quality, characterized natural product libraries. This strategy suffers from poorly characterized culture collections, which makes the choice of organisms for screening difficult, and from the inability to control the expression of metabolic potential.

Recombinant Drugs

Recombinant DNA techniques coupled with the typing of human genes now allows the development of a new class of drugs, ranging from expression of proteins to individualized drugs that would show specific response in patients. Once a biotechnological target has been identified, decisions must be made regarding the choice of the expression organism and the screening procedures to evaluate efficiency. Rules of thumb include relying on proven organisms such as actinomycetes, using taxon-chemistry and taxon-property databases to predict antibiotic potential, focusing on novel and overlooked taxa. An example of the last consideration is the study of cyanobacteria to isolate the human immunodeficiency (HIV)-inactivating protein cyanovirin-N. There is a strong view that biopharmaceutical leads are more likely to be detected in cell function assays than in in vitro assays. In this context, construction of surrogate host cells for in vivo drug screening is an interesting development. For example, the ability of *S. cerevisiae* to express heterologous proteins makes it an attractive option; it is used in screens based on substitution assays, differential expression assays, and transactivation assays, which are proved to be an effective route for drug discovery.

Considerable effort has been and is being expended in the development of screening assays, particularly as a response to the need for the evaluation of large numbers of samples in HTS and the expectation that many new targets will be identified in the wake of genome sequencing projects. HTS involves the robotic handling of very large numbers of candidate samples, the registering of appropriate signals from the assay system, and data management and interpretation. However, the advent of HTS, whereby lead discoveries may be identified in a matter of days from libraries of 103–105 compounds, may be limited by the provision of sufficient quantities of the assay components. The development of surrogate hosts, such as enzyme inhibition, receptor binding, and cell function assays provide possible means of alleviating such bottlenecks.

A large number of drugs manufactured through recombinant methods are coming off patents, creating markets for biogeneric medicines.

REFERENCES

1. http://www.elan.com/AboutUs/.
2. http://www.nektar.com.
3. http://www.phrma.org.
4. http://nihroadmap.nih.gov/newpathways/.
5. http://pubchem.ncbi.nlm.nih.gov.
6. http://nature.com/nrd/index.html.
7. http://cmm.info.nih.gov/modeling/guide_documents/about.html.
8. http://PredictionCenter.llnl.gov/.
9. http://www.expasy.org/prosite/.
10. http://www.herbal-ahp.org/index.html.
11. http://www.escop.com/.
12. http://www.ex.ac.uk/phytonet/phytojournal/.
13. http://www.promedmail.org.
14. http://www.fda.gov.
15. http://www.centerwatch.com/patient/drugs/druglist.html#.
16. http://www.accessdata.fda.gov/scripts/cder/drugsatfda/.

BIBLIOGRAPHY

Abu-Qare AW, Elmasry E, Abou-Donia MB. A role for P-glycoprotein in environmental toxicology. J Toxicol Environ Health B Crit Rev 2003; 6(3):279–288.

Agrafiotis DK, Lobanov VS, Salemme FR. Combinatorial informatics in the post-genomics ERA. Nat Rev Drug Discov 2002; 1(5):337–346.

Alayash AI. Oxygen therapeutics: can we tame haemoglobin? Nat Rev Drug Discov 2004; 3(2):152–159.

Anzali S, Barnickel G, Cezanne B, Krug M, Filimonov D, Poroikov V. Discriminating between drugs and nondrugs by prediction of activity spectra for substances (PASS). J Med Chem 2001; 44(15):2432–2437.

Apostolakis J, Caflisch A. Computational ligand design. Comb Chem High Throughput Screen 1999; 2(2):91–104.

Bailey D, Zanders E, Dean P. Site-specific molecular design and its relevance to pharmacogenomics and chemical biology. Pharmacogenomics J 2001; 1(1):38–47.

Bailey DG, Dresser GK. Interactions between grapefruit juice and cardiovascular drugs. Am J Cardiovasc Drugs 2004; 4(5):281–297.

Barreca ML, Rao A, De Luca L, et al. Efficient 3D database screening for novel HIV-1 IN inhibitors. J Chem Inf Comput Sci 2004; 44(4):1450–1455.

Barril X, Hubbard RE, Morley SD. Virtual screening in structure-based drug discovery. Mini Rev Med Chem 2004; 4(7):779–791.

Benet LZ, Cummins CL, Wu CY. Transporter-enzyme interactions: implications for predicting drug-drug interactions from in vitro data. Curr Drug Metab 2003; 4(5):393–398.

Bleicher KH, Bohm HJ, Muller K, Alanine AI. Hit and lead generation: beyond high-throughput screening. Nat Rev Drug Discov 2003; 2(5):369–378.

Bogman K, Zysset Y, Degen L, et al. P-glycoprotein and surfactants: effect on intestinal talinolol absorption. Clin Pharmacol Ther 2005; 77(1):24–32.

Breier A, Barancik M, Sulova Z, Uhrik B. P-glycoprotein—implications of metabolism of neoplastic cells and cancer therapy. Curr Cancer Drug Targets 2005; 5(6):457–468.

Briz O, Serrano MA, MacIas RI, Gonzalez-Gallego J, Marin JJ. Role of organic anion-transporting polypeptides, OATP-A, OATP-C and OATP-8, in the human placenta-maternal liver tandem excretory pathway for foetal bilirubin. Biochem J 2003; 371(Pt 3):897–905.

Bugrim A, Nikolskaya T, Nikolsky Y. Early prediction of drug metabolism and toxicity: systems biology approach and modeling. Drug Discov Today 2004; 9(3):127–135.

Butte A. The use and analysis of microarray data. Nat Rev Drug Discov 2002; 1(12): 951–960.

Carpenter D, Ting MM. Essay: the political logic of regulatory error. Nat Rev Drug Discov 2005; 4(10):819–823.

Carson SW, Ousmanou AD, Hoyler SL. Emerging significance of P-glycoprotein in understanding drug disposition and drug interactions in psychopharmacology. Psychopharmacol Bull 2002; 36(1):67–81.
Chalmers DT, Behan DP. The use of constitutively active GPCRs in drug discovery and functional genomics. Nat Rev Drug Discov 2002; 1(8):599–608.
Chandra P, Brouwer KL. The complexities of hepatic drug transport: current knowledge and emerging concepts. Pharm Res 2004; 21(5):719–735.
Chen HF, Dong XC, Zen BS, Gao K, Yuan SG, Panaye A, Doucet JP, Fan BT. Virtual screening and rational drug design method using structure generation system based on 3D-QSAR and docking. SAR QSAR Environ Res 2003; 14(4):251–264.
Chien A, Foster I, Goddette D. Grid technologies empowering drug discovery. Drug Discov Today 2002; 7(20 Suppl):S176–S180.
Chin-Dusting J, Mizrahi J, Jennings G, Fitzgerald D. Outlook: finding improved medicines: the role of academic-industrial collaboration. Nat Rev Drug Discov 2005; 4(11):891–897.
Chiou WL, Chung SM, Wu TC, Ma C. A comprehensive account on the role of efflux transporters in the gastrointestinal absorption of 13 commonly used substrate drugs in humans. Int J Clin Pharmacol Ther 2001; 39(3):93–101.
Crowe A, Wong P. pH Dependent uptake of loperamide across the gastrointestinal tract: an in vitro study. Drug Dev Ind Pharm 2004; 30(5):449–459.
Cummins CL, Jacobsen W, Benet LZ. Unmasking the dynamic interplay between intestinal P-glycoprotein and CYP3A4. J Pharmacol Exp Ther 2002; 300(3):1036–1045.
Cvetkovic M, Leake B, Fromm MF, Wilkinson GR, Kim RB. OATP and P-glycoprotein transporters mediate the cellular uptake and excretion of fexofenadine. Drug Metab Dispos 1999; 27(8):866–871.
Cvetkovic RS, Goa KL. Lopinavir/ritonavir: a review of its use in the management of HIV infection. Drugs 2003; 63(8):769–802.
Davis AM, Keeling DJ, Steele J, Tomkinson NP, Tinker AC. Components of successful lead generation. Curr Top Med Chem 2005; 5(4):421–439.
de Julian-Ortiz JV. Virtual Darwinian drug design: QSAR inverse problem, virtual combinatorial chemistry, and computational screening. Comb Chem High Throughput Screen 2001; 4(3):295–310.
Dey S, Gunda S, Mitra AK. Pharmacokinetics of erythromycin in rabbit corneas after single-dose infusion: role of P-glycoprotein as a barrier to in vivo ocular drug absorption. J Pharmacol Exp Ther 2004; 311(1):246–255. Epub June 2, 2004. Erratum in: J Pharmacol Exp Ther 2005; 314(1):483.
Doherty MM, Charman WN. The mucosa of the small intestine: how clinically relevant as an organ of drug metabolism? Clin Pharmacokinet 2002; 41(4):235–253.
Doppenschmitt S, Spahn-Langguth H, Regardh CG, Langguth P. Role of P-glycoprotein-mediated secretion in absorptive drug permeability: an approach using passive membrane permeability and affinity to P-glycoprotein. J Pharm Sci 1999; 88(10):1067–1072.
Dorsett Y, Tuschl T. siRNAs: applications in functional genomics and potential as therapeutics. Nat Rev Drug Discov 2004; 3(4):318–329.
Dresser GK, Bailey DG. The effects of fruit juices on drug disposition: a new model for drug interactions. Eur J Clin Invest 2003; 33(suppl 2):10–16.
Dresser GK, Bailey DG, Leake BF, et al. Fruit juices inhibit organic anion transporting polypeptide-mediated drug uptake to decrease the oral availability of fexofenadine. Clin Pharmacol Ther 2002; 71(1):11–20.
Dresser GK, Kim RB, Bailey DG. Effect of grapefruit juice volume on the reduction of fexofenadine bioavailability: possible role of organic anion transporting polypeptides. Clin Pharmacol Ther 2005; 77(3):170–177.
Dube DH, Bertozzi CR. Glycans in cancer and inflammation—potential for therapeutics and diagnostics. Nat Rev Drug Discov 2005; 4(6):477–488.
DuBuske LM. The role of P-glycoprotein and organic anion-transporting polypeptides in drug interactions. Drug Saf 2005; 28(9):789–801.
Eagling VA, Profit L, Back DJ. Inhibition of the CYP3A4-mediated metabolism and P-glycoprotein-mediated transport of the HIV-1 protease inhibitor saquinavir by grapefruit juice components. Br J Clin Pharmacol 1999; 48(4):543–552.

Ekins S, Berbaum J, Harrison RK. Generation and validation of rapid computational filters for cyp2d6 and cyp3a4. Drug Metab Dispos 2003; 31(9):1077–1080.
Ekins S, Rose J. In silico ADME/Tox: the state of the art. J Mol Graph Model 2002; 20(4):305–309.
Ellis C, Smith A. Highlighting the pitfalls and possibilities of drug research. Nat Rev Drug Discov 2004; 3(3):238–278. No abstract available.
Elsinga PH, Hendrikse NH, Bart J, Vaalburg W, van Waarde A. PET Studies on P-glycoprotein function in the blood-brain barrier: how it affects uptake and binding of drugs within the CNS. Curr Pharm Des 2004; 10(13):1493–1503.
Engels MF, Venkatarangan P. Smart screening: approaches to efficient HTS. Curr Opin Drug Discov Devel 2001; 4(3):275–283.
Engels MF, Wouters L, Verbeeck R, Vanhoof G. Outlier mining in high throughput screening experiments. J Biomol Screen 2002; 7(4):341–351.
Entzeroth M. Emerging trends in high-throughput screening. Curr Opin Pharmacol 2003; 3(5):522–529.
Ereshefsky L, Riesenman C, Lam YW. Antidepressant drug interactions and the cytochrome P450 system. The role of cytochrome P450 2D6. Clin Pharmacokinet 1995; 29(suppl 1):10–18; discussion 18–19.
Evans AM. Influence of dietary components on the gastrointestinal metabolism and transport of drugs. Ther Drug Monit 2000; 22(1):131–136.
Fagan R, Swindells M. Bioinformatics, target discovery and the pharmaceutical/biotechnology industry. Curr Opin Mol Ther 2000; 2(6):655–661.
Fejzo J, Lepre C, Xie X. Application of NMR screening in drug discovery. Curr Top Med Chem 2003; 3(1):81–97.
Fischer HP. Towards quantitative biology: integration of biological information to elucidate disease pathways and to guide drug discovery. Biotechnol Annu Rev 2005; 11:1–68.
Fischer HP, Heyse S. From targets to leads: the importance of advanced data analysis for decision support in drug discovery. Curr Opin Drug Discov Devel 2005; 8(3):334–346.
Fischer WJ, Altheimer S, Cattori V, Meier PJ, Dietrich DR, Hagenbuch B. Organic anion transporting polypeptides expressed in liver and brain mediate uptake of microcystin. Toxicol Appl Pharmacol 2005; 203(3):257–263.
Fitzgerald GA. Opinion: anticipating change in drug development: the emerging era of translational medicine and therapeutics. Nat Rev Drug Discov 2005; 4(10):815–818.
Frank R, Hargreaves R. Clinical biomarkers in drug discovery and development. Nat Rev Drug Discov 2003; 2(7):566–580.
Fromm MF. P-glycoprotein: a defense mechanism limiting oral bioavalability and CNS accumulation of drugs. Int J Clin Pharmacol Ther 2000; 38(2):69–74.
Funakoshi S, Murakami T, Yumoto R, Kiribayashi Y, Takano M. Role of organic anion transporting polypeptide 2 in pharmacokinetics of digoxin and beta-methyldigoxin in rats. J Pharm Sci 2005; 94(6):1196–1203.
Funakoshi S, Murakami T, Yumoto R, Kiribayashi Y, Takano M. Role of P-glycoprotein in pharmacokinetics and drug interactions of digoxin and beta-methyldigoxin in rats. J Pharm Sci 2003; 92(7):1455–1463.
Gao B, Huber RD, Wenzel A, et al. Localization of organic anion transporting polypeptides in the rat and human ciliary body epithelium. Exp Eye Res 2005; 80(1):61–72.
Glen RC, Allen SC. Ligand-protein docking: cancer research at the interface between biology and chemistry. Curr Med Chem 2003; 10(9):763–767.
Gombar VK, Silver IS, Zhao Z. Role of ADME characteristics in drug discovery and their in silico evaluation: in silico screening of chemicals for their metabolic stability. Curr Top Med Chem 2003; 3(11):1205–1225.
Good A. Structure-based virtual screening protocols. Curr Opin Drug Discov Devel 2001; 4(3):301–307.
Good AC, Cheney DL. Analysis and optimization of structure-based virtual screening protocols (1): exploration of ligand conformational sampling techniques. J Mol Graph Model 2003; 22(1):23–30.
Gozalbes R, Doucet JP, Derouin F. Application of topological descriptors in QSAR and drug design: history and new trends. Curr Drug Targets Infect Disord 2002; 2(1): 93–102.

Green DV. Virtual screening of virtual libraries. Prog Med Chem 2003; 41:61–97.
Greiner B, Eichelbaum M, Fritz P, et al. The role of intestinal P-glycoprotein in the interaction of digoxin and rifampin. J Clin Invest 1999; 104(2):147–153. Erratum in: J Clin Invest 2002; 110(4):571.
Hagenbuch B, Meier PJ. Organic anion transporting polypeptides of the OATP/SLC21 family: phylogenetic classification as OATP/SLCO superfamily, new nomenclature and molecular/functional properties. Pflugers Arch 2004; 447(5):653–665. Epub 2003, Oct 25.
Harris RZ, Jang GR, Tsunoda S. Dietary effects on drug metabolism and transport. Clin Pharmacokinet 2003; 42(13):1071–1088.
Hochman JH, Chiba M, Nishime J, Yamazaki M, Lin JH. Influence of P-glycoprotein on the transport and metabolism of indinavir in Caco-2 cells expressing cytochrome P-450 3A4. J Pharmacol Exp Ther 2000; 292(1):310–318.
Hochman JH, Pudvah N, Qiu J, et al. Interactions of human P-glycoprotein with simvastatin, simvastatin acid, and atorvastatin. Pharm Res 2004; 21(9):1686–1691.
Hochman JH, Yamazaki M, Ohe T, Lin JH. Evaluation of drug interactions with P-glycoprotein in drug discovery: in vitro assessment of the potential for drug-drug interactions with P-glycoprotein. Curr Drug Metab 2002; 3(3):257–273.
Hong H, Tong W, Xie Q, Fang H, Perkins R. An in silico ensemble method for lead discovery: decision forest. SAR QSAR Environ Res 2005; 16(4):339–347.
Hopfinger AJ, Duca JS. Extraction of pharmacophore information from high-throughput screens. Curr Opin Biotechnol 2000; 11(1):97–103.
Hou T, Xu X. Recent development and application of virtual screening in drug discovery: an overview. Curr Pharm Des 2004; 10(9):1011–1033.
Huth JR, Sun C. Utility of NMR in lead optimization: fragment-based approaches. Comb Chem High Throughput Screen 2002; 5(8):631–643.
Inui KI, Masuda S, Saito H. Cellular and molecular aspects of drug transport in the kidney. Kidney Int 2000; 58(3):944–958.
Issa AM. Ethical perspectives on pharmacogenomic profiling in the drug development process. Nat Rev Drug Discov 2002; 1(4):300–308.
Jacoby E, Davies J, Blommers MJ. Design of small molecule libraries for NMR screening and other applications in drug discovery. Curr Top Med Chem 2003; 3(1):11–23.
Jacoby E, Schuffenhauer A, Floersheim P. Chemogenomics knowledge-based strategies in drug discovery. Drug News Perspect 2003; 16(2):93–102.
Jacoby E, Schuffenhauer A, Popov M, et al. Key aspects of the Novartis compound collection enhancement project for the compilation of a comprehensive chemogenomics drug discovery screening collection. Curr Top Med Chem 2005; 5(4):397–411.
Jain AN. Virtual screening in lead discovery and optimization. Curr Opin Drug Discov Devel 2004; 7(4):396–403.
Johnson BM, Charman WN, Porter CJ. Application of compartmental modeling to an examination of in vitro intestinal permeability data: assessing the impact of tissue uptake, P-glycoprotein, and CYP3A. Drug Metab Dispos 2003; 31(9):1151–1160.
Johnson BM, Charman WN, Porter CJ. The impact of P-glycoprotein efflux on enterocyte residence time and enterocyte-based metabolism of verapamil. J Pharm Pharmacol 2001; 53(12):1611–1619.
Jonsdottir SO, Jorgensen FS, Brunak S. Prediction methods and databases within chemoinformatics: emphasis on drugs and drug candidates. Bioinformatics 2005; 21(10):2145–2160. Epub 2005, 15 Feb.
Jorgensen WL. The many roles of computation in drug discovery. Science 2004; 303(5665):1813–1818.
Kajosaari LI, Niemi M, Neuvonen M, Laitila J, Neuvonen PJ, Backman JT. Cyclosporine markedly raises the plasma concentrations of repaglinide. Clin Pharmacol Ther 2005; 78(4):388–399.
Kamath AV, Yao M, Zhang Y, Chong S. Effect of fruit juices on the oral bioavailability of fexofenadine in rats. J Pharm Sci 2005; 94(2):233–239.
Kantardjieff K, Rupp B. Structural bioinformatic approaches to the discovery of new antimycobacterial drugs. Curr Pharm Des 2004; 10(26):3195–3211.

Karnachi PS, Brown FK. Practical approaches to efficient screening: information-rich screening protocol. J Biomol Screen 2004; 9(8):678–686. Erratum in: J Biomol Screen 2005; 10(1):93.

Kellogg GE, Semus SF. 3D QSAR in modern drug desin. EXS 2003; 93:223–241. No abstract available.

Kenakin T. New concepts in drug discover: collateral efficacy and permissive antagonism. Nat Rev Drug Discov 2005; 4(11):919–927.

Kharasch ED, Hoffer C, Altuntas TG, Whittington D. Quinidine as a probe for the role of p-glycoprotein in the intestinal absorption and clinical effects of fentanyl. J Clin Pharmacol 2004; 44(3):224–233.

Kharasch ED, Hoffer C, Whittington D, Sheffels P. Role of P-glycoprotein in the intestinal absorption and clinical effects of morphine. Clin Pharmacol Ther 2003; 74(6): 543–554.

Kim RB. Organic anion-transporting polypeptide (OATP) transporter family and drug disposition. Eur J Clin Invest 2003; 33(suppl 2):1–5.

Kitchen DB, Decornez H, Furr JR, Bajorath J. Docking and scoring in virtual screening for drug discovery: methods and applications. Nat Rev Drug Discov 2004; 3(11): 935–949.

Kitchen DB, Stahura FL, Bajorath J. Computational techniques for diversity analysis and compound classification. Mini Rev Med Chem 2004; 4(10):1029–1039.

Kivisto KT, Niemi M, Fromm MF. Functional interaction of intestinal CYP3A4 and P-glycoprotein. Fundam Clin Pharmacol 2004; 18(6):621–626.

Klebe G. Recent developments in structure-based drug design. J Mol Med 2000; 78(5):269–281.

Kobayashi D, Nozawa T, Imai K, Nezu J, Tsuji A, Tamai I. Involvement of human organic anion transporting polypeptide OATP-B (SLC21A9) in pH-dependent transport across intestinal apical membrane. J Pharmacol Exp Ther 2003; 306(2):703–708. Epub 2003, 30 Apr.

Kramer R, Cohen D. Functional genomics to new drug targets. Nat Rev Drug Discov 2004; 3(11):965–972.

Krejsa CM, Horvath D, Rogalski SL, et al. Predicting ADME properties and side effects: the BioPrint approach. Curr Opin Drug Discov Devel 2003; 6(4):470–480.

Kusuhara H, Sugiyama Y. Active efflux across the blood–brain barrier: role of the solute carrier family. NeuroRx 2005; 2(1):73–85.

Kusuhara H, Sugiyama Y. Efflux transport systems for organic anions and cations at the blood–CSF barrier. Adv Drug Deliv Rev 2004; 56(12):1741–1763.

Kusuhara H, Sugiyama Y. Role of transporters in the tissue-selective distribution and elimination of drugs: transporters in the liver, small intestine, brain and kidney. J Control Release 2002; 78(1–3):43–54.

Lanctot JK, Putta S, Lemmen C, Greene J. Using ensembles to classify compounds for drug discovery. J Chem Inf Comput Sci 2003; 43(6):2163–2169.

Langer T, Hoffmann RD. Virtual screening: an effective tool for lead structure discovery? Curr Pharm Des 2001; 7(7):509–527.

Langer T, Krovat EM. Chemical feature-based pharmacophores and virtual library screening for discovery of new leads. Curr Opin Drug Discov Devel 2003; 6(3): 370–376.

Lengauer C, Diaz LA Jr, Saha S. Cancer drug discovery through collaboration. Nat Rev Drug Discov 2005; 4(5):375–380.

Lengauer T, Lemmen C, Rarey M, Zimmermann M. Novel technologies for virtual screening. Drug Discov Today. 2004; 9(1):27–34. PMID: 14761803 [PubMed—indexed for MEDLINE].

Lennernas H. Clinical pharmacokinetics of atorvastatin. Clin Pharmacokinet 2003; 42(13):1141–1160.

Leung S, Bendayan R. Role of P-glycoprotein in the renal transport of dideoxynucleoside analog drugs. Can J Physiol Pharmacol 1999; 77(8):625–630.

Lewell XQ, Judd DB, Watson SP, Hann MM. RECAP—retrosynthetic combinatorial analysis procedure: a powerful new technique for identifying privileged molecular fragments with useful applications in combinatorial chemistry. J Chem Inf Comput Sci 1998; 38(3):511–522.

Ley SV, Baxendale IR. New tools and concepts for modern organic synthesis. Nat Rev Drug Discov 2002; 1(8):573–586.

Lin J, Sahakian DC, de Morais SM, Xu JJ, Polzer RJ, Winter SM. The role of absorption, distribution, metabolism, excretion and toxicity in drug discovery. Curr Top Med Chem 2003; 3(10):1125–1154.

Lin JH. Drug–drug interaction mediated by inhibition and induction of P-glycoprotein. Adv Drug Deliv Rev 2003; 55(1):53–81.

Lin JH. How significant is the role of P-glycoprotein in drug absorption and brain uptake? Drugs Today (Barc) 2004; 40(1):5–22.

Lin JH, Lu AY. Inhibition and induction of cytochrome P450 and the clinical implications. Clin Pharmacokinet 1998 Nov; 35(5):361–390.

Lin JH, Yamazaki M. Role of P-glycoprotein in pharmacokinetics: clinical implications. Clin Pharmacokinet 2003; 42(1):59–98.

Lindsay MA. Target discovery. Nat Rev Drug Discov 2003; 2(10):831–838.

Littman BH, Williams SA. The ultimate model organism: progress in experimental medicine. Nat Rev Drug Discov 2005; 4(8):631–638.

Loscher W, Potschka H. Role of drug efflux transporters in the brain for drug disposition and treatment of brain diseases. Prog Neurobiol 2005; 76(1):22–76.

Lundqvist T. The devil is still in the details—driving early drug discovery forward with biophysical experimental methods. Curr Opin Drug Discov Devel 2005; 8(4): 513–519.

Lyall A. Bioinformatics in the pharmaceutical industry. Trends Biotechnol 1996; 14(8):308–312.

Lyne PD. Structure-based virtual screening: an overview. Drug Discov Today 2002; 7(20):1047–1055.

Marzolini C, Tirona RG, Kim RB. Pharmacogenomics of the OATP and OAT families. Pharmacogenomics 2004; 5(3):273–282.

Mason JS, Cheney DL. Library design and virtual screening using multiple 4-point pharmacophore fingerprints. Pac Symp Biocomput 2000; xx:576–587.

Masuda S, Inui K. [Molecular mechanisms on drug transporters in the drug absorption and disposition] Nippon Rinsho 2002; 60(1):65–73.

Matsushima S, Maeda K, Kondo C, et al. Identification of the hepatic efflux transporters of organic anions using double-transfected Madin-Darby canine kidney II cells expressing human organic anion-transporting polypeptide 1B1 (OATP1B1)/multidrug resistance-associated protein 2, OATP1B1/multidrug resistance 1, and OATP1B1/breast cancer resistance protein. J Pharmacol Exp Ther 2005; 314(3):1059–1067. Epub 2005, 18 May.

Merlot C, Domine D, Church DJ. Fragment analysis in small molecule discovery. Curr Opin Drug Discov Devel 2002; 5(3):391–399.

Mestres J. Computational chemogenomics approaches to systematic knowledge-based drug discovery. Curr Opin Drug Discov Devel 2004; 7(3):304–313.

Michelson S. Assessing the impact of predictive biosimulation on drug discovery and development. J Bioinform Comput Biol 2003; 1(1):169–177.

Mikkaichi T, Suzuki T, Tanemoto M, Ito S, Abe T. The organic anion transporter (OATP) family. Drug Metab Pharmacokinet 2004; 19(3):171–179.

Mizutani MY, Itai A. Efficient method for high-throughput virtual screening based on flexible docking: discovery of novel acetylcholinesterase inhibitors. J Med Chem 2004; 47(20):4818–4828.

Molnar L, Keseru GM. A neural network based virtual screening of cytochrome P450 3A4 inhibitors. Bioorg Med Chem Lett 2002; 12(3):419–421.

Moore J, Abdul-Manan N, Fejzo J, et al. Leveraging structural approaches: applications of NMR-based screening and X-ray crystallography for inhibitor design. J Synchrotron Radiat 2004; 11(Pt 1):97–100. Epub 2003, 28 Nov.

Moore WR Jr. Maximizing discovery efficiency with a computationally driven fragment approach. Curr Opin Drug Discov Devel 2005; 8(3):355–364.

Muchmore SW, Hajduk PJ. Crystallography, NMR and virtual screening: integrated tools for drug discovery. Curr Opin Drug Discov Devel 2003; 6(4):544–549.

Muegge I. Selection criteria for drug-like compounds. Med Res Rev 2003; 23(3): 302–321.

Narayanan R, Gunturi SB. In silico ADME modelling: prediction models for blood–brain barrier permeation using a systematic variable selection method. Bioorg Med Chem 2005; 13(8):3017–3028.

Nicholson JK, Connelly J, Lindon JC, Holmes E. Metabonomics: a platform for studying drug toxicity and gene function. Nat Rev Drug Discov 2002; 1(2):153–161.
Nicholson JK, Wilson ID. Opinion: understanding "global" systems biology: metabonomics and the continuum of metabolism. Nat Rev Drug Discov 2003; 2(8):668–676. No abstract available.
Nozawa T, Imai K, Nezu J, Tsuji A, Tamai I. Functional characterization of pH-sensitive organic anion transporting polypeptide OATP-B in human. J Pharmacol Exp Ther 2004; 308(2):438–445. Epub 2003, 10 Nov.
Nozawa T, Sugiura S, Nakajima M, et al. Involvement of organic anion transporting polypeptides in the transport of troglitazone sulfate: implications for understanding troglitazone hepatotoxicity. Drug Metab Dispos 2004; 32(3):291–294.
O'Connor KA, Roth BL. Finding new tricks for old drugs: an efficient route for public-sector drug discovery. Nat Rev Drug Discov 2005; 4(12):1005–1014.
Oprea TI, Matter H. Integrating virtual screening in lead discovery. Curr Opin Chem Biol 2004; 8(4):349–358.
Patel J, Buddha B, Dey S, Pal D, Mitra AK. In vitro interaction of the HIV protease inhibitor ritonavir with herbal constituents: changes in P-gp and CYP3A4 activity. Am J Ther 2004; 11(4):262–277.
Patel J, Mitra AK. Strategies to overcome simultaneous P-glycoprotein mediated efflux and CYP3A4 mediated metabolism of drugs. Pharmacogenomics 2001; 2(4):401–415.
Pea F, Furlanut M. Pharmacokinetic aspects of treating infections in the intensive care unit: focus on drug interactions. Clin Pharmacokinet 2001; 40(11):833–868.
Perkins R, Fang H, Tong W, Welsh WJ. Quantitative structure-activity relationship methods: perspectives on drug discovery and toxicology. Environ Toxicol Chem 2003; 22(8):1666–1679.
Perri D, Ito S, Rowsell V, Shear NH. The kidney—the body's playground for drugs: an overview of renal drug handling with selected clinical correlates. Can J Clin Pharmacol 2003; 10(1):17–23.
Petri N, Tannergren C, Rungstad D, Lennernas H. Transport characteristics of fexofenadine in the Caco-2 cell model. Pharm Res 2004; 21(8):1398–1404.
Petricoin EF, Zoon KC, Kohn EC, Barrett JC, Liotta LA. Clinical proteomics: translating benchside promise into bedside reality. Nat Rev Drug Discov 2002; 1(9):683–695.
Pirard B. Computational methods for the identification and optimisation of high quality leads. Comb Chem High Throughput Screen 2004; 7(4):271–280.
Pirard B. Knowledge-driven lead discovery. Mini Rev Med Chem 2005; 5(11):1045–1052.
Pizzagalli F, Hagenbuch B, Stieger B, Klenk U, Folkers G, Meier PJ. Identification of a novel human organic anion transporting polypeptide as a high affinity thyroxine transporter. Mol Endocrinol 2002; 16(10):2283–2296.
Pottorf RS, Player MR. Process technologies for purity enhancement of large discovery libraries. Curr Opin Drug Discov Devel 2004; 7(6):777–783.
Renfrey S, Featherstone J. Structural proteomics. Nat Rev Drug Discov 2002; 1(3):175–176.
Renwick AG. The metabolism of antihistamines and drug interactions: the role of cytochrome P450 enzymes. Clin Exp Allergy 1999; 29(suppl 3):116–124.
Roberts PM, Hayes WS. Advances in text analytics for drug discovery. Curr Opin Drug Discov Devel 2005; 8(3):323–328.
Roberts SA. High-throughput screening approaches for investigating drug metabolism and pharmacokinetics. Xenobiotica 2001; 31(8–9):557–589.
Roche O, Guba W. Computational chemistry as an integral component of lead generation. Mini Rev Med Chem 2005; 5(7):677–683.
Roden DM, George AL Jr. The genetic basis of variability in drug responses. Nat Rev Drug Discov 2002; 1(1):37–44.
Rollinger JM, Haupt S, Stuppner H, Langer T. Combining ethnopharmacology and virtual screening for lead structure discovery: COX-inhibitors as application example. J Chem Inf Comput Sci 2004; 44(2):480–488.
Sams-Dodd F. Target-based drug discovery: is something wrong? Drug Discov Today 2005; 10(2):139–147.

Sauvant C, Holzinger H, Gekle M. Short-term regulation of basolateral organic anion uptake in proximal tubular OK cells: EGF acts via MAPK, PLA(2), and COX1. J Am Soc Nephrol 2002; 13(8):1981–1991.
Savchuk NP, Balakin KV, Tkachenko SE. Exploring the chemogenomic knowledge space with annotated chemical libraries. Curr Opin Chem Biol 2004; 8(4):412–417.
Schellens JH, Malingre MM, Kruijtzer CM, et al. Modulation of oral bioavailability of anticancer drugs: from mouse to man. Eur J Pharm Sci 2000; 12(2):103–110.
Schiffer R, Neis M, Holler D, et al. Active influx transport is mediated by members of the organic anion transporting polypeptide family in human epidermal keratinocytes. J Invest Dermatol 2003; 120(2):285–291.
Schneider G. Trends in virtual combinatorial library design. Curr Med Chem 2002; 9(23):2095–2101.
Schuffenhauer A, Popov M, Schopfer U, Acklin P, Stanek J, Jacoby E. Molecular diversity management strategies for building and enhancement of diverse and focused lead discovery compound screening collections. Comb Chem High Throughput Screen 2004; 7(8):771–781.
Schwardt O, Kolb H, Ernst B. Drug discovery today. Curr Top Med Chem 2003; 3(1):1–9.
Seeberger PH, Werz DB. Automated synthesis of oligosaccharides as a basis for drug discovery. Nat Rev Drug Discov 2005; 4(9):751–763.
Shitara Y, Hirano M, Sato H, Sugiyama Y. Gemfibrozil and its glucuronide inhibit the organic anion transporting polypeptide 2 (OATP2/OATP1B1:SLC21A6)-mediated hepatic uptake and CYP2C8-mediated metabolism of cerivastatin: analysis of the mechanism of the clinically relevant drug-drug interaction between cerivastatin and gemfibrozil. J Pharmacol Exp Ther 2004; 311(1):228–236. Epub 2004, 11 June.
Shitara Y, Sato H, Sugiyama Y. Evaluation of drug–drug interaction in the hepatobiliary and renal transport of drugs. Annu Rev Pharmacol Toxicol 2005; 45:689–723.
Shoichet BK. Virtual screening of chemical libraries. Nature 2004; 432(7019):862–865.
Siraki AG, Chevaldina T, Moridani MY, O'Brien PJ. Quantitative structure-toxicity relationships by accelerated cytotoxicity mechanism screening. Curr Opin Drug Discov Devel 2004; 7(1):118–125.
Smith PA, Sorich MJ, Low LS, McKinnon RA, Miners JO. Towards integrated ADME prediction: past, present and future directions for modelling metabolism by UDP-glucuronosyltransferases. J Mol Graph Model 2004; 22(6):507–517.
So SS, Karplus M. Evaluation of designed ligands by a multiple screening method: application to glycogen phosphorylase inhibitors constructed with a variety of approaches. J Comput Aided Mol Des 2001; 15(7):613–647.
Soldner A, Christians U, Susanto M, Wacher VJ, Silverman JA, Benet LZ. Grapefruit juice activates P-glycoprotein-mediated drug transport. Pharm Res 1999; 16(4):478–485.
Stahura FL, Bajorath J. Partitioning methods for the identification of active molecules. Curr Med Chem 2003; 10(8):707–715.
Stahura FL, Bajorath J. Virtual screening methods that complement HTS. Comb Chem High Throughput Screen 2004; 7(4):259–269.
Stoughton RB, Friend SH. How molecular profiling could revolutionize drug discovery. Nat=Rev Drug Discov 2005; 4(4):345–350.
St-Pierre MV, Hagenbuch B, Ugele B, Meier PJ, Stallmach T. Characterization of an organic anion-transporting polypeptide (OATP-B) in human placenta. J Clin Endocrinol Metab 2002; 87(4):1856–1863.
Su Y, Zhang X, Sinko PJ. Human organic anion-transporting polypeptide OATP-A (SLC21A3) acts in concert with P-glycoprotein and multidrug resistance protein 2 in the vectorial transport of Saquinavir in Hep G2 cells. Mol Pharm 2004; 1(1):49–56.
Sugiyama D, Kusuhara H, Taniguchi H, et al. Functional characterization of rat brain-specific organic anion transporter (Oatp14) at the blood-brain barrier: high affinity transporter for thyroxine. J Biol Chem 2003; 278(44):43489–43495. Epub 2003, 15 Aug. Erratum in: J Biol Chem 2003; 278(49):49662.
Sugiyama Y, Kusuhara H, Suzuki H. Kinetic and biochemical analysis of carrier-mediated efflux of drugs through the blood-brain and blood-cerebrospinal fluid barriers: importance in the drug delivery to the brain. J Control Release 1999; 62(1–2): 179–186.

Sumner-Smith M. Beginning to manage drug discovery and development knowledge. Curr Opin Drug Discov Devel 2001; 4(3):319–324.

Sun H, Dai H, Shaik N, Elmquist WF. Drug efflux transporters in the CNS. Adv Drug Deliv Rev 2003; 55(1):83–105.

Sun H, Frassetto L, Benet LZ. Effects of renal failure on drug transport and metabolism. Pharmacol Ther 2006; 109(1–2):1–11. Epub 2005, 8 Aug.

Suzuki H, Sugiyama Y. Role of metabolic enzymes and efflux transporters in the absorption of drugs from the small intestine. Eur J Pharm Sci 2000; 12(1):3–12.

Suzuki H, Sugiyama Y. Transport of drugs across the hepatic sinusoidal membrane: sinusoidal drug influx and efflux in the liver. Semin Liver Dis 2000; 20(3):251–263.

Swaan PW, Ekins S. Reengineering the pharmaceutical industry by crash-testing molecules. Drug Discov Today 2005; 10(17):1191–1200.

Tahara H, Kusuhara H, Fuse E, Sugiyama Y. P-glycoprotein plays a major role in the efflux of fexofenadine in the small intestine and blood-brain barrier, but only a limited role in its biliary excretion. Drug Metab Dispos 2005; 33(7):963–968. Epub 2005, 8 Apr.

Tan DS. Current progress in natural product-like libraries for discovery screening. Comb Chem High Throughput Screen 2004; 7(7):631–643.

Thomas SA. Drug transporters relevant to HIV therapy. J HIV Ther 2004; 9(4):92–96.

Tirona RG, Kim RB. Pharmacogenomics of organic anion-transporting polypeptides (OATP). Adv Drug Deliv Rev 2002; 54(10):1343–1352.

Tirona RG, Leake BF, Wolkoff AW, Kim RB. Human organic anion transporting polypeptide-C (SLC21A6) is a major determinant of rifampin-mediated pregnane X receptor activation. J Pharmacol Exp Ther 2003; 304(1):223–228.

Treiber A, Schneiter R, Delahaye S, Clozel M. Inhibition of organic anion transporting polypeptide-mediated hepatic uptake is the major determinant in the pharmacokinetic interaction between bosentan and cyclosporin A in the rat. J Pharmacol Exp Ther 2004; 308(3):1121–1129. Epub 2003, 14 Nov.

van de Waterbeemd H, Gifford E. ADMET in silico modelling: towards prediction paradise? Nat Rev Drug Discov 2003; 2(3):192–204.

van der Greef J, McBurney RN. Innovation: Rescuing drug discovery: in vivo systems pathology and systems pharmacology. Nat Rev Drug Discov 2005; 4(12):961–967.

van Dongen M, Weigelt J, Uppenberg J, Schultz J, Wikstrom M. Structure-based screening and design in drug discovery. Drug Discov Today 2002; 7(8):471–478.

van Montfoort JE, Hagenbuch B, Groothuis GM, Koepsell H, Meier PJ, Meijer DK. Drug uptake systems in liver and kidney. Curr Drug Metab 2003; 4(3):185–211.

Varma MV, Sateesh K, Panchagnula R. Functional role of P-glycoprotein in limiting intestinal absorption of drugs: contribution of passive permeability to P-glycoprotein mediated efflux transport. Mol Pharm 2005; 2(1):12–21.

Vavricka SR, van Montfoort J, Ha HR, Meier PJ, Fattinger K. Interactions of rifamycin SV and rifampicin with organic anion uptake systems of human liver. Hepatology 2002; 36(1):164–172.

Verdonk ML, Hartshorn MJ. Structure-guided fragment screening for lead discovery. Curr Opin Drug Discov Devel 2004; 7(4):404–410. PMID: 15338949 [PubMed— indexed for MEDLINE].

Veselovsky AV, Ivanov AS. Strategy of computer-aided drug design. Curr Drug Targets Infect Disord 2003; 3(1):33–40.

Viswanadhan VN, Balan C, Hulme C, Cheetham JC, Sun Y. Knowledge-based approaches in the design and selection of compound libraries for drug discovery. Curr Opin Drug Discov Devel 2002; 5(3):400–406.

von Moltke LL, Weemhoff JL, Perloff MD, et al. Effect of zolpidem on human cytochrome P450 activity, and on transport mediated by P-glycoprotein. Biopharm Drug Dispos 2002; 23(9):361–367.

Walters WP, Namchuk M. Designing screens: how to make your hits a hit. Nat Rev Drug Discov 2003; 2(4):259–266.

Wandel C, Kim RB, Guengerich FP, Wood AJ. Mibefradil is a P-glycoprotein substrate and a potent inhibitor of both P-glycoprotein and CYP3A in vitro. Drug Metab Dispos 2000; 28(8):895–898.

Wang E, Lew K, Barecki M, Casciano CN, Clement RP, Johnson WW. Quantitative distinctions of active site molecular recognition by P-glycoprotein and cytochrome P450 3A4. Chem Res Toxicol 2001; 14(12):1596–1603.
Waszkowycz B. Structure-based approaches to drug design and virtual screening. Curr Opin Drug Discov Devel 2002; 5(3):407–413.
Weber A, Teckentrup A, Briem H. Flexsim-R: a virtual affinity fingerprint descriptor to calculate similarities of functional groups. J Comput Aided Mol Des 2002; 16(12):903–916.
Weinshilboum R, Wang L. Pharmacogenomics: bench to bedside. Nat Rev Drug Discov 2004; 3(9):739–748.
Wenk MR. The emerging field of lipidomics. Nat Rev Drug Discov 2005; 4(7):594–610. Erratum in: Nat Rev Drug Discov 2005; 4(9):725.
Wienkers LC, Heath TG. Predicting in vivo drug interactions from in vitro drug discovery data. Nat Rev Drug Discov 2005; 4(10):825–833.
Wilke RA, Reif DM, Moore JH. Combinatorial pharmacogenetics. Nat Rev Drug Discov 2005; 4(11):911–918.
Winkler DA. Neural networks as robust tools in drug lead discovery and development. Mol Biotechnol 2004; 27(2):139–168.
Xu C, Li CY, Kong AN. Induction of phase I, II and III drug metabolism/transport by xenobiotics. Arch Pharm Res 2005; 28(3):249–268.
Xu H. Retrospect and prospect of virtual screening in drug discovery. Curr Top Med Chem 2002; 2(12):1305–1320.
Xue L, Bajorath J. Molecular descriptors in chemoinformatics, computational combinatorial chemistry, and virtual screening. Comb Chem High Throughput Screen 2000; 3(5):363–372.
Yamazaki M, Suzuki H, Sugiyama Y. Recent advances in carrier-mediated hepatic uptake and biliary excretion of xenobiotics. Pharm Res 1996; 13(4):497–513.
Yasuda K, Lan LB, Sanglard D, Furuya K, Schuetz JD, Schuetz EG. Interaction of cytochrome P450 3A inhibitors with P-glycoprotein. J Pharmacol Exp Ther 2002; 303(1):323–332.
Yi B, Hughes-Oliver JM, Zhu L, Young SS. A factorial design to optimize cell-based drug discovery analysis. J Chem Inf Comput Sci 2002; 42(5):1221–1229.
Yu H, Adedoyin A. ADME-Tox in drug discovery: integration of experimental and computational technologies. Drug Discov Today 2003; 8(18):852–861.
Yumoto R, Murakami T, Sanemasa M, Nasu R, Nagai J, Takano M. Pharmacokinetic interaction of cytochrome P450 3A-related compounds with rhodamine 123, a P-glycoprotein substrate, in rats pretreated with dexamethasone. Drug Metab Dispos 2001; 29(2):145–151.
Zavodszky MI, Sanschagrin PC, Korde RS, Kuhn LA. Distilling the essential features of a protein surface for improving protein-ligand docking, scoring, and virtual screening. J Comput Aided Mol Des 2002; 16(12):883–902.
Zernov VV, Balakin KV, Ivaschenko AA, Savchuk NP, Pletnev IV. Drug discovery using support vector machines. The case studies of drug-likeness, agrochemical-likeness, and enzyme inhibition predictions. J Chem Inf Comput Sci 2003; 43(6):2048–2056.
Zhang Y, Benet LZ. The gut as a barrier to drug absorption: combined role of cytochrome P450 3A and P-glycoprotein. Clin Pharmacokinet 2001; 40(3):159–168.
Zhang Y, Guo X, Lin ET, Benet LZ. Overlapping substrate specificities of cytochrome P450 3A and P-glycoprotein for a novel cysteine protease inhibitor. Drug Metab Dispos 1998; 26(4):360–366.
Zhou S, Yung Chan S, Cher Goh B, et al. Mechanism-based inhibition of cytochrome P450 3A4 by therapeutic drugs. Clin Pharmacokinet 2005; 44(3):279–304.

2 Intellectual Property Considerations

INTRODUCTION

The development of new pharmaceutical or biotechnology-derived products undergoes a lengthy and expensive cycle, which is adequately described elsewhere in the book. Given the cost of development now hovering around the billion dollar mark for each new drug, amortized over all other molecules under development, the only way to protect this investment is to create intellectual property claims to not only the active molecule but where possible, each and every step of its production, processing, and testing. Almost 4% of the development cost is spent on filing and prosecuting patents. Preformulation scientists play an important role in creating intellectual property because at this stage the specifications of the new drug entity are defined, and often newer lead compounds are substituted.

Typically, the research-based pharmaceutical companies spend about 15–20% of their total sales revenue on developing new drugs, as compared with less than 4% for the industry as a whole; the cost of patent protection ranges between 1% and 3% of the R&D expenditure. Successful patenting, patent protection, and exploitation of expired patents involve a complex interaction among scientists and lawyers. Generally, the development teams should have a basic understanding of the patenting process to be able to make the best use of legal expertise in the field; the more complex the field is, the more input from scientists becomes valuable. In this chapter, I describe the fundamentals of patenting that I consider important enough for scientists to understand well.

PATENTING STRATEGIES

Pharmaceutical and biotechnology scientists face several major challenges in designing their research:

- Does the research create a product that is prohibited by statute to be patented; this includes a thought process, a law of physics, an object of little utility or another prohibited area of patentability? Obviously, the end goal is not always to secure a patent for a product; a proprietary process need not be patented if it is possible to keep it protected, something which is becoming very difficult to assure as the information flow becomes easier between individuals and around the world.
- Is the research likely to lead to a novel product or process meaning something that has never existed before? Novel is not necessarily patentable as we go through the legality of the patenting process; this is, however, one of the fundamental requirements.

- Is the novel research unobvious to those with ordinary skill in the art? This aspect of patenting is most confusing and often leads to the most rejections of patent applications. Researchers need to study existing art (what is called prior art) before designing experiments to assure that they can create sufficient features in the invention to take it out of the obviousness arena. What is needed here is a demonstration that there was an inventive step, or in ordinary words, something was actually discovered that was not something that could have been easily discovered. A lot of experimentation goes into demonstrating unusual results to obviate the assertion of obviousness by the patent examiners.
- Can the claim made withstand court challenge? Even though these areas of interpretations are beyond what a scientist would be expected to have any expertise, a keen understanding of the scope of claims is essential; a patent with too broad a claim is not necessarily a good patent nor is a patent with too narrow a claim. The patent disclosure must support the claim adequately.
- Does the disclosure in the patent meet the legal requirements of patent allowance? In the United States, the patentee must disclose a best mode, the best formula, and the best approach to use the research outcome. There is no need to disclose an industrial model but there should be a workable model. Keeping the information out of a patent can be a double-edge sword.
- Is it possible to create a sequel to a successful research product? It is not unusual for the companies to keep some portions of the research aside and out of a patent to be able to claim it later. This can be a dangerous practice. If prior art appears in the interim, then all research is lost and even one's own patent will be taken as a prior art.
- Is it possible to continue the patent coverage by designing products around an issued patent? This is most critical to the pharmaceutical and biotechnology industries. Good examples of this practice include changes in the formulation to improve the product, such as Abbott did when its patent on calcitriol expired; the company changed the specification on the level of oxygen in the solution. What it meant was that there would be no generic equivalent to Abbott's calcitriol, though the generic forms may not be any less active. Smith Kline Beecham's remarkable patenting of the combination of amoxicillin and clavulanic is a good example of intelligent patenting; you will find not only patents for a specific combination of the two antibiotics, but a score of different dosage forms; one patent is based only on the hardness of the tablet. This type of research requires a keen understanding of the patenting process.
- Is it possible to invent around a patent or a publication? When sildenafil citrate became a success and statins became the choice of treatment for lowering cholesterol, the drug companies rushed to make similar products; obviously, there were problems in prior art. What is already disclosed cannot be obviated. The researchers are then faced with a challenge to find solutions novel enough to be patented; some degree of reverse engineering is required. However, as we see the results, they did succeed. But, with each patent issued the field narrows as to what can be patented. Scientists often get bogged down with the science and do not realize that to receive a patent on an invention there is no need to explain how it works. In fact, there need not even be an explanation about how it came into existence. Just the fact that there is a working example is sufficient. In many instances even a working example is not required. A new

drug may be subject to Food and Drug Administration (FDA) comments before the United States Patent Office allows the claims; however, it is not required of the patent office. On the other hand, the patent office need not accept the approval by the regulatory authorities as a proof of the utility, a key requirement for patentability.

Obviously, a productive researcher contributes to the company's profits. To achieve marketability and profitability from the research, the product must be either patentable or of a type that could not be reverse-engineered if proprietary techniques are relied upon. The challenges to scientists are great, with expanding technology and stiffer competition to produce new products. One way to meet these challenges head on is to understand the art and the science of patenting. Keeping abreast of the knowledge is critical, and scientists are strongly urged to hone their skills in the use of computers to search the literature.

PATENTING SYSTEMS

A patent is a grant by the government of exclusive rights for a limited time with respect to a new and useful invention, rights to prevent others from making or selling the invention in a limited territory as defined by the patent-issuing authority. A patent is not a right to sell the invention. The word "patent" means "open," and the words, "letter patent," mean an announcement made to all that the inventor has been awarded rights to the invention. Open here means without having to break any seal, as was the custom in the decrees issued in the past by the sovereign governments. Note that "letters patent" are not restricted to inventions, as even today these are issued to appoint judges in the United Kingdom. Historically, patents were issued on elaborate stationary, as they still are to some extent. The point about the rights to exercise an invention, vis-à-vis to prevent others from practicing, needs further elaboration. The new chemical entity patent for sildenafil citrate expired in March 2003; the use and composition patents listed for sildenafil citrate in the Orange Book (FDA) extended to 2012 and 2019; 17 U.S. patents were issued to Pfizer on the various aspects of sildenafil citrate. A total of 44 patents making claims for sildenafil citrate covered its use in Tourette's syndrome, its chewing gum formulations that require chewing the gum for not less than two minutes, and other uses. The example of erythropoietin demonstrates how the patent laws can create great difficulties for biogeneric offerings. A classical example of exploiting patenting systems is the patent on interferon, held by both Roche and Biogen; when the patent was filed, the patent office declared interference, meaning similar inventions submitted by two companies. Both Roche and Biogen fought it for decades while agreeing to cross license while the patent was pending; decades later when the patent was issued, they agreed to cross license as well, resulting in an unprecedented 35+ years of protection; this will likely be challenged in the courts.

WHAT IS A PATENT?

It is noteworthy that patents are awarded for things or *"res,"* which must have some use; there need not be any rationale provided as to why an invention works; obviously it must meet all statutory requirements to be eligible for patenting. The patent laws are extremely complex, often inexplicably irrational, and in almost all instances questionable in their enforcement.

The following definitions and terms are commonly used in describing patent laws:

- *Invention*: An invention is the conception of a new and useful article, machine, composition, or process.
- *Patent application*: A document describing an invention in detail, which is to be submitted to a patent office with the aim of obtaining a patent on the invention.
- *Patent*: Right of ownership granted by the government to a person that gives the owner the right to exclude others from making, selling, or using the claimed invention.
- *Reduction to practice*: An in-depth description of how the invention works, described in concrete terms.
- *Prior art*: The existing or public knowledge available before the date of an invention or more than one year prior to the first patent application date.
- *Utility*: This is the most common type of patent. It includes inventions that operate in a new and useful manner.
- *Design*: The emphasis of this type of patent is on the design of the invention and not on its functionality. What are important with this type of patent are the invention's unique ornamental and aesthetic properties.
- *Plant*: This type of patent includes new varieties of asexually reproduced plants.
- *Actual reduction to practice*: Constructing the machine or article, synthesizing the composition, or performing the method and testing sufficiently to demonstrate that the invention works for its intended purpose. Testing is NOT required if one with ordinary skills in the art would recognize that it will work.
- *Constructive reduction to practice*: Filing U.S. patent application—in compliance with the first paragraph of §112.
- *Diligence*: Working on reducing the invention to practice, or else it is considered abandoned.

PATENT MYTHS

The patenting of inventions is a complex process that has historically been confusing, particularly to the inventors who may not be practitioners of patent law. For example:

- Patents are valuable only if they can be used to protect a profit stream by excluding others from making, using, or selling whatever is covered by the patent's claims.
- A patent does not mean that the invention works as verified by the government; that is left for the licensors to evaluate; it is suspected that as many as 10% of all issued patents are invalid for being nonfunctional as claimed.
- A Provisional Application is not "just describing the idea," "it is a complete application except for the required claim(s)." You may not change anything in the body of the application when you file the regular application if you want to take priority advantage.
- You cannot get a patent for an idea or mere suggestion. Patents are granted to people who (claim to) "invent or discover any new and useful process, machine, manufacture, or composition of matter, or any new and useful improvement thereof," to quote the essence of the U.S. statute governing patents. Complete and enabling disclosure is also required.

- A patent can be enforceable from the time it is issued till it expires, not necessarily 20 years. New rules provide some guarantee that the enforceable term of a utility patent will be at least 17 years, and that some royalties may be collectable when a patent is published before it is issued. Design patents are only good for 14 years and only cover the ornamental appearance of the item, and not its structure or functionality.
- A patent does not give the owner the exclusive right to make, use, and sell the invention; it gives its owner the right to EXCLUDE others from making, using, and selling exactly what is covered by their patent claims. A holder of a prior patent with broader claims may prevent the inventor whose patent has narrower claims from using the inventor's own patent. A patent right is exclusory only.
- A U.S. patent is only enforceable in the United States. It can be used to stop others from importing what the patent covers into the United States, but people in other countries are free to make, use, and sell the invention anywhere else in the world that the inventor does not also have a patent. This is the reason why one must consider filing a Patent Cooperative Treaty and follow it up with either individual state filings or consortium filing, such as European Patent Office.
- A patent does not protect an invention because only a patent in conjunction with a legal opinion of infringement will give the owner(s) of the patent the right to sue in a civil case against the alleged infringer. The U.S. Government does not enforce patents (however, the Customs Service can help block infringing imports) and infringement of a patent is not a crime. The responsibility, and all expenses, for enforcing the rights granted by a patent (and securing Customs Service help) lie with the patent owner(s).
- Filing for a patent is not the only way to protect an invention. When properly used, the United States Patent and Trademark Office Disclosure Document Program ($10), non-Disclosure Agreements (free), and Provisional Applications for Patent ($80) along with maintaining good records and diligent pursuit can keep your patenting rights intact until you do file.
- A patent attorney or agent is not needed to file your patent; an inventor may choose to go pro se (on his own). However, given the complexity of the law, advice from professionals always proves invaluable.

Of great importance to the pharmaceutical industry is the Drug Price Competition and Patent Restoration Act of 1984, commonly known as the Hatch/Waxman Act, which provided for extensions of patent terms for human drugs, food additives, and medical devices. The commercialization of this act had been delayed by regulatory procedures in the FDA, making registration easier for competitors when patent protection expired, and allowing that testing for regulatory approval involving a patented drug did not amount to patent infringement.

There are three principal requirements for an invention to be patented as set out in the European Patent Convention:

- that the invention must be new;
- that it must involve an inventive step; and
- that it must be capable of industrial application.

The same three requirements must be met in one form or another in the United States, Japan, and indeed in practically every country that has a patent system at all. Some countries and conventions exclude certain inventions but these exclusions may be forbidden under the Trade Related Aspects of Intellectual Property section (TRIPS) of the World Trade Organization regulations.

WHAT IS NOT PATENTABLE?

There are certain specific exceptions to patentability, which apply whether or not the invention is capable of industrial application. Artistic works and esthetic creations are not patentable, and are generally not industrially applicable either; but scientific theories and mathematical methods, the presentation of information, business methods, and computer programs are also unpatentable, although they may very well be applied in industry.

Animal and plant varieties are not patentable in countries adhering to the European Patent Commission, although in the United States plants may be protected either by normal utility patents or by special plant patents for plant varieties. In the United Kingdom and certain other European countries new plant varieties, although not patentable, can be protected by plant breeders' rights granted under the UPOV (The International Union for the Protection of New Varieties of Plants) convention. Transgenic plants and animals are in principle patentable only if they do not constitute a variety that might create a difficult and uncertain situation. A further exception applies to offensive, immoral, or antisocial inventions, which need not just be an article prohibited by the law; if it is abhorrent to society, it is unpatentable, though it may be legal. A portable nuclear device would be such an example. Also excluded are inventions contrary to well-establishednatural laws, for example, a perpetual motion machine, though the European Patent Commission does not spell it out like this.

PATENT SEARCH

The patent offices worldwide have opened their databases to the public; there is no better place to start the search for patentability than with these free databases; the same databases that provide additional services and literature search are packaged by other vendors. The United States Patent and Trademark Office (1) has created one of the world's largest electronic databases that includes every patent issued; recently, published applications are also available in the database. Scientists are strongly urged to develop expert skills in interacting with the database of the United States Patent and Trademark Office. The search at United States Patent and Trademark Office can be most beneficial if the scientist learns how to use the patent classification system. (Tutorials are available at the United States Patent and Trademark Office website; alternately, please consult *Filing Patents Online: A Professional Guide* by Sarfaraz K. Niazi, CRC Press 2002).

The U.S. Patent Office Classification 435 includes the following subcategories related to therapeutic proteins.

Class 435: Chemistry: Molecular Biology and Microbiology, provides for methods of purifying, propagating, or attenuating a microorganism; for example, a virus, bacteria, etc., except for propagating a microorganism in an animal for the purpose of producing an antibody-containing sera. Class 435 provides for methods of propagating animal organs, tissues, or cells; for example, blood, sperm, etc., and culture media. Class 435 is the generic home for processes of: (*i*) analyzing or testing that involve a fermentation step or (*ii*) qualitative or quantitative testing for fermentability, or fermentative power.

435: Chemistry: Molecular Biology and Microbiology, see appropriate subclasses for processes in which a material containing an enzyme or microorganism is used to perform a qualitative or quantitative measurement or test; for compositions or test strips for either of the stated processes; for the processes of making such compositions or test strips; for the processes of using micro-organisms or enzymes to synthesize a chemical product; for the processes of treating a material with microorganisms or enzymes to separate, liberate, or purify a pre-existing substance or to destroy hazardous or toxic wastes; for processes of propagating microorganisms; for processes of genetically altering a microorganism; for processes of tissue, organ, blood, sperm, or microbial maintenance; for processes of malting or mashing; for microorganisms, per se, and subcellular parts thereof; for recombinant vectors and their preparation; for enzymes, per se, compositions containing enzymes not otherwise provided for and processes of preparing and purifying enzymes; for compositions for microbial propagation; for apparatus for any of the processes of the class; for composting apparatus; and subclasses 4+ for in vitro processes in which there is a direct or indirect, qualitative or quantitative, measurement or test, or of a material that contains an enzyme or microorganism [for the purposes of Class 435, microorganisms include bacteria, actinomycetales, cyanobacteria (unicellular algae), fungi, protozoa, animal cells, plant cells, and virus]. Class 424 definition contains controlling statements on the class lines.

93.2: Genetically modified microorganisms, such as a cell or a virus (e.g., transformed, fused, hybrid, and the like): This subclass is indented under subclass 93.1. Subject matter involving a microorganism, cell, or virus, which (*i*) is a product of recombination, transformation, or transfection with a vector or a foreign or exogenous gene or (*ii*) is a product of homologous recombination if it is directed rather than spontaneous or (*iii*) is a product of fused or hybrid cell formation. (1) Note. Examples of subject matter included in this and the indented subclass are compositions containing microorganisms, cells, or viruses resulting from: (*i*) a process in which the cellular matter of two or more fusing partners is combined producing a cell, which initially contains the genes of both the fusing partners or (*ii*) a process in which a cell is treated with an immortalizing agent which results in a cell that proliferates in long-term culture or (*iii*) a process involving recombinant DNA methodology. (2) Note. Excluded from this subclass are products of unidentified or noninduced mutations; products of microbial conjugation wherein specific genetic material is not identified and controlled; and products of natural, spontaneous, or arbitrary conjugation or recombination events. These products are not considered genetically modified for this subclass and therefore will be classified as unmodified microorganisms, cells, or viruses.

93.21: Eukaryotic cell: This subclass is indented under subclass 93.2. Subject matter involving a eukaryotic cell, such as an animal cell, plant cell, fungus, protozoa, or higher algae, which has been genetically modified. (1) Note. A eukaryotic cell has a nucleus defined by a nuclear membrane wherein the nucleus contains chromosomes that comprise the genome of the cell.

93.3: Intentional mixture of two or more microorganisms, cells, or viruses of different genera: This subclass is indented under subclass 93.1. Subject matter involving a mixture consisting of two or more different microbial, cellular, or viral genera. (1) Note. A mixture of *Escherichia coli* and *Pseudomonas* or a mixture of *Aspergillus* and *Bacillus* would be considered proper for this subclass while a mixture of *Bacillus cereus* and *Bacillus brevis* would be classified under *Bacillus* rather than in this subclass as they are both in the genus, *Bacillus*. (2) Note. Rumen, intestinal, vaginal, and other microflora mixtures are mixtures appropriate for this subclass unless mixture constituents are disclosed and are found to be contrary to the subclass definition.

133.1: Structurally modified antibody, immunoglobulin, or fragment thereof (e.g., chimeric, humanized, complementarity determining region-grafted, mutated, and the like): This subclass is indented under subclass 130.1. Subject matter involving an antibody, immunoglobulin, or fragment thereof that is purposely altered with respect to its amino acid sequence or glycosylation, or with respect to its composition of heavy and light chains or immunoglobulin regions or domains, as compared with that found in nature; or wherein the antibody, immunoglobulin, or fragment thereof is part of a larger, synthetic protein. (1) Note. Structurally modified antibodies may be made by chemical alteration or recombination of existing antibodies, or by various cloning techniques involving recombinant DNA or hybridoma technology. (2) Note. Structurally modified antibodies may be chimeric (i.e., comprising amino acid sequences derived from two or more nonidentical immunoglobulin molecules, such as interspecies combinations, and so on). (3) Note. Structurally modified antibodies may have domain deletions or substitutions (e.g., deletions of particular constant-region domains or substitutions of constant-region domains from other classes of immunoglobulins). (4) Note. Structurally modified antibodies may have deletions of particular glycosylated amino acids, or may have their glycosylation otherwise altered, which may alter their function. (5) Note. While expression of cloned antibody genes in cells of species other than from which they originated may result in altered glycosylation of the product, compared with that found in nature, this subclass and indented subclasses are not meant to encompass such antibodies or fragments thereof, unless such cloning is a deliberate attempt to alter their glycosylation. However, such antibodies or fragments thereof may still be classified here or in indented subclasses if they are structurally modified in other ways (e.g., if they are single chain, and the like). (6) Note. It is suggested that the patents of this subclass and indented subclasses be cross-referenced to the appropriate subclass(es) that provide for the binding specificities of these antibodies, if disclosed.

141.1: Monoclonal antibody or fragment thereof (i.e., produced by any cloning technology): This subclass is indented under subclass 130.1. Subject matter involving an antibody or fragment thereof produced by a clone of cells or cell line, which are derived from a single antibody-producing cell or antibody fragment-producing cell, wherein the said antibody or fragment thereof is identical to all other antibodies or fragments thereof produced by that clone of cells or cell line. (1) Note. This and the indented subclasses provide for bioaffecting and body-treating compositions of antibodies or fragments thereof as well as bioaffecting and body-treating methods of using said compositions, said antibodies, or said fragments, which are produced by any cloning technology that yields identical molecules (e.g., hybridoma technology, recombinant DNA technology, and so on). (2) Note. Monoclonal antibodies, per se, are considered compounds and are provided for elsewhere. See the search notes that follow. (3) Note. Monoclonal antibodies are sometimes termed monoclonal receptors or immunological binding partners.

1.49 and 1.53, for methods of using radiolabeled monoclonal antibodies or compositions thereof for bioaffecting or body-treating purposes and said compositions, per se.

9.1+ for methods of using monoclonal antibodies or compositions thereof for in vivo testing or diagnosis and said compositions, per se.

178.1+ for bioaffecting or body-treating methods of using monoclonal antibodies or fragments thereof that are conjugated to or complexed with nonimmunoglobulin material; bioaffecting or body-treating methods of using compositions of monoclonal antibodies or fragments thereof, which are conjugated to or complexed with nonimmunoglobulin material; and said compositions, per se.

199.1: Recombinant virus encoding one or more heterologous proteins or fragments thereof: This subclass is indented under subclass 184.1. Subject matter involving a virus into whose genome is integrated one or more nucleic acid sequences encoding one or more heterologous proteins or fragments thereof. (1) Note. A heterologous protein is one derived from another species (e.g., another viral species). (2) Note. Such genetically modified viruses may be used as multivalent vaccines.

200.1: Recombinant or stably transformed bacterium encoding one or more heterologous proteins or fragments thereof: This subclass is indented under subclass 184.1. Subject matter involving a bacterium into whose genome is integrated one or more nucleic acid sequences encoding one or more heterologous proteins or fragments thereof; or involving a bacterium that carries stable, replicative plasmids that include one or more nucleic acid sequences encoding one or more heterologous proteins or fragments thereof. (1) Note. A heterologous protein is one derived from another species (e.g., another bacterial species). (2) Note. Such genetically modified bacteria may be used as multivalent vaccines.

201.1: Combination of viral and bacterial antigens (e.g., multivalent viral and bacterial vaccines, and so on): This subclass is indented under subclass 184.1. Subject matter involving a combination of viral and bacterial antigens, such as that found in a multivalent viral and bacterial vaccine.

202.1: Combination of antigens from multiple viral species (e.g., multivalent viral vaccine, and so on): This subclass is indented under subclass 184.1. Subject matter involving a combination of antigens from multiple viral species, such as that found in a multivalent viral vaccine. (1) Note. A combination of antigens from multiple variants of the same viral species should be classified with that viral species.

203.1: Combination of antigens from multiple bacterial species (e.g., multivalent bacterial vaccine, and so on): This subclass is indented under subclass 184.1. Subject matter involving a combination of antigens from multiple bacterial species, such as that found in a multivalent bacterial vaccine. (1) Note. A combination of antigens from multiple variants of the same bacterial species should be classified with that bacterial species.

801: *Involving Antibody or Fragment thereof Produced by Recombinant DNA Technology*: This subclass is indented under the class definition. Subject matter involves an antibody or fragment thereof produced by recombinant DNA technology.

A search under CCL/"435/69.1" yields 8898 patents including the earliest patents wherein insulin was produced by genetically modified fungi from the University of Minnesota and the two classic patents from Stanford and Columbia.

Second to the United States Patent Office, the largest database is accessed through the European Patent Office, where one should conduct a similar classification search as suggested previously for the United States Patent and Trademark Office (2). The World Intellectual Property Organization (3) offers many useful features including complete details of the Patent Cooperative Treaty and its gazette. The Canadian Patent Office can be reached at the website mentioned in Ref. 4.

Internet Search Engines

Though intended for the lay public, the Internet can turn up remarkable information particularly as it pertains to prior art. The following sites are recommended:

www.searchenginecolossus.com/
www.searchengineguide.com/
www.google.com/
www.lycos.com/
www.yahoo.com/
www.altavista.com/
www.alltheweb.com/
www.webcrawler.com/
www.excite.com/
www.infoseek.com/
www.msn.com/
www.infospace.com/

Information Portals

http://lcweb.loc.gov	The Library of Congress is the best place to start as it is the world's largest library.
www.firstgov.gov/	Gateway to all government information.
http://spireproject.com/	A guide to what is coming online.
http://scout.cs.wisc.edu/	The Scout Report is one of the Internet's longest-running weekly publications, offering a selection of new and newly discovered online resources.
http://infomine.ucr.edu/Main.html	A University of California Library service on what is available on the Internet in the sciences.
www.ipl.org/ref/RR/	Internet Public Library. An annotated collection of high-quality Internet resources for providing accurate, factual information on a particular topic or topics.
www.invent.org/	A highly artistic website on the patenting process with much support for independent inventors.
www.access.gpo.gov/getcfr.html	This page describes the HTML coding necessary to link directly to documents contained in the *Code of Federal Regulation* (CFR) databases resident on Government Printing Office's wide area information system servers.
http://vlib.org/	The WWW Virtual Library.
www.invent1.org/	Minnesota Inventors Congress; inventor resources and the oldest convention center.
www.uww.edu/business/innovate/innovate.htm	
www.innovationcentre.ca	The Canadian Innovation Center.
www.sul.stanford.edu/depts/swain/patent/pattop.html	Selected Resources for Patents, Inventions, and Technology Transfer selected by Stanford University.
www.bl.uk/	The 150-million volume British Library is a good source; search here with the key word "patent."
www.rand.org/radius	The most comprehensive listing of federally funded research projects, free to federal employees.

www.knowledgeexpress.com/Knowledge	Express provides business development and competitive intelligence resources—including intellectual property, technology transfer, and corporate partnering opportunities—to organizations involved with science/technology research and new inventions.
http://searchlight.cdlib.org/cgi-bin/searchlight	Publicly available databases and other resources.
http://productnews.com	New product information on thousands of products weekly.
www.techexpo.com	Aimed at manufacturers and inventors.
www.pubcrawler.ie/	Health and medical information.
http://medlineplus.gov/	National Library of Medicine consumer site on health.
http://clinicaltrials.gov/	Information on all U.S. clinical trials underway and you can branch out to learn about how to do clinical trials.
www.pubmedcentral.nih.gov/	A major archive of free online life science journals.
http://highwire.stanford.edu/lists/freeart.dtl	Free online journals in science, technology, and medicine.
www.healthfinder.gov/	Reliable health information from government agencies.
www.cdc.gov/	The Center for Disease Control.
www.fda.gov	Food and Drug Administration regulations.
www.nal.usda.gov/fnic/	USDA food and nutrient information center.
www.mdtmag.com	Medical design technology for devices.
www.medscape.com/	Medical information from WebMD.
http://bmn.com	Biomedical information portal.
www.intelihealth.com	Consumer healthcare information from Harvard.
www.bmj.com	British Medical Journal; prestige medical issues.
www.devicelink.com/	Medical devices.
www.ornl.gov/hgmis/	The human genome project.

Technical Databases

The National Institutes of Health (NIH; Ref. 5) offers over 12 million research papers mainly in the biomedical sciences that are available through the National Library of Medicine. This free database allows downloads of abstracts in an ASCII format for direct placement into programs like Microsoft Word to develop a comprehensive bibliography.

Derwent (6) is one of the most widely used databases, from which the United States Patent and Trademark Office examiners benefit as well.

Dialog (7) is another large database that allows you to search without having to register an account; you pay as you go along using your credit card. You cannot do this if you are searching for Trademarks.

Delphion (formerly IBM Intellectual Property database; Ref. 8) allows for more in-depth search of patents and access to many consolidated databases. It is a low-cost solution.

American Chemical Society offers large databases (9). This is one of the premier scientific databases where you will be able to get original papers faxed to you if you need them urgently.

Nerac (10): Numerous online databases.

Patent Search and Intellectual Property Services
- Patent Café (11): a hangout for patent product vendors.
- Investor's Digest (12): Resources for inventors. Very active site.
- The Massachusetts Institute of Technology (MIT) (13) has provided an elaborate and detailed website for invention development; worth a detailed look.

Patent Copies, Search Facilities
- Micropatent (14): This is perhaps the most comprehensive service available at a very reasonable cost and is an easy website to navigate. This is a highly recommended site;
- Questel (15);
- Faxpat (16);
- Patentec (17);
- Lexis-Nexus (18); and
- Mayall (19).

COMPONENTS OF A PATENT APPLICATION
A patent specification (not specifications) is a legal document that gets published as the patent, if granted. Great care and detail goes into writing this document in defining the scope of invention, deciding what is claimed, and wording the claims (which are part of specification) such that they can withstand challenges.

Deciding on what is "invented" is the job of the research scientist, but the decision is made in light of prior art; for example, if there is a discovery of a new group of chemicals, the breadth of the group should be ascertained in light of the prior art available, what can be reasonably predicted regarding the structure of chemicals, and the ability to synthesize representative chemicals, if not all. Where a completely new molecular structure has been invented, a broad scope, including all kinds of derivatives of the basic structure which the inventor thinks may be useful, is possible. It is surprising how many times scientists who are sure of the novelty of a structure, composition, or application find out otherwise when a thorough search is made of the possible prior art. It is worth realizing that with the availability of databases in electronic format, the Internet, and generally a faster access to remove publications (even brochures for promotion in remote countries), what a thorough search would reveal; know that the patent examiners have access to these same channels of information and even more. The author has been humbled more than once by what the patent examiner can dig out. So, scientists are advised not to jump the gun in making very broad claims until so advised by the patent attorneys after conducting a thorough patent search. Obviously, the scope is narrowed down gradually as more and more prior art emerges. The goal is not to narrow it down to a point where it loses its commercial importance. The scope of protection, which is commercially important to achieve, varies from one field to another. In extreme situations, it is sufficient to have a scope which includes a singular compound, if that is what the company wishes to market. The strength in this approach comes from the regulatory control of pharmaceutical products. Once a company receives marketing authorization from the FDA, at a great expense, imitators would like only to reproduce the invention, which they cannot do through the course of the patent term and any extensions granted by the FDA. So, while there may be other molecules, perhaps better ones, available, imitators are unlikely to invest in their development as they are unprotected and there is no guarantee that the FDA will approve them. So, in the field of new drug development, a single chemical entity does have substantial value. Such is not the case in other industries where regulatory costs are not involved, such as in the chemical industry.

Once a decision is made about what is invented, how much to claim, and what specifically to claim, the process of drafting the specification begins. One method is to draft the main claim, defining the scope of the invention in the form of the statement of invention, which is the heart of the patent specification, then the rest of the specification and claims will be drafted.

Specification begins with a title. Newcomers to the field of patenting would be amazed or perhaps amused at the choice of patent titles and even the language used to describe an invention. Historically, inventors kept the titles vague to keep the searchers (who then did manual searches) from finding out about their inventions. Today, as most patent offices have gone electronic, this is no long an issue, nevertheless the practice continues.

Patent applications have fixed formats that often vary between patent offices but nevertheless require a similar information submission: a background, summary, details of invention, and so on. The patent application must be comprehensive to demonstrate novelty and the inventive step in light of prior art; it should be understood that the purpose is not to fool the patent examiner into allowance, but to protect the invention from competitors who will challenge it, should it be worth anything. A full disclosure is required to keep the infringers out, to decrease the chance of their success in knocking out a patent. Additional statements are included defining the features of the invention for use as a basis for specific claims. For example, stating that: "In one aspect the invention provides...." "In another aspect the invention provides...." The described widgets are new and form part of the invention. Attorneys have their word preferences and standard statements to fill the specification are written quickly.

After the statements of invention, there is a description to indicate the preferred parts of the scope, and one or more formulae may be given defining narrower subgeneric scopes. This section fulfills the requirement of adequacy of description. The specification must also describe how the invention is to be carried out, an essential part of a patent application. This is a critical stage in deciding how and what to disclose. As discussed earlier, often at this stage a decision may be made not to file a patent application, for the disclosure will inevitably cause the invention to escape from the hands of inventors, and if there were no certain ways to determine infringement, this would make patenting useless. It is also not a smart move to be deceptive when it comes to describing how the invention works; many a patent application has been declared invalid after the companies have made significant investments in marketing the invention, because a competitor was able to demonstrate that the inventors hid certain critical facts. It must be understood that the disclosure need not be for a commercial model of the invention and thus need not include many fine details generally required for a large-scale production of the invention, such as in-process specifications, certain handling conditions, the grade of excipients used, and so on, which may be material to produce a product fit for a particular purpose, such as human consumption. As long as the competitor can manufacture the article, not necessarily for the commercial production, using the details provided, the requirement of sufficiency of disclosure is met. This becomes more important in the discovery of new chemical entities where chemical synthesis can be described adequately, but not necessarily for the grade of material required; for example, the impurity profile of a new chemical entity (NCE: FDA) is critical for the purpose of a new drug application (NDA: FDA); manufacturers often are able to produce a product that would meet FDA's requirement for quality, yet not report this method in the patent, which would allow the competitor to manufacture the product only with impurities. The reason companies are able to get away with this trick is that the patent claims a chemical compound, not necessarily what would be suitable for ingestion by humans. Obviously, if the molecule turns out to be a blockbuster, many will imitate the process and may challenge the patent; case law on this aspect is silent. It is well known as a result, that once an NCE comes to the end of its patent cycle, new sources of active pharmaceutical ingredient (API) are developed with greater difficulty than what would be anticipated from the disclosures in the patent.

Next is an indication of what the invention is useful for, a part of the specification usually called the utility statement. Whereas in the case of mechanical inventions it is often obvious, it requires explanation for chemical inventions along with any peculiar or particular advantages. It is important to know that there is no requirement to explain how and why the invention works; thus, it is best not to offer any hypothesis about the invention. However, if a theory must be given, one should leave room for a change of mind later, for example, by wording as, "while we do not intend to be bound to any particular theory, it is believed that . . . ." In a chemical case, a number of examples are given with detailed instructions for the preparation of at least one of the compounds within the scope, and for the use of the compounds.

After these is the heart of the patent, the claims. There is no limitation on how many claims are made; however, redundant and superfluous claims are frowned upon by the examiners and should be avoided. It is important to know that all dependent claims are narrower in scope and written exclusively for the purpose of protecting the invention, or any part of it, should the broader claim or claims be knocked out in court proceedings.

Other parts of a patent application, such as priority dates, affidavit requirements, assignment, appointment of attorney or agents are best left to the patent practitioners and the company's legal department to worry about; however, there may be some interaction with the inventor in filling out certification documents. The filing of an application is followed by numerous communications and office actions from the patent office, the responses to which are drafted in full consultation with the scientists and their approval for accuracy of information and its interpretation. A word of caution is needed here. In the United States, there is a clause of "estoppel," under which admissions made in the specification or in responses to office action about what was actually taught by the prior art may be binding upon the applicant. A wrong statement about the acceptance of an article as prior art may be reversed later. Court proceedings will have the entire text of correspondence available for examination, and the "file wrapper" becomes part of the patent. The safest rule is to admit nothing and say as little as possible about the prior art. Of course, all relevant prior arts known to the applicant or his attorney must be brought to the attention of the United States Patent and Trademark Office, but this does not mean that it has to be mentioned in the specification.

There must be at least one claim in each patent application. Claims define the scope of subject matter for which protection is sought. A competitor does not infringe and cannot be stopped unless he makes, sells, offers for sale, imports, or does something that falls within the scope of at least one claim of the granted patent, in other words, if the infringing object "reads onto claim," then it is an infringement. How claims are interpreted keeps the courts filled with opportunities to create case laws. Claims are always read in light of the published specification and thus the issue of prior art comes up again; had there been a mistake in allowing a claim in the light of prior art, the claim will be thrown out and in some cases, the entire patent is rendered invalid.

All claims fall into one of two broad categories: they claim either a product (a mechanical device, a machine, a composition of matter, and the like), or a process (a method of making, using, or testing something). For chemical patents, this may include the chemical per se as a useful intermediate, a composition in a pharmaceutical product, a specific form (optical isomer, crystal form, and so on), or for direct use. The process claims would include the processes of synthesis, isolation, or purification as the case may be; the methods of use may be the first use or a subsequent use, such as a method of medical treatment or diagnosis or testing and analysis methods.

Drawing(s) (§113)

When necessary, drawings must be submitted, most likely in mechanical or electrical and some chemical applications. Filing date is not assigned if drawings are not provided at the office of initial patent examination (OIPE) level of evaluation; examiner may require drawings but filing date is not affected; drawings MUST show all of the claimed elements; drawings may be added later by amendment if already described in the specification or claim as originally filed; no need for manufacturing drawings (such as tolerances or in-process controls).

Specification (§112 ¶1)

The written description, the manner and process of making and using, in such full, clear, concise, and exact terms as to enable any person skilled (with ordinary skills) in the art to which it pertains, or with which it is most nearly connected, to make and use the invention and setting forth the best mode contemplated by the inventor for carrying out his invention. There are three requirements of disclosure that are dealt with in the following.

Description
- Must describe what is claimed clearly;
- focus is on the claimed invention only;
- scope commensurate with scope of claim(s): disclosure of a single species may or may not support a generic claim;
- critical or essential element MUST be recited in claims;
- the vantage point is one of ordinary skill in the art;
- inventors may be their own lexicographers by so stating; however, the term can not be used in a contrary manner to what is commonly acceptable;
- theory need not be set forth; if theory is wrong, the error is not fatal (unless theory is claimed invention);
- manner of invention is not important (how the invention was made).

Enablement
- To one of ordinary skills to make and use;
- without undue experimentation;
- not necessarily for commercial production;

- this requirement is different from §101's requirement of being useful;
- claim not reciting essential matter may be rejected for lack of enablement or failing to claim the subject matter, which the applicant considers as the invention;
- publications after filing date may not be used to support enablement, but may be used to defeat enablement (such as by examiner);
- scope of enablement must be commensurate with the scope of claims(s);
- amount of disclosure required depends on the state of the art and predictability—the more is known and the greater is the predictability, the less is the required disclosure.

Best Mode
- What inventors consider as the best mode is not what anyone or everyone else considers and not what is objectively the best.
- After an application is filed, it need not and cannot be updated by amendment (as it will be considered new matter) even in a division or continuation, but can be updated in continuation-in-part if it pertains to a new claim made.
- Must be disclosed though not necessarily identified as such; embodiment disclosed is automatically considered as the best mode; several embodiments may be disclosed without identifying which one is the best.

Paragraph 1 35 USC §112 Requirements
Description, enablement, and best mode must exist in each claim as filed or else it renders claims invalid if contested. All of these are intended to be understood by a person of ordinary skills in the art, not a layperson. Need to disclose what is considered as required knowledge of one of ordinary skill. Any one or more of the following can satisfy each of these three requirements: specification, drawing(s), and claims as originally filed. Unclaimed inventions need not satisfy this requirement.

Paragraph 2 35 USC §112 Requirement: Parts of a Claim
- Preamble sets the scope of the invention's technical environment and class (composition, process or apparatus, and the like: A method of..., Apparatus for..., A composition...). It is not limiting if it merely states the purpose of the invention; however, if it breathes life and meaning into the claim (such as if it is essential to tell what is claimed or if the body of claim refers to it as antecedent support) it can become limiting.
- Transitional phrase connects the body of the claim to the preamble: comprising, consisting of, consisting essentially of, and the like. Three types: open-ended: comprising, including, containing, characterized by, and the like; closed: consisting of, also composed of, having, being, and so on—some of these can be interpreted differently; partially closed: consisting essentially of... wherein it allows only those additional elements that do not affect the basic and novel characteristics of invention. No synergism. The applicant has burden of proof to show that additional elements in prior art would materially change the characteristics of the invention. If ABCD is known and ABC is claimed; absence of D must be demonstrated to affect materially the invention.

- Body is a list of elements, such as ingredients of a composition and components of the apparatus; all elements must be interconnected. (There must a reason why a component is recited; not just to list it.)

UNDERSTANDING CLAIMS
The heart of a patent is within the boundaries defined by the claims.

Reading a Claim
Determination of infringement is done by reading (or matching) the claim to prior art to prove the validity of the claim. A device or process to indicate infringement or own specification satisfies §112 ¶1 requirements. Claim does not read on prior art with elements ABCD if the claim is ABC and closed (consisting of), but it reads if the transitional phrase is open such as "comprising" and may or may not read if it is "consisting essentially of."

Punctuation of Claim
A claim is written as one sentence, contains a comma after preamble, a colon after introducing another element, and each element gets its own paragraph. There is a semicolon at the end of each paragraph and between the last two elements. If more than one period is used this would be rejected as indefinite claim (¶2 §112).

Definiteness of Claim
Without proper antecedent basis a claim is rendered indefinite. "A" or "an" introduces an element for the first time except in a means-plus-function format. "Said" or "the" refers back to previously introduced elements or limitations or refers to inherent properties (not required to be recited for antecedent purpose; for example, "the surface of said element" when "surface" is not defined earlier). Inferential claiming where interconnectivity of elements is not certain—does not tell if the element is part of combination or not.

Narrowing of Claim
Narrowed by adding an element or limitation to a previously recited element; narrow claim can be dependent or independent. Adding a step narrows method claims. Adding an element to a closed (such as Markush Group) claim broadens, not narrows claim.

Dependent Claims (§112 ¶3 ¶4)
- Claim can be dependent or independent; a dependent claim incorporates by reference all the limitations of the claims to which it refers and is always narrower; must depend on a preceding claim and not on a following claim (numbering of claims is readjusted during prosecution).
- "Further comprising" or "further including" are used to narrow a claim by adding another element or step.
- Claims narrowed by further defining an element or the relationship between elements. Transitional element "wherein" is used to add limitation. Narrowing can be both adding an element and further defining its relationship.
- Defining a step further narrows a method claim.

Multiple Dependent Claims (§112 ¶5)
- A claim referring to more than one previously set forth claim, but only in the alternative ("or") narrows the claim on which it depends.
- Cannot serve as a basis for another multiple dependent claim; may refer to other dependent claims, and a dependent claim may depend on a multiple dependent claim.
- Incorporates by reference all the limitations of the particular claim in relationship to which it is being considered (individually and not collectively).
- It takes the place of writing several dependent claims—in its spirit.
- A flat special fee is charged at the time of filing application if multiple dependent claim or claims are included.

Dominant-Subservient Claims
Dominant claims involve subcombination or genus while combination means subsurvient species. Need two members (species) to have genus, which is illustrated by the selection of species; genus is an inherent commonality among embodiments (species).

Means-Plus-Function Clauses §112 ¶6
Claim defines an element by its function, not what it is. It is a means for performing a function. Interpreted by the literal function recited and corresponding structure or materials described in specification and equivalents thereof. Does not cover all structures for performing the recited function. A claim reciting only a single means-plus-function clause without any other element is impossible. Must have the phrase, "means for," which then must be modified by functional language, but not modified by the recitation of structure sufficient to accomplish the specific function. If specification does not adequately disclose the structure corresponding to the "means" claimed, the claims fail to comply with paragraph two of the patent law requirement for "particularly pointing out and distinctly claiming" the invention. If disclosure is implicit (for those skilled in the art), an amendment may be required or stated on record what structure performs the function. *Equivalents*: examiner must explain rationale; prior art must perform, not excluded by explicit definition in the specification for an equivalent, prior art supported by:

- Identical function, substantially same way, substantially same results
- art-recognized interchangeability
- insubstantial differences
- structural equivalency

Process Claims
These include a method for making a product, comprising the steps of . . . ; a method of using a specified or known material, comprising the steps of . . . ; recitation of at least one step required and a single step method claim is proper.

Step-Plus-Function Clauses
- Functional method claims reciting a particular result but not the specified act—that is, techniques used to achieve results—adjusting pH, raising temperature, reducing friction, and so on.
- No recital of acts in support required.

- Typically introduced by words like "whereby," "so that," or "for."
- Addition of a functional description alone is not sufficient to differentiate claim—rejected under §102.
- Functional language without recitation of structure, which performs the function, may render the claim broader (rather than narrower), and rejected under §112 ¶1.

Ranges
Commonly used for temperature, pressure, time, and dimensional limitations. "Up to" means from zero to the top limit; "at least" means not less than (does not set upper limit, which must be fully disclosed in specification); specification must support eventual ranges. A dependent claim cannot broaden the range. Range within range is indefinite in a claim, but acceptable in specification.

Negative Limitations
Permissible if boundaries are set forth definitely, such as free of an impurity or a particular element or incapable of performing a certain function. Absence of structures cannot be claimed—holes, channels, and the like—as structural elements.

Relative and Exemplary Terminology
Imprecise language may satisfy definiteness requirement (for one of ordinary skill). "So dimensioned" or "so spaced" can be definite if it is as accurate as the subject would permit; "about" is clear and flexible, but rendered indefinite if specification or prior art does not provide indication about the dimensions anticipated. "Essentially," "substantially," "effective amount" are definite if one with ordinary skills would understand. Exemplary terminology is always indefinite: such as, of like material, similar—all rejected.

Markush Group
Closed form. Two forms: wherein P is a material selected from a group consisting of A, B, C, and D; or wherein P is A, B, C, or D. Members must belong to a recognized class, possess properties in common as disclosed in the specification, and these properties are mainly responsible for their function or the grouping is clear from their nature or the prior art that all members possess the property. Adding members broadens claim. Prior art with one of the members anticipates the claim.

Markush Alternates
"Or" terminology used if choices are related: one or several pieces; made entirely or part of; red, blue, or white. If unrelated choices, the use of "or" will lead to indefinite interpretation. "Optionally" used if definite or if there are no ambiguities in the scope of the claim as a result of the choices offered.

Jepson-Type Claims—Improvement Claims
Preamble defines what is conventional; transitional phrase, "wherein the improvement comprises;" body builds on preamble; can add element or modify element in preamble. Preamble is limiting.

Mixed-Class Claims

Mixed elements are improper: methods claims should have no structural elements; apparatus claims should have no step elements. Limitations can be mixed, such as method step may include a structural limitation, and an apparatus may include a process limitation.

Product-by-Process Claims

A product claim that defines the claimed product in terms of the process by which it is made: A product made by the process comprising of steps . . . Patentability based on product itself and NOT on method of production. If the product is the same (as prior art), using another process does not make it patentable. If examiner shows that product appears to be the same or similar the burden shifts to the applicant; United States Patent and Trademark Office bears lesser burden of proof in making out a case of prima facie obviousness. One-step method claims are acceptable, but claims where body consists of single "means" elements are not acceptable.

Patent Term Adjustment

The United States Congress passed legislation known as the Hatch-Waxman Act in 1984 that weakened patent law for pharmaceuticals, making it easier for generic copies to enter the market based on the innovator's safety and effectiveness data. Under the Act, pharmaceutical research companies lost nearly all of their rights to defend their unexpired patents before generic copies entered the market. Patent holders can sue to defend their unexpired patents *only* when a generic drug manufacturer submits a filing to the FDA seeking to bring the generic copy to market. The Act also created a 30-month stay procedure to allow patent holders the opportunity to obtain a court ruling on whether the generic copy infringes their patent. Thirty-month stays do not extend patents—they are triggered *before* the patent expires, and provide a period of time during which patent infringement cases can be resolved.

Patent lawsuits based on the Act are rare because generally challenges to patents on prescription medicines are rare. FDA reports that of 8,259 generic applications filed between 1984 and 2001, only 6% raised a patent issue, the necessary condition for patent litigation. According to the Federal Trade Commission, more than one-quarter of patent challenges studied did not result in a lawsuit by the innovator company. Since enactment of the law, generic company share of prescription medicine use has increased from 19% of prescription units in 1984 to 50% today.

The average effective patent life for prescription medicines under the Hatch-Waxman Act is 11–12 years, compared with an average of 18.5 years for other products.

With effect from August 18, 2003, FDA revised its regulations as follows:

> Permit only one 30-month stay in the approval process for a generic drug pending resolution of patent litigation. Past regulations acquiesced to the delayed launch of generic versions beyond 30 months when there were multiple, consecutive patent challenges that were made against the launch of the generic versions, even if the challenges were frivolous.
>
> Clarify the types of patents that may be listed in the "Orange Book," which is the Food and Drug Administration's official register of approved pharmaceutical products that provides notice to generic drug makers of name brand patent rights. Patents may

no longer be listed that cover drug packaging or other minor matters not related to effectiveness. Patents are to be listed that pertain to active ingredients, drug formulations/compositions and approved uses of a drug. A more detailed, signed attestation will be required to accompany a patent submission. False statements in the attestation can lead to criminal charges.

For patents that are granted after the drug application is filed, the brand name drug maker has 30 days to list the patent(s) from the grant date.

To seek approval for a generic, the generic drug maker must certify to the Food and Drug Administration that either (*i*) there are no Orange Book listed patents for the name brand drug or (*ii*) the patent(s) has (have) expired or (*iii*) will expire by the time approval is sought or (*iv*) the listed patent(s) is(are) invalid or will not be infringed. If the latter, notice of the certification is given to the patent owner and to the brand name drug maker with an explanation as to why the patent(s) is(are) invalid or not infringed. If the patent owner does not bring a patent infringement suit against the generic drug maker within 45 days, the Food and Drug Administration may approve the generic version. Otherwise, the approval process is stayed for the shorter of 30 months or the date when a court concludes the patent(s) is(are) either invalid or not infringed.

Require generic manufacturers to demonstrate to the Food and Drug Administration that their generic drug is therapeutically equivalent to an approved brand name drug. That is, equivalence in terms of safety, strength, quality, purity, performance, intended use and other characteristics. Review drug applications for generics more quickly. The Food and Drug Administration is hiring 40 generic drug experts to expedite the approval process and to institute targeted research to expand the range of generic drugs available to consumers.

Improve the review process for generic drugs by instituting internal reforms. The reforms include making early communications with generic drug manufacturers who submit applications and guiding generic manufacturers in preparing and submitting quality, complete applications.

The recent decision in the U.S. patent infringement case Madey versus Duke is very important to academic researchers and the industry. Duke University had challenged the general assumption that academic research using a patented device or method cannot constitute infringement. The subject matter was a laser device, which had originally been developed and patented by Duke University. When the inventor left the University to pursue commercial applications for the laser, Duke University continued to use their model for research purposes. Duke claimed that it was entitled to continue using the laser for noncommercial purposes under the experimental use exception in U.S. patent law. However, the court held that Duke University's use of the laser "unmistakably" furthered its commercial goals, including facilitating the education of students. The court further held that research using the laser had helped the University to obtain research grants. The equivalent provision in English law is section 60(5)(b) of the Patents Act 1977, which states that an act relating to the subject matter of a patent, which is done for "experimental purposes" will not constitute infringement. The provision does not set out whether the exemption is available to those whose experimental purposes have a commercial element. There is no United Kingdom equivalent; however, of the United States exemption, which permits the unauthorized use of a patented device or method by a person seeking FDA approval to market a new product. The exemption applies only while the application is pending, but extends to the use of patented devices or drugs in clinical trials, their sale for use in trials, demonstrations at trade shows, and the reporting of clinical data to potential investors.

The United States Patent Office prescribes specific regulations regarding patent term adjustment:

- Application filed prior to June 8, 1995: 17 years from the date of issuance regardless of the length of prosecution.
- Application filed June 8, 1995-May 28, 2000: The Uruguay Round Agreements Act (URAA): 20 years from filing date but with up to five years extension for delays resulting from secrecy orders, interferences, and/or successful appears.
- Application pending or patent in force on June 8, 1995: 17 years from issue date or the period between the issue date and the 20th anniversary of the filing date, whichever is greater.
- Application filed on or after May 29, 2000: American Inventors Protection Act (AIPA) may be entitled to patent term adjustment (PTA) in a continuing application including continued prosecuting application (CPA), request for continued examination (RCE) filed after May 29, 2000 in an application filed before May 29, 2000 does NOT provide PTA eligibility; Patent Cooperative Treaty's eligibility depends on its filing date, not its national stage entry date (Patent Cooperative Treaty must be filed on or after May 29, 2000 to be eligible for PTA).
- PTA: termination date (20th anniversary from filing date) is extended by number of days Patent and Trademark Office delays minus the number of days that the applicant delays.

Patent and Trademark Office Delays: Guaranteed Adjustment Basis

Guaranteed adjustment basis (GAB)1: Patent and Trademark Office failure to take certain actions within 14 months from filing date and four months from other events: Patent and Trademark Office must mail an examination notification (first Office action including Quayle action or notice of allowability, restriction requirement and request for information, but NOT OIPE notice of incompleteness of application or other such notices) to applicant within 14 months of the filing date; Patent and Trademark Office must also respond within four months to the applicant's reply to an office action or applicant's opening appeal brief; Patent and Trademark Office must act within four months of a board of patent appeals and interferences (BPAI) or court decision, where allowable claims remain in the application; Patent and Trademark Office must issue the patent within four months of date on which the issue fee is paid, and all outstanding requirements are satisfied.

GAB2: Patent and Trademark Office delays due to interference, secrecy order, or successful appellate review (where BPAI or court reverses determination of patentability of at least one claim [allowance by examiner after a remand from BPAI is not a final decision]. GAB2 was also the basis of PTA under URAA, but for a maximum of five years, AIPA removes five-year limit.

GAB3: Patent and Trademark Office fails to issue a patent within three years excluding time consumed in RCE, secrecy order, interference, or appellate review (whether successful or not), time consumed by applicant-requested delays (e.g., suspension of action up to six months for "good and sufficient cause," up to three-month delay request at time of filing RCE or CPA, up to three-year deferral of examination requested by applicant. Filing an RCE for an application filed on or after May 29, 2000 cuts off any additional PTA due to failure to issue patent within three years, but it does NOT eliminate PTA in GAB1 and 2.

Required Reduction Basis

Applicant's delay for failure to engage in reasonable effort to conclude prosecution of the application is subtracted from GAB1-3: failure to reply within three months to any notice from the office making any rejection, objection, argument, or other requests (even though the applicant pays for and receives extension), days in excess of three months are deducted; reinstatement of deduction of up to three months can be made by applicant showing that "in spite of all due care, the applicant was unable to reply," due perhaps to testing to demonstrate unexpected results, death of applicant's sole practitioner, or a natural disaster. (Do not confuse the three months concession with three months required to respond.) Additional required reduction bases (RRBs) are generated because of: suspension of action under Rule 1.103, deferral of issuance under Rule 1.3114, abandonment or late payment of issue fee, petition to revise more than two months after notice of abandonment, conversion of provisional to nonprovisional, preliminary amendment within one month of office action that requires supplemental office action (i.e., a response is sent when an office action is to come within one month)*, inadvertent omission in reply to office action*, supplemental reply not requested by examiner*, submission filed after BPAI or court decision within one month of office action that requires supplemental office action*, submission filed after notice of allowance*, and filing a continuing application to continue prosecution. Note that in instances marked with an asterisk (*) information disclosure statement submission will not create reduction if information is received from foreign patent office within the last 30 days (i.e., the applicant responds within 30 days of receiving such information).

Summary: Patent term begins on the day of patent issuance; terminal disclaimer date ends patent term; failure to pay a postissuance maintenance fee ends patent term (notice: no such fee required for design and plant patents); term extension beyond statutory period only through private congressional legislation or by showing government agency delays (e.g., FDA); 20-year term begins from the earliest ancestral application from which priority is claimed (does not include provisional application or a foreign application for term running purpose); design applications excluded from URAA and AIPA as they have fixed 14-year term from issue.

FOOD AND DRUG ADMINISTRATION

A listing of drugs for which the patent term had been extended by the US FDA is available at the website mentioned in Ref. 20. The longest patent term extension given by the U.S. FDA to any drug belongs to U.S. Patent 3,737,433 (Trental) for 2494 days.

The point in the development program at which a patent application is filed will vary somewhat from company to company, but will normally be at an early stage in the process, when the substance has been made and been shown to be active in early screening. For a patent with a nominal term of 20 years from filing, the effective term during which the patentee has exclusive rights to a marketed product is only 8 to 12 years. This explains the importance attached by the pharmaceutical industry to provisions to extend the patent term, whether directly, as in United States and Japan, or indirectly by way of the supplementary protection certificate (SPC) in Europe, in order to compensate for this loss of effective patent term. It also explains the importance to the industry of the minimum 20-year term guaranteed by the

TRIPS Agreement. The SPC for medicinal and plant protection products by the United Kingdom Patent office does not extend the entire scope of the patent on which it is based, but is limited to the product covered by the marketing authorization, and for any medicinal use of the product that has been authorized before the expiration of the certificate. Thus, sales of the product for nonmedicinal uses do not infringe, but the SPC would be infringed by sales of a medicinal product by a third party even if that party had a marketing authorization for a different indication. Apart from this, the SPC confers the same rights as the basic patent, and is subject to the same limitations and obligations to allow existing licenses under the basic patent to continue under the SPC. The scope of protection given by a U.S. patent during the Hatch/Waxman extension period is essentially the same.

REFERENCES
1. http://www.uspto.gov.
2. http://ep.espacenet.com; http://register.epoline.org/espacenet/ep/en/srch-reg.htm.
3. http://ipdl.wipo.int/.
4. http://patents1.ic.gc.ca/srch_bool-e.html.
5. http://www4.ncbi.nlm.nih.gov/PubMed.
6. http://www.derwent.com.
7. http://www.dialog.com.
8. http://www.delphion.com/.
9. http://stneasy.cas.org/html/english/login1.html.
10. http://www.nerac.com.
11. http://www.patentcafe.com.
12. http:/www.inventorsdigest.com/.
13. http://web.mit.edu/invent/.
14. http://www.1790.com/0/patentweb9809.html.
15. http://www.questel.orbit.com/.
16. http://www.faxpat.com.
17. http://www.patentec.com.
18. http://www.lexis-nexis.com.
19. http://www.mayallj.freeserve.co.uk/.
20. http://www.uspto.gov/web/offices/pac/dapp/opla/term/156.html.

BIBLIOGRAPHY
Grubb PW. Patents for Chemicals, Pharmaceuticals and Biotechnology: Fundamentals of Global Law, Practice and Strategy (Hardcover). 4th ed. USA: Oxford University Press, February 10, 2005.
Niazi S. Filing Patents OnLine. Boca Raton, FL: CRC Press, 2003.
United States Patent and Trade Mark Office Manual of Patent Examination Practices http://www.uspto.gov/web/offices/pac/mpep/mpep.htm.
Voet MA. The Generic Challenge: Understanding Patents, FDA and Pharmaceutical Life-Cycle Management [DOWNLOAD: ADOBE READER] (Digital), Brown Walker Press, January 1, 2005.

3 The Scope of Preformulation Studies

INTRODUCTION
A detailed understanding of the properties of the drug substance is essential to minimize formulation problems in later stages of drug development, reduce drug development costs, and decrease the product's time to market (i.e., from drug substance to drug product). The goals of preformulation studies are to choose the correct form of the drug substance, evaluate its physical properties, and generate a thorough understanding of the material's stability under the conditions that will lead to development of an optimal drug delivery system. This chapter examines the continuously evolving scope of preformulation studies. These changes are driven by three forces: the regulatory requirements, the market requirements, and the technological development. However, prior to reviewing the various requirements that determine the scope of preformulation studies, it is important to review how the drug discovery models are rapidly changing and why there is a need for not just one, but several levels of preformulation studies.

New drugs are discovered through serendipity, by chemical synthesis, by extraction from natural sources or more recently through biotechnology processes. The drug discovery is rarely a linear process and the many probes involved often run concurrently. Chapter 1 described in detail the various phases and stages of development; whereas the industry has developed its own jargon about preclinical phases and clinical phases, the entire process of development goes through a highly orchestrated undertaking that requires very close association among all groups.

PREFORMULATION TESTING CRITERIA
The classical preformulation studies include the physicochemical characterization of the solid and solution properties of compounds that would be useful in formulating the drug into a suitable delivery system. It is in this critical decision-making about what constitutes the "suitable" that many a lead compound falls through. A good pharmacological and toxicological profile alone does not suffice. The drug delivery system must be able to take the molecules to the site of action, at a cost and convenience commensurate with the treatment trends. An excellent remedy for headache that requires intravenous injection would not go past this stage; on the other hand, for those drugs where there are no alternatives, such as in cancer treatment or other diseases where the patient is hospitalized and critically ill, any dosage form would be acceptable if cost and reimbursement issues do not impair the commercial projections of sales. Besides delivering the drug, the dosage form must provide a stable environment through a reasonable shelf life, preferably at room temperature and sufficient bioavailability without any food or other drug interactions. Now, it is clearly seen as to why decision-making can be very difficult at the preformulation stages of drug development.

Whereas most preformulation studies start during the lead optimization (LO) phase, the involvement of some nevertheless begins much earlier, even at the lead identification stage to rule out undesirable features, such as chirality (though at times it can be the desired target), polymorphism, hygroscopicity, and extreme stability problems. LO taking about two years to complete narrows the choice to no more than about three to four compounds based on the fine balance of pharmacology, toxicity, and biopharmaceutic compatibility. An optimally available oral drug would be a jackpot. Given the small quantity of sample available at this stage, agreements must be reached between the preformulation and drug discovery group to obviate redundant testing. The drug discovery group may take on nuclear magnetic resonance (NMR), mass spectra, and elemental analysis whereas the preformulation group may use almost two-thirds of the supply (generally less than 10 mg) to perform Karl Fischer, pK_a, log P/log D, initial solubility, crystal structure, hygroscopicity, stability in solution and high-performance liquid chromatography (HPLC), and other spectroscopic data. For salt forms, additional testing of dynamic vapor sorption (DVS), X-ray, differential scanning calorimetry (DSC), solubility/stability tests, polymorphism studies using DSC/differential thermal analysis (DTA)/hot stage microscopy (HSM), crystal habit using microscopy, both light and scanning electron microscopies (SEM), stability using temperature and humidity stress to rule out hydrate or solvate status often using circular dichroism require a larger sample quantity, around 100 mg.

REGULATORY REQUIREMENTS
Small Molecules/General
Regulatory agencies are continuously pushing the quality systems that appear in the early phases of drug development. The ICH (International Conference on Harmonization) offers several guidelines for the characterization of the drug substance. Notably, these include:

1. Q1A(R2) Stability Testing of New Drug Substances and Products (Issued 11/2003, Posted 11/20/2003);
2. Q1B Photostability Testing of New Drug Substances and Products (Issued 11/1996, Reposted 7/7/1998);
3. Q1C Stability Testing for New Dosage Forms (Issued 5/9/1997, Posted 3/19/1998);
4. Q1D Bracketing and Matrixing Designs for Stability Testing of New Drug Substances and Products (Issued 1/2003, Posted 1/15/2003);
5. Q3A Impurities in New Drug Substances (Issued 2/10/2003, Posted 2/10/2003);
6. Q3B(R) Impurities in New Drug Products (Issued 11/2003, Posted 11/13/2003);
7. Q3C Impurities: Residual Solvents (Issued 12/24/1997, Posted 12/30/1997);
8. Q3C Tables and Lists (Posted 11/12/2003);
9. Q6A International Conference on Harmonization; Guidance on Q6A Specifications: Test Procedures and Acceptance Criteria for New Drug Substances and New Drug Products: Chemical Substances (12/29/2000);
10. Q7A Good Manufacturing Practice Guidance for Active Pharmaceutical Ingredients (Issued 8/2001, Posted 9/24/2001).

The physicochemical and biological properties of the drug substance that can influence the performance of the drug product and its manufacturability that are specifically designed into the drug substance (e.g., crystal engineering) should be identified and discussed. Examples of physicochemical and biological properties that might need to be examined include solubility, water content, particle size, crystal properties, biological activity, and permeability. These properties could be inter-related and might need to be considered in combination. Some of these properties can change with time and require time studies.

To evaluate the potential effect of the physicochemical properties of the drug substance on the performance of the drug product, studies on drug product might be warranted. For example, the ICH *Q6A Specifications: Test Procedures and Acceptance Criteria for New Drug Substances and New Drug Products: Chemical Substances* describes some of the circumstances in which drug product studies are recommended. The knowledge gained from the studies investigating the potential effect of drug substance properties on drug product performance can be used, as appropriate, to justify elements of the drug substance specification.

One purpose of these comprehensive guidelines it to prepare for compliance with process analytical technology (PAT), a recent initiative of the Food and Drug Administration (FDA; Ref. 1). PAT is intended to encourage drug makers to build quality into their development processes so they can anticipate the impact of changes on a final formulation. Although PAT is voluntary, the initiative is designed to promote a better understanding, among drug manufacturers, of the mechanics of their processes so that they can avoid failures and minimize the amount of testing required at the end of production. Preformulation studies support PAT by providing more information on an active pharmaceutical ingredient's (API) characteristics to facilitate downstream efficiency and success. Drug manufacturers can eventually submit their documents to a special PAT group within the FDA, which can expedite regulatory approval. Preformulation studies also support reference standard characterization. The regulations of the FDA require that the drug manufacturers establish a primary reference standard at a certain stage in drug development, whereby a compound is characterized as thoroughly and precisely as possible. Subsequent tests and analyses must be based on samples that meet this standard.

The FDA considers PAT to be a system for designing, analyzing, and controlling manufacturing through timely measurements (i.e., during processing) of critical quality and performance attributes of raw and in-process materials and processes, with the goal of ensuring final product quality. It is important to note that the term *analytical* in PAT is viewed broadly to include chemical, physical, microbiological, mathematical, and risk analysis conducted in an integrated manner. The goal of PAT is to enhance the understanding and control the manufacturing process, which is consistent with our current drug quality system: *quality cannot be tested into products; it should be built-in or should be by design*. Consequently, the tools and principles described in this guidance should be used for gaining process understanding and can also be used to meet the regulatory requirements for validating and controlling the manufacturing process.

Quality is built into pharmaceutical products through a comprehensive understanding of:

- The intended therapeutic objectives; patient population; route of administration; and pharmacological, toxicological, and pharmacokinetic characteristics of a drug.
- The chemical, physical, and biopharmaceutic characteristics of a drug.

- Design of a product and selection of product components and packaging based on drug attributes listed previously.
- The design of the manufacturing processes by using principles of engineering, material science, and quality assurance to ensure acceptable and reproducible product quality and performance throughout a product's shelf life.

Effective innovation in development, manufacture, and quality assurance would be expected to answer the following questions:

- What are the mechanisms of degradation, drug release, and absorption?
- What are the effects of product components on quality?
- What sources of variability are critical?
- How does the process manage variability?

This guidance facilitates innovation in development, manufacture, and quality assurance by focusing on process understanding. These concepts are applicable to all manufacturing situations.

Phytomedicines

In January 2004, the U.S. FDA issued a guideline for botanical products (2). The information discussed in section VII.A.1 of the guideline pertains to the initiation of characterization of the drug substance. Also, it should be provided for all products. It is important for the safe conduct of clinical trials to ensure the proper identity of botanical raw materials used in the trials. As there is no history of experience in United States with botanical raw materials marketed only outside the United States, a certificate of authenticity of the plant and plant parts should be provided for such materials. A trained professional who is competent to determine authenticity should sign this certificate. This information should also be provided, if available, for a botanical raw material marketed in the United States.

The general method of preparation (e.g., pulverization, decoction, expression, aqueous extraction, or ethanolic extraction) is provided under §312. 23(a)(7)(iv)(a). This is especially important where more than one process exists in the literature onwhich the safety of the botanical drug substance is based.

The EMEA provides the following guidelines for herbal (botanical as listed in United States) products:

- committee for proprietary medicine products (CPMP)/quality working party (QWP)/2819/00 [European Medicines Evaluation Agency (EMEA)/committee on veterinary medicinal products (CVMP)/814/00] Note for Guidance on Quality of Herbal Medicinal Products (CPMP/CVMP adopted July 01).
- CPMP/QWP/2820/00 (EMEA/CVMP/815/00) Note for Guidance on Specifications: Test procedures and Acceptance Criteria for Herbal Drugs, Herbal Drug Preparations and Herbal Medicinal Products (CPMP/CVMP adopted July 01).

A comprehensive specification for each herbal drug must be submitted, even if the starting material is an herbal drug preparation. This also applies if the applicant is not the manufacturer of the preparation. In the case of fatty or essential oils used as active substances of herbal medicinal products, a specification for the herbal drug is required unless fully justified. The scientific name of the parent plant and its part(s) also need to be stated.

If no monograph for the herbal drug is given in a Pharmacopoeia referred to in Directives 75/318/European Economic Community (EEC) and 81/852/EEC,

Annex 1, a comprehensive specification on the herbal drug must be supplied and should be set out in the same way, where practicable, as the monographs on herbal drugs in the European Pharmacopoeia. This should include the botanical name and authority and the common name, if used for labeling purposes. Information on the site of collection, the time of harvesting and stage of growth, treatment during growth with pesticides, and so on, and drying and storage conditions should be included if possible. The comprehensive specification should be established on the basis of recent scientific data. In the case of herbal drugs with constituents of known therapeutic activity, assays of their content (with test procedure) are required. The content must be included as a range, so as to ensure reproducibility of the quality of the finished product. In the case of herbal drugs where constituents of known therapeutic activity are not known, assays of marker substances (with test procedure) are required. The choice of the markers should be justified.

As a general rule, herbal drugs must be tested for microbiological quality and for residues of pesticides and fumigation agents, toxic metals, likely contaminants and adulterants, and others, unless otherwise justified. Radioactive contamination should be tested for if there are reasons for concerns. Specifications and descriptions of the analytical procedures must be submitted, together with the limits applied. Analytical procedures not given in a Pharmacopoeia should be validated in accordance with the ICH guideline "Validation of analytical procedures: methodology" (CPMP/ICH/281/95) and registration of veterinary products (VICH) guideline (CVMP/VICH/591/98).

If the herbal medicinal product does not contain the herbal drug itself but a preparation, the comprehensive specification on the herbal drug must be followed by a description and validation of the manufacturing process for the herbal drug preparation. The information may be supplied either as part of the marketing authorization application or with the help of the European Drug Master File procedure.

For each herbal drug preparation, a comprehensive specification must be submitted. This must be established on the basis of recent scientific data and must give particulars of the characteristics, identification tests, and purity tests. This has to be done, for example, by the appropriate chromatographic methods. If deemed necessary by the results of the analysis of the starting material, tests on microbiological quality, residues of pesticides, fumigation agents, solvents, and toxic metals have to be carried out. Radioactivity should be tested if there are reasons for concerns. Quantitative determination (assay) of markers or of substances with known therapeutic activity is required. The content must be indicated with the lowest possible tolerance. The test methods must be described in detail.

If preparations from herbal drugs with constituents of known therapeutic activity are standardized (i.e., adjusted to a defined content of constituents with known therapeutic activity), the mode of achievement of the standardization must be stated. If another substance is used for these purposes, it is necessary to specify the quantity, as a range, that can be added.

This section should be in accordance with the "Note for guidance on stability testing of new active substances and medicinal products" (CPMP/ICH/380/95 and CVMP/VICH/899/99) and the "Note for guidance on stability testing of existing active substances and related finished products" (CPMP/QWP/556/96 and EMEA/CVMP/846/99).

Because the herbal drug or the herbal drug preparation in its entirety is regarded as the active substance, a mere determination of the stability of the constituents with known therapeutic activity will not suffice. It must also be shown, as far as possible, for example, by means of appropriate fingerprint chromatograms, that other substances present in the herbal drug or in the herbal drug preparation are likewise stable and that their proportional content remains constant.

If an herbal medicinal product contains several herbal drugs or preparations of several herbal drugs and if it is not possible to determine the stability of each active substance, the stability of the medicinal product should be determined by appropriate fingerprint chromatograms, appropriate overall methods of assay and physical and sensory tests, or other appropriate tests. The appropriateness of the tests should be justified by the applicant.

In the case of an herbal medicinal product containing an herbal drug or herbal drug preparation with constituents of known therapeutic activity, the variation in content during the proposed shelf life should not exceed $\pm 5\%$ of the initial assay value, unless justified. In the case of an herbal medicinal product containing an herbal drug or herbal drug preparation where constituents with known therapeutic activity are unknown, a variation in content during the proposed shelf life of $\pm 10\%$ of the initial assay value can be accepted, if justified by the applicant. These criteria also apply to the stability testing of active substances in a similar manner.

Large Molecule Drugs

A biopharmaceutical drug can go into development before anyone knows much about how it works. The protein may be identified through genomics or proteomics activities or through more traditional medical research. It may initially be associated with a particular disease process or a certain metabolic event. In any case, its mechanism of action—as well as many of its structural characteristics and biochemical properties—may be unknown. One of the more challenging aspects of developing protein pharmaceuticals is dealing with and overcoming the inherent physical and chemical instabilities of proteins. This inherent instability has the potential to alter the state of the protein from the desired (native) form to an undesirable form (upon storage), compromising patient safety and drug efficacy. Marketing concerns come up earlier in the development of protein drugs. Route of administration is determined by the target product profile and whether the product will treat a chronic or acute disorder, or if it will need specific targeting—a broad or narrow therapeutic window, or if it will be administered at home or in the clinic or hospital. For example, marketing considerations arise early in product development for monoclonal antibodies (MAbs). Typically, MAbs are needed at high doses (hundreds of milligrams per dose) and are normally delivered intravenously. The drive to reduce healthcare costs has created a need to administer MAb therapeutics more conveniently, at home, subcutaneously. Thus, MAbs must be available at high concentrations ($\sim$200 mg/mL) in the vial. At these high concentrations, MAb-containing solutions are viscous, making them difficult to administer conveniently. Hence, a preformulation activity that needs to be considered is a concentration study investigating solubility behavior, effect of concentration on viscosity, and increased potential for aggregation. These studies have the potential to strongly influence the target product profile and the design of the clinical trial.

All of these questions can affect the optimal formulation of a drug. For example, an early formulation question is whether the product will be lyophilized (freeze-dried) or sold as a liquid. The advantages of a liquid include timesaving, lower cost, and ease-of-use for patients and clinicians, all of which are good sales points. But stability questions often make freeze-drying necessary for protein and peptide pharmaceuticals. Freeze-dried drugs have a longer shelf life and better stability for shipping and storage, even if they cost more and take longer to make.

The seemingly endless variation of polypeptides makes them interesting as potential therapeutics, but it also makes them a challenge to develop into products. Each protein is unique, and just as variation from protein to protein affects biologic production and purification, so it is central to the formulation development process. Methods developed for one biopharmaceutical are not always directly applicable to others. Similarly, it is quite likely that a formulation developed for one biopharmaceutical may not provide the same level of stability for a different biopharmaceutical.

While there are numerous ways for a protein to lose its stability, the three most commonly encountered modes of denaturation and degradation are aggregation, oxidation, and deamidation. The commonly accepted strategy for rational formulation development relies on identifying mechanisms of denaturation and degradation in order to develop effective countermeasures. Once the specifics of any particular degradation pathway are understood, a more informed choice regarding excipients and formulation can be made, accelerating the product development.

International disagreement over preservatives in food and drugs may present a problem at the preformulation stage. The United States, European Union, and Japanese compendia standards differ regarding the timing of antimicrobial tests and which preservatives and excipients are allowed. Japan, for example, does not accept phenol, a preservative used commonly in the United States. So at this early stage, companies must decide if and where their products will be distributed outside the United States. The European Union is known to have the toughest acceptance criteria for preservatives. The ICH is working toward a common standard, but many formulators have criticized its slow progress. This is applicable if preformulation studies would conduct compatibility studies.

In the development of proteins, physical stability is of prime importance. Four important preformulation stress tests are the shake test (agitation), surfactant test, freeze-thaw test, and heating experiment. Each formulation configuration is shaken in a vial to determine whether it forms aggregates. Then a surfactant (usually a polysorbate detergent, such as Tween) may be selected to prevent formation of precipitants by making it harder for proteins to aggregate. Most proteins are stable around 2–8°C, but few are stable at room temperature. Heating experiments help scientists examine degradation at temperature extremes by heating them to 30°C (about 86°F), and maybe even 45°C (about 113°F). At high temperatures, different mechanisms of protein denaturation may arise.

Recombinant DNA Products

The U.S. FDA provides a detailed description of the characterization of the substances obtained by recombinant DNA technique. The details are provided in the website mentioned in Ref. 3. In addition, several guidelines of the ICH and other guidelines at the U.S. FDA provide additional information on stability testing of biological products. What follows in the successive paragraphs is an outline of what constitutes the required minimum studies. A drug substance is defined by the U.S. FDA as the unformulated active substance that may be subsequently formulated with excipients to produce the drug product.

A clear description of the drug substance should be provided. This description may include, but not be limited to, any of the following: chemical structure, primary and subunit structure, molecular weight, molecular formula, established the US Adapted Names, antibody class/subclass (if appropriate), and so on.

A description and the results of all the analytical testing performed on the manufacturer's reference standard lot and qualifying lots to characterize the drug substance should be included. Information from specific tests regarding identity, purity, stability, and consistency of manufacture of the drug substance should be provided. Examples of analyses for which information may be submitted include, but are not necessarily limited to the following:

- amino acid analysis;
- amino acid sequencing, entire sequence or amino- and carboxy-terminal sequences;
- peptide mapping;
- determination of disulfide linkage;
- sodium dodecyl sulfate-polyacrylamide gel electrophoresis (SDS-PAGE) (reduced and nonreduced);
- isoelectric focusing;
- conventional chromatography and HPLC, for example, reverse-phase, size exclusion, ion-exchange, etc.
- mass spectroscopy (MS);
- assays to detect product-related proteins including deamidated, oxidized, cleaved, and aggregated forms and other variants, for example, amino acid substitutions and adducts/derivatives;
- assays to detect residual host proteins, DNA, and reagents;
- immunochemical analyses;
- assays to quantitate bioburden, endotoxin.

Additional physicochemical characterization may be required for products undergoing post-translational modifications, for example, glycosylation, sulfation, phosphorylation, or formylation. Additional physicochemical characterization may also be required for products derivatized with other agents, including other proteins, toxins, drugs, radionuclides, or chemicals. The information submitted should include the degree of derivatization or conjugation, the amount of unmodified product, removal of free materials (e.g., toxins, radionuclides, linkers, and others), and the stability of the modified product. All test methods should be fully described and the results provided. The application should also include the actual data, such as legible copies of chromatograms, photographs of SDS-PAGE or agarose gel, spectra, and the like.

A description and results of all relevant in vivo and in vitro biological testing performed on the manufacturer's reference standard lot to show the potency and activity(ies) of the drug substance should be provided. Results of relevant testing performed on lots other than the reference standard lot, which might have been used in establishing the biological activity of the product, should also be included. The description and validation of the bioassays should include the methods and standards used, the inter- and intra-assay variability, and the acceptable limits of the assay.

A description of the storage conditions, study protocols, and results supporting the stability of the drug substance should be submitted in this section. (ICH document Stability Testing of Biotechnological/Biological Products or other FDA documents, such as Guideline for Submitting Documentation for the Stability of Human Drug and Biologics for specific information.) Data from tests to monitor the biological activity and degradation products, such as aggregated, deamidated, oxidized, and cleaved forms should be included, as appropriate. Data supporting any proposed storage of intermediate(s) should also be provided.

TESTING SYSTEMS

The significant downturn in the number of drugs approved by the US FDA has prompted newer models of drug discovery that promise to produce a larger number of possible leads. All promising leads must be put through some level of preformulation testing, creating a large burden on preformulation groups to produce results in shorter times. The technologic developments in analytical methodologies allow greater understanding of drug substances and most companies would rather quickly adopt these new techniques, particularly if they automate the testing methods. Some of the most recent introductions of new techniques include liquid chromatography (LC) or gas chromatography (GC)-MS/MS systems, use of Sirius GLpK_a and lipophilicity and pION pSOL instruments, CheqSol® (chasing equilibrium solubility) measurements, nanocrystal technology, and the modulating role of solubilizers on drug efflux by P-glycoprotein (Pgp), and in silico prediction of the effect of solubilizers on Pgp are some of the newer goals of preformulation studies. There are scores of new innovations in dissolution instrumentation, drug substance stability study, and identification of degradation products. One test at the preformulation level pertains to biological transportability of the new drug substance; much of this information is gained from differential solubility analysis. Many methods like the Caco-2 cells model are now routinely used to provide initial estimates about the biological activity potential of new compounds. All of the changing regulatory requirements and technological developments place a significant burden on the preformulation team to stay alert and stay abreast of the technology.

Polymorph Screening

Some of the most significant information comes from identifying the optimal crystal form and the corresponding behavior in different humidity conditions. This information will have implications for a drug's stability and solubility and guide decisions concerning the appropriate dosage form or formulation and how it is packaged, handled, and stored. If the compound is a crystal, the next step is to identify its shape or the different shapes it can take. This information is crucial for several reasons, the most obvious of which is to ensure uniform synthesis, manufacturing, and testing of the compound within a formulation.

Drug manufacturers need to confirm that each batch of API has the same crystalline structure, and that this crystal form remains constant throughout the formulation and life of the drug product (particularly for solid oral dosage forms, powders, creams, ointments, and suspensions). The crystalline structure can also place physical constraints on the ability to manufacture a particular dosage form. For example, needle-shaped crystals tend to entangle and often do not flow well in manufacturing equipment. This can cause formulation of "hot spots," with high concentrations of the API in some areas and deficits in others. If the compound comes in different crystal shapes, then formulators will prefer the shape that is most conducive to the physical manufacture of the desired dosage form—other things being equal. For example, if drug manufacturers prefer a tablet or capsule, we may recommend that they synthesize more spherically shaped crystals rather than flat plates or needles. The crystalline structure also affects a compound's stability and solubility, which again has important implications for formulating, manufacturing, packaging, and storing pharmaceutical products and API. A trade-off may often present itself when selecting a crystal form. For example, crystalline structures that are more desirable from the standpoint of synthesis or formulation manufacture may be less advantageous when considering stability or solubility.

pK_a, Partitioning, and Solubility

Critical variables that should be considered when making formulation decisions are pK_a, lipophilicity, and solubility. The pK_a and lipophilicity can be measured using Sirius GLpK_a and a pION pSOL instrument is used to measure the intrinsic solubility of the compound. The pK_a value is the pH at which acidic or basic groups attached to molecules exist as 50% ionized and 50% nonionized in aqueous solution. The pK_a value provides valuable data on the interaction of an ionizable drug with charged biological membranes and receptor sites and information on where the drug may be absorbed in the digestive tract. Knowing the pK_a also enables the scientist to know how much to alter the pH to drive a compound to its fully ionized or nonionized form for analytical and other purposes, such as formulation, solubility, and stability. Formulators need to know where a drug will dissolve in the digestive tract and whether that corresponds to the optimal region for absorption, especially if they are planning to create a dosage form that will be taken orally. If the drug dissolves too early, it may reprecipitate in a form that is poorly absorbed. But if a drug does not dissolve until after it travels through the stomach or small intestine, it is not likely to get absorbed. In the first case, scientists may want to create a formulation that slows the dissolution and in the second case, a formulation that speeds it up. Another option would be to formulate a dosage that could be administered by injection. Often it is preferred to use a traditional, manual test process to evaluate solubility. For example, one may place samples into three buffer solutions at different pH, shake them mechanically overnight, and then measure how much of the compound has dissolved into the solutions. The measure of the intrinsic solubility of a compound (i.e., the fundamental solubility at which the compound is completely unionized) is useful for formulators in many ways. Working over a pH range from 2 to 11, the pSOL instrument can typically determine intrinsic solubility across a range of 5–50 mg/mL. The use of the Sirius GLpK_a to create lipophilicity profiles is very useful. Drugs that can be taken orally must fall into a fairly narrow window between extreme lipophilicity and extreme hydrophilicity. Many drugs cross biological lipid membranes by passive transport, and there is an optimum value of lipophilicity for each type of membrane. For example, drugs that are highly lipophilic may be easily transported or absorbed, but may get trapped inside fat storage regions, where they will be ineffective. On the other hand, a drug that is extremely hydrophilic may not penetrate the membrane and therefore, has no pharmacological effect. Hence, formulators often find lipophilicity profiles very valuable.

Poorly soluble compounds represent an estimated 60% of compounds in development and many major marketed drugs. It is important to measure and predict solubility and permeability accurately at an early stage, and interpret these data to help assess the potential for development of candidates. This requires developing an effective strategy to select the most appropriate tools to examine and improve solubility in each phase of development, and optimization of solid-state approaches to enhance solubility including the use of polymorphs, co-crystals, and amorphous solids.

Poor solubility can hinder—or even prevent—drug development. Yet the volume and level of poorly soluble compounds is dramatically increasing, leaving gaps in development pipelines. Currently only 8% of new drug candidates have both high solubility and permeability. It is important to know the solubility of drugs as it helps in the identification of potential screening and bioavailability issues. It is valuable in planning chemistry changes during biopharmaceutical evaluation, is important for the confirmation of bioavailability issues, and is also useful in early development of formulations. In drug development, solubility knowledge is needed for biopharmaceutical classification, biowaivers and bioequivalence; it is also required for formulation optimization and salt selection. In manufacture, solubility also affects the optimization of the manufacturing processes.

With this trend of increasingly insoluble drugs stretching resources, many companies are now re-evaluating their strategy. They know that there are many available technologies to measure, predict and improve solubility, and several new emerging techniques. Studies that encompass this scope would include how membrane permeation of drugs can be enhanced by means of solubilizing agents, how the solid state is characterized and modified to improve solubility and drug performance, how salt screening and selection can impact dissolution rate and oral absorption, application of nanocrystal technology to increase dissolution rate, and analysis of the use of pharmaceutical co-crystals in enhancing drug properties.

There are several new emerging methods to measure the solubility of ionizable drugs, such as using the method called CheqSol. CheqSol is a software product that processes data, and controls Sirius' existing GLpK_a, PCA200, and D-PAS instrumentation. Not only does CheqSol measure equilibrium and kinetic solubility rapidly and accurately, it also provides insights into compound behavior that will be of value for the better understanding of drug bioavailability, modeling of precipitation processes, and for investigating changes of crystalline form in suspensions. Pharmaceutical scientists need to know the solubility of drug molecules during drug discovery, as well as in confirmation of bioavailability issues, human formulation design, and Biopharmaceutical Classification, which is required by the FDA. CheqSol is much faster than shake-flask methods, and, it measures both the equilibrium and the turbidimetric (or kinetic) solubilities in the same experiment. CheqSol works by monitoring the pH, as hydrochloric acid (HCl) or potassium hydroxide (KOH) solutions are carefully added to a 10-mL solution of the ionized drug until it precipitates, as detected by an abrupt decrease in the amount of light transmitted through the solution. The concentration at this point is equivalent to a kinetic solubility. Chasing equilibrium then begins—HCl and KOH are added sequentially to force the solution to become supersaturated or subsaturated, and the state of saturation is determined from subsequent small changes in the pH reading. The concentration of unionized species at the crossing points, when the pH change is zero and the sample is neither super nor subsaturated, is equal to the intrinsic solubility. For "chasers," such as diclofenac that supersaturate and chase equilibrium, CheqSol often finds an equilibrium solubility result within 20 minutes, and confirms it several times during a 60-minute experiment. For "nonchasers," such as chlorpromazine that do not chase equilibrium, the pH after precipitation follows the Precipitation Bjerrum Curve and the software calculates the result from the shape of the curve.

Predicting aqueous solubility with in silico tools is a key drug property. It is however, difficult to measure accurately, especially for poorly soluble compounds, and thus numerous in silico models have been developed for its prediction. Some in silico models can predict aqueous solubility of simple, uncharged organic chemicals reasonably well; however, solubility prediction for charged species and drug-like chemicals is not very accurate. However, extrapolating solubility data to intestinal absorption from pharmacokinetic and physicochemical data, elucidating crucial parameters for absorption, and assessing the potential for improvement of bioavailability are important at the preformulation stages.

Solubilizers (e.g., organic solvents, detergents, Pluronics) are often used to solubilize drugs in aqueous solution without considering their effects on biological systems, such as (*i*) lipid membranes and (*ii*) multidrug resistance (MDR) efflux transporters (e.g., Pgp or MDR1).

The modulatory role of solubilizers on drug efflux by Pgp and in silico prediction of the effect of solubilizers on Pgp are some of the newer goals of preformulation studies.

Liposomal solubilization is an effective approach for the delivery of potent, insoluble drug candidates. However, careful consideration of the various lipid and drug properties along with an emphasis on manufacturing conditions is needed for the successful development of a marketable formulation.

Increasing dissolution rates using nanocrystal technologies is becoming common. The NanoCrystal Technology was developed by Elan Corporation (Dublin 2, Ireland). For poorly water-soluble compounds, Elan's proprietary NanoCrystal technology can enable formulation and improve compound activity and final product characteristics. The NanoCrystal technology can be incorporated into all dosage forms, both parenteral and oral, including solid, liquid, fast-melt, pulsed release, and controlled release dosage forms. Poor water solubility correlates with slow dissolution rate, and decreasing particle size increases the surface area, which leads to an increase in dissolution rate. This can be accomplished predictably and efficiently using NanoCrystal technology (4). NanoCrystal particles are small particles of the drug substance, typically less than 1000 nm in diameter, which are produced by milling the drug substance using a proprietary wet milling technique (Fig. 1). The NanoCrystal particles of the drug are stabilized against agglomeration by surface adsorption of selected GRAS (generally regarded as safe) stabilizers. The result is an aqueous dispersion of the drug substance that behaves like a solution—a NanoCrystal colloidal dispersion, which can be processed into finished dosage forms for all routes of administration.

Nanonization is a formulation technology that can universally be applied to all drugs—each drug can be transferred to drug nanocrystals. The main production technologies available to produce drug nanocrystals have their advantages and limitations. The reduction of solid particles to nanoparticles is achieved by high-pressure homogenization.

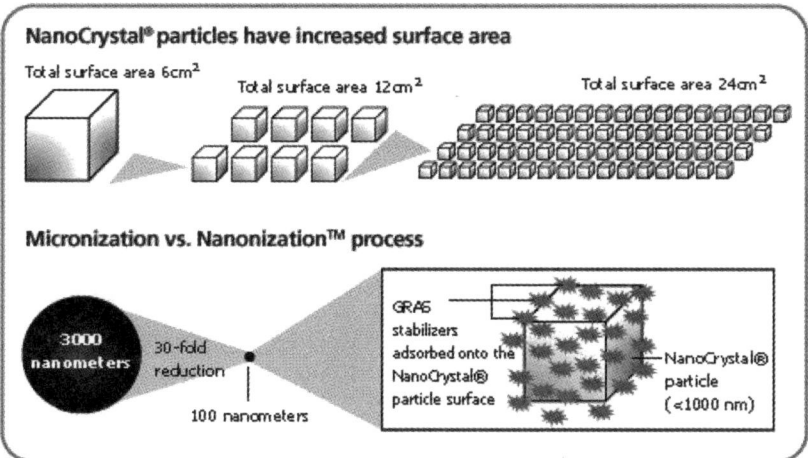

Figure 1 The NanoCrystal® technology. *Source*: Courtesy of Elan Corporation, Dublin 2, Ireland.

Salt Screening

Recent trends in combinatorial chemistry have resulted in the synthesis of large molecular weight (MW) lipophilic drugs. Converting the free acid/base form to a salt is an important option to explore when trying to improve solubility and oral bioavailability. Of the 21 new molecular entities approved by the FDA in 2003, 10 were salt forms. Selection of the right counter ion with optimum physiochemical characteristics is crucial to drug development. Consideration of the new compound's physical–chemical properties, processability under various manufacturing conditions, and bioavailability must be made. A complete range of characterization tools for a complete salt screen would include:

- X-ray powder diffraction analysis (XRPD)
- thermal analysis [DSC, thermogravimetric analyzer (TGA), thermo-mechanical analyzer (TMA)]
- microscopy (light and polarized)
- dynamic vapor sorption—moisture absorption and desorption
- density (intrinsic and bulk)
- NMR analysis
- solubility analysis in various media
- dissolution (including intrinsic dissolution testing)
- particle size analysis (optical, laser light, and light obscuration)

Scheme 1 describes a salt screening decision-making tree.

SOLID-STATE CHARACTERIZATION
Powder Properties

Powders are masses of solid particles or granules surrounded by air (or other fluid) and it is the solid plus fluid combination that significantly affects the bulk properties of the powder. It is perhaps the most complicating characteristic because the amount of fluid can be highly variable.

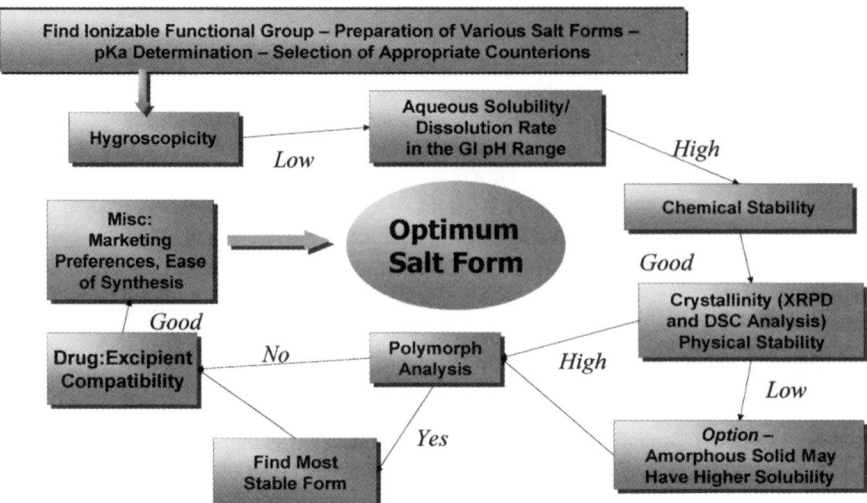

Scheme 1 Salt selection tree. *Source*: Courtesy of Cardinal Health, Dublin, Ohio, U.S.A.

Powders are probably the least predictable of all materials in relation to flow ability because of the large number of factors that can change their rheological properties. Physical characteristics of the particles, such as size, shape, angularity, size variability, and hardness will all affect flow properties. External factors, such as humidity, conveying environment, vibration, and perhaps most importantly, aeration, will compound the problem. The more common variables would include:

- Powder or particle variables:
  - particle size
  - size distribution
  - shape
  - surface texture
  - cohesivity
  - surface coating
  - particle interaction
  - wear or attrition characteristic
  - propensity to electrostatic charge
  - hardness
  - stiffness
  - strength
  - fracture toughness
- External factors influencing powder behavior:
  - flow rate
  - compaction condition
  - vibration
  - temperature
  - humidity
  - electrostatic charge

- aeration
- transportation experience
- container surface effects
- storage time

Another characteristic of powders is that they are often inherently unstable in relation to their flow performance. This instability is most obvious when a free flowing material ceases to flow. This transition may be initiated by the formation of a bridge in a bin, by adhesion to surfaces, or by any event that may promote compaction of the powder. The tendency to switch in this way varies greatly from one powder to another, but can even be pronounced between batches of the same material.

The nature of powders therefore is such that an adverse combination of environmental factors can cause an otherwise free flowing powder to block or flow with difficulty. Conversely, a very cohesive powder may be processed satisfactorily if the handling conditions are optimized.

Given the complex nature of powders, it is not surprising that processing difficulties are very common. Being able to predict flow performance would bring about many operational advantages, such as reducing stoppages and improving product quality. To achieve this, we need to know how a given powder is affected by the variables mentioned earlier, and also to have a reliable indicator of the potential instability of the powder. Predicting flow ability performance in a particular plant therefore requires knowledge of the handling and processing conditions as well as the flow ability characteristics of the material under these conditions. It means that the process conditions relevant to flow ability need to be determined. These might include the level of static and dynamic head produced in a storage bin or hopper, the amount of aeration that occurs, the opportunity to adsorb moisture, become electrostatically charged or be consolidated due to vibration. Another factor that can affect powder flow includes an increase in the amount of finer particles as a result of attrition. All or at least the most important of these factors then needs to be quantified regarding how they affect flow ability.

Slight compaction, a small vibration, or the smallest amount of aeration can significantly affect flow ability. This is the main reason why traditional methods of flow ability measurement have not been suitable as a basis for repetitive testing. In all traditional techniques, the packing condition and the air content are largely unknown quantities, and so the results will vary accordingly. When making an assessment, it is essential to know what was tested and the condition of the powder when tested. In addition to the packing problem, traditional flow ability measurements are prone to operator error, have poor repeatability and, for the most part, are very time-consuming. An automated test and analysis system is needed that takes only minutes, is very repeatable, and is independent of the operator.

The most important innovation required in relation to traditional techniques is a way of classifying powders so that flow ability performance of each powder can be measured and recorded along with its processing experience. Eventually, such a database of information could remove much of the uncertainty from processing and provide a reference base for the development of new powders. It would allow each production machine to be classified in terms of the powders that could be efficiently processed.

Ideally, the classification of powders would provide more than just flow ability data, such as flow rate and compaction indices. It would also include data describing the robustness and stability of the powder, for example, vulnerability to segregation, attrition, and vibration. Given this, then the two key issues of powder processing could be addressed. First, will the powder flow satisfactory—does it have flowability properties that suit the process? And second, is the powder robust—will it be adversely affected by being processed?

Freeman Technology and the FT4 Powder Rheometer offer real benefits to all users of powders (5). These include:

- the more efficient use of powder handling systems by reducing stoppages and optimizing throughput;
- improved product quality by introducing quality conformance checks at all stages of production; and
- overall—improved competitiveness.

Microscopy

Significant advances have been made in the field of microscopy over the past decade, allowing study of nanocrystals and elemental analysis using small samples. Some of the spectroscropic and microscopic methods available include:

- Energy dispersive X-ray spectrometry (EDS) for quick and easy elemental analysis of samples in the SEM. Minimum detection limit of 0.1% by weight.
- Wavelength dispersive X-ray spectrometry (WDS) for a more detailed elemental analysis of samples in the SEM. JEOL Four-Crystal Spectrometer attached to the JSM-35C SEM can be used for 1-μm spot analysis, digital and analog line scans, and X-ray image mapping, elements detection from Be to U, minimum detection limit of 0.01% by weight, fully quantitative results by extended $\varphi-\rho-z$.
- Inductively coupled plasma-atomic emission spectroscopy (ICP) provides trace level and bulk elemental analyses of solid and liquid samples. Using Varian Analytical Instruments Liberty 100 air pass inductively coupled plasma atomic emission sequential spectrometer, minimum detection limits better than 1 ppb by weight (element/line dependent) bulk solid acid digestion (for powders, residue, ingots, and so on) and liquid analyses can be performed. Analysis of all elements from Li to U (excluding N, O, F, S, and noble gases), 0.75-m Czerny Turner monochromator with holographic grating allows high intensity spectra up to four peak orders with 0.006-nm resolution through a wavelength range of 189–900 nm.
- Surface analysis (AES/XPS): Electron spectroscopy for elemental analysis of surfaces, sensitive to as low as two atomic layers. Physical electronics model PHI-570 Auger Electron Spectroscopy/X-ray Photoelectron Spectroscopy System is a double pass cylindrical mirror energy analyzer with dual anode (Mg/Al) X-ray source and has a rapid sample introduction probe. It can detect elements at the first five to ten atomic layers of sample and detect all elements except H and He.
- Scanning electron microscopy for high resolution and high magnification photographs. Can also perform elemental analysis with EDS and WDS attachments. JEOL JSM6320F & JSM-35C research-grade SEM can provide imaging from 10× to 400,000×. Using analytical electron microscopy very high magnification images with excellent depth of focus can be obtained. This is especially

important when rough surface structures are being examined. In addition, information about the chemical composition at the microlevel and the phase composition of the sample under study can be directly obtained. Using SEM, such as the Fraunhofer Institut Für Fertigungstechnik und Angewandte Material for Schung it is possible to magnify structures up to 500,000 times with high depth of focus. A finely focused electron beam allows structures down to 0.001 mm to be resolved. The acceleration voltage of the electron beam directed at the sample surface can be varied between 300 V and 30 kV. The emitted secondary and back-scattered electrons give information about the topology of the sample. Back-scattering electrons can also be used to produce material contrast images.

- X-ray diffraction (XRD) for phase analysis, crystallographic information, residual stress, texture analysis, and reflectometry on powders, bulk, or thin films. Philips X'Pert PRO, and a second Philips dual diffractometer system with automated PC control, independent theta/2θ, sample spinner, and 21 sample changer can be used for crystallography and Rietveld analysis of samples; flat, irregular, thin films, or in glass capillaries.
- Fourier transform infrared spectroscopy (FT-IR) is useful for identifying organic and inorganic compounds by comparison with library references. Perkin Elmer System 2000 offers near IR, mid IR, far IR: 15,000–15,030 cm, transmittance (T), specular reflectance (SR; Ref. 6) and diffuse reflectance (DR), horizontal and vertical attenuated total reflectance (ATR) microscope (>10-μm spot, 10,000–10,580 cm)$^{-1}$.

Thermal Analysis

Materials characterization requires the measurement of molecular and macroscopic properties. Thermal analysis techniques determine calorimetric and mechanical properties, such as heat capacity, mechanical modulus, sample mass, and dimensional changes in temperature ranges between $-150°C$ and $1600°C$. Thermal analysis utilizes DSC, TGA, TMA, and dynamic mechanical analysis instrumentation supplemented by software products, accessories, consumables, and documentation. Applications are frequently found in research and QC environments. They cover the characterization of materials, process development, and evaluation as well as safety investigations. All METTLER TOLEDO thermal analysis products belong to the latest generation STARe family. The associated METTLER TOLEDO FP900 series includes instruments for the rapid determination of physical properties, such as melting, boiling, dropping, or softening points.

Microthermal analysis is a recently introduced thermoanalytical technique that combines the principles of scanning probe microscopy with thermal analysis via replacement of the probe tip with a thermistor. This allows samples to be spatially scanned in terms of both topography and thermal conductivity, whereby placing the probe on a specific region of a sample and heating, it is possible to perform localized thermal analysis experiments on those regions.

Molecular Spectroscopy

The foundations for fluorescence correlation spectroscopy (FCS) were already laid in the early 1970s, but this technique did not become widely used until single-molecule detection was established almost 20 years later with the use of diffraction-limited confocal volume element. The analysis of molecular noise from the GHz- to the Hz-region facilitates measurements over a large dynamic range covering photophysics, conformational transitions and interactions as well as transport properties of fluorescent biomolecules. From the Poissonian nature of the noise spectrum the absolute number of molecules is obtainable. Originally used for the analysis of molecular interactions in solutions, the strength of FCS lies also in its applicability to molecular processes at either the surface or interior of single cells. Examples of the analysis of surface kinetics including on and off rates of ligand–receptor interactions will be given. The possibility of obtaining this type of information by FCS will be of particular interest for cell-based drug screening.

Recrystallization, grinding, compaction, and freeze-drying are frequently used in the pharmaceutical industry to obtain a desirable crystalline form of bulk powder and excipients. These processes affect not only the surface area, but also the crystalline disorder of the powder materials. Because both these parameters may affect the bioavailability of a drug through the rate of dissolution, it is necessary to control the conditions under which the pharmaceutical drug powders are produced. The extent of disorder in a crystalline solid may induce the hygroscopicity of the drug in addition to the flow, mechanical properties, and chemical stability. Because the qualities of a pharmaceutical preparation depend on the characteristics of the bulk powders and excipients, controlling the production process is important. An amorphous solid-state powder may determine the bioavailability of a slightly water-soluble drug because the property affects solubility and hence, absorption of the drug in the gastrointestinal tract. However, the amorphous form has problems regarding stability and hygroscopicity, resulting in transformation to a more stable crystalline form during preservation.

Therefore, in order to control the quality of pharmaceutical solid dosage products, techniques for the evaluation of crystallinity of the bulk powders and/or excipients are needed. XRD, DSC, FT-Raman spectroscopy, and microcalorimetry are currently the most widely used methods to evaluate crystallinity.

Near-infrared (NIR) spectroscopy is becoming an important technique for pharmaceutical analysis. This spectroscopy is simple and easy because no sample preparation is required and samples are not destroyed. In the pharmaceutical industry, NIR spectroscopy has been used to determine several pharmaceutical properties, and a growing literature exists in this area. A variety of chemoinfometric and statistical techniques have been used to extract pharmaceutical information from raw spectroscopic data. Calibration models generated by multiple linear regression (MLR) analysis, principal component analysis, and partial least squares regression analysis have been used to evaluate various parameters.

X-Ray Diffraction

The determination of the average morphology is often a "bottle neck" in elucidating other important behaviors of large quantities of crystalline powders used in pharmaceutical development and processing.

X-rays are electromagnetic radiation of wavelength about 1 Å (10^{-10} m), which is about the same size as an atom. They occur in that portion of the electromagnetic spectrum between gamma rays and the ultraviolet. The discovery of X-rays in 1895 enabled scientists to probe crystalline structure at the atomic level. X-ray diffraction has been in use in two main areas: for the fingerprint characterization of crystalline materials and the determination of their structure. Each crystalline solid has its unique characteristic X-ray powder pattern, which may be used as a "fingerprint" for its identification. Once the material has been identified, X-ray crystallography may be used to determine its structure, that is, how the atoms pack together in the crystalline state and what the interatomic distance and angle are. X-ray diffraction is one of the most important characterization tools used in solid-state chemistry and materials science. The size and the shape of the unit cell for any compound can be most easily determined using the diffraction of X-rays.

It is possible to use XRD techniques to estimate the average shape and "habit" of organic crystalline material using a single crystal. The relative intensities of the peaks in an XRPD pattern from a sample exhibiting a "standard" preferred orientation correlates with the shape of the crystallites present. Models have been developed to yield a quantitative "enhancement" factor for each face. The combined simple forms morphology (CSM) of the material can be produced by indexing the observed faces and modifying the simulated Bravais-Friedel-Donnay-Harker (BFDH) morphology (7). The average shape of crystallites can be estimated from the CSM by multiplying each face by its enhancement factor.

Stability Testing

The regulatory authorities clearly define the protocols for the testing of drug products for stability during the shelf life. However, testing of drug substances at the preformulation level for stability evaluation offers several advantages and opportunities once the drug substances enter the formulation stage. First, it provides a clear idea about which types of dosage forms can be used. A highly unstable protein drug cannot be placed in anything but a highly preserved and protected parenteral form, as an example. The development of stability testing protocols start with the development of stability indicating methods, the details of which can be readily found in any pharmaceutical analysis text or through the website of the US FDA. The Q1A R2 (Stability Testing of New Drug Substances and Products) is a good starting place (8). Similar guidelines are provided for biotechnology and botanical products.

Moisture Isotherm

The crystalline structure can significantly affect a compound's tendency to absorb moisture, which can impact sample handling for analytical testing, formulation, stability, and product shelf life in areas of varying humidity. For example, if a compound attracts water as it is exposed to rising humidity levels, each kilogram of material will contain more water and less of the compound as the humidity level increases. This could carry over to the formulation, where, as a consequence of the higher moisture content in the API, a subpotent formulation could be manufactured. Hence, drug manufacturers need this information in order to adequately control humidity and ensure uniform moisture to compound ratios across batches. They also need the information to design specific formulations and packaging that will maintain the stability of a finished product when it is shipped and stored in different environments. A compound may remain stable in the controlled humidity and temperature of a lab or manufacturing facility, but degrade if it acquires water when trucked across country or stored on a pharmacy shelf. For example, compound A may exist in two salt forms—a nonhygroscopic monohydrochloride form and a hygroscopic dihydrochloride form. The client needs to use the dihydrochloride form for a solid oral dosage.

Increasing moisture levels in the formulation leads to localized areas of high concentrations of hydrochloric acid in the vicinity of the API molecules. The decision must then be made whether to develop a capsule formulation or a tablet formulation, as each have their own unique set of manufacturing and packaging challenges. While the basic generation of a capsule formulation may be the quickest route, initially, the need for extensive coating and or packaging may slow the entire process. These measures would be required to protect the product from moisture, as the generation of hydrochloric acid in the formulation could result in partial digestion of the capsule shell. For a tablet formulation, the application of a moisture impermeable coating may be a simpler process and more options may be available to package the material in a protective environment. Knowing this information at the beginning of the process may save the client considerable time while providing some early direction to the development of an appropriate and successful formulation. Experimental data can generate a lot of information regarding the characteristic behavior of the molecule. In addition to understanding the hygroscopicity of the compound, drug manufacturers should know whether their compound's vapor sorption is "reproducible" as humidity levels go up and down. One of our most useful instruments is the vapor sorption analyzer, which, among other things, subjects samples to increases and decreases in relative humidity (RH).

These studies involve weighing out a small sample and exposing it to a very dry atmosphere at low or high temperatures. Once the sample dries to a prespecified level, the humidity is raised in increments, for example, from 5% to 95%, while keeping the temperature constant. By weighing the sample at each increment, one can determine how much moisture the compound acquires at a given percentage of RH. This is very important information for predicting stability and shelf life. We can then determine whether vapor desorption is reproducible by weighing the sample as we decrease the humidity in the same increments. The kinetics of absorption can also be studied by subjecting samples to the same incremental humidity changes, but by varying the times spent by different samples at each percentage of RH. In addition, the percentage of RH at which a transition occurs between an amorphous and crystalline structure can be determined. An API may be more stable, more processable, or more soluble (giving the impression of different pharmacological activity) in one form or another and hence, must be formulated, packaged, and handled to maintain that form. In general, four possible basic vapor sorption profiles are observed—the compound can be found to be non-hygroscopic, vapor sorption is reproducible, vapor sorption curves demonstrate a degree of hysteresis, or the vapor sorption is nonreproducible either because of deliquescence, development of nonreversible hydrates, or other reasons.

Vapor sorption is reproducible when the rate at which a compound acquires moisture during humidity increases is matched by the rate at which it loses moisture during humidity decreases. If the rates are the same, then scientists can control the moisture/compound ratio simply by controlling the humidity levels, without having to consider the specific history of the material. Hence, if a batch of material has a certain water/compound ratio in facility A and acquires more water as it is moved across a humid environment to facility B, one can restore the original ratio by insuring that B's humidity level is the same as A's (and waiting for an adequate amount of time). If vapor sorption is not reproducible, then one will need to know how the absorption rate differs from the desorption rate, and the precise humidity conditions that the material undergoes as it is transported from A to B, as the history of the material will affect the amount of moisture present in the material. Of course, whether vapor sorption is reproducible or not, it takes time to raise or lower the water content under certain humidity and temperature conditions. What is considered to be an adequate time can be ascertained experimentally by preformulation studies of the kinetics of vapor sorption for a particular compound, which focuses not only on the amount of water absorbed or released, but also on the time it takes for the processes to occur.

While the moisture content of the sample at any given RH is dependent on the history of the sample, all the moisture gained by the sample in the adsorption phase is eventually lost in the desorption phase.

While the majority of experiments conducted using the vapor sorption analyzer involve monitoring weight changes at constant temperatures and varying humidity levels, the instrument can also be used to measure changes in weight when incrementally altering temperature, while maintaining a constant humidity.

Other methods and instruments to test the effects of temperature include the DSC and the TGA. DSC measures the amount of energy (as heat) absorbed or released as a sample is heated, cooled, or held at constant temperature. These measurements provide information on effects such as glass transition, crystallization, and melting point, and provide quantitative insight into the composition of a sample. When the sample is heated, inorganic salts first split off their water of crystallization, and then other volatile components evaporate. The weight loss indicates the amount of water or volatile components in the sample. TGA also helps formulators to understand decomposition behavior. Although some similar information can be obtained with the vapor sorption analyzer, TGA is a specialized instrument that allows users to measure effects at much higher temperatures. They can even set the starting and ending temperatures and control the speed at which the temperature rises or falls.

Excipient Compatibility

Whereas the choice of excipients starts with the stages of formulation, some excipients are historically used in specific drug formulations; for example, if the newly discovered drug is a cephalosporin for use as an intravenous product, compatibility with arginine or sodium carbonate would be advised as these are the most commonly used active excipients used for solubilization. Similarly, for drugs that are likely to be compressed, compatibility with common ingredients of compression and disintegration are plausible choices at this stage. The relative emphasis on excipient interaction would depend on how the company research is planned; in many situations, the preformulation group is more closely aligned with the drug discovery group and many of these studies are left to the formulation group.

TRANSPORT ACROSS BIOLOGICAL MEMBRANES
Drug Efflux and Multidrug Resistance Studies

The problem of MDR has gained increasing importance in recent years, particularly in the fields of tumor therapy and treatment of bacterial and fungal infections. One of the major mechanisms responsible for development of MDR is overexpression of drug efflux pumps. These membrane-bound, ATP-driven transport proteins efflux a wide variety of natural product toxins and chemotherapeutic drugs out of the cells and, give rise to decreased intracellular accumulation of these compounds. Thus, inhibition of efflux pumps is a versatile approach for overcoming MDR, and several compounds are in clinical Phase III studies. The main target is Pgp, which is responsible for MDR in tumor cells, and transport systems in *Staphylococcus aureus, Pseudomonas aerugiosa* and *Escherichia coli*. Due to the fact that 3D structures of the proteins at atomic resolution were not available, drug development was performed solely on the basis of ligand design. However, electron microscopy studies as well as X-ray structures of three bacterial efflux pumps may open the door to target-based drug design in the near future.

The lipophilicity of drug molecules (represented as the logarithm of the *n*-octanol/water partition coefficient) often strongly correlates with their pharmacological and toxic activities. It is therefore, not surprising that there is considerable interest in developing mathematical models capable of accurately predicting their value for new drug candidates. The key importance of lipophilicity in bio-studies is discussed for β-blockers. Examples of their lipophilicity-dependent pharmacological properties including pharmacokinetic, pharmacodynamic, and clinical aspects are reviewed. Comprehensive lipophilicity compilations of β-blockers are not available so far. Log *P* calculations with 10 programs for 30 clinically relevant β-blockers are presented for the first time in this review.

Modulators and inhibitors of multidrug efflux transporters, such as Pgp, are used to reduce or inhibit MDR, which leads to a failure of the chemotherapy of, for example, cancers, epilepsy, bacterial, parasitic, and fungal diseases. Binding and transport of first-, second-, and third-generation modulators and inhibitors of Pgp take into account the properties of the drug (H-bonding potential, dimensions, and pK_a values) as well as the properties of the membrane.

Gram-positive lactic acid bacteria possess several MDRs that excrete out of the cell a wide variety of mainly cationic lipophilic cytotoxic compounds as well as many clinically relevant antibiotics. These MDRs are either proton/drug antiporters belonging to the major facilitator superfamily of secondary transporters or ATP-dependent primary transporters belonging to the ATP-binding cassette superfamily of transport proteins.

It is increasingly recognized that efflux transporters play an important role, not only in chemo protection, for example, MDR, but also in the absorption, distribution, and elimination of drugs. The modulation of drug transporters through inhibition or induction can lead to significant drug–drug interactions by affecting intestinal absorption, renal secretion, and biliary excretion, thereby changing the systemic or target tissue exposure of the drug. Few clinically significant drug interactions that affect efficacy and safety are due to a single mechanism and there is considerable overlap of substrates, inhibitors, and inducers of efflux transporters and drug metabolizing enzymes, such as CYP3A. As well, genetic polymorphisms of efflux transporters have been correlated with human disease and variability of drug exposure.

In Vitro–In Vivo Correlation

The in vitro–in vivo correlation (IVIVC) is an extremely useful exercise at the preformulation level that determines how scale up and post approval changes or Biowaiver principles would be exploited. Conceptually, IVIVC describes a relationship between the in vitro dissolution/release versus the in vivo absorption. This relationship is an important item of research in the development of drug delivery systems. In vitro dissolution testing serves as a guidance tool to the formulator regarding product design and in quality control. Especially, it is of specific importance for modified release dosage forms, which are intended for the purpose of prolonging, sustaining, or extending the release of drugs. By applying mathematical principles, such as linear system analysis or moment analysis, data describing in vitro and in vivo processes can be obtained. Developing a predictable IVIVC depends upon the complexity of the delivery system, its formulation composition, method of manufacture, physicochemical properties of the drug, and the dissolution method. Several sophisticated commercial dissolution methods are available along with the software to develop IVIVC models; these will bediscussed elsewhere in the book.

Caco-2 Cell Studies

Caco-2 monolayer, a model for human drug intestinal permeability, is of great interest. Kinetics of intestinal drug absorption, permeation enhancement, chemical moiety, structure–permeability relationships, dissolution testing, in vitro/in vivo correlation, and bioequivalence are studied using Caco-2. The Caco-2 cell line is heterogeneous and is derived from a human colorectal adenocarcinoma. Caco-2 cells are used as in vitro permeability models to predict human intestinal absorption because they exhibit many features of absorptive intestinal cells. This includes their ability to spontaneously differentiate into polarized enterocytes that express high levels of brush border hydrolases, and form well-developed junctional complexes. Consequently, it becomes possible to determine whether passage is transcellular or paracellular based on a compound's transport rate. Caco-2 cells also express a variety of transport systems including di-peptide transporters and Pgp. Due to these features, drug permeability in Caco-2 cells correlates well with human oral absorption, making Caco-2 an ideal in vitro permeability model. Additional information can be gained on metabolism and potential drug–drug interactions as the drug undergoes transcellular diffusion through the Caco-2 transport model. The Millipore MultiScreen Caco-2 assay system is a reliable 96-well platform for predicting human oral absorption of drug compounds (using Caco-2 cells or other cell lines whose drug transport properties have been well characterized).

REFERENCES

1. http://www.fda.gov/cder/OPS/PAT.htm.
2. http://www.fda.gov/cder/guidance/4592fnl.htm.
3. http://www.fda.gov/cder/cmcdna.txt.
4. http://www.elan.com/EDT/drug_delivery/nanocrystal_technology.asp.
5. http://www.freemantech.co.uk/.
6. http://www.specac.com/diffuse_specular_reflectance.htm.
7. http://www.scripps.edu/rc/softwaredocs/msi/cerius45/proppred/Morphology.doc.html.
8. http://www.fda.gov/cber/gdlns/ichstab.pdf.

BIBLIOGRAPHY

Adessi C, Soto C. Converting a peptide into a drug: strategies to improve stability and bioavailability. Curr Med Chem 2002 May; 9(9):963–978.

Adibi SA. The oligopeptide transporter (Pept-1) in human intestine: biology and function. Gastroenterology 1997 Jul; 113(1):332–340.

Adibi SA. Regulation of expression of the intestinal oligopeptide transporter (Pept-1) in health and disease. Am J Physiol Gastrointest Liver Physiol 2003 Nov; 285(5): G779–G788.

Anderle P, Huang Y, Sadee W. Intestinal membrane transport of drugs and nutrients: genomics of membrane transporters using expression microarrays. Eur J Pharm Sci 2004 Jan; 21(1):17–24.

Artursson P, Borchardt RT. Intestinal drug absorption and metabolism in cell cultures: Caco-2 and beyond. Pharm Res 1997 Dec; 14(12):1655–1658.

Artursson P, Palm K, Luthman K. Caco-2 monolayers in experimental and theoretical predictions of drug transport. Adv Drug Deliv Rev 2001, 1 Mar; 46(1–3):27–43.

Avdeef A. Physicochemical profiling (solubility, permeability and charge state). Curr Top Med Chem 2001 Sep; 1(4):277–351.

Barrett D. From natural products to clinically useful antifungals. Biochim Biophys Acta 2002 18 Jul; 1587(2–3):224–233.

Barthe L, Woodley J, Houin G. Gastrointestinal absorption of drugs: methods and studies. Fundam Clin Pharmacol 1999; 13(2):154–168.

Basson MD. In vitro evidence for matrix regulation of intestinal epithelial biology during mucosal healing. Life Sci 2001 9 Nov; 69(25–26):3005–3018.

Bauer M. [Crystallization and solid state properties of molecules of pharmaceutical interest]. Ann Pharm Fr 2002 May; 60(3):152–160. French.

Benet LZ, Cummins CL. The drug efflux–metabolism alliance: biochemical aspects. Adv Drug Deliv Rev 2001 Oct 1; 50 Suppl 1:S3–11

Benet LZ, Cummins CL, Wu CY. Transporter–enzyme interactions: implications for predicting drug–drug interactions from in vitro data. Curr Drug Metab 2003 Oct; 4(5):393–398.

Benet LZ, Cummins CL, Wu CY. Unmasking the dynamic interplay between efflux transporters and metabolic enzymes. Int J Pharm 2004 Jun 11; 277(1–2):3–9.

Berger V, Gabriel AF, Sergent T, Trouet A, Larondelle Y, Schneider YJ. Interaction ofochratoxin A with human intestinal Caco-2 cells: possible implication of a multidrug resistance-associated protein (MRP2). Toxicol Lett 2003 11 Apr; 140–141:465–476.

Bergstrom CA. In silico predictions of drug solubility and permeability: two rate-limiting barriers to oral drug absorption. Basic Clin Pharmacol Toxicol 2005 Mar; 96(3): 156–161.

Bertrand M, Jackson P, Walther B. Rapid assessment of drug metabolism in the drug discovery process. Eur J Pharm Sci 2000 Oct; 11(suppl 2):S61–S72.

Bettinetti GP. [X-ray diffractometry in the analysis of drugs and pharmaceutical forms]. Boll Chim Farm 1989 May; 128(5):149–162. Italian.

Blanchfield J, Toth I. Lipid, sugar and liposaccharide based delivery systems 2. Curr Med Chem 2004 Sep; 11(17):2375–2382.

Bohets H, Annaert P, Mannens G, et al. Strategies for absorption screening in drug discovery and development. Curr Top Med Chem 2001 Nov; 1(5):367–383.

Boyle EC, Finlay BB. Bacterial pathogenesis: exploiting cellular adherence. Curr Opin Cell Biol 2003 Oct; 15(5):633–639.

Brandau DT, Jones LS, Wiethoff CM, Rexroad J, Middaugh CR. Thermal stability of vaccines. J Pharm Sci 2003 Feb; 92(2):218–231.

Braun A, Hammerle S, Suda K, et al. Cell cultures as tools in biopharmacy. Eur J Pharm Sci 2000 Oct; 11(suppl 2):S51–S60.

Braun K, Pipkorn R, Waldeck W. Development and characterization of drug delivery systems for targeting mammalian cells and tissues. Curr Med Chem 2005; 12(16):1841–1858.

Brewster ME, Loftsson T. The use of chemically modified cyclodextrins in the development of formulations for chemical delivery systems. Pharmazie 2002 Feb; 57(2): 94–101.

Brot-Laroche E. Differential regulation of the fructose transporters GLUT2 and GLUT5 in the intestinal cell line Caco-2. Proc Nutr Soc 1996 Mar; 55(1B):201–208.

Brown LR. Commercial challenges of protein drug delivery. Expert Opin Drug Deliv 2005 Jan; 2(1):29–42.

Bruguera JL, Chulia D, Verain A, Note D, Rovira G, Salles JP. [The therapeutic utilization of magnesium: medical consequences]. Pharm Acta Helv 1992; 67(7): 189–194. French.

Bugay DE. Characterization of the solid-state: spectroscopic techniques. Adv Drug Deliv Rev 2001, 16 May; 48(1):43–65.

Burke CJ, Hsu TA, Volkin DB. Formulation, stability, and delivery of live attenuated vaccines for human use. Crit Rev Ther Drug Carrier Syst 1999; 16(1):1–83.
Burton PS, Conradi RA, Ho NF, Hilgers AR, Borchardt RT. How structural features influence the biomembrane permeability of peptides. J Pharm Sci 1996 Dec; 85(12):1336–1340.
Carriere V, Chambaz J, Rousset M. Intestinal responses to xenobiotics. Toxicol In Vitro 2001 Aug–Oct; 15(4–5):373–378.
Carstea ED, Hough S, Wiederholt K, Welch PJ. State-of-the-art modified RNAi compounds for therapeutics. IDrugs 2005 Aug; 8(8):642–647.
Castoria G, Lombardi M, Barone MV, et al. Rapid signalling pathway activation by androgens in epithelial and stromal cells. Steroids 2004 Aug; 69(8–9):517–522.
Christopher R, Dhiman A, Fox J, et al. Data-driven computer simulation of human cancer cell. Ann NY Acad Sci 2004 May; 1020:132–153.
Cross TA. Solid-state nuclear magnetic resonance characterization of gramicidin channel structure. Methods Enzymol 1997; 289:672–696.
Curstedt T, Johansson J. New synthetic surfactants—basic science. Biol Neonate 2005; 87(4):332–337. Epub 2005, 1 Jun.
Debnam ES, Grimble GK. Methods for assessing intestinal absorptive function in relation to enteral nutrition. Curr Opin Clin Nutr Metab Care 2001 Sep; 4(5):355–367.
Delie F, Rubas W. A human colonic cell line sharing similarities with enterocytes as a model to examine oral absorption: advantages and limitations of the Caco-2 model. Crit Rev Ther Drug Carrier Syst 1997; 14(3):221–286.
Donowitz M, Janecki A, Akhter S, et al. Short-term regulation of NHE3 by EGF and protein kinase C but not protein kinase A involves vesicle trafficking in epithelial cells and fibroblasts. Ann NY Acad Sci 2000; 915:30–42.
Drobny GP, Long JR, Karlsson T, et al. Structural studies of biomaterials using double-quantum solid-state NMR spectroscopy. Annu Rev Phys Chem 2003; 54:531–571. Epub 2002, 21 Mar.
Duffy LC. Interactions mediating bacterial translocation in the immature intestine. J Nutr 2000 Feb; 130(2S suppl):432S–436S.
During A, Harrison EH. Intestinal absorption and metabolism of carotenoids: insights from cell culture. Arch Biochem Biophys 2004, 1 Oct; 430(1):77–88.
Fairweather-Tait SJ. Iron the Journal is J Nutr 2001 Apr; 131(4 suppl):1383S–1386S.
Favennec L. Physiopathologic and therapeutic studies in in vitro and in vivo models of Cryptosporidium parvum infection. J Eukaryot Microbiol 1997 Nov–Dec; 44(6):69S–70S.
Ferruzza S, Sambuy Y, Rotilio G, Ciriolo MR, Scarino ML. The effect of copper on tight junctional permeability in a human intestinal cell line (Caco-2). Adv Exp Med Biol 1999; 448:215–222.
Fiese EF. General pharmaceutics—the new physical pharmacy. J Pharm Sci 2003 Jul; 92(7):1331–1342.
Ford D. Intestinal and placental zinc transport pathways. Proc Nutr Soc 2004 Feb; 63(1):21–29.
Friis LM, Pin C, Pearson BM, Wells JM. In vitro cell culture methods for investigating Campylobacter invasion mechanisms. J Microbiol Methods 2005 May; 61(2):145–160. Erratum in: J Microbiol Methods 2005 Oct; 63(1):104.
Gabor F, Bogner E, Weissenboeck A, Wirth M. The lectin–cell interaction and its implications to intestinal lectin-mediated drug delivery. Adv Drug Deliv Rev 2004, 3 Mar; 56(4):459–480.
Garg S, Kandarapu R, Vermani K, et al. Development pharmaceutics of microbicide formulations. Part I: preformulation considerations and challenges. AIDS Patient Care STDS 2003 Jan; 17(1):17–32.
Garg S, Tambwekar KR, Vermani K, et al. Development pharmaceutics of microbicide formulations. Part II: formulation, evaluation, and challenges. AIDS Patient Care STDS 2003 Aug; 17(8):377–399.
Gautherot I, Sodoyer R. A multi-model approach to nucleic acid-based drug development. BioDrugs 2004; 18(1):37–50.
Gouyer V, Leteurtre E, Zanetta JP, Lesuffleur T, Delannoy P, Huet G. Inhibition of the glycosylation and alteration in the intracellular trafficking of mucins and other glycoproteins by GalNAcalpha-O-bn in mucosal cell lines: an effect mediated through the intracellular

synthesis of complex GalNAcalpha-O-bn oligosaccharides. Front Biosci 2001, 1 Oct; 6:D1235–D1244.
Gregory PA, Lewinsky RH, Gardner-Stephen DA, Mackenzie PI. Regulation of UDP glucuronosyltransferases in the gastrointestinal tract. Toxicol Appl Pharmacol 2004, 15 Sep; 199(3):354–363.
Harris ED, Qian Y, Reddy MC. Genes regulating copper metabolism. Mol Cell Biochem 1998 Nov; 188(1–2):57–62.
Harris ED, Reddy MC, Qian Y, Tiffany-Castiglioni E, Majumdar S, Nelson J. Multiple forms of the Menkes Cu-ATPase. Adv Exp Med Biol 1999; 448:39–51.
Hauri HP, Matter K. Protein traffic in intestinal epithelial cells. Semin Cell Biol 1991 Dec; 2(6):355–364.
Hidalgo IJ. Assessing the absorption of new pharmaceuticals. Curr Top Med Chem 2001 Nov; 1(5):385–401.
Honda T. [Enteropathogenicity of Providencia alcalifaciens, a newly recognized pathogen causing food-poisoning]. Nippon Rinsho 2002 Jun; 60(6):1228–1232. Japanese.
Huang LF, Tong WQ. Impact of solid state properties on developability assessment of drug candidates. Adv Drug Deliv Rev 2004, 23 Feb; 56(3):321–334.
Huang N, Wu GD. Short chain fatty acids inhibit the expression of the neutrophil chemoattractant, interleukin 8, in the Caco-2 intestinal cell line. Adv Exp Med Biol 1997; 427:145–153.
Huebert ND, Dasgupta M, Chen Y. Using in vitro human tissues to predict pharmacokinetic properties. Curr Opin Drug Discov Devel 2004 Jan; 7(1):69–74.
Hussain MM. A proposed model for the assembly of chylomicrons. Atherosclerosis 2000 Jan; 148(1):1–15.
Ingels FM, Augustijns PF. Biological, pharmaceutical, and analytical considerations with respect to the transport media used in the absorption screening system, Caco-2. J Pharm Sci 2003 Aug; 92(8):1545–1558.
Jepson MA, Clark MA. Studying M cells and their role in infection. Trends Microbiol 1998 Sep; 6(9):359–365.
Jimenez-Barbero J, Asensio JL, Canada FJ, Poveda A. Free and protein-bound carbohydrate structures. Curr Opin Struct Biol 1999 Oct; 9(5):549–555.
Kawakami K, Pikal MJ. Calorimetric investigation of the structural relaxation of amorphous materials: evaluating validity of the methodologies. J Pharm Sci 2005 May; 94(5):948–965.
Kerneis S, Caliot E, Stubbe H, Bogdanova A, Kraehenbuhl J, Pringault E. Molecular studies of the intestinal mucosal barrier physiopathology using cocultures of epithelial and immune cells: a technical update. Microbes Infect 2000 Jul; 2(9):1119–1124.
Klaenhammer TR, Kullen MJ. Selection and design of probiotics. Int J Food Microbiol 1999, 15 Sep; 50(1–2):45–57.
Kozlowski A, Charles SA, Harris JM. Development of pegylated interferons for the treatment of chronic hepatitis C. BioDrugs 2001; 15(7):419–429.
Kumar GN, Surapaneni S. Role of drug metabolism in drug discovery and development. Med Res Rev 2001 Sep; 21(5):397–411.
Kuo SM. Flavonoids and gene expression in mammalian cells. Adv Exp Med Biol 2002; 505:191–200.
le Bouguenec C, Bertin Y. AFA and F17 adhesins produced by pathogenic Escherichia coli strains in domestic animals. Vet Res 1999 Mar–Jun; 30(2–3):317–342.
le Ferrec E, Chesne C, Artusson P, et al. In vitro models of the intestinal barrier. The report and recommendations of ECVAM Workshop 46. European Centre for the Validation of Alternative methods. Altern Lab Anim 2001 Nov–Dec; 29(6):649–668.
Lemieux P, Vinogradov SV, Gebhart CL, et al. Block and graft copolymers and NanoGel copolymer networks for DNA delivery into cell. J Drug Target 2000; 8(2):91–105.
Lennernas H. Human intestinal permeability. J Pharm Sci 1998 Apr; 87(4):403–410.
Lennernas H. Human jejunal effective permeability and its correlation with preclinical drug absorption models. J Pharm Pharmacol 1997 Jul; 49(7):627–638.
Levy E, Bendayan M. Use of immunoelectron microscopy and intestinal models to explore the elaboration of apolipoproteins required for intraenterocyte lipid transport. Microsc Res Tech 2000, 15 May; 49(4):374–382.

Levy E, Mehran M, Seidman E. Caco-2 cells as a model for intestinal lipoprotein synthesis and secretion. FASEB J 1995 May; 9(8):626–635.

Li S, He H, Parthiban LJ, Yin H, Serajuddin AT. IV-IVC considerations in the development of immediate-release oral dosage form. J Pharm Sci 2005 Jul; 94(7):1396–1417.

Li Y, Shin YG, Yu C, et al. Increasing the throughput and productivity of Caco-2 cell permeability assays using liquid chromatography-mass spectrometry: application to resveratrol absorption and metabolism. Comb Chem High Throughput Screen 2003 Dec; 6(8):757–767.

Luzio JP, Jackman MR, Ellis JA. Endocytic and transcytic pathways in Caco-2 cells. Biochem Soc Trans 1992 Nov; 20(4):717–719.

MacDonald RS, Guo J, Copeland J, et al. Environmental influences on isoflavones and saponins in soybeans and their role in colon cancer. J Nutr 2005 May; 135(5):1239–1242.

Mahato RI, Narang AS, Thoma L, Miller DD. Emerging trends in oral delivery of peptide and protein drugs. Crit Rev Ther Drug Carrier Syst 2003; 20(2–3):153–214.

Malik A. Advances in sol–gel based columns for capillary electrochromatography: sol–gel open-tubular columns. Electrophoresis 2002 Nov; 23(22–23):3973–3992.

Marshall SA, Lazar GA, Chirino AJ, Desjarlais JR. Rational design and engineering of therapeutic proteins. Drug Discov Today 2003, 1 Mar; 8(5):212–221.

Martin Y, Vermette P. Bioreactors for tissue mass culture: design, characterization, and recent advances. Biomaterials 2005 Dec; 26(35):7481–7503.

Masimirembwa CM, Bredberg U, Andersson TB. Metabolic stability for drug discovery and development: pharmacokinetic and biochemical challenges. Clin Pharmacokinet 2003; 42(6):515–528.

Mato TK, Davis PA, Odin JA, Coppel RL, Gershwin ME. Sidechain biology and the immunogenicity of PDC-E2, the major autoantigen of primary biliary cirrhosis. Hepatology 2004 Dec; 40(6):1241–1248.

McCauley R, Kong SE, Heel K, Hall JC. The role of glutaminase in the small intestine. Int J Biochem Cell Biol 1999 Mar–Apr; 31(3–4):405–413.

McDermott AE. Structural and dynamic studies of proteins by solid-state NMR spectroscopy: rapid movement forward. Curr Opin Struct Biol 2004 Oct; 14(5):554–561.

Mehnert W, Mader K. Solid lipid nanoparticles: production, characterization and applications. Adv Drug Deliv Rev 2001, 25 Apr; 47(2–3):165–196.

Mesonero J, Mahraoui L, Matosin M, Rodolosse A, Rousset M, Brot-Laroche E. Expression of the hexose transporters GLUT1-GLUT5 and SGLT1 in clones of Caco-2 cells. Biochem Soc Trans 1994 Aug; 22(3):681–684.

Meunier V, Bourrie M, Berger Y, Fabre G. The human intestinal epithelial cell line Caco-2; pharmacological and pharmacokinetic applications. Cell Biol Toxicol 1995 Aug; 11(3–4):187–194.

Milovic V, Bauske R, Turchanowa L, Stein J. Epidermal growth factor, polyamines, and epithelial remodeling in Caco-2 cells. Ann NY Acad Sci 2000; 915:279–281.

Moribe K, Maruyama K. Pharmaceutical design of the liposomal antimicrobial agents for infectious disease. Curr Pharm Des 2002; 8(6):441–454.

Morissette SL, Almarsson O, Peterson ML, et al. High-throughput crystallization: polymorphs, salts, co-crystals and solvates of pharmaceutical solids. Adv Drug Deliv Rev 2004, 23 Feb; 56(3):275–300.

Murakami T, Yumoto R, Nagai J, Takano M. Factors affecting the expression and function of P-glycoprotein in rats: drug treatments and diseased states. Pharmazie 2002 Feb; 57(2):102–107.

Nagao A. Oxidative conversion of carotenoids to retinoids and other products. J Nutr 2004 Jan; 134(1):237S–240S.

Nail SL, Jiang S, Chongprasert S, Knopp SA. Fundamentals of freeze-drying. Pharm Biotechnol 2002; 14:281–360.

Nath SK, Desjeux JF. Human intestinal cell lines as in vitro tools for electrolyte transport studies with relevance to secretory diarrhoea. J Diarrhoeal Dis Res 1990 Dec; 8(4):133–142.

Nebuloni M. [Thermoanalytic techniques for the study of pharmaceutical products]. Boll Chim Farm 1990 Mar; 129(3):87–96. Italian.

Neckers L. Development of small molecule Hsp90 inhibitors: utilizing both forward and reverse chemical genomics for drug identification. Curr Med Chem 2003 May; 10(9): 733–739.

Newman AW, Byrn SR. Solid-state analysis of the active pharmaceutical ingredient in drug products. Drug Discov Today 2003, 1 Oct; 8(19):898–905.

Ng AW, Lukic T, Pritchard PH, Wasan KM. Development of novel water-soluble phytostanol analogs: disodium ascorbyl phytostanyl phosphates (FM-VP4): preclinical pharmacology, pharmacokinetics and toxicology. Cardiovasc Drug Rev 2003 Fall; 21(3):151–168.

Nicoletti C. Unsolved mysteries of intestinal M cells. Gut 2000 Nov; 47(5):735–739.

Nishimura N. [Effects of Chinese herbal medicines on intestinal drug absorption]. Yakugaku Zasshi 2005 Apr; 125(4):363–369. Japanese.

Oguchi S, Shinohara K, Yamashiro Y, Walker WA, Sanderson IR. Growth factors in breast milk and their effect on gastrointestinal development. Zhonghua Min Guo Xiao Er Ke Yi Xue Hui Za Zhi 1997 Sep–Oct; 38(5):332–337.

Oldfield E. Chemical shifts and three-dimensional protein structures. J Biomol NMR 1995 Apr; 5(3):217–225.

Oprea TI, Matter H. Integrating virtual screening in lead discovery. Curr Opin Chem Biol 2004 Aug; 8(4):349–358.

Paine MF, Leung LY, Watkins PB. New insights into drug absorption: studies with sirolimus. Ther Drug Monit 2004 Oct; 26(5):463–467.

Pan M, Choudry HA, Epler MJ, et al. Arginine transport in catabolic disease states. J Nutr 2004 Oct; 134(10 suppl):2826S–2829S; discussion 2853S.

Park JW, Benz CC, Martin FJ. Future directions of liposome- and immunoliposome-based cancer therapeutics. Semin Oncol 2004 Dec; 31(6 suppl 13):196–205.

Pauluhn J. Overview of inhalation exposure techniques: strengths and weaknesses. Exp Toxicol Pathol 2005 Jul; 57(suppl 1):111–128.

Peppas NA, Leobandung W. Stimuli-sensitive hydrogels: ideal carriers for chronobiology and chronotherapy. J Biomater Sci Polym Ed 2004; 15(2):125–144.

Perrin MA. Crystallography of drug polymorphism: emergence of new resolution methods and prediction of crystalline structures]. Ann Pharm Fr 2002 May; 60(3): 187–202.

Pifferi G, Santoro P, Pedrani M. Quality and functionality of excipients. Farmaco 1999 Jan–Feb; 54(1–2):1–14.

Prasad JV, Lunney EA, Para KS, et al. Nonpeptidic potent HIV-1 protease inhibitors. Drug Des Discov 1996 Apr; 13(3–4):15–28.

Pritchard JF, Jurima-Romet M, Reimer ML, Mortimer E, Rolfe B, Cayen MN. Making better drugs: Decision gates in non-clinical drug development. Nat Rev Drug Discov 2003 Jul; 2(7):542–553.

Rice-Evans C, Spencer JP, Schroeter H, Rechner AR. Bioavailability of flavonoids and potential bioactive forms in vivo. Drug Metabol Drug Interact 2000; 17(1–4):291–310.

Roberts SA. High-throughput screening approaches for investigating drug metabolism and pharmacokinetics. Xenobiotica 2001 Aug–Sep; 31(8–9):557–589.

Rochu D, Masson P. [Capillary electrophoresis for monitoring stability of pharmaceutical proteins]. Ann Pharm Fr 2003 May; 61(3):185–195. French.

Rousset M. The human colon carcinoma cell lines HT-29 and Caco-2: two in vitro models for the study of intestinal differentiation. Biochimie 1986 Sep; 68(9):1035–1040.

Royall PG, Gaisford S. Application of solution calorimetry in pharmaceutical and biopharmaceutical research. Curr Pharm Biotechnol, 2005 Jun; 6(3):215–222.

Said HM. Cellular uptake of biotin: mechanisms and regulation. J Nutr 1999 Feb; 129(2S suppl):490S–493S.

Saito H. [Molecular and cell biological analyses for intestinal absorption and renal excretion of drugs]. Yakugaku Zasshi 1997 Aug; 117(8):522–541. Japanese.

Sambuy Y, De Angelis I, Ranaldi G, Scarino ML, Stammati A, Zucco F. The Caco-2 cell line as a model of the intestinal barrier: influence of cell and culture-related factors on Caco-2 cell functional characteristics. Cell Biol Toxicol 2005 Jan; 21(1):1–26.

Sanders ME, Klaenhammer TR. Invited review: the scientific basis of Lactobacillus acidophilus NCFM functionality as a probiotic. J Dairy Sci 2001 Feb; 84(2):319–331.

Sanderson IR, He Y. Nucleotide uptake and metabolism by intestinal epithelial cells. J Nutr 1994 Jan; 124(1 suppl):131S–137S.
Shih JC, Chen K. Regulation of MAO-A and MAO-B gene expression. Curr Med Chem 2004 Aug; 11(15):1995–2005.
Shire SJ, Shahrokh Z, Liu J. Challenges in the development of high protein concentration formulations. J Pharm Sci 2004 Jun; 93(6):1390–1402.
Sitrin MD, Bissonnette M, Bolt MJ, et al. Rapid effects of 1,25(OH)2 vitamin D3 on signal transduction systems in colonic cells. Steroids 1999 Jan–Feb; 64(1–2):137–142.
Spahn-Langguth H, Langguth P. Grapefruit juice enhances intestinal absorption of the P-glycoprotein substrate talinolol. Eur J Pharm Sci 2001 Feb; 12(4):361–367.
Steimer A, Haltner E, Lehr CM. Cell culture models of the respiratory tract relevant to pulmonary drug delivery. J Aerosol Med 2005 Summer; 18(2):137–182.
Stephenson GA, Forbes RA, Reutzel-Edens SM. Characterization of the solid state: quantitative issues. Adv Drug Deliv Rev 2001, 16 May; 48(1):67–90.
Stevenson CL. Characterization of protein and peptide stability and solubility in non-aqueous solvents. Curr Pharm Biotechnol 2000 Sep; 1(2):165–182.
Stutzmann J, Bellissent-Waydelich A, Fontao L, Launay JF, Simon-Assmann P. Adhesion complexes implicated in intestinal epithelial cell-matrix interactions. Microsc Res Tech 2000, 15 Oct; 51(2):179–190.
Sun D, Yu LX, Hussain MA, Wall DA, Smith RL, Amidon GL. In vitro testing of drug absorption for drug 'developability' assessment: forming an interface between in vitro preclinical data and clinical outcome. Curr Opin Drug Discov Devel 2004 Jan; 7(1):75–85.
Suzuki H, Sugiyama Y. Role of metabolic enzymes and efflux transporters in the absorption of drugs from the small intestine. Eur J Pharm Sci 2000 Nov; 12(1):3–12.
Suzuki YA, Lonnerdal B. Characterization of mammalian receptors for lactoferrin. Biochem Cell Biol 2002; 80(1):75–80.
Szabo C. Nitric oxide, intracellular calcium overload, and cytotoxicity. Shock 1996 Jul; 6(1):25–26.
Taglialatela M. DPC-423 Bristol-Myers Squibb. Curr Opin Investig Drugs 2002 Feb; 3(2):252–254.
Tang L, Persky AM, Hochhaus G, Meibohm B. Pharmacokinetic aspects of biotechnology products. J Pharm Sci 2004 Sep; 93(9):2184–2204.
Telko MJ, Hickey AJ. Dry powder inhaler formulation. Respir Care 2005 Sep; 50(9):1209–1227.
Thanou M, Verhoef JC, Junginger HE. Chitosan and its derivatives as intestinal absorption enhancers. Adv Drug Deliv Rev 2001, 1 Oct; 50(suppl 1):S91–S101.
Tirucherai GS, Yang C, Mitra AK. Prodrugs in nasal drug delivery. Expert Opin Biol Ther 2001 Jan; 1(1):49–66.
Tishmack PA, Bugay DE, Byrn SR. Solid-state nuclear magnetic resonance spectroscopy—pharmaceutical applications. J Pharm Sci 2003 Mar; 92(3):441–474.
Tuomola E, Crittenden R, Playne M, Isolauri E, Salminen S. Quality assurance criteria for probiotic bacteria. Am J Clin Nutr 2001 Feb; 73(2 suppl):393S–398S.
Tycko R. Biomolecular solid state NMR: advances in structural methodology and applications to peptide and protein fibrils. Annu Rev Phys Chem 2001; 52:575–606.
Urakami K. Characterization of pharmaceutical polymorphs by isothermal calorimetry. Curr Pharm Biotechnol 2005 Jun; 6(3):193–203.
van de Waterbeemd H, Smith DA, Beaumont K, Walker DK. Property-based design: optimization of drug absorption and pharmacokinetics. J Med Chem 2001, 26 Apr; 44(9):1313–1333.
van der Merwe SM, Verhoef JC, Verheijden JH, Kotze AF, Junginger HE. Trimethylated chitosan as polymeric absorption enhancer for improved peroral delivery of peptide drugs. Eur J Pharm Biopharm 2004 Sep; 58(2):225–235.
Vippagunta SR, Brittain HG, Grant DJ. Crystalline solids. Adv Drug Deliv Rev 2001, 16 May; 48(1):3–26.
Volkin DB, Sanyal G, Burke CJ, Middaugh CR. Preformulation studies as an essential guide to formulation development and manufacture of protein pharmaceuticals. Pharm Biotechnol 2002; 14:1–46.

Wacher VJ, Salphati L, Benet LZ. Active secretion and enterocytic drug metabolism barriers to drug absorption. Adv Drug Deliv Rev 2001, 1 Mar; 46(1–3):89–102.

Wagner D, Spahn-Langguth H, Hanafy A, Koggel A, Langguth P. Intestinal drug efflux: formulation and food effects. Adv Drug Deliv Rev 2001, 1 Oct; 50(suppl 1):S13–S31.

Wang W. Instability, stabilization, and formulation of liquid protein pharmaceuticals. Int J Pharm 1999, 20 Aug; 185(2):129–188.

Waterhouse RN. Determination of lipophilicity and its use as a predictor of blood–brain barrier penetration of molecular imaging agents. Mol Imaging Biol 2003 Nov–Dec; 5(6):376–389.

Wienk KJ, Marx JJ, Beynen AC. The concept of iron bioavailability and its assessment. Eur J Nutr 1999 Apr; 38(2):51–75.

Wilson G. Cell culture techniques for the study of drug transport. Eur J Drug Metab Pharmacokinet 1990 Apr–Jun; 15(2):159–163.

Wood RJ, Han O. Recently identified molecular aspects of intestinal iron absorption. J Nutr 1998 Nov; 128(11):1841–1844.

Yamamoto A. [Improvement of intestinal absorption of peptide and protein drugs by chemical modification with fatty acids]. Nippon Rinsho 1998 Mar; 56(3):601–607. Japanese.

Zanetti D, Poli G, Vizio B, Zingaro B, Chiarpotto E, Biasi F. 4-hydroxynonenal and transforming growth factor-beta1 expression in colon cancer. Mol Aspects Med 2003 Aug–Oct; 24(4–5):273–280.

Zweibaum A. [Differentiation of human colon cancer cells: a new approach to cancer of the colon]. Ann Gastroenterol Hepatol (Paris) 1993 Oct; 29(5):257–261; discussion 261–262. French.

Zweibaum A. [Differentiation of human colon cancer cells: a new approach to colon cancer]. Bull Acad Natl Med 1993 Jan; 177(1):63–71; discussion 71–73.

4 Dissociation, Partitioning, and Solubility

INTRODUCTION

For a newly discovered molecule to become an active drug it must traverse through a multitude of physiologic barriers, both aqueous and nonaqueous; these barriers exist to protect our body from the noxious agents that can be toxic to our body. The system by which Nature chose to protect us is based on the solubility of compounds. A compound highly soluble in water or highly insoluble in water would not be able to penetrate the deeper tissues and thus rendered ineffective. Neutral compounds without any polarizable centers often prove to be inert pharmacologically; for example, fluorinated hydrocarbons, such as perfluorodecalin, which is a hexane structure with full fluorination. Fluorine is so highly electronegative that it pulls the electrons from the parent structure, making it an inert compound. Interactions at the site of action are often electrically driven and as a result, it is more likely that we will discover a compound that has weak acid or base properties as an active entity. This necessitates studies that would yield information on how well the compound will distribute throughout the body tissues and the lipophilic/hydrophilic balance of the molecular structure becomes the focus of studies at an early stage in preformulation.

Compounds that ionize in the aqueous phase are rendered water-soluble, because they can polarize the medium and can create solute–solvent electrostatic bonding to increase their solubility. The ionization of a compound depends on the strength of binding of the ionizable group to the core of the molecule, a property that is determined by the value of the dissociation constant; once ionized, the molecule acquires new solubility characteristics; when placed between aqueous and nonaqueous phases, the distribution between these two phases, generally called partitioning, will change. It is this partitioning behavior of drugs that makes them useful as drugs; without a significant degree of partition between aqueous and non-aqueous phases of body tissues, no molecule can become active. This ionization also determines the quantity of a solute that is eventually contained in a medium, aqueous or nonaqueous—the solubility of compound. So, what starts with dissociation affects both partitioning and the solubility of the compound, the two most important parameters that will determine if a newly discovered molecule will end up as an active drug or not. This chapter describes these three inter-related properties that form the first step in any preformulation evaluation.

THE IONIZATION PRINCIPLE

Chemical moieties are known to attract to each other and under appropriate conditions, disassociate. When this process is driven by the electrical charges on the components of the moiety, this phenomenon is known as ionization. The physicochemical properties of dissociated species differ significantly from the undissociated species, and form a basis not only of the physicochemical stability, but also of the physiological activity of molecules and ions. A detailed description of ionization principle is provided here as a refresher for scientists and to emphasize the relative importance of this property.

The Acid–Base Theory

For hundreds of years, substances that behaved like vinegar have been classified as acids, while those that have properties like the ash from a wood fire have been referred to as alkalies or bases. The name "acid" comes from the Latin *acidus*, which means "sour," and refers to the sharp odor and sour taste of many acids. Vinegar tastes sour because it is a dilute solution of acetic acid in water; lemon juice is sour because it contains citric acid; milk turns sour when it is spoilt because of the formation of lactic acid; and the sour odor of rotten meat can be attributed to carboxylic acids, such as butyric acid formed when fat spoils.

Arrhenius in 1887 was the first person to give a definition of an acid and a base. According to him, an acid is one that gives rise to excess of H^+ in aqueous solution, whereas a base gives rise to excess of OH^- in solution. This was modified by Bronsted-Lowry in 1923 such that a proton donor was defined as an acid and a proton acceptor as a base. They also introduced the familiar concept of the conjugate acid–base pair. The final refinement to the acid–base theory was completed by Lewis in 1923, who extended the concept that acid is an acceptor of electron pairs while base is a donor of electron pairs.

Bronsted-Lowry Theory

According to this theory, all acid–base reactions involve the transfer of an H^+ ion, or a proton. Water reacts with itself, for example, by transferring an H^+ ion from one molecule to another to form an H_3O^+ ion and an OH^- ion.

$$H_2O + H_2O \longrightarrow H_3O^+ + OH^- \tag{1}$$

According to this theory, an acid is a "proton donor" and a base is a "proton acceptor." Acids are often divided into categories, such as "strong" and "weak." One measure of the strength of an acid is the acid-dissociation equilibrium constant, K_a, of the acid.

$$K_a = \frac{[H_3O^+][A^-]}{[HA]} \tag{2}$$

When K_a is relatively large, we have a strong acid, which is mostly present in an ionized form:

HCl: $K_a = 1 \times 10^3$ (here the ratio of ionized to unionized species is 1000 to 1).

When K_a is small, we have a weak acid, which is mostly present in an unionized form:

CH_3CO_2H: $K_a = 1.8 \times 10^{-5}$ (here the ratio of ionized to unionized species is 0.000018 to 1).

When K_a is very small, we have a very weak acid, such as water:

H_2O: $K_a = 1.8 \times 10^{-16}$ (barely ionized)

As shown, the range of ratio of ionized and unionizes species can be very large; to manage it mathematically, the values of K_a and the ionized forms are expressed in a logarithmic form, since 1909 (suggested by Sorenson).

$$pH = -\log[H_3O^+] \quad (3)$$
$$pOH = -\log[OH^-] \quad (4)$$

The "p" in pH and pOH is an operator that indicates that the negative of the logarithm should be calculated for any quantity to which it is attached. Thus, pK_a is the negative of the logarithm of the acid-dissociation equilibrium constant.

$$pK_a = -\log K_a \quad (5)$$

The only disadvantage of using pK_a as a measure of the relative strengths of acids is the fact that large numbers now describe weak acids, and small (negative) numbers describe strong acids.

HCl: $pK_a = -3$
CH_3CO_2H: $pK_a = 4.7$
H_2O: $pK_a = 15.7$

An important feature of the Bronsted theory is the relationship it creates between acids and bases. Every Bronsted acid has a conjugate base, and vice versa.

$$\begin{array}{ccccccc} HCl & + & H_2O & \longrightarrow & H_3O^+ & + & Cl^- \\ Acid & + & Base & & Acid & + & Base \end{array} \quad (6)$$

$$\begin{array}{ccccccc} NH_3 & + & H_2O & \longrightarrow & NH_4^+ & + & OH^- \\ Base & + & Acid & & Acid & + & Base \end{array} \quad (7)$$

Just as the magnitude of K_a is a measure of the strength of an acid, the value of K_b reflects the strength of its conjugate base. Consider what happens when we multiply the K_a expression for a generic acid (HA) by the K_b expression for its conjugate base (A^-).

$$\frac{[H_3O^+][A^-]}{[HA]} \times \frac{[HA][OH^-]}{[A^-]} = [H_3O^+][OH^-] \tag{8}$$

If we now replace each term in this equation by the appropriate equilibrium constant, we get the following equation:

$$K_a K_b = K_w = 1 \times 10^{-14} \tag{9}$$

Because the product of K_a and K_b is a relatively small number, either the acid or its conjugate base can be "strong." But if one is strong, the other must be weak. Thus, a strong acid must have a weak conjugate base.

$$\underset{\substack{\text{Strong} \\ \text{acid}}}{HCl} + H_2O \longrightarrow H_3O^+ + \underset{\substack{\text{Weak} \\ \text{base}}}{Cl^-} \tag{10}$$

A strong base, on the other hand, must have a weak conjugate acid.

$$\underset{\substack{\text{Strong} \\ \text{base}}}{NH_4^+} + OH^- \longrightarrow NH_3 + \underset{\substack{\text{Weak} \\ \text{acid}}}{H_2O} \tag{11}$$

Water has a limiting effect on the strength of acids and bases. All strong acids behave the same in water; 1 M solutions of the strong acids all behave as 1 M solutions of the H_3O^+ ion, and very weak acids do not act as acids in water. However, the acid–base reactions can take place in any solvent and ionization in water is not a requirement to meet the definition. Figure 1 shows the inverse relationship that exists between pK_a and pK_b values of typical acids and bases.

The strongest acids appear on the left side of the figure and the strongest bases on the right side of the figure. Any base can deprotonate any acid on the left side of it, a weaker base. Acetic acid, a weak acid will ionize (or get deprotonated)

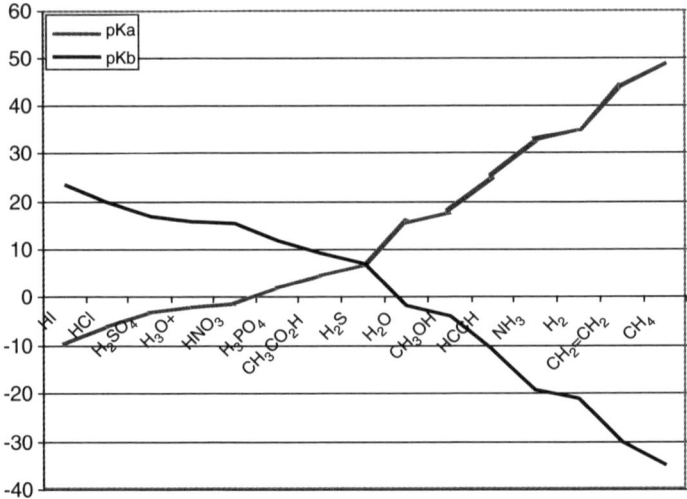

Figure 1 Typical Bronsted acids and their conjugate bases.

water, methanol, or ammonia. However, methanol will not deprotonate in water, but it will in ammonia.

Lewis Theory

The theory of acids and bases, the Bronsted-Lowry theory, was dependent on the presence of proton (H^+) to show acidic or basic properties, which may not always be the case; hence in 1923, G. N. Lewis introduced a theory of acids and bases with a more general definition of acid–base reactions by examining what happens when an H^+ ion combines with an OH^- ion to form water.

$$H^+ + :\!\overset{\cdot\cdot}{\underset{\cdot\cdot}{O}}\!\!-\!H^- \longrightarrow H\!-\!\overset{\cdot\cdot}{\underset{\cdot\cdot}{O}}\!\!-\!H \tag{12}$$

In this observation, the H^+ ion picks up (or accepts) a pair of electrons from the OH^- ion to form a new covalent bond. As a result, any substance that can act as an acceptor of electron pair is a Lewis acid (such as the H^+ ion).

The pair of electrons that went into the new covalent bond were donated by the OH^- ion. Lewis therefore argued that any substance that can act as a donor of electron pair is a Lewis base (such as the OH^- ion).

Whereas the Lewis acid–base theory does not contradict Bronsted theory, as "bases" in Bronsted theory must have a pair of nonbonding electrons in order to accept a proton, it expands the family of compounds that can be called "acids": any compound that has one or more empty valence-shell orbital and provides an explanation for the instantaneous reaction of boron triflouride (BF_3) with ammonia (NH_3). The nonbonding electrons on the nitrogen in NH_3 are donated into an empty orbital on the boron to form a new covalent bond, as shown in Eq. 13.

$$\begin{array}{c} F\!\!-\!\!B\!\!-\!\!F \\ F \\ \uparrow \\ :\!N\!: \\ H\ |\ H \\ H \end{array} \longrightarrow \begin{array}{c} F\ \ F \\ \diagdown\!/ \\ B \\ | \\ N \\ H\ |\ H \\ H \end{array} \tag{13}$$

It also explains why Cu^{2+} ions pick up NH_3 to form the four-coordinate $Cu(NH_3)_4^{2+}$ ion.

$$Cu^{2+}(aq) + 4NH_3(aq) \longrightarrow Cu(NH_3)_4^{2+}(aq) \tag{14}$$

In this case, a pair of nonbonding electrons from each of the four NH_3 molecules is donated into an empty orbital on the Cu^{2+} ion to form a covalent Cu—N bond.

$$\left[\begin{array}{c} NH_3 \\ | \\ NH_3\!-\!Cu\!-\!NH_3 \\ | \\ NH_3 \end{array}\right]^{2+} \tag{15}$$

$$\text{H}_2\text{N}^+\text{H}_2 + \text{H}-\text{C}\equiv\text{C}-\text{H} \longrightarrow \text{H}_2\text{N}^{++}-\text{H} + {}^{\ddagger}\text{C}\equiv\text{C}-\text{H} \qquad (16)$$

The flow of electrons from a Lewis base to a Lewis acid is often indicated with a curved arrow.

Henderson-Hasselbach Equation

The Henderson-Hasselbach equation defines the relationship between ionization and pH (Eq. 1). This equation relates the pK_a to the pH of the solution and the relative concentrations of the dissociated and undissociated parts of a weak acid.

$$pH = pK_a + \log[A^-]/[HA] \qquad (17)$$

or

$$pH = pK_a + \log[\text{salt}]/[\text{acid}] \qquad (18)$$

where $[A^-]$ is the concentration of the dissociated species, and $[HA]$ is the concentration of the undissociated species.

This equation can be manipulated into the form given by Eq. 2 to yield the percentage of a compound that will be ionized at any particular pH.

$$\text{Percent ionized} = \frac{100}{1 + 10^{[\text{charge}(pH-pK_a)]}} \qquad (19)$$

One simple point to note about Eq. 19 is the 50% dissociation (or ionization), $pK_a = pH$. It should also be noted that, usually, pK_a values are preferred for bases instead of the pK_b values ($pK_w = pK_a + pK_b$).

The pH Scale

The pH concept was introduced in 1909 by the Danish chemist S. P. L. Sorenson. The pH is defined by the negative logarithm of the hydrogen ion activity:

$$pH = -\log a_H \qquad (20)$$

where a_H is the activity of the hydrogen ion.

The pH scale is derived from the characteristics of the auto-dissociation of water. Pure water has a low conductivity and is only slightly ionized:

$$2H_2O = H_3O^+ + OH^- \quad \text{or} \quad H_2O = H^+ + OH^- \qquad (21)$$

The concentration of H^+ and OH^- ions, which are equal, is 1×10^{-7} ions/L. The equilibrium constant (or ion product) for the dissociation of water, K_w, is

$$K_w = \{H^+\}\{OH^-\} = 1.01 \times 10^{-14} \text{ at } 25°C \qquad (22)$$

By taking log of both sides, we get:

$$-\log\{H^+\} + -\log\{OH^-\} = 14 \qquad (23)$$

Using the standard abbreviation p for $\{-\log 10\}$, we get:

$$pH + pOH = 14 \qquad (24)$$

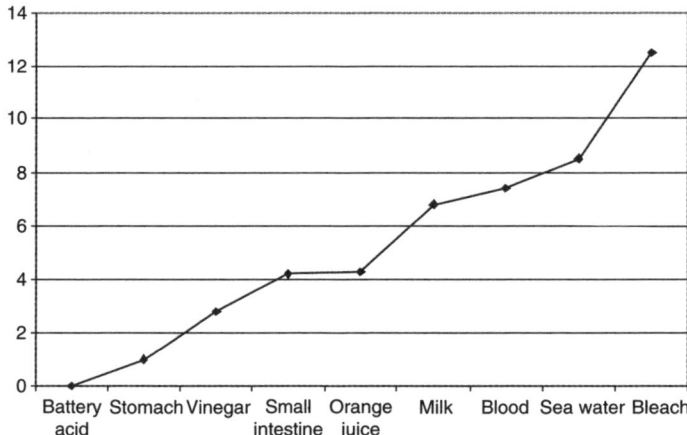

Figure 2 pH of common fluids.

This equation sets the pH scale to 0–14, which gives a convenient way to express 14 orders of magnitude of [H$^+$]. Any solution with pH > 7 contains excessive hydroxyl ions, and is alkaline; those with pH < 7 are acidic, containing excess hydrogen ions. Figure 2 shows pH values of common fluids.

It should be noted that the above definitions of pH are based on an assumption that the solution is behaving in an ideal nature meaning that the thermodynamic activity is equal to concentration (e.g., what happens when the dilution is infinite). However, as the concentration increases, ionic attraction and incomplete hydration results in a decrease in the effective concentration (or the activity). This activity is defined as the "apparent concentration" of an ionic species, which is due to the attraction that ions exert on one another, and the incomplete hydration of ions in solutions that are too concentrated. The lower the concentration, the less is the interaction. At infinite dilution, activity coefficients approach unity.

The activity of a species X is equal to the product of its concentration and its activity coefficient, f_x:

$$\{X\} = f_x[X] \tag{25}$$

The pH from an electrode relates to $\{H^+\}$, and not [H$^+$]; though the value of ionic strength is below 0.01, these terms have values that are very close between pH 2 and 10. A pH electrode consists of a pH sensor, which varies in proportion to the $\{H^+\}$ of the solution and a reference electrode, which provides a stable constant voltage. The output is in mV, which needs to be converted to pH units (Fig. 3).

A plot of electrode potential against pH at different temperature shows the effect of temperature on activity and should be corrected (Fig. 4).

This results in the following equation correlating the reference potential:

$$E_{obs} = \text{constant} + \text{slope} \cdot \text{pH} \tag{26}$$
$$E_{obs} = E_c + N_f \log \{H^+\} \tag{27}$$

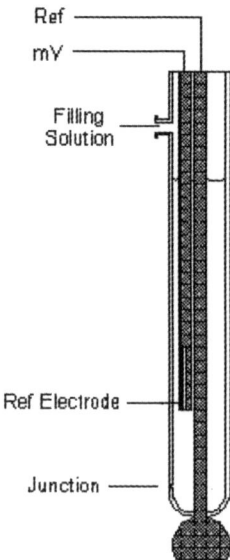

Figure 3 pH probe design.

where E_c is the reference potential; N_f is the Nernstian slope factor = $2.3RT/nF$ = 59.1 at 25°C; where R is the gas constant, T is the absolute temperature in Kelvin, F is the Faraday's constant, and n is the valance factor.

The slope factor is temperature-dependent and thus, the pH is derived from:

$$pH = pH_{std} - (E - E_{std})/slope \qquad (28)$$

At pH 7, where $\{H^+\} = \{OH^-\}$, the voltage from the electrode is zero. This is called the isopotential point (Fig. 4). In theory, this point is temperature-independent. The International Union for Physical and Applied Chemistry (IUPAC) (1) operational pH scale is defined as the pH relative to a standard buffer measured using a hydrogen electrode. In practice, a pH electrode is calibrated with standard buffers of pH 7.00 and pH 4 or 9 to determine the isoelectric point and slope, respectively. Conventional pH meters will read accurately over a range of 2.5–11 and beyond these ranges, accuracy cannot be assured. However, recently, instruments have become available that carry out calibration to allow correction for nonideal electrode behavior allowing accurate measurements between ranges of pH 1 and 13.

Compendia Specification for pH Measurement

For compendial purposes, pH is defined as the value given by a suitable, properly standardized, potentiometric instrument (pH meter) capable of reproducing pH values to 0.02 pH unit using an indicator electrode sensitive to hydrogen ion activity, the glass electrode, and a suitable reference electrode. The instrument should be capable of sensing the potential across the electrode pair and, for pH standardization purposes, applying an adjustable potential to the circuit by manipulation of "standardization," "zero," asymmetry, or "calibration" control,

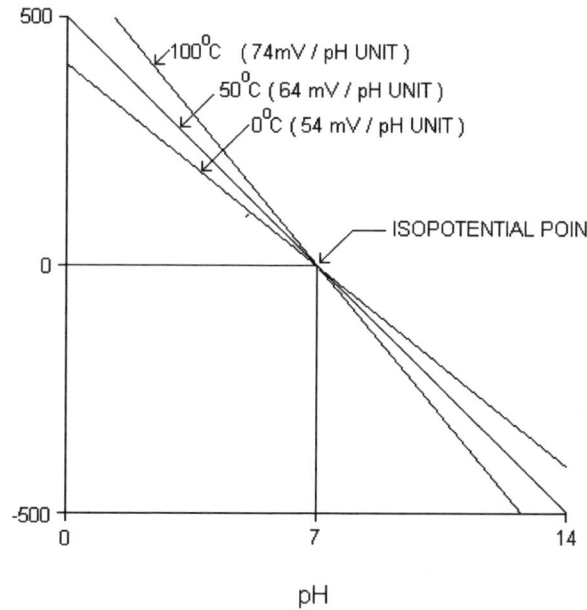

Figure 4 Relationship between pH and electrode potential as a function of temperature.

and should be able to control the change in millivolts per unit change in pH reading through a "temperature" and/or "slope" control.

Measurements are made at 25 ± 2°, unless otherwise specified in the individual monograph. The pH scale is defined by the equation:

$$\text{pH} = \text{pHs} + (E - ES)/k, \tag{29}$$

where E and ES are the measured potentials where the galvanic cell contains the solution under test, represented by pH, and the appropriate buffer solution for standardization, represented by pHs, respectively. The value of k is the change in potential per unit change in pH and is theoretically [0. 05916 + 0. 000198 $(t - 25°)$] V at any temperature, t. This operational pH scale is established by assigning rounded pH values to the buffer solutions for standardization from the corresponding National Institute of Standards and Technology (NIST) molal solutions.

It should be emphasized that the definitions of pH, the pH scale, and the values assigned to the buffer solutions for standardization are for the purpose of establishing a practical, operational system so that results may be compared between laboratories. The pH values thus measured do not correspond exactly to those obtained by the definition, $\text{pH} = -\log_a \text{H}^+$. As long as the solution being measured is sufficiently similar in composition to the buffer used for standardization, the operational pH corresponds fairly closely to the theoretical pH. Although no claim is made with respect to the suitability of the system for measuring hydrogen ion activity or concentration, the values obtained are closely related to the activity of the hydrogen ion in aqueous solutions.

Where a pH meter is standardized by use of an aqueous buffer and then used to measure the "pH" of a nonaqueous solution or suspension, the ionization constant of the acid or base, the dielectric constant of the medium, the liquid-junction potential (which may give rise to errors of approximately 1 pH unit), and the hydrogen ion response of the glass electrode are all changed. For these reasons, the values so obtained with solutions that are only partially aqueous in character can be regarded only as apparent pH values.

Because of variations in the nature and operation of the available pH meters, it is not practicable to give universally applicable directions for the potentiometric determinations of pH. The general principles to be followed in carrying out the instructions provided for each instrument by its manufacturer are set forth in the following paragraphs. Examine the electrodes and, if present, the salt bridge prior to use. If necessary, replenish the salt bridge solution, and observe other precautions indicated by the instrument or the electrode manufacturer. Commercially available buffer solutions for pH meter standardization, standardized by methods traceable to the NIST (2), labeled with a pH value accurate to 0.01 pH unit may be used. Solutions prepared from American Chemical Society reagent grade materials or other suitable materials, in the stated quantities, may be used provided the pH of the resultant solution is the same as that of the solution prepared from the NIST certified material.

To standardize the pH meter, select two buffer solutions for standardization whose difference in pH does not exceed 4 units, and such that the expected pH of the material under test falls between them. Fill the cell with one of the buffer solutions for standardization at the temperature at which the test material is to be measured. Set the control "temperature" at the temperature of the solution, and adjust the calibration control to make the observed pH value identical with that tabulated. Rinse the electrodes and the cell with several portions of the second buffer solution for standardization, then fill the cell with it, at the same temperature as the material to be measured. The pH of the second buffer solution is within ± 0.07 pH unit of the tabulated value. If a larger deviation is noted, examine the electrodes and, if they are faulty, replace them. Adjust the "slope" or "temperature" control to make the observed pH value identical with that tabulated. Repeat the standardization until both buffer solutions for standardization give observed pH values within 0.02 pH unit of the tabulated value without further adjustment of the control. When the system functions satisfactorily, rinse the electrodes and cell several times with a few portions of the test material, fill the cell with the test material, and read the pH value. Use carbon dioxide-free water for solution or dilution of test material in pH determinations. In all pH measurements, allow a sufficient time for stabilization.

Where approximate pH values suffice, indicators and test papers may be suitable for use instead of pH meters.

Dissociation

At a given temperature, the thermodynamic ionization constants are independent of concentration and at a pH value equal to pK_a, the activity of ionized and neutral forms, are equal. In many measurement techniques, we measure concentration rather than activity, such as in the use of spectroscopic methods. In such instances,

$$K_c a = [H^+][A^-]/[HA] \tag{30}$$

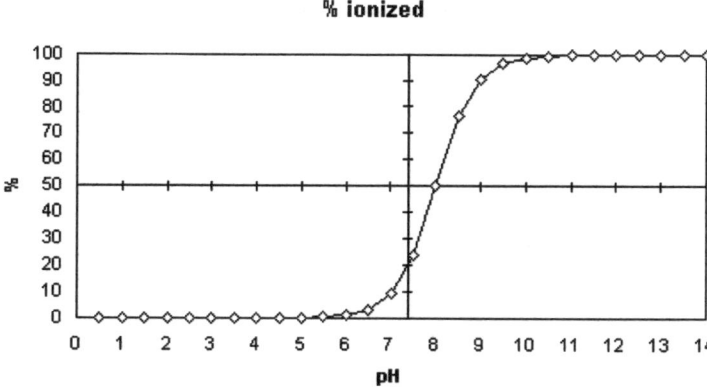

Figure 5 Percent ionization of an acid with pK_a of 8.0.

where values in brackets are the observed concentrations from spectroscopic measurements based on the Beer-Lambert law. The "thermodynamic" ionization coefficient is related to the "concentration" ionization coefficient by:

$$K_a = K_c a \cdot (f_A f_H / f_{HA}) \tag{31}$$

where f is the activity coefficient.

The pK_a values are also temperature-dependent, often in a nonlinear and unpredictable way. Samples measured by potentiometry are therefore held at a constant temperature bath and therefore, pK_a value should be quoted at a specific temperature. Often a temperature of 25°C is chosen to reflect room temperature whereas this may be quite different from the body temperature. Percent ionization at different temperatures can be calculated as:

$$\text{Percent ionized} = \frac{100}{1 + 10^{(\text{charge}(pH - pK_a))}} \tag{32}$$

where charge is 1 for bases and −1 for acids. Percent ionization is 50% when the pH equals pK_a (Figs. 5 and 6).

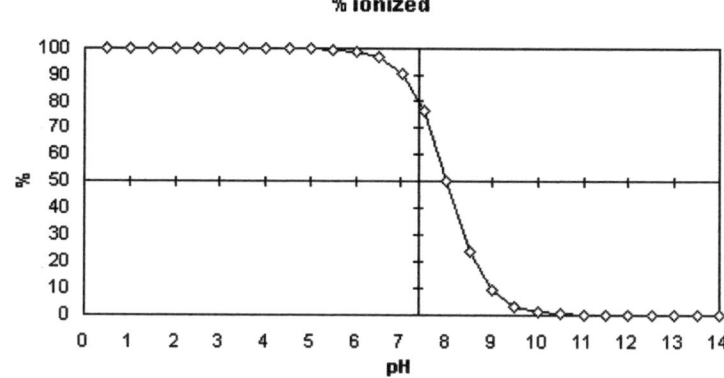

Figure 6 Percent ionization of a base with pK_a of 8.0.

QUANTITATIVE STRUCTURE–ACTIVITY RELATIONSHIPS

The relationship between chemical structure, lipophilicity, and its disposition in vivo has been extensively studied. These include solubility, absorption potential, membrane permeability, plasma protein binding, volume of distribution, and renal and hepatic clearance. Activities used in quantitative structure–activity relationships (QSAR) include chemical measurements and biological assays. QSAR currently are applied in many disciplines, with many pertaining to drug design and environmental risk assessment.

QSAR studies date back to the nineteenth century. In 1863, A. F. A. Cros at the University of Strasbourg observed that toxicity of alcohols to mammals increased as the water solubility of the alcohols decreased. In the 1890s, Hans Horst Meyer of the University of Marburg and Charles Ernest Overton of the University of Zurich, working independently, noted that the toxicity of organic compounds depended on their lipophilicity.

Little additional development of QSAR occurred until the work of Louis Hammett (1894–1987), who correlated electronic properties of organic acids and bases with their equilibrium constants and reactivity. Consider the dissociation of benzoic acid:

$$\text{C}_6\text{H}_5\text{-COOH} \rightleftharpoons \text{C}_6\text{H}_5\text{-COO}^- + \text{H}^+ \quad K=6.27\times10^{-5} \tag{33}$$

Hammett observed that adding substituents to the aromatic ring of benzoic acid had an orderly and quantitative effect on the dissociation constant. For example,

$$m\text{-O}_2\text{N-C}_6\text{H}_4\text{-COOH} \rightleftharpoons m\text{-O}_2\text{N-C}_6\text{H}_4\text{-COO}^- + \text{H}^+ \quad K=32.1\times10^{-5} \tag{34}$$

a nitro group in the meta position increases the dissociation constant, because the nitro group withdraws electron, thereby stabilizing the negative charge that develops. Consider now the effect of a nitro group in the para position:

$$p\text{-O}_2\text{N-C}_6\text{H}_4\text{-COOH} \rightleftharpoons p\text{-O}_2\text{N-C}_6\text{H}_4\text{-COO}^- + \text{H}^+ \quad K=37.0\times10^{-5} \tag{35}$$

The equilibrium constant is even larger than for the nitro group in the meta position, indicating even greater withdrawal of electrons. Now, consider the case in which an ethyl group is in the para position:

$$p\text{-CH}_3\text{CH}_2\text{-C}_6\text{H}_4\text{-COOH} \rightleftharpoons p\text{-CH}_3\text{CH}_2\text{-C}_6\text{H}_4\text{-COO}^- + \text{H}^+ \quad K=4.47\times10^{-5} \tag{36}$$

In this case, the dissociation constant is lower than for the unsubstituted compound, indicating that the ethyl group donates electrons, thereby destabilizing the negative charge that arises upon dissociation.

Hammett also observed that substituents have a similar effect on the dissociation of other organic acids and bases. Consider the dissociation of phenylacetic acids:

$$\text{C}_6\text{H}_5\text{-CH}_2\text{COOH} \rightleftharpoons \text{C}_6\text{H}_5\text{-CH}_2\text{COO}^- + \text{H}^+ \quad K=5.20\times10^{-5}$$

$$m\text{-O}_2\text{N-C}_6\text{H}_4\text{-CH}_2\text{COOH} \rightleftharpoons m\text{-O}_2\text{N-C}_6\text{H}_4\text{-CH}_2\text{COO}^- + \text{H}^+ \quad K=10.7\times10^{-5}$$

$$p\text{-O}_2\text{N-C}_6\text{H}_4\text{-CH}_2\text{COOH} \rightleftharpoons p\text{-O}_2\text{N-C}_6\text{H}_4\text{-CH}_2\text{COO}^- + \text{H}^+ \quad K=14.1\times10^{-5}$$

$$p\text{-CH}_3\text{CH}_2\text{-C}_6\text{H}_4\text{-CH}_2\text{COOH} \rightleftharpoons p\text{-CH}_3\text{CH}_2\text{-C}_6\text{H}_4\text{-CH}_2\text{COO}^- + \text{H}^+ \quad K=4.27\times10^{-5}$$

(37–40)

Withdrawal of electrons by the nitro group increases dissociation, with the effect being less for the meta than for the para substituent, just as was observed with benzoic acid. The electron-donating ethyl group decreases the equilibrium constant, as expected.

Data for these equilibria typically are graphed in Figure 7:

K_0 or K_0' represent equilibrium constants for unsubstituted compounds and K or K', for substituted compounds. Values for the abscissa are calculated from the dissociation constants of unsubstituted and substituted benzoic acid. Values for the ordinate are obtained from another organic acid or base with identical patterns of substitution, in this case phenylacetic acid.

Because this relationship is linear, the following equation can be written:

$$\log\frac{K}{K_0} = \rho\log\frac{K'}{K_0'} \quad (41)$$

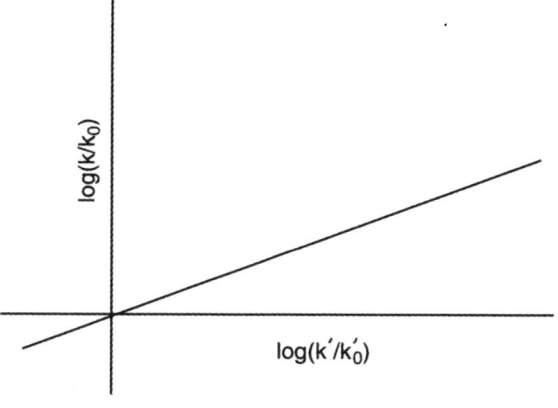

Figure 7 Example of a graph for a linear free-energy relationship.

where ρ is the slope of the line. The values for the abscissa in Figure 7 are always those for benzoic acid and are given the symbol, σ. Therefore, we can write:

$$\log \frac{K}{K_0} = \rho\sigma \tag{42}$$

where ρ, the slope of the line, is a proportionality constant pertaining to a given equilibrium. It relates the effect of substituents on this equilibrium to the effect of those substituents on the benzoic acid equilibrium. That is, if the effect of substituents is proportionally greater than on the benzoic acid equilibrium, then $\rho > 1$; if the effect is less than on the benzoic acid equilibrium, then $\rho < 1$. By definition, ρ for benzoic acid is equal to one.

σ is a descriptor of the substituents. The magnitude of σ gives the relative strength of the electron-withdrawal and -donating properties of the substituents. σ is positive if the substituent is electron-withdrawing and negative if it is electron-donating.

These relationships as developed by Hammett are termed linear free-energy relationships, which relate free energy to an equilibrium constant:

$$\Delta G = -RT \ln K \tag{43}$$

The free energy is proportional to the logarithm of the equilibrium constant. These linear free-energy relationships are termed "extrathermodynamic." Although they can be stated in terms of thermodynamic parameters, no thermodynamic principle states that the relationships should be true.

To develop a better understanding of these relationships, Table 1 gives some values of ρ.

In the aniline and phenol equilibria, the hydrogen ion dissociates and removes one atom from the phenyl ring, whereas in the benzoic acid equilibrium two atoms are removed. Thus, substituents are able to exert a greater effect on the dissociation in aniline and phenol than in benzoic acid and the value of ρ is > 1. In phenylacetic and phenylpropionic acids, the hydrogen ion dissociates and removes three and four atoms, respectively, from the phenyl ring. Substituents exert a lesser effect on the equilibrium in this case than on the benzoic acid

Table 1 ρ Values of Various Substituents

Compound	ρ
Ph—NH_3^+	$\rho = 2.90$
Ph—OH	$\rho = 2.23$
Ph—CH_2COOH	$\rho = 0.49$
Ph—CH_2CH_2COOH	$\rho = 0.21$

Table 2 σ Values at Meta and Para Substitution Positions

Substituent	σ_m	σ_p
—H	0.0	0.0
—NO$_2$	0.71	0.78
—Cl	0.37	0.23
—OCH$_3$	0.12	−0.27

equilibrium, and ρ is <1. Examples of σ for substituents in the meta and para positions are given in Table 2.

By definition, σ for hydrogen is 0. The positive values of σ for the nitro group indicate that it withdraws electrons. In understanding the magnitudes of the σ values for the nitro group in meta versus para positions, consider the mechanisms of electron withdrawal or donation. For a nitro group in the meta position, withdrawal of electrons is due to an inductive effect produced by the electronegativity of the constituent atoms. If only induction were operative, one would expect the electron withdrawal effect of a nitro group in the para position to be less than that in the meta position. The larger value for a para-substituted nitro group results from the combination of both inductive and resonance effects. The resonance structures for para-nitrobenzoate are shown below:

(44)

For chlorine, the electronegativity of the atom produces an inductive effect on electron withdrawal, with the magnitude of the effect in the para position being less than that in the meta position. For chlorine, only the inductive effect is possible. The methoxy group can donate or withdraw electrons, depending on the position of substitution. In the meta position, the electronegativity of the oxygen produces an inductive effect on withdrawal of electrons. In the para position, only a small inductive effect is expected. Moreover, an electron-donating resonance effect, as illustrated in Eq. 45, occurs for the methoxy group in the para position, giving an overall electron-donating effect.

(45)

Table 3 Hammett σ Constants[a]

Group	σ_{meta}	σ_{para}	σ_I	σ_v	π	E_s	MR
H	0.00	0.00	0.00	0.00	0.00	0.00	1.03
CH_3	−0.07	−0.17	−0.04	0.52	0.56	−1.24	5.65
C_2H_5	−0.07	−0.15	−0.05	0.56	1.02	−1.31	10.30
n-C_3H_7	−0.07	−0.13	−0.03	0.68	1.55	−1.60	14.96
i-C_3H_7	−0.07	−0.15	−0.03	0.76	1.53	−1.71	14.96
n-C_4H_9	−0.08	−0.16	−0.04	0.68	2.13	−1.63	19.61
t-C_4H_9	−0.10	−0.20	−0.07	1.24	1.98	−2.78	19.62
$H_2C=CH$[b]	0.05	−0.02	0.09	2.11	0.82		10.99
C_6H_5[b]	0.06	−0.01	0.10	2.15	1.96	−3.82	25.36
CH_2Cl	0.11	0.12	0.15	0.60	0.17	−1.48	10.49
CF_3	0.43	0.54	0.42	0.91	0.88	−2.40	5.02
CN	0.56	0.66	0.53	0.40	−0.57	−0.51	6.33
CHO	0.35	0.42	0.25		−0.65		6.88
$COCH_3$	0.38	0.50	0.29	0.50	−0.55		11.18
CO_2H[b]	0.37	0.45	0.39	1.45	−0.32		6.93
$Si(CH_3)_3$	−0.04	−0.07	−0.13	1.40	2.59		24.96
F	0.34	0.06	0.52	0.27	0.14	−0.46	0.92
Cl	0.37	0.23	0.47	0.55	0.71	−0.97	6.03
Br	0.39	0.23	0.50	0.65	0.86	−1.16	8.88
I	0.35	0.18	0.39	0.78	1.12	−1.40	13.94
OH	0.12	−0.37	0.29	0.32	−0.67	−0.55	2.85
OCH_3	0.12	−0.27	0.27	0.36	−0.02	−0.55	7.87
OCH_2CH_3	0.10	−0.24	0.27	0.48	0.38		12.47
SH	0.25	0.15	0.26	0.60	0.39	−1.07	9.22
SCH_3	0.15	0.00	0.23	0.64	0.61	−1.07	13.82
NO_2[b]	0.71	0.78	0.76	1.39	−0.28	−2.52	7.36
NO	0.62	0.91	0.37		−0.12		5.20
NH_2	−0.16	−0.66	0.12		−1.23	−0.61	5.42
NHCHO	0.19	0.00	0.27		−0.98		10.31
$NHCOCH_3$	0.07	−0.15	0.26		−0.37		16.53
$N(CH_3)_2$	−0.15	−0.83	0.06	0.43	0.18		15.55
$N(CH_3)_3^+$	0.88	0.82	0.93	1.22	−5.96		21.20

[a] σ_{meta}, σ_{para} are the Hammett constants; σ_I is the inductive σ constant; σ_v is the Charton's v (size) values; π is the hydrophobicity parameter; E_s is the Taft size parameter; and MR is the molar refractivity (polarizability) parameter.
[b] The group is in the most sterically hindered conformation.
Abbreviation: MR, molar refractivity.

Tables of σ values for numerous substituents available in the literature are shown in Table 3.

In some cases, the σ values are generally applicable to much different equilibrium. In other cases, σ values have been derived for specific equilibria, which is particularly true when one considers σ values for ortho substituents. A good example of the application of Hammett's electronic descriptors in a QSAR relating the inhibition of bacterial growth is in the series of sulfonamides,

$$H_2N-\text{C}_6\text{H}_4-\underset{\underset{O}{\overset{\overset{O}{\|}}{\|}}}{S}-\underset{H}{N}-\text{C}_6\text{H}_4-X$$

where X represents various substituents. A QSAR was developed based on the values of the substituents,

$$\log\left(\frac{1}{C}\right) = 1.05\sigma - 1.28 \tag{46}$$

where C is the minimum concentration of compound that inhibited growth of *Escherichia coli*. From this relationship, it is seen that electron-withdrawing substituents favor inhibition of growth.

Hansch Analysis

The QSARs based on Hammett's relationship utilize electronic properties as the descriptors of structures. Difficulties were encountered when investigators attempted to apply Hammett-type relationships to biological systems, indicating that other structural descriptors were necessary. Robert Muir, a botanist at Pomona College, studied the biological activity of compounds that resembled indoleacetic acid and phenoxyacetic acid, which function as plant growth regulators. In attempting to correlate the structures of these compounds with their activities, he consulted his colleague in chemistry, Corwin Hansch, while using Hammett σ parameters to account for the electronic effect of substituents that did not lead to meaningful QSAR. However, Hansch recognized the importance of the lipophilicity, expressed as the octanol–water partition coefficient, on biological activity. This parameter is now recognized to provide a measure of the bioavailability of compounds, which will determine, in part, the amount of the compound that gets to the target site.

Relationships were developed to correlate a structural parameter (i.e., lipophilicity) with activity. In some cases, a univariate relationship correlating structure and activity was adequate. The form of the equation is:

$$\log\left(\frac{1}{C}\right) = a \log P + b \tag{47}$$

where C is the molar concentration of the compound that produces a standard response (e.g., LD50, ED50). With other data, it was observed that correlations were improved by combining Hammett's electronic parameters and Hansch's measure of lipophilicity using an equation, such as

$$\log\left(\frac{1}{C}\right) = k_1 \pi + k_2 \sigma + k_3 \tag{48}$$

where σ is the Hammett substituent parameter and π is defined analogously to σ. That is,

$$\pi = \log\left(\frac{P_X}{P_H}\right) \tag{49}$$

The significance of the slopes in univariate QSAR that involve the correlation of log P with toxicity has been considered. For example, an analysis of data in which the lysis of erythrocytes (hemolysis) by various neutral organic compounds (e.g., alcohols, carboxylic acids, amines, phenols, esters, and so on) was studied, which yielded the following general equation:

$$\log\left(\frac{1}{C}\right) = 0.93(\pm 0.17) \log P + 0.09(\pm 0.23) \tag{50}$$

Note that the slope of this equation is approximately one and that the intercept is approximately zero. Many other QSARs that involve nonspecific toxicity also show correlations of $\log P$ with toxicity with a slope near one. However, a number of QSARs involving neutral organic compounds have slopes considerably less than one. The reasons for this phenomenon are not entirely clear. One analysis of the problem involves a consideration of the meaning of hydrophobicity at the molecular level. That is, hydrophobicity can be considered to be due largely to the free energy change associated with the desolvation of a compound as it moves from an aqueous phase to the biological phase with which it interacts to produce toxicity (e.g., entering a membrane, binding to a protein, and so on). It appears that the slope of the regression equation is related to the desolvation of the compound that occurs when it interacts at its target site. When the slope of the equation is approximately one, the environment of the biological phase appears to be similar to that of octanol; partitioning of a compound from water into octanol would require complete desolvation of the compound. For some of the examples of QSAR with a slope less than one, the measured effect is a result of ligands binding to proteins or DNA.

While most of the QSAR observations involve linear relationships between $\log P$ and toxicity, there are other relationships, such as parabolic seen between biological response and hydrophobicity. One interpretation to account for this observation is that many membranes may have to be traversed for compounds to get to the target site, and compounds with the greatest hydrophobicity will become localized in the membranes they encounter initially, thereby slowing their transit to the target site.

Another factor that alters QSAR involves steric effects. For studies that involve reactivity of organic compounds, a steric parameter, E_s, was defined by Taft as

$$E_s = \log\left(\frac{K_x}{K_H}\right)_A \tag{51}$$

where k is the rate constant for the acid hydrolysis of esters of the type

$$X-CH_2-\overset{\overset{O}{\|}}{C}-OR$$

The transition state for this hydrolysis can be represented as:

$$X-CH_2-\overset{\overset{OH}{|}}{\underset{HOH}{C}}-OR$$

Assuming that the electronic effects of substituent X can be ignored, the size of X will affect the transition state and hence, the rate of the reaction. By definition, $E_s = 0$ for $X = H$. Table 3 lists values of E_s for other substituents.

Another parameter that is related to molecular volume and steric effects is the molar refractivity (MR). Experimentally, it is obtained from the equation

$$MR = \frac{n^2 - 1}{n^2 + 2} \cdot \frac{MW}{d} \tag{52}$$

where n is the index of refraction, d is the density, and MW is the molecular weight.

Steric effects can be particularly difficult to define in complex biological systems. Quantitative structure–activity relationships in biological systems have been developed using parameters, such as E_s and MR. In addition, factors, such as van der Waals radii, standard bond angles and lengths, and conformational flexibility have been applied as a way to define the space occupied by molecules. However, it is often difficult to define a single parameter that can account for all of these factors. A more recent treatment of steric effects that is applied to biological systems is comparative molecular field analysis (CoMFA). This approach, which examines and superimposes the conformations of molecules of interest, is an extension of the ligand-based drug design.

One of the largest QSAR database is offered by the Pomona College (3); for advanced search functions, the user needs a license from Biobyte (4). These large databases represent large values for preformulation and drug discovery groups; most of the development in this area came about during the past few years with the availability of high-speed computational devices.

PARTITIONING

The partition coefficient is a measure of the extent a substance partitions between two phases, generally an oil phase and an aqueous phase. This ratio is often expressed as log P (logarithm of partition ratio). Both pK_a and log P measurements are useful parameters for understanding the behavior of drug molecules at the preformulation stage. The pK_a will determine the species of molecules, which is likely to be present at the site of action and how quickly or completely would the species cross a large number of transport barriers in the body, regardless of the route of administration. Factors, such as absorption, excretion, and penetration of the central nervous system (CNS) are also related to the log P value of a drug and in certain cases predictions can be made; these are important in assessing the endogenous toxicity of compounds and their activity.

Partition coefficient is a ratio of the concentration in two immiscible solvents.

Partition coefficient, $P = $ [organic]/[aqueous]   (53)

where the values in brackets describe measured concentrations.

$\log P = \log_{10}$ (partition coefficient)   (54)

In practical terms, the uncharged or neutral molecule exists for bases >2 pK_a units above the pK_a and for acids >2 pK_a units below the pK_a. In practice, the log P will vary according to the conditions under which it is measured and the choice of the partitioning solvent.

It is worth noting that this is a logarithmic scale; therefore, log $P = 0$ means that the compound is equally soluble in water and in the partitioning solvent. If the compound has a log $P = 5$, then the compound is 100,000 times more soluble in the partitioning solvent. A log $P = -2$ means that the compound is 100 times more soluble in water, that is, it is quite hydrophilic.

Log P values have been studied in approximately 100 organic liquid-water systems. As it is virtually impossible to determine lop P in a realistic biological medium, the octanol-water system has been widely adopted as a model of the lipid phase. While there has been much debate about the suitability of this system, it is the most widely used in pharmaceutical studies. Octanol and water are immiscible, but some water does dissolve in octanol in a hydrated state. This hydrated state contains 16 octanol aggregates, with the hydroxyl head groups surrounded by trapped aqueous solution. Lipophilic (unionized) specices dissolve in the aliphatic regions.

Generally, compounds with log P values between one and three show good absorption, whereas those with log Ps greater than six or less than three often have poor transport characteristics. Highly nonpolar molecules have a preference to reside in the lipophilic regions of membranes, and highly polar compounds show poor bioavailability because of their inability to penetrate membrane barriers. Thus, there is a parabolic relationship between log P and transport, that is, candidate drugs that exhibit a balance between these two properties will probably show the best oral bioavailability.

Distribution Coefficient

The partition coefficient refers to the intrinsic lipophilicity of the drug, in the context of the equilibrium of unionized drug between the aqueous and organic phases. If the drug has more than one ionization center, the distribution of species present will depend on the pH. The concentration of the ionized drug in the aqueous phase will therefore have an effect on the overall observed partition coefficient. This leads to the definition of the distribution coefficient (log D) of a compound, which takes into account the dissociation of weak acids and bases.

As in the aqueous phase, the total concentration may comprise of both ionized and unionized forms, the distribution is given as:

$$\text{Distribution coefficient}, D = [\text{unionized}]_{(o)}/[\text{unionized}]_{(aq)}$$
$$+ [\text{ionized}]_{(aq)} \tag{55}$$

$$\log D = \log 10 \text{ (distribution coefficient)} \tag{56}$$

Log D is related to log P and the pK_a by the following equations:

$$\log D_{(pH)} = \log P - \log[1 + 10^{(pH - pK_a)}]_{\text{for acids}} \tag{57}$$
$$\log D_{(pH)} = \log P - \log[1 + 10^{(pH_a - pH)}]_{\text{for bases}} \tag{58}$$

Log D is the log distribution coefficient at a particular pH. This is not constant and will vary according to the protogenic nature of the molecule. Log D at pH 7.4 is often quoted to give an indication of the lipophilicity of a drug at the pH of blood plasma. Figures 8–10 show the distribution profiles of various acids and bases.

It is important to understand that the species that partition are primarily the neutral molecules or molecules that appear neutral through interactions, such as ion pairing, which allows transport of ionic species and thus complicates the calculations of log P and log D. However, instruments like the Sirius PCA200 or G-pK_a can isolate the ion pairing effect that requires several titrations. It is noteworthy that ion pairing also affects the readings taken using spectrophotometers. As a result, comparisons of values obtained from different methods may not corroborate, because of differences in the concentrations or the differences in the use of the counter ion involved. Generally, a 0.1 M solution of a background electrolyte is used, which is close to the biological level of 0.16 M. The type of electrolyte can make much difference, for example, 0.15 M potassium chloride in place of sodium chloride to obviate the "sodium effect" on the electrode at high

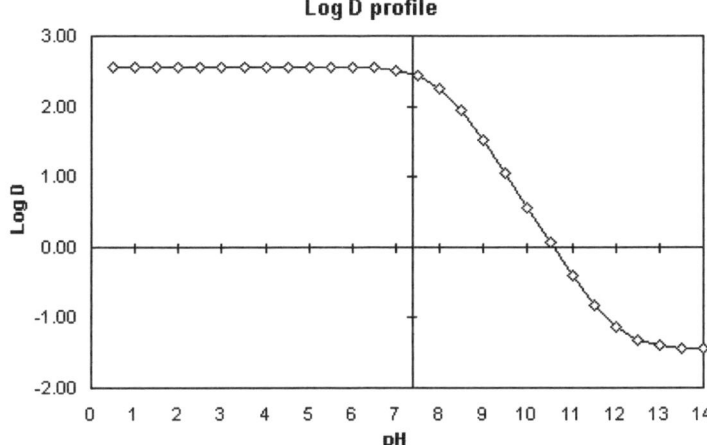

Figure 8 Log D profile of an acid with p$K_a = 8$.

pH with similar profile as obtained using sodium chloride. Choice of a proper background electrolyte can be very important in many instances, especially where some significance to biological systems is desired.

Log D membrane, or log D_{mem}, is another way of measuring the lipophilicity of a compound, which is less frequently used than the octanol/water partition system. Log D_{mem} utilizes liposomes prepared from the synthetic phospholipids dimyristoylphosphatidylcholine (DMPC), and although the system is more physiologically relevant than octanol, it suffers from a lack of predictability. This method requires 1 mg of the compound, whereby a solution of the compound is equilibrated with a solution of DMPC liposomes at 37°C for two hours. The free and liposome-bound compounds are then separated by centrifugation, and the solutions are analyzed by high-performance liquid chromatography (HPLC).

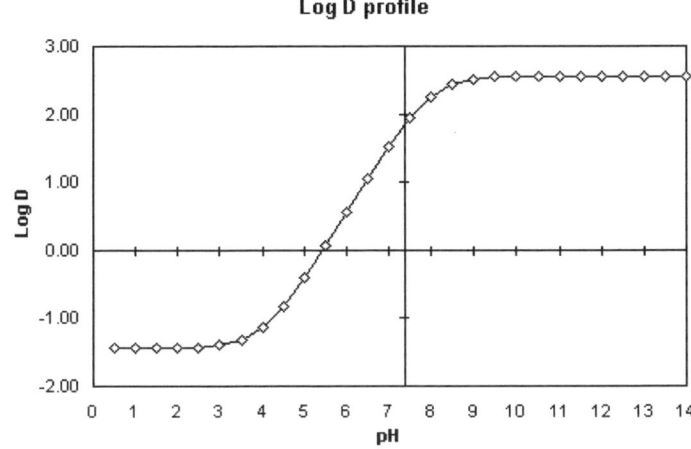

Figure 9 Log D profile of a base with p$K_a = 8$.

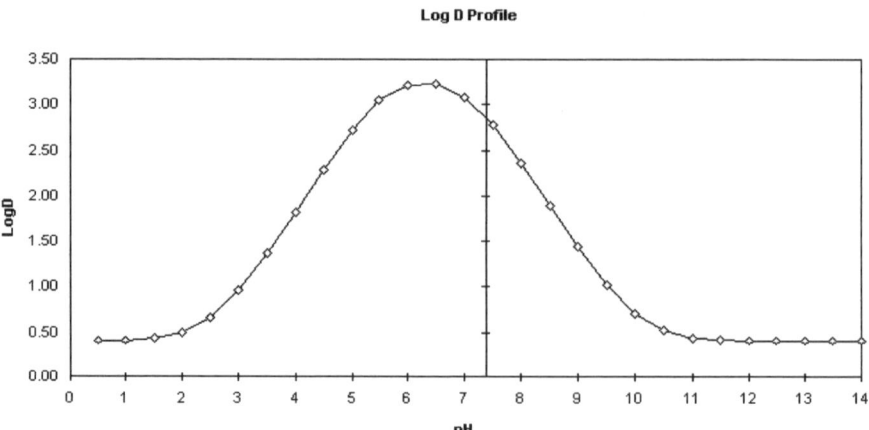

Figure 10 Log D profile of a zwitter ion (base) with pK_a = 5.6 and 7.0 for acid and base, respectively.

Log D_{mem} is then calculated as the log of the ratio of the concentration of the compound in the liposome phase to that of the compound in the aqueous phase.

More recently, computer methods have been devised to calculate these values. The molecule is broken down into fragments of known lipophilicity, and the log P is calculated using various computer routines. Alternatively, there are atom-based methods, and lipophilicity is measured, and calculated log Ps (c log Ps) agreement is reasonably good.

Partitioning Solvent

The choice of the partitioning solvent can have significant effects on the results obtained and also their relevance to biological systems. The most commonly used solvent is octan-1-ol that simulates the phospholipid membrane. To simulate the blood–brain barrier and skin penetration, octanol, chloroform, cyclohexane, and propylene glycol dipelargonate (PGDP) are more suitable. The log P values measured in these solvents show differences principally due to hydrogen bonding effects. Octanol can donate and accept hydrogen bonds making it amphiprotic; cyclohexane is inert; chloroform is a proton donor; and PGDP is an electron acceptor that can only accept them. Despite scores of very good publications and hundreds of studies on the choice of the solvent to be used, the choice remains very fluid. Differences in the log P values between different solvents of solvent mixture systems have often proven useful in QSAR studies. The differences in the partitioning value between PGDP–water and octanol–water or between octanol–water and alkali–water point to differences in hydrogen bonding capacity that affects skin penetration. Compounds with high log P values and low H bonding capacity can readily get past ester/phosphate groups in skin membranes.

Partitioning experiments have been carried out using liposomes, where neutral log P values from liposomes tend to be very similar to those measured in octanol, but the log P values of the ion pair differ. The "surface ion pair" log P is found to be much higher in bases, zwitter ions, and amphophiles. The values for acids tend to be similar to the octanol values. This reflects the increased potential for partitioning of molecules with basic groups into membranes. Quantitative structure–activity relationship studies have found improved correlations with liposome-derived "surface ion pair" log P values.

The measurement of partitioning is only practical if the compound shows some solubility; insoluble compounds are difficult to characterize and often prove less valuable anyway. A relationship between log P and the observed biology is frequently found in a series where structural modifications do not significantly affect the pK_a values. The classical work of Hansch showed that these relationships were often parabolic; hence, the relationship often leads to an optimum value for the log P for a desired activity or selective distribution. Relationships of this type include:

Linear: Activity $= m \log P + k'$ (59)

Parabolic: Activity $= m \log P - c(\log P)2 - k$ (60)

Rectilinear: Activity $= m \log P - c(\log P + 1) - k$ (61)

where m, k, and c are constants generated using regression analysis to correlate the observed biological data with measured partition coefficients. The mathematical techniques most commonly used to develop this correlation involve use of multivariate analysis, principal component analysis, and partial least square regression. Standard textbooks of statistics should be consulted to learn more about the applications and limitations of each of these approaches to data analysis.

The use of organic solvents to model complex bilipids is very simplistic. While there have been some successes in modeling the response of compounds, large differences in the activity between molecules of different structures or the activity between enantiomers cannot be easily understood. In these cases, it is very useful to combine physical measurements with molecular modeling, molecular property, and spectroscopic data and use multivariate analysis. For both CNS penetration and gastric absorption, the relationship appears to be parabolic with an optimum log P value of around 2 ± 1. Evidence for this comes from a wide variety of experiments in the literature from brain concentration of radiolabeled compounds to behavioral studies.

Other methods of analysis used include molecular properties, such as partial charged surface area (PSA) and study of the effects of hydrogen bonding on drug absorption.

Besides projecting the solubilitythe log P value has several important applications providing greater insight into how the molecule crosses various biological barriers and hence, proves effective as a prospective new lead compound. In general, where passive absorption is assumed, log P can be related to various fixed value ranges (Fig. 11).

Generally, a low log P (below 0) is desirable for injectable products, whereas a medium (0–3) range is suitable for oral administration; transdermal administration requires a higher value (3–4), but once the range of four to seven is reached, we risk the accumulation of the drug in the body fat that can be prove toxic due to accumulation of drug in multiple dosing situations. The renal clearance of drugs with log D (measured at pH 7.4) above zero will decrease renal clearance and increase metabolic clearance; the pK_a of drugs also plays an important role here as highly ionized drugs are kept out of cells and thus out of systemic toxicity;

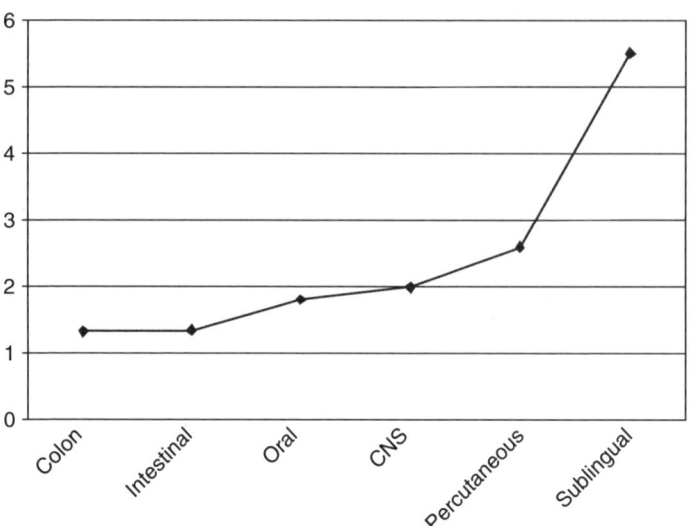

Figure 11 Optimal log P values for absorption from different parts of the gastrointestinal tract.

generally, a pK_a of six to eight will be most optimal for transport across various biological membranes.

When making a choice, generally a drug with lowest log P will be desirable; however, it might be required to make a choice between a high versus a low MW molecule; it is known that high molecular weight drugs are generally more allergenic. The goal should be to achieve a minimum hydrophobicity using a combination of log P, pK_a, and molecular size. The principle of minimum hydrophobicity keeps the drugs out of CNS that might produce side effects like depression, and son on, which means that most molecules should have a log P lower than 2.0; this technique was used in the design of the new generation of nonsedative antihistamines. A very high lipophilicity should also be avoided because of adverse effects on protein binding and on drug absorption, including solubility.

The ideal drug candidate, going into human studies, should have already been designed with the idea of keeping lipophilicity as low as possible, provided this can be done without loss of affinity to the target receptor.

Solubility

The discussions of pK_a, log P, and log D above all pertain to the end point analysis of the solubility of a drug candidate. A correlation exists between log P of neutral immiscible liquids with their solubility in water; however, for solids solubility also depends on the energy required to break the crystal lattice where log P is related to solubility. It is therefore possible to have compounds with high log P values, which are still soluble on account of their low melting point. Similarly, it is possible to have a low log P compound with a high melting point, which is very insoluble.

There are instances when titrating a basic compound that there appears a precipitation; in such cases, the solubility of the free base may be calculated

Table 4 The United States Pharmacopeia Solubility Classification

Descriptive term	Parts of solvent required for one part of solute
Very soluble	Less than 1
Freely soluble	From 1 to 10
Soluble	From 10 to 30
Sparingly soluble	From 30 to 100
Slightly soluble	From 100 to 1000
Very slightly soluble	From 1000 to 10,000
Practically insoluble or insoluble	10,000 and over

using the equation:

$$St_x = S_0 \cdot [1 + 10^{pK_a - pH_x}] \tag{62}$$

where St_x is the solubility at pH_x and S_0 is the solubility of the free base.

It is this parameter that ultimately determines the activity, toxicity, stability and dosage form, and route of administration. The United States Pharmacopoeia (USP) classifies drugs based on their solubility (Table 4).

High solubility is defined as the highest dose strength that is soluble in 250 mL or less of aqueous media across the physiological pH range. Poorly soluble drugs can be defined as those with an aqueous solubility of less than 100 μg/mL. If a drug is poorly soluble, then it will only slowly dissolve, perhaps leading to incomplete absorption. Some general observations about the behavior of solutes in solution systems include:

- Electrolytes dissolve in conducting solvents.
- Solutes containing hydrogen capable of forming hydrogen bonds dissolve in solvents capable of accepting hydrogen bonds and vice versa.
- Solutes having significant dipole moments dissolve in solvents having significant dipole moments.
- Solutes with low or zero dipole moments dissolve in solvents with low or zero dipole moments.

There are always exceptions to these rules, but a good rule of thumb "like dissolves like" mostly applies. Therefore, solvents fall into three classes:

- Protic solvents, such as methanol and formamide, which are hydrogen bond donors.
- Dipolar aprotic solvents (e.g., acetonitrile nitrobenzene) with dielectric constants greater than 15, but which cannot form hydrogen bonds with the solute and
- Aprotic solvents in which the dielectric constant is weak and the solvent is nonpolar, for example, pentane or benzene.

If solubility of a compound is accompanied by degradation, the estimation of solubility becomes difficult and must be quoted with the degradant found. Many candidate drugs are ionizable compounds, and thus there are a number of parameters that determines the solubility of a compound. These parameters include, for example, molecular size and substituent groups on the molecule, degree of ionization, ionic strength, salt form, temperature, crystal properties, and complexation.

As compound supply is likely to be limited, only a few solvent systems can be investigated during initial solubility studies. Typically, the solubility of the compounds in water, 0.9% w/v saline, 0.1 M HCl and 0.1 M NaOH will be determined. If there is sufficient compound, then the solubility in other systems may be considered (e.g., co-solvents, such as PEG400 and propylene glycol). In addition, the solubility in oils and surfactants systems (e.g., Tween 80) may be considered. Automated instruments allow ready measure of solubility, such as by using the pION's pSOL (5) whereas as little as 0.1 mg of the compound is sufficient to establish a complete solubility profile.

Molecular Size

Large organic molecules have a smaller aqueous solubility than smaller molecules. This is due to interactions between the nonpolar groups and water, that is, solubility is dependent on the number of solvent molecules that can pack around the solute molecule.

Additives

Additives might increase or decrease the solubility of a solute in a given solvent. In the case of salt, those that increase the solubility are said to "salt in" the solute, and those that decrease the solubility "salt out" the solute. The effect of the additive depends very much on the influence it has on the structure of the water or its ability to compete with solvent water molecules. Both effects are described by the empirically derived Setschenow equation:

$$\log^{S_0/S} = kM \tag{63}$$

This equation describes the relationship between the aqueous solubility of sparingly soluble salts (S_0) and the empirical Setschenow salting-out constant $k = 0.217/S_0$. This relationship and the Setschenow equation are valid only at low concentrations of added salt. As the concentration of added salt increases, the apparent k value is not constant, but is dependent on the solubility and the rate of change of solubility with added salt concentration. It was concluded that the Setschenow treatment is generally inappropriate for description and analysis of common ion equilibria.

Another aspect of the effect of electrolytes on the solubility of a salt is the concept of the solubility product for poorly soluble substances. The experimental consequences of this phenomenon are that if the concentration of a common ion is high, then the other ion becomes low in a saturated solution of the substance, that is, precipitation occurs. Conversely, the effect of foreign ions on the solubility of sparingly soluble salts is just the opposite, and the solubility increases. This is called the salt effect.

Temperature

As dissolution is usually an endothermic process, an increase in the solubility of solids with a rise in temperature is the general rule. Therefore, most graphs of solubility plotted against temperature show a continuous rise, but there are exceptions, for example, the solubility of sodium chloride is almost invariant, while that for calcium hydroxide falls slightly from a solubility of 0.185 g/mL at 0°C to 0.077 g/mL at 100°C.

MEASUREMENT STRATEGIES

The understanding of the factors that affect the values of pK_a or log P obtained as discussed previously is essential in making the best use of the multitude of techniques available to measure these parameters and then interpret them to the drug discovery group. The limitations faced by the preformulation group at this stage include:

- *Solubility of the compound.* There must be some solubility. The pK_a of poorly soluble compounds must be measured in aqueous methanol solution. If several titrations are carried out with different ratios of methanol:water, the Yesuda-Shedlovsky equation can reveal the theoretical pK_a in purely aqueous solution. Similarly, for poorly soluble compounds, provided the log P is high enough, the compound may be determined by titration, by addition of the sample to the octanol first. The compound will then back partition into the aqueous layer. If this fails, then spectroscopic methods have to be employed as more dilute solutions may be used.
- *Stability of the compound.* The compound must be able to withstand the rigors of testing, such as not breaking down during the time it takes to establish equilibrium between the two phases.
- *Purity of the compound.* Substances submitted for pK_a and log P studies need to be pure, of accurately known composition, and be submitted as free bases or inorganic acid salts. In general, no reliable measurements can be made on organic acid salts.
- *Single compound.* It is preferred to prepare a series of compounds to validate methods of testing and to validate a trend and to obviate many experimental problems.

Table 5 summarizes the advantages and disadvantages of the most commonly used methods to measure pK_a and log P. These are presented together because in many instances both parameters can be obtained from unified experimental runs, a strategy that is very useful when the quantity of the substance available is limited, as is the case in most instances.

Most companies would do well by studying the available equipment from Sirius (6). Table 6 lists the available equipments and their applications.

Ion Pair Log *P*

Ion pair log *P*s might be determined by at least two, or preferably three titrations in different ratios of octanol to water. The apparent pK_a in the presence of octanol, the poK_a, can be used to determine the presence of ion pair partitioning according to the equations:

$$P_{XH} = \frac{r_2 10^{P_a K_a(2) - pK_a} - r_1 10^{P_a K_a(1) - pK_a} - (r_2 - r_1) \times 10^{P_a K_a(1) + pK_a + P_a K_a(2) - 2pK}}{r_1 r_2 (10^{P_a K_a(1) - pK_a} - 10^{P_a K_a(2) - pK_a})} \tag{64}$$

$$P_X = \frac{r_1 10^{P_a K_a(2) - pK_a} - r_2 10^{P_a K_a(1) - pK_a} + r_2 - r_1}{r_1 r_2 (10^{P_a K_a(1) - pK_a} - 10^{P_a K_a(2) - pK_a})} \tag{65}$$

where r is the octanol:water ratio.

Table 5 Advantages and Disadvantages of pK_a and Log P Measurement Techniques

Method	Measures	Advantage	Disadvantage	Concentration required	Sample size
Sirius Potentiometric pK_a/log P	pK_a, log P, log P_{app}	Rapid, convenient	Insoluble or neutral samples cannot be measured	0.0001 M (0.1 mM)	1–5 mg
Sirius Yesuda-Shedlovsky	pK_a	pK_a for insoluble samples	Takes three or more titrations	0.0005 M (0.5 mM)	5 mg
Sirius ion pair log P	log P, log P (ip)	Predict log D more accurately	Takes three or more titrations	0.0001 M (0.1 mM)	3–15 mg
Manual potentiometric pK_a	pK_a	Simple, rapid	Not for low or overlapping pK_as; precipitation when titrated to neutral species; multiple experiments required	>0.0025 M (2.5 mM)	50 mg
pK_a by UV	pK_a	pK_a for poorly soluble or scarce compounds	Slow	0.000025 M (25 uM)	6 mg
pK_a by solubility	pK_a	pK_a for highly insoluble compounds	Slow, low accuracy	Below 0.0005 M (0.5 mM)	10 mg
Log P by filter probe	Log P	Log P for poorly soluble compounds, reliable > log P of 0.2	Messy, slow to set up, requires care. Inaccurate below log P of 0.2	0.000025 M (25 uM)	6 mg
Log D by filter probe	pK_a, log P, log D	Can determine log P_{app} at any pH	Only possible with compounds possessing isobestic point[a]	0.000025 M (25 uM)	6 mg
Log P by shake flask	Log P, log D at chosen pH	Low log P values. Can investigate surface effects	Slow, tedious, messy	0.000025 M (25 uM)	6 mg
Log P by HPLC	Log D at pH 7	Many compounds can be measured at once. Small sample size	Inaccurate, generally only carried out at pH 7	~2.5 mM	0.5 mg

[a] A wavelength, wavenumber, or frequency at which the total absorbance of a sample does not change during a chemical reaction or a physical change of the sample, such as when one molecular entity is converted into another, which has the same molar absorption coefficient at a given wavelength.
Abbreviations: HPLC, high-performance liquid chromatography; UV, ultraviolet.

Table 6 Equipment Available for Sirius for pK_a, Log P and Solubility Measurement, and Data Handling

Instrument	Features
GLpK_a: simultaneous pK_a, log P and solubility determination	• pH range of 1.8–12.2 (for standard titrations) • up to 30 samples in 24 hr • automated operation • measurements below 10^{-4} M • built-in traceability and reminders • supports D-PAS for spectrophotometric pK_a
D-PAS: spectrophotometric pK_a	• concentrations as low as 10^{-6} M • sample weight down to 50 µg • single- or multiprotic molecules • macro- and microconstants • easily fitted to Sirius GLpK_a or PCA200
PCA200: entry-level pK_a, log P, and solubility	• measures pK_a, log P, and intrinsic solubility • ideal for smaller workloads or academic research • uses the pH-metric technique and RefinementPro2 software • automated addition of acid and base titrants • supports D-PAS (UV method)
ProfilerSGA: rapid pK_a measurement	• fully automated physicochemical profiling • measures pK_a in just 2 min • multiwavelength spectrophotometric technique • open access option • automated data handling and processing • full computer control • microplate format • measures below 10^{-5} M concentration • 300 assays in 24 hr • MDM mixed solvent system
ProfilerLDA: log D/P octanol/water partitioning	• fully automated log D determination • multiwavelength spectrophotometric technique • open access (administrator and user profiles) • automated data handling and processing • full computer control • microplate format (96- and 384-well) • remote network monitoring • performance monitoring
ProfilerLDA: log D/P octanol/water partitioning	• fully automated log D determination • multiwavelength spectrophotometric technique • open access (administrator and user profiles) • automated data handling and processing • full computer control • microplate format (96- and 384-well) • remote network monitoring • performance monitoring
GLpH: automated pH measurement	• high degree of accuracy • easy to use • automatic calibration and reminders • available in single-sample or autosampler versions • one-button operation for most measurements • built-in GLP • "normal" and "supervisor" access modes

(Continued)

Table 6 Equipment Available for Sirius for pK_a, Log P and Solubility Measurement, and Data Handling (*Continued*)

Instrument	Features
RefinementPro2: automatic clipping of data	• automatic selection of reference ranges for D-PAS assay • dataset quality marker • improved "Approx" function with Auto-Bjerrum analysis • GLpK_a /D-PAS to be used with a laptop • turbidity sensing with D-PAS probe (optional) • assay planner (optional) • autorefinement of pH-metric and D-PAS data
CheqSol: solubility on GLpK_a, with results in less than 1 h	• equilibrium and kinetic result • sample does not require chromophore • result confirmed within same experiment • most samples measured without cosolvent • no filtration or separation required • no need to speculate whether sample is at equilibrium • wide measurement range
Fast D-PAS: pK_a values in just 4 min	• fast D-PAS is a new way to measure pK_a values • it combines the power of UV spectroscopy with the flexibility of pH-metric titration • gains speed by titrating in the presence of a unique linear buffer solution • fast D-PAS assays take about 4 min—including the time required to fill dispensers, move probes, titrate, read spectra, and clean up after assay

Note: Aqueous titrations using Sirius instruments is the easiest method for pK_a and log P measurements, and provides detailed information on the partitioning characteristics of a sample at all pH values. The PCA200 and GLpK_a, pK_a/log P analyzers (Table 2 in chapter 3) are based on a potentiometric titration method. The basic principle of operation is to determine the pK_a by titration followed by a back titration to determine the apparent pK_a in the presence of octanol. Any partitioning by the compound will shift the equilibrium and cause a change in the apparent pK_a. From this shift, the log P may be calculated. Sophisticated software allows detailed iterative calculations to be made and values to be carefully refined. If a sample is soluble, then it is possible to determine all its pK_a values, its log P, and the apparent log P at every pH. In addition, log P values of ionized species where they occur may also be calculated. The technique can be performed on samples at a concentration of 0.0001 M or above, the ideal concentration being 0.0005 M. Using the PCA200 for a suitable molecule, the analysis time would be a few hours, including calculation time. The GLpK_a has an autosampler and can also do multiple titrations on each sample. If the sample is highly insoluble, then the log P cannot be measured. The pK_a can be measured by either partial titration or by a Yesuda Shedlovsky experiment where three titrations in aqueous methanol is performed, each with a different proportion of methanol. From the results, a corrected extrapolation gives the theoretical aqueous value; the technique can be performed on samples at a concentration of 0.0005 M (6).
Abbreviations: CheqSol, chasing equilibrium solubility; UV, ultraviolet.
Source: From Ref. 7.

Both the log P and the log D values may be severely affected if one or more of the charged species is partitioned. Ion pairing effects may be fully determined with the Sirius Instruments, even in cases of polyprotic compounds where any of the charged or neutral species may partition.

Manual Titration

While this technique has mostly been replaced by automated techniques described previously, the principle remains the same wherein a pH electrode and a volumetric pipette are used. The technique may be carried out on compounds with reasonable aqueous solubilities (>0.0025 M), and that are available in amounts greater than 30 mg. This method is rapid, simple, and accurate; however, very low pK_as

($pK_a < 3$) and overlapping pK_as cannot be determined. A computer program or spreadsheet to calculate pK_a values from the experimental data can be written to speed up the calculations. Analysis time is a few hours. The reader may refer to *Excel for Chemists* by E. Joseph Billo, John Wiley & Sons, New York (2001) for help on setting up the spreadsheets.

Spectroscopy

Where the compounds are poorly soluble or where the quantity of compound available is small, pK_a values are calculated from ultraviolet (UV) spectrophotometric measurements. The method simply relies on the change in UV spectra at different pH values. An adaptation of the filter probe method (described next) is used. Sample concentrations down to 4 mg/400 mL may be determined (approx. 0.000025 M). Sirius Analytical Instruments supplies the D-PAS probe, which enables spectrophotometric determinations using the $GLpK_a$ instrument (Table 6).

The UV spectroscopy often proves to be a very useful tool if the new compound contains a chromophore (e.g., a group containing a double bond or other electron-rich zones; most sutiable for compounds with an ionizing group close to or within an aromatic ring, as the shift is large in such instances upon ionization). A simple plot of UV absorption as a function of change in pH gives instant results (provided the compound has sufficient solubility to yield a stable UV absorption reading). If the compound contains an UV chromphore that changes with the extent of ionization, then a method involving UV spectroscopy can be used. This method involves measuring the UV spectrum of the compound as a function of pH. Mathematical analysis of the spectral shifts can then be used to determine the pK_a or pK_as of the compound. The pK_a of a compound may be estimated using ACDpKa software, which also contains a large database of measured pK_a data. The use of the computer program SPARC (scalable processor architecture) for estimating the pK_a of pharmaceuticals is also used (8).

Generally, the quantity of substance required when the spectroscopy method is used is about 1 mg (traditional titrimetry requires about 3 mg). To maximize the utility of the small quantities available, the preformulation scientist would perform a multipurpose experiment; the compound is partitioned between a mixture of water and octanol (or a buffer), separating the two layers, centrifuged to separate any emulsification, both layers analyzed by HPLC method, and then using the aqueous layer to titrate or using UV spectroscopy to determine the pH shift at pK_a. These microscale techniques form the strong basis of preformulation studies. With newer techniques of quantitation becoming available, such as TOF mass spectroscopy (MS), liquid chromatography-tandem mass spectroscopy (LC-MS/MS), the burden to produce significant amounts of the lead compound is substantially reduced. This has also broadened the involvement of preformulation scientists at the earliest stages of drug discovery.

Solubility Method

Where the compounds have extremely low solubility, pK_a values may be measured by the solubility method. An aqueous solution of a substance is titrated in the direction of its neutral species until the free base or free acid is precipitated. pK_a can then be calculated from the solubility product. This method is not very accurate but may be used on very dilute solutions.

Filter Probe Method

The log P determinations using the filter probe method is essentially a variation of the shake-flask method, except that it is rapid, and relies on continuous sampling. An aqueous solution of the sample under test is placed in a reaction vessel and circulated through a spectrophotometric flow cell. The absorbance of the aqueous solution is measured before and after the addition of octanol. A solvent inlet filter prevents any octanol from passing through to the detector. The method is rapid and reliable for log P values from around 0.2 upward; but low log P values are difficult to measure due to the insignificant change in absorbance, which results. The reason for this is that below a phase volume ratio of 40 (400 mL water/10 mL octanol), the octanol tends to break through the filter. In cases of low log P, the shake-flask method has to be adopted. Sample concentrations down to 4 mg/400 mL may be determined (approx. 0.000025 M).

Log D profiles may be obtained by performing the filter probe experiment over a range of pH values. The critical part of the experiment is to discover whether the compound of interest has an isobestic point in its UV spectrum. If it does not, the experiment cannot be performed; if it does, both log P and pK_a values may be determined. This method is similar to the potentiometric method described earlier except that the amount of un-partitioned substance is determined by spectroscopy rather than by potentiometry. The advantage of this method is that only one experiment needs to be performed to yield log P, log P_{app}, and pK_as, but not all compounds have an isobestic point. Sample concentrations down to 4 mg/400 mL may be determined (approx. 0.000025 M).

Shake-Flask Method

The most common method for determining partition and distribution coefficients is the shake-flask method. In this technique, the candidate drug is shaken between octanol (previously shaken together or presaturate each phase with the other) and water layers, from which an aliquot is taken and analyzed using UV absorption, HPLC, or titration. The shake-flask method is the oldest and the most tedious way of measuring log P values. The UV absorbance of an aqueous solution is measured before and after being shaken with a known volume of octanol. Despite its disadvantages, this remains the only method that can be used in cases of very low log P values. One advantage of this method is that the appearance of the compound in the octanol may be checked against the disappearance from the aqueous phase to see if any surface effects have occurred. Some molecules may form effective surfactants! It is very important to presaturate the solvents in prolonged shake-flask experiments. The experiment must be performed over three days or more to ensure equilibrium is reached, although the actual time taken for the experiment is a few hours.

High-Pressure Liquid Chromatography

This chromatography may be used to estimate log P values. Compounds with known log Ps are injected onto a C18 reverse-phase HPLC column and their capacity factors are used to create a calibration curve. Unknown compounds are then injected and their capacity factors are used to predict log P. Strictly, this technique is only valid for neutral molecules. Charged molecules have more complex retention behavior than simple partition. Some liquid–liquid experiments have been reported using an octanol saturated column and an aqueous mobile phase; however, the method is messy and requires frequent re-generation of the column.

The chromatographic methods suffer the disadvantage that the retention time is linearly related to the partition coefficient, that is, for a doubling of the log P, there is a 10-fold increase in the retention. This often requires different length columns to be used, short ones for high log P values and long ones for low values. Additionally, where strong charged molecules are involved, the data become less reliable. However, reasonable correlation is seen with neutral compounds or those that are uncharged at pH 7.4. In cases where the molecule is charged and the pK_a is known, a correction factor may be added to correct the log D measurement to log P.

The discrepancies with the HPLC method are probably due to the imperfect nature of the C18-silica columns. Some of the new-generation reverse-phase materials, such as C18-alumina, polymeric C18, ultra-high carbon-loaded C18, and porous graphitic carbon may overcome these problems. A recent development is an immobilized artificial membrane (IAM) column, which should more closely model the biological membranes.

The main advantage of the HPLC method is that a range of compounds may be determined at the same time. A new rapid technique has been reported where all compounds and standards are simultaneously injected and the identity of each peak is determined by mass spectroscopy. A refinement of this technique is the determination of log k'_0. This is achieved by measuring log k' in several different concentrations of aqueous methanol mobile phases and extrapolating back to 0% methanol. The resultant log k'_0 values have been correlated to log P values more successfully. The concerns about polar interactions and the charge present on the analytes still remain.

Capillary Zone Electrophoresis

Capillary electrophoresis (CE) has emerged as a method of choice for determining compound pK_a values, as it possesses many favorable qualities as outlined in the following:

- potential impurities and degradants can be separated from the target compound;
- knowledge of sample concentration is not required for analysis;
- sparingly soluble compounds with a suitable UV chromophore can be analyzed;
- no changes in spectral properties are required for detection of a pK_a value;
- minimal sample amounts are required for analysis (<1 mg).

Single CE-UV systems possess a throughput of approximately one-compound/hr when analyzing 12 pH points per compound. Using the 96-capillary cePRO 9600™ system (9), it is possible to analyze 12 compounds/hr over 24 pH points. This breakthrough results in a significant increase in sample throughput in combination with an extended pH range and improved data quality.

High-performance liquid chromatography techniques have also been used in the determination of log P values. A potential problem with the use of HPLC retention data is that it is not a direct method and thus requires calibration. Furthermore, there may be problems with performing experiments above pH 8.

Plate Method for Solubility Testing

The multiscreen solubility filter plate method is a screening method that provides a fast, convenient, automation-compatible, high-throughput means to estimate the aqueous solubility of hundreds of compounds per day. It correlates well with standard shake-flask methodology, and can be implemented readily for this method in the typical drug discovery laboratory. Using a single-point calibration, the screening ratio is simply and quickly derived, and compound solubility is approximated easily. Multiple samples, each requiring approximately 200 nmol ($\sim$100 µg) per result, can be run in parallel. This method allows for the analysis of approximately 45 compounds (duplicate determinations) per plate with the capability of completing four or more plates in a standard eight-hour day. The assay is inherently compatible with the method by which most compound libraries are produced (e.g., as stock solutions in dimethylsulfoxide—DMSO), and is integrated easily into existing chemical profiling and early ADME (absorption, distribution, metabolism, and excretion) workflows. The method's resultant filtrate quickly provides a particulate-free and known soluble compound that can be used with confidence for other downstream ADME analysis.

Using this method, entire libraries can be screened for solubility. This narrows the number of candidates to only those that can meet solubility levels consistent with predicted oral bioavailability levels (especially when combined with permeability results) necessary to proceed down the pipeline.

High-throughput screening (HTS) assays are a routine for preformulation group; where the solubility is low, highly unreliable data are often seen particularly if there are molecules that skew the correlation significantly. Though compound aggregation has been determined as the cause of many false-positives, compound solubility often is not addressed until late in the ADME process. Screening compounds earlier (pharmacologic profiling) in the drug discovery process can minimize these issues, saving time and unnecessary expense. The standard method for solubility testing, the shake-flask method, has significant drawbacks, especially as an early screening tool. Long incubation times, tedious analysis requirements, and the need for large quantities of solid compounds preclude its usefulness in screening high numbers of candidates. A HTS method using a 96-well filter plate (Fig. 12) (MultiScreen® Solubility Filter Plate, Millipore, Billerica, Massachusetts, U.S.A.; Ref. 10) is a suitable substitute for the classical flask method. Studies report good correlation between this method and the classical flask method. An additional benefit is that the resultant filtrate can be used for downstream ADME analysis because it contains both a known concentration and known soluble compounds. The 96-well plate method can be easily introduced into the drug discovery workflow because the new entities are typically stored in DMSO, a condition that is readily integrated with this methodology. This eliminates the need for the large quantities of solid compounds required in the shake-flask method. The 96-well filter plate design meets new ANSI/SBS 2004 standards and is compatible with all standard laboratory robotics and analytical equipments. Incubation time for this assay is less than two hours, and hundreds of compounds can be evaluated on a given day.

This automatable assay can provide aqueous solubility data (in triplicate) for up to 120 drug compounds per (eight hours) day. pK_a data on compounds with ionizing groups can also be obtained when solubility is determined versus pH. There are however some important limitations to this method. Compounds must remain soluble, especially standards, over the duration of the assay, in whatever

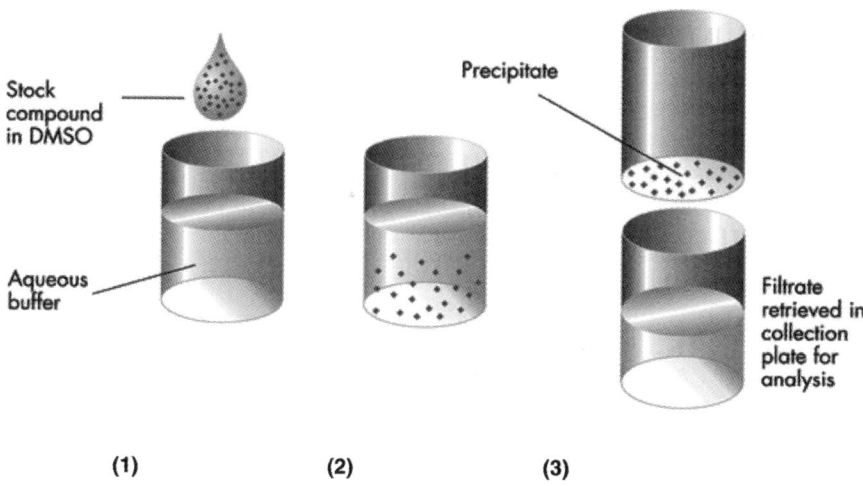

Figure 12 Filter plate method for solubility testing. (1) Add compound dissolved in organic solvent to aqueous buffer. (2) Shake for 90 minutes to allow insoluble compound to precipitate. (3) Apply vacuum to filter solution into collection plate. Precipitates remain on membrane. Analyze filtrate in collection plate to quantify the amount of the compound still in the solution. *Source*: Courtesy of Millipore Corporation, Billerica, Massachusetts, U.S.A.

solvent system is used. Certain compounds, if not soluble in 20% aqueous acetonitrile (ACN), may produce visible precipitate and cloudiness, which can interfere with the UV spectroscopic analysis. If the precipitate is found, other analytical methods, such as HPLC or LC/MS/MS may be used. Some compounds, upon visual inspection, may contain color-producing chromophores, so the spectral range should be increased beyond 500 nm. The fact that the sample is made up in a 5% (v/v) DMSO solution may result in an overestimation of the compound's solubility in a purely aqueous solution. Lowering the amount of DMSO (e.g., 0.5%) may improve the correlation between the aqueous solubility method and the shake-flask method. If the compound is less than 95% pure, the UV spectroscopy method may not be suitable. A complex mixture would require some sort of chromatographic separation prior to analysis. It is essential that the compounds being investigated have sufficient UV spectroscopic molar absorptivities (extinction coefficients) to provide the requisite analytical sensitivity.

With these limitations in mind, the assay is still well suited as a high-throughput tool for a number of compound screening applications—including the determination of structure–solubility relationships and establishing appropriate dosing concentration ranges for subsequent in vitro testing programs.

This method also allows use of multliple wavelength measurement and estimation of solubility. By analyzing the compound using spectroscopy at six wavelengths, the relative solubility, in the form of a screening ratio, can be calculated using the ratio of the pre- and postfiltered test samples. The calculated screening ratio provides a fast method for identifying compounds that are highly, moderately, or marginally soluble in aqueous solutions. As the screening ratio approaches unity, the sample approaches the upper limit of solubility—500 μM—as measured by the assay. If the screening ratio has a value less than 1 but greater than 0.5, the solubility of the compound is known to be between 100 and 500 µM. A screening ratio of less than 0.5 indicates that the compound's solubility is likely to be less than 100 µM.

If: $\dfrac{(\Sigma AU \text{ at } 280, 300, 320, 340, 360 \text{ nm}) - (AU \text{ at } 800 \text{ nm}) \text{ Filtrate}}{(\Sigma AU \text{ at } 280, 300, 320, 340, 360 \text{ nm}) - (AU \text{ at } 800 \text{ nm}) \text{ Standard}} \approx 1.00$

Then: Aqueous Solubility ≥ 500 µM

If: $\dfrac{(\Sigma AU \text{ at } 280, 300, 320, 340, 360 \text{ nm}) - (AU \text{ at } 800 \text{ nm}) \text{ Filtrate}}{(\Sigma AU \text{ at } 280, 300, 320, 340, 360 \text{ nm}) - (AU \text{ at } 800 \text{ nm}) \text{ Standard}} \leq 0.5$

Then: Aqueous Solubility ≤ 100 µM

(66–68)

If: $\dfrac{(\Sigma AU \text{ at } 280, 300, 320, 340, 360 \text{ nm}) - (AU \text{ at } 800 \text{ nm}) \text{ Filtrate}}{(\Sigma AU \text{ at } 280, 300, 320, 340, 360 \text{ nm}) - (AU \text{ at } 800 \text{ nm}) \text{ Standard}} < 1.00 \text{ and } > 0.5$

Then: 100 µM < Aqueous Solubility ≤ 500 µM

REFERENCES
1. http://www.iupac.org/index_to.html.
2. http://www.nist.gov.
3. http://www.cqsar.com/medchem/chem/qsar-db/.
4. http://www.biobyte.com.
5. http://www.pion-inc.com/pSOL.htm.
6. Avdeef A. pH-metric log P. II: Refinement of partition coefficients and ionization constants of multiprotic substances. J Pharm Sci 1993 Feb; 82(2):183–190.
7. http://www.sirius-ai.com/index.htm.
8. http://ibmlc2.chem.uga.edu/sparc/.
9. http://www.combisep.com/pka.html.
10. http://www.millipore.com/userguides.nsf/docs/p36523.

BIBLIOGRAPHY
Agbada Co O, York P. Dehydration of theophylline monohydrate powder-effects of particle size and sample weight. Int J Pharma 1994; 106:33–40.

Ahlneck C, Zografi G. The molecular basis of moisture effects on the physical and chemical stability of drugs in the solid state. Int J Pharm 1990; 62:87–95.

Ahmed S, Owen CP, James K, Patel CK, Patel M. Acid dissociation constant, a potential physicochemical factor in the inhibition of the enzyme estrone sulfatase (ES). Bioorg Med Chem Lett 2001, 9 Apr; 11(7):899–902.

Akers MJ. Preformulation testing of solid oral dosage forms. Methodology, management and evaluation. Can J Pharm Sci 1976; 11:1–10.

Albert A, Seargent EP. The Determination of Ionisation Constants—A Laboratory Manual. 3rd ed. London: Chapman and Hall, 1984.

Albini A, Fashini E., eds. Drugs: Photochemistry and Photostability. Cambridge, UK: The Royal Society of Chemistry, 1998.

Alverez Nunex FA, Lenneras VP Shah, Crison JR. A theoretical basis for a biopharmaceutic drug classification: the correlation of in vitro drug product dissolution and in vivo bioavailability. Pharma Res 1995; 12:413–420.

Anderson BD, Conradi RA. Predictive relationship in the water solubility of salts of a nonsteridal anti-inflammatory drug. J Pharm Sci 1985; 74:815–820.

Anderson NH, Johnston D, McLelland MA, Munden P. Photostability testing of drug substances and drug products in UK pharmaceutical laboratories. J Pharm Biomed Anal 1991; 9:443–449.

Andersson A, Hedenmalm H, Elfsson B, Ehrsson H. Determination of the acid dissociation constant for cis-diammineaquachloroplatinum(II) ion. A hydrolysis product of cisplatin. J Pharm Sci 1994 Jun; 83(6):859–862.

Anwair MA, Karolyhazy L, Szabo D, et al. Lipophilicity of aminopyridazinone regioisomers. J Agric Food Chem 2003, 27 Aug; 51(18):5262–5270.

ASTM: E 1148-02, Standard Test Method for Measurements of Aqueous Solubility, Book of Standards, Vol. 11.05.

Asuero AG, Navas MJ, Herrador MA, Recamales AF. Spectrophotometric evaluation of acidity constants of isonicotinic acid. Int J Pharma 1986; 34:81–92.

Austin RP, Davis AM, Manners CN. Partitioning of ionizing molecules between aqueous buffers and phospholipids vesicles. J Pharma Sci 1995; 84:1180–1183.

Avdeef A. pH-metric log P I: Difference plots for determining ion-pair octanol-water partition coefficients of multiprotic substances. Quant Struct-Act Relat 1992; 11: 510–517.

Avdeef A. pH-metric log P. II: Refinement of partition coefficients and ionization constants of multiprotic substances. J Pharm Sci 1993 Feb; 82(2):183–190.

Avdeef A. Weighting scheme for regression analysis using pH data from acid base titrations. Anal Chim Acta 1983; 148:237–244.

Avdeef A, Berger CM, Brownell C. pH-metric solubility. 2. Correlation between the acid-base titration and the saturation shake-flask solubility-pH methods. Pharm Res 2000; 17:85–89.

Avdeef A, Box KJ, Comer JE, et al. pH-metric log P 11. pKa determination of water-insoluble drugs in organic solvent-water mixtures. J Pharm Biomed Anal 1999 Aug; 20(4):631–641.

Avdeef A, Comer JEA, Thomson SJ. pH-metric log P 3. Glass electrode calibration in methanol-water applied to pKa determination of water insoluble substances by potentiometric titration. Anal Chem 1993; 65:42–49.

Bannerjee, Yalkowsky, Valvoni. Pharmacologists and environmental chemists make use of two measures which must...73–83, 1993. Envir Sci Tech 1980; 14:1227.

Barnett SP, Hill AP, Livingstone DJ, Wood J. A new method for the calculation of partition coefficients from experimental data for both mixture and pure compounds. Quant Struct-Act Relat 1992; 11:505–509.

Bartolini M, Bertucci C, Gotti R, et al. Determination of the dissociation constants (pKa) of basic acetylcholinesterase inhibitors by reversed-phase liquid chromatography. J Chromatogr A 2002, 7 Jun; 958(1–2):59–67.

Bartolomei M, Ramusino MC, Ghetti P. Solid state investigation of flucinolone acetonide. J Pharm Biomed Anal 1997; 15:1813–1820.

Beall HD, Getz JJ, Sloan KB. The estimation of relative water solubility for prodrugs that are unstable in water. Int J Pharm 1993; 93:37–47.

Benjamin EJ, Lin LH. Preparation and in vitro evaluation of salts of an antihypertensive agent to obtain slow release. Drug Dev Ind Pharm 1985; 11:771–790.

Berge SM, Bighley LD, Monkhouse DC. Pharmaceutical salt. J Pharm Sci 1977; 66:1–18.

Bogardus JB. Common ion equilibria of hydrochloride salts and the Setschenow equation. J Pharm Sci 1982 May; 71(5):588–590.

Borman S. New QSAR techniques eyed for environmental assessments. Chem Eng News 1990; 68:20–23.

Bray ML, Janhansouz H, Kaufman MJ. Selection of optimal hydrate/solvate forms of a fibrinogen receptor antagonist for solid dosage development. Pharm Dev Tech 1999; 4:81–87.

Briggner LE, Buckton G, Bystrom K, Dacy P. Use of isothermal microcalorimetry in the study of changes in crystallinity induced during the processing of powders. Int J Pharm 1994; 105:125–135.

Brittian HG. Spectral methods for the characterization of polymorphs and solvates. J Pharm Sci 1997; 86:405–412.

Burger A, Brox W, Ratz AW. Polymorphic und pseudopolymorphic von ceil prololhydrochlorid. Acta Pharm Technol 1985; 31:230–235.

Busetta B, Courseille C, Hospital M. Crystal and molecular structure of three polymorphous forms of estrone. Acta Cryst 1973; B29:298–313.

Byrn SR. Solid State Chemistry of Drugs. New York: Academic Press, 1982.

Byrn SR, Pfeiffer R, Ganey M, Hoiberg C, Poochikian G. Pharmaceutical solids: A strategic approach to regulatory considerations. Pharm Res 1995; 12:945–954.

Callaghan JC, Clearly GW, Elefant M, Kaplan G, Kensler T, Nash RA. Equilibrium moisture content of pharmaceutical excipients. Drug Dev Ind Pharm 1982; 8:355–369.

Cartensen JT, Danjo K, Yoshioka S, Uchiyama M. Limits to the concept of solid-state stability. J Pharm Sci 1987; 76:548–550.

Cartensen JT, Li Wan Po A. The state of water in drug decomposition in the moist solid state. Description and modeling. Int J Pharm 1992; 83:87–94.

Caselli M, Mangone A, Paolillo P, Traini A. Determination of the acid dissociation constant of bromocresol green and cresol red in water/AOT/isooctane reverse micelles by multiple linear regression and extended principal component analysis. Ann Chim 2002 May–Jun; 92(5–6):501–512.

Chan HK, Doelker E. Polymorphic transformation of some drugs under compression. Drug Dev Ind Pharm 1985; 11:315–332.

Chauret N, Lloyd DK, Levorse D, Nicoll-Griffith DA. Automated pKa determination of soluble and sparingly soluble drugs by capillary zone electrophoresis. Pharm Sci 1995; 1:59–62.

Chen MJ, Chen HS, Lin CY, Chang HT. Indirect detection of organic acids in non-aqueous capillary electrophoresis. 16: J Chromatogr A 1999, 20 Aug; 853(1–2):171–180.

Chowan, ZT. pH solubility profiles of organic carboxylic acids and their salts. J Pharm Sci 1978; 67:1257–1260.

Clarke FH. Ionization constants by curve fitting: application to the determination of partition coefficients. J Pharm Sci 1984 Feb; 73(2):226–230.

Clarke MJ, Potter UJ, Gilpin C, Tobyn MJ, Staniforth JN. Imaging of hygroscopic ultrafine pharmaceutical powders using low temperature and environmental scanning electron microscopy. Pharm Pharmacol Commun 1998; 4:419–425.

Coleman NJ, Craig DQM. Modulated temperature differential scanning calorimetry: A novel approach to pharmaceutical thermal analysis. Int J Pharm 1996; 135:13–29.

Connors KA, Amidon GL, Stella VJ. Chemical Stability of Pharmaceuticals. A Handbook for Pharmacists. 2nd ed. New York: John Wiley and Sons, 1986.

Cotton ML, Lamarache P, Motola S, Vadas EB. L—649, 923—The selection of an appropriate salt form and preparation of a stable oral formulation. Int J Pharm 1994; 109:237–249.

Craig DQM, Royall PG. The use of modulated temperature DSC for the study of pharmaceutical systems: potential uses and limitations. Pharm Res 1998; 15:1152–1153.

Danjo K, Kinoshita K, Kitagawa K, Iida K, Sunada H, Otsuka A. Effect of particle shape on the compaction and flow properties of powders. Chem Pharm Bull 1989; 37:3070–3073.

Davey, RJ, Carder PT, McEwan D, Sadler DE. Rate processes in solven-mediated phase transformations. J Cryst Growth 1986; 79:648–653.

David WIF, Shankland K, Shankland N. Routine determination of molecular crystal structures from powder diffraction data. Chem Commun 1998; 931–932.

de Villiers MM, van der Watt JG, Lotter AP. Kinetic study of the solid-state photolytic degradation of two polymorphic forms of furosemide. Int J Pharm 1992; 88:275–283.

Dearden JC, Bresen GM. The measurement of partition coefficients. Quant Struct–Act Relat 1988; 7:133–144.

DesJarlais RL, Sheridan RP, Seibel GL, Dixon JS, Kuntz ID, Venkataraghavan R. Using shape complementarity as an initial screen in designing ligands for a receptor binding site of known three-dimensional structure. J Med Chem 1988; 31:722–729.

Dissociation Constants of Inorganic Acids. Butterworths, 1969 (reprint of Pure and Applied Chemistry, Vol. 20, No. 2, 1969).

Dunn III W, Block JH, Pearlman RS. Partition Coefficient Determination and Estimation. Pergamon: St. Louis Missouri, 1986.

Elliot SR, Rao CNR, Thomas JM. The chemistry of the non-crystalline state. Agnew Chem Int Ed Engl 1986; 25:31–46.

El-Tayar N, Tsai RS, Test B, Carrupt PA, Leo A. Partitioning of solutes in different solvent systems: contribution of hydrogen-bonding capacity and polarity. J Pharm Sci 1991; 80:590–598.

Engel GE, Wilke S, Konig O, Harris KDM, Leussen FJJ. PowderSolve—A complete package for crystal structure solution from powder diffraction patterns. J Appl Cryst 1999; 32:1169–1179.

Eros D, Kovesdi I, Orfi L, Takacs-Novak K, Acsady G, Keri G. Reliability of log P predictions based on calculated molecular descriptors: a critical review. Curr Med Chem 2002 Oct; 9(20):1819–1829.

Fang X, Fernando Q, Ugwu SO, Blanchard J. An improved method for determination of acid dissociation constants of peptides. Pharm Res 1995 Oct; 12(10):1423–1429.
Fini A, de Maria P, Guarnieri A, Varoli L. Acidity constants of sparingly water-soluble drugs from potentiometric determinations in aqueous dimethyl sulfoxide. J Pharm Sci 1987 Jan; 76(1):48–52.
Fini A, Fazio G, Orienti I, Zecchi V, Rapaport I. Chemical properties-dissolution relationship. Part 4. Behaviour in solution of the diclofenac-N-(2-hydroxyethyl)pyrrolidine salt (DHEP). Pharm Acta Helv 1991; 66:201–203.
Franks NP, Abraham MH, Lieb WR. Molecular organization of liquid n-octanol: an X-ray diffraction analysis. J Pharm Sci 1993; 82:466–470.
Fuhrer C. Crystal engineering. Acta Pharm Technol 1986; 32:161–163.
Fyhr P, Hogstrom C. A preformulation study on the kinetics of the racemization of ropivacaine hydrochloride. Acta Pharm Suec 1988; 25:121–132.
Ganellin CR. Uses of Partition Coefficients by Brain Penetration Applied to the Design of H2-receptor Histamine Antagonists. Elsevier: St. Louis Missouri, 1991:103–110.
Gavezotti A, Filppini G. Polymorphic forms of organic crystals at room conditions—thermodynamic and structural implications. J Am Chem Soc 1995; 17:12299–12305.
Ghetti P, Ghedini A, Stradi R. Analytical potential of FT-IR microscopy. I. Applications to the drug polymorphism study. Boll Chin Farm 1994; 133:689–697.
Giachetti C, Assandri A, Mautone G, Tajana E, Palumbo B, Palumbo R. Pharmacokinetics and metabolism of N-(2-hydroxymethyl)-2, 5-[14C]-pyrolidine (HEP, epolamine) in male healthy volunteers. Eur J Drug Metab Pharmcokinet 1996; 21:261–268.
Giron D. Thermal analysis and calorimetric methods in the characterization of polymorphs and solvates. Thermochim Acta 1995; 248:1–59.
Giron D, Goldbronn C. Place of DSC purity analysis in pharmaceutical development. J Thermal Anal 1995; 44:217–251.
Giron-Forest D, Goldbronn Ch, Piechon P. Thermal analysis methods for pharmacopeial materials. J Pharm Biomed Anal 1989; 7:1421–1433.
Goa D, Rytting JH. Use of solution calorimetry to determine the extent of crystallinity of drugs and excipients. Int J Pharm 1997; 151:183–192.
Gould PL. Salt selection for basic drugs. Int J Pharm 1986; 33:201–217.
Grant DJW. Report and recommendation of the USP advisory panel on physical test methods: crystallinity determination by solution calorimetry. Pharmacop Forum 1999; 25:9266–9268.
Grant DJW, Mehdizadeh, Chow AHL, Fairbrother JE. Nonlinear van't Hoff solubility-temperature plots and their pharmaceutical interpretation. Int J Pharm 1984; 18:25–38.
Green DA, Meenan P. Acetaminophen crystal habit:Solvent effects. In: Myerson AS, Green DA, Meenan P, eds. Crystal Growth of Organic Materials. ACS Conference Proceedings, 1996:78–84.
Griesser UJ, Burger A. The effect of water vapor pressure on desolvation kinetics of caffeine 4/5-hydrate. Int J Pharm 1995; 120:83–93.
Gu L, Stickley RG. Preformulation salt selection–physical property comparisons of the tris(hydroxymethyl)aminomethane (THAM) salt of four analgesic/anti-inflammatory agents with the sodium salts and the free acids. Pharm Res 1987; 4:255–257.
Guillory JK, Erb DM. Using solution calorimetry to quantitate binary mixtures of three crystalline forms of sulfamethoxazole. Pharm Manuf 1985; 2:29–33.
Gustavo Gonzalez, Strickley A. Practical digest for the evaluation of acidity constants of drugs by reversed phase high performance liquid chromatography. Int J Pharm 1987; 91:R1–R5.
Heller H, Schaeffer M, Schulten K. Molecular dynamics simulation of a bilayer of 200 lipids in the gel and in the liquid-crystal phases. J Phys Chem 1993; 97:8343–8360.
Haleblian J, McCrone W. Pharmaceutical applications of polymorphism. J Pharm Sci 1969; 58:911–929.
Haleblian JK. Characterization of habits and crystalline modification of solids and their pharmaceutical applications. J Pharm Sci 1975; 64:1269–1288.
Hanai T. Simulation of chromatography of phenolic compounds with a computational chemical method. J Chromatogr A 2004, 20 Feb; 1027(1–2):279–287.

Hancock BC, Zografi G. Characteristics and significance of the amorphous state in pharmaceutical systems. J Pharm Sci 1997; 86:1–12.
Hansch C. A quantitative approach to biochemical structure-activity relationships. Acct Chem Res 1969; 2:232–239.
Hansch C. Drug research or the luck of the draw. J Chem Ed 1974; 51:360–365.
Hansch C. The QSAR paradigm in the design of less toxic molecules. Drug Metab Rev 1984–1985; 15:1279–1294.
Hansch C, Leo A, Hoekman D. Exploring QSAR—Hydrophobic, Electronic, and Steric Constants. Washington, D.C.: American Chemical Society, 1995.
Hansch C, Leo A, Taft RW. A survey of hammett substituent constants and resonance and field parameters. Chem Rev 1991; 91:165–195.
Hansch C, McClarin J, Klein T, Langridge R. A quantitative structure-activity relationship and molecular graphics study of carbonic anhydrase inhibitors. Mol Pharmacol 1985; 27:493–498.
Hatheway GJ, Hansch C, Kim KH, et al. Antitumor 1-(X-Aryl)-3,3-dialkyltriazenes. 1. Quantitative structure-activity relationships vs. L1210 leukemia in mice. J Med Chem 1978; 21:563–574.
Hilal SH, El-Shabrawy Y, Carreira LA, Karickhoff SW, Toubar SS, Rizk M. Estimation of the ionization pKa of pharmaceutical substances using the computer program SPARC. Talanta 1996; 43:607–619.
Hirsch CA, Messenger RJ, Brannon JL. Fenoprofen: drug form selection and preformulation stability studies. J Pharm Sci 1978; 67:231–236.
Holgard A, Møller N. Hydrate formation of metronidazole benzoate in aqueous suspensions. Int J Pharm 1983; 15:213–221.
Hong-guang W, Ru-hua Z. Compaction behaviour of paracetamol powders of different crystal shapes. Drug Dev Ind Pharm 1995; 21:863–868.
Horter D, Dressman JB. Influence of physicochemical properties on dissolution of drugs in the gastrointestinal tract. Adv Drug Devel Rev 1997; 25:3–14.
Jakobsen DF, Frokjer S, Larsen C, Niemann H, Buur A. Application of isothermal microcalorimetry in preformulation. J. Hygroscopicity of drug substances. Int J Pharm 1997; 156:67–77.
James KC. Solubility and Related Phenomena. New York: Marcel Dekker, 1986.
Jashnani RK, Byron PR. Dry powder aerosol generation in different environments: performance comparisons of albuterol, albuterol suphate, albuterol adipate and albuterol stearate. Int J Pharm 1996; 130:13–24.
Jashnani RK, Byron PR, Dalby RN. Validation of an improved Wood's rotating disk dissolution apparatus. J Pharm Sci 1993; 82:670–671.
Jenkins R, Fawcett TG, Smith DK, Visser JW, Morris MC, Frevel LK. JCPDS—International centre for diffraction data sample preparation method in X-ray powder diffraction. Powder Diff 1986; 1:15–63.
Jenkins R, Snyder RL. Introduction to X-ray powder diffractometry. New York: John Wiley & Sons, 1996.
Joseph Billo E. Excel for Chemists. New York, NY: John Wiley & Sons, 2001.
Jozwiakowski MJ, Williams SO, Hathaway RD. Relative physical stability of the solid forms of amiloride hydrochloride. Int J Pharm 1993; 91:195–207.
Karfunkel HR, Wu ZJ, Burkan A, et al. Crystal packing calculations and Rietveld refinement in elucidating the crystal structures of two modifications of 4-amidoindance guanylhydrazone. Acta Cryst B 1996: 52:555–561.
Kariuki BM, Psallidas K, Harris KDM, et al. Structure determination of a steroid directly from powder diffraction data. Chem Commun 1999; 17:1677–1678.
Kilday MV. Systematic errors in an isoperibol solution calorimeter measure with standard reference reactions. J Res Natl Bur Stand 1980; 85:449–465.
Koehler MG, Grigoras S, Dunn III WJ. The relationship between chemical structure and the logarithm of the partition coefficient. Quant Struct–Act Relat 1988; 7:150–159.
Komatsu H, Yoshii K, Okada S. Application of thermogravimetry to water-content determinations of drugs. Chem Pharm Bull 1994; 42:1631–1635.

Kontny MJ, Grandolfi GP, Zografi G. Water vapor sorption of water-soluble substance: Studies of crystalline solids below their critical relative humidities. Pharm Res 1987; 4:104–112.
Koort E, Herodes K, Pihl V, Leito I. Estimation of uncertainty in pKa values determined by potentiometric titration. Anal Bioanal Chem 2004 Jun; 379(4):720–729. Epub 2004, 22 Apr.
Kortum G, Vogel W, Andrussow K. Dissociation constants of Organic Acids in Aqueous Solution. Butterworths, 1961 (reprint of Pure and Applied Chemistry, Vol. 1, No. 2–3, 1961).
Krahn FU, Mielck JB. Relations between several polymorphic forms and the dehydrate of carbamazepine. Pharm Acta Helv 1987; 62:247–254.
Krezel A, Bal W. A formula for correlating pKa values determined in D_2O and H_2O. J Inorg Biochem 2004 Jan; 98(1):161–166.
Kuhnert-Brandstatter M, Gasser P. Solvates and polymorphic modifications of steroid hormones. I. Microchem J 1971; 16:419–428.
Kumar K, King RW, Carey PR. Carbonic anhydrase—aromatic sulfonamide complexes: a resonance Raman study. FEBS Lett 1974; 48:283–287.
Lambert WJ, Wright LA, Stevens JC. Development of a preformulation screen utilizing a C-18-derivitized polystyrene–divinylbenzene high-performance liquid chromatographic (HPLC) column. Pharm Res 1990; 7:577–586.
Leahy DE et al. QSAR: rational Approaches to the Design of Bioactive Compounds. Elsevier, St. Louis Missouri, 1991:75–82.
Ledwidge MT, Corrigan OI. Effects of surface active characteristics and solid state forms on the pH solubility profiles of drug–salt systems. Int J Pharm 1998; 174:187–200.
Lehto V-P, Salonen J, Laine E. Real time detection of photoreactivity in pharmaceutical solids and solutions with isothermal microcalorimetry. Pharm Res 1999; 16:368–373.
Leo A, Hansch C, Elkins D. Partition coefficients and their uses. Chem Rev 1971 Dec; 71(6): 525–616.
Levich VG. Physicochemical Hydrodynamics. Englewood Cliffs, N.J.: Prentice Hall, 1962.
Lipinski CA, Lombardo F, Dominy BW, Feeney PJ. Experimental and computational approaches to estimated solubility and permeability in drug discovery in drug discovery and development settings. Adv Drug Del Rev 1997; 23:2–25.
Lipnick RL. Charles Ernest Overton: narcosis studies and a contribution to general pharmacology. Trends Pharmacol Sci 1986; 7:161–164.
Loudon GC. Mechanistic interpretation of pH-rate profiles. J Chem Ed 1991; 68:973–984.
Marshall PV, Cook PA, Williams DR. A new analytical technique for characterising the water vapour sorption properties of powders. Stockholm: Int Symp Solid Dosage Forms, 1994.
Martinez V, Maguregui MI, Jimenez RM, Alonso RM. Determination of the pKa values of beta-blockers by automated potentiometric titrations. J Pharm Biomed Anal 2000, 15 Aug; 23(2–3):459–468.
Matoga M, Laborde-Kummer E, Langlois MH, et al. Determination of pKa values of 2-amino-2-oxazolines by capillary electrophoresis. J Chromatogr A 2003, 17 Jan; 984(2):253–260.
Mayers CL, Jenke DR. Stabilization of oxygen-sensitive formulations via a secondary oxygen scavenger. Pharm Res 1993; 10:445–448.
McCrone WC. Fusion Methods in Chemical Microscopy. New York: Interscience Publishers Inc., 1957.
McGovern SL, Caselli E, Grigorieff N, Shoichet BK. A common mechanism underlying promiscuous inhibitors from virtual and high-throughput screening. J Med Chem 2002; 45:1712–1722.
Mehta AC. Anayltical issues in the chemical stability testing of drugs in solution. Anal Proc 1995; 32:67–69.
Meng EC, Shoichet BK, Kuntz ID. Automated docking with grid-based energy evaluation. J Comput Chem 1992; 13:505–524.
Mengelers MJ, Hougee PE, Janssen LH, Van Miert AS. Structure-activity relationships between antibacterial activities and physicochemical properties of sulfonamides. J Vet Pharmacol Ther 1997 Aug; 20(4):276–283.
Merrofield DR, Carter PL, Clapham D, Sanderson FD. Addressing the problem of light instability during formulation development. In: Tønnesen HH, ed., Photostab Drugs Drug Formulation. London: Taylor and Francis, 1996:141–154.

Miller JL, Shea D, Khaledi MG. Separation of acidic solutes by nonaqueous capillary electrophoresis in acetonitrile-based media. Combined effects of deprotonation and heteroconjugation. J Chromatogr A 2000, 4 Aug; 888(1–2):251–266.

Millipore MultiScreen Solubility Filter Plate: 96-well filter plate designed and optimized for aqueous solubility assays, Millipore Lit. No. PF1315EN00, 2003.

Millipore MultiScreen Solubility Filter Plate: Automated screening of aqueous compound solubility in drug discovery, Millipore Lit. No. PS2153EN00, 2003.

Millipore MultiScreen Solubility Filter Plate: Determination of aqueous compound solubility using a 96-well filter plate to remove precipitated solids prior to UV/Vis spectroscopic analysis, Millipore Lit. No. PC2445EN00, 2003.

Misture ST, Chatfield L, Snyder RL. Accurate powder using zero background holders. Powder Diff 1994; 9:172–179.

Miyazaki S, Oshiba M, Nadai T. Precaution on the use of HCl salts in pharmaceutical formulation. J Pharm Sci 1981; 70:594–596.

Monkhouse DC, van Campen L. Solid state reations—theoretical and experimental aspects. Drug Dev Ind Pharm 1984; 10:1175–1276.

Morgan ME, Lui K, Anderson BD. Microscale titrimetric and spectrophotometric methods for determination of ionization constants and partition coefficients of new drug candidates. J Pharm Sci 1998 Feb; 87(2):238–245.

Morris, KR, Fakes MG. Thakur AB, et al. An intergrated approach of the selection of optimal salt form for a new drug candidate. Int J Pharm 1994; 105:209–217.

Nakanishi T, Suzuki M, Mashiba A, Ishikawa K, Yokotsuka T. Synthesis of NK109, an anticancer benzo [c] phenanthridine alkaloid. J Org Chem 1998; 63:4235–4239.

Navia MA, Chaturvedi PR. Design principles for orally bioavailable drugs. DDT 1996; 1:179–189.

Nichols G. Optical properties of polymorphic forms I and II of paracetamol. Microscope 1998; 46:117–122.

Nururkar AN, Purkaystha AR, Sheen PC. Effect of various factors on the corrosion and rusting of tooling material used in tablet manufacturing. Drug Dev Ind Pharm 1985; 11:1487–1495.

Nyqvist H. Saturated salt solutions for maintaining specified relative humidities. Int J Pharm Tech Prod Mfr 1983; 4:47–48.

Paulson B, ed. Polymorph changes important to FDA. Gold Sheet 1985; 19:1–20.

Payne RS, Roberts RJ, Rowe RC, Docherty R. Examples of successful crystal structure prediction: polymorphs of primidone and progesterone. Int J Pharm 1999; 177:231–245.

Perrin DD. Dissociation Constants of Organic Bases in Organic Solution. London: Butterworths, 1965.

Pikal MJ, Lang JE, Shah S. Desolvation kinetics of cefamandole sodium methonalate: effect of water vapour. Int J Pharm 1983; 17:237–262.

Pikal MJ, Lukes AL, Lang JE, Gaines K. Quantitative crystallinity determinations for β-actam antibiotics by solution calorimetry: correlations with stability. J Pharm Sci 1978; 67:767–772.

Pudipeddi M, Sokoloski TD, Duddu SP, Carstensen JT. Quantitative characterization of adsorption isotherms using isothermal microcalorimetry. J Pharm Sci 1996; 85:381–386.

Quarterman CP, Banham NM, Irwan AK. Improving the odds—high throughput techniques in new drug selection. Eur Pharm Rev 1998; 3:27–31.

Rabel SR, Jona JA, Maurin MB. Applications of modulated differential scanning calorimetry in preformulation studies. J Pharm Biomed Anal 1999; 21:339–345.

Rodergues C, Bugey DE. Characterization of pharmaceutical solvates by combined thermogravimetric and infrared analysis. J Pharm Sci 1997; 86:263–266.

Roman L, Mirel S, Florean E, Oprean R. The potentiometric and spectrophotometric determination of dissociation constants for same 2-mercapto-5-R-amino-1,3,4-thiadiazole derivatives. J Pharm Biomed Anal 1998 Oct; 18(1–2):137–144.

Rosenberg LS, Wagenknecht DM. pKa determination of sparingly soluble compounds by difference potentiometry. Drug Dev Ind Pharm 1986; 12:1449–1467.

Roy SD, Flynn GL. Solubility behaviour of narcotic analgesic in aqueous media: Solubilities and dissociation constants of morphine, fentanyl and sufentanil. Pharm Res 1989; 6:147–151.
Royall PG, Craig DQM, Doherty C. Characterization of the glass transition of an amorphous drug using modulated DSC. Pharm Res 1998; 15:1117–1130.
Rubin JT. Solubilities and solid state properties of the sodium salts of drugs. J Pharm Sci 1989; 78:485–489.
Rubino JT, Thomas E. Influence of solvent composition on the solubilities and solid state properties of the sodium salts of some drugs. Int J Pharm 1990; 65:141–145.
Ruiz R, Rafols C, Roses M, Bosch E. A potentially simpler approach to measure aqueous pKa of insoluble basic drugs containing amino groups. J Pharm Sci 2003 Jul; 92(7):1473–1481.
Saesmaa T, Halmekoski J. Slightly water-soluble salts of β-lactam antibiotics. Acta Pharm Fenn 1987; 96:65–78.
Saesmaa T, Makela T, Tannienen VP. Physical studies on the benzathine and emanate salts of some β-lactam antibiotics. Part 1. X-ray powder diffractometric study. Acta Pharm Fenn 1990; 99:157–162.
Serrajudin ATM, Mufson D. pH solubility profiles of organic bases and their hydrochloride salts. Pharm Res 1985; 2:65–68.
Seydel JK. Prediction of in vitro activity of sulfonamides, using hammett constants or spectrophotometric data of the basic amines for calculation. Mol Pharmacol 1966; 2:259–265.
Shah JC, Maniar M. pH-dependent solubility and dissolution of bupivacaine and its relevance to the formulation of a controlled release system. J Controlled Rel 1993; 23:261–270.
Simmons DL, Ranz RJ, Gyannchandani ND, Picotte P. Polymorphism in pharmaceuticals. II. Tolbutamide. Can J Pharm Sci 1972; 7:1436–1442.
Sirius Analytical Instruments Ltd. Applications and Theory Guide to pH-Metric pKa and logP determination, 1993.
Sirius Analytical Instruments Ltd. STAN Sirius Technical Application notes, Vol. 1, 1994.
Smith A. Pharmaceutical and biological applications of thermal analysis studies in polymorphism and hydration. Anal Proc 1986; 23:338–389.
Stewart PJ, Tucker IG. Prediction of drug stability—Part 2: hydrolysis. Aust J Hosp Pharm 1985; 15:11–16.
Stoltz M, Lotter AP, van der Watt JG. Physical characterization of two oxyphenbutazone pseudopolymorphs. J Pharm Sci 1988; 77:1047–1049.
Stylli C, Theobald AE. Determination of ionisation constants of radiopharmaceuticals in mixed solvents by HPLC. Int J Rad Appl Instrum [A] 1987; 38(9):701–708.
Subramaniam B, Rajewski RA, Snavely K. Pharmaceutical processing with supercritical carbon dioxide. J Pharm Sci 1997; 86:885–890.
Suleiman MS, Najib NM. Isolation and physicochemical pKa determination of water-insoluble compounds: validation study in methanol/water mixtures. Int J Pharm 1989; 151:235–248.
Takacsne Novak K. Practical aspects of partition measurements according to GLP rule. Acta Pharm Hung 1997 Sep; 67(5):179–191.
Takacs-Novak K, Avdeef A. Interlaboratory study of log P determination by shake-flask and potentiometric methods. J Pharm Biomed Anal 1996 Aug; 14(11):1405–1413.
Takacs-Novak K, Nagy P, Jozan M, Orfi L, Dunn WJ 3rd, Szasz G. Relationship between partitioning properties and (calculated) molecular surface. SPR investigation of imidazoquinazolone derivatives. Acta Pharm Hung 1992 Jan–Feb; 62(1–2):55–64.
Takacs-Novak K, Tam KY. Multiwavelength spectrophotometric determination of acid dissociation constants part V: microconstants and tautomeric ratios of diprotic amphoteric drugs. J Pharm Biomed Anal 2000 Jan; 21(6):1171–1182.
Takla PG, Dakas CJ. Infrared study of tautomerism in acetohexamide polymorphs. J Pharm Pharmacol 1989; 41:227–230.
Tam KY, Takacs-Novak K. Multiwavelength spectrophotometric determination of acid dissociation constants: Part II. First derivative vs. target factor analysis. Pharm Res 1999 Mar; 16(3):374–381.
Tanaka H, Tachibana T. Determination of acid/base dissociation constants based on a rapid detection of the half equivalence point by feedback-based flow ratiometry. Anal Sci 2004 Jun; 20(6):979–981.

Thomas E, Rubino J. Solubility, melting point and salting-out relationships in a group of secondary amine hydrochloride salt. Int J Pharm 1996; 130:179–185.
Threlfall TL. The analysis of organic polymorphs. Analyst 1995; 120:2435–2460.
Tomlinson E. Filter probe extractor: a tool for the rapid determination of oil-water partition coefficients. J Pharm Sci 1982; 71:602–604.
Tong W-Q, Vhitesell G. In situ salt screening—a useful technique for discovery support and preformulation studies. Pharm Dev Tech 1998; 3:215–223.
Tønneson HH. Photochemical degradation of components in drug formulation. Part I. an approach to the standardization of degradation studies. Pharmazie 1991; 46: 263–265.
Torniainen K, Tammilehto S, Ulvi V. The effect of pH, buffer type and drug concentration on the photodegradation of ciprofloxacin. Int J Pharm 1996; 132:53–61.
Umprayn K, Mendes RW. Hygroscopicity and moisture adsorption kinetics of pharmaceutical solids: a review. Drug Dev Ind Pharm 1987; 13:653–693.
van de Waterbeemed H, Mannhold R. Programms and methods for calculation of log P values. Quant Struct–Act Relat 1996; 15:410–412.
van Dooren AA, Muler BW. Effects of heating rates and particle sizes on DSC peaks. Thermochimica, Acta, 1982; 54:115–129.
Venger BH, Hansch C, Hatheway GJ, Amrein YU. Ames test of 1-(X-Phenyl)-3,3-dialkyltriazenes. A quantitative structure–activity study. J Med Chem 1979; 22:473–476.
Verwer P, Leusen FJJ. Computer simulation to predict possible crystal polymorphs. Rev Comput Chem 1998; 12:327–365.
Vitez IM, Newman AC, Davidovich M, Kiesnowski C. The evolution of hot-stage microscopy to aid solid-state characterization of pharmaceutical solids. Thermochim Acta 1998; 324:187–196.
Wadsten, T, Simon L, Talpas GS. Crystal habit as the basis of the change in bioactivity of pentetrazole. Pharmazie 1990; 45:66–67.
Walkling, WD, Reynolds BE, Fegely BJ, Janicki CA. Xilobam: effect of salt form on pharmaceutical properties. Drug Dev Ind Pharm 1983; 9:809–819.
Wang DX, Yang GL, Song XR, Kou SR. Determination of the pKa value of icariin and its content in the Chinese herb medicine epimedium grandiflorum morr. by capillary electrophoresis. Se Pu 2001 Jan; 19(1):64–67.
Watanabe A, Yamaoka Y, Kuroda K. Study of crystalline drugs by means of polarizing microscope. III. Key refractive indices of some crystalline drugs and their measurement using an improved immersion method. Chem Pharm Bull 1980; 28:372–378.
Watanabe A, Yamaoka Y, Takada K. Crystal habits and dissolution behaviours of aspirin. Chem Pharm Bull 1982; 30:2958–2963.
Wells JI. Pharmaceutical preformulation. The physicochemical properties of drug substances. Chichester, UK: Ellis Horwood Ltd, 1988:219.
Wilson RJ, Beezer AE, Mitchell JC. Determination of thermodynamic and kinetic parameters from isothermal heat conduction microcalorimetry: applications to long term reaction studies. J Phys Chem 1996; 99:7108–7113.
Wu LE, Gerard C, Hussain MA. Thermal analysis and solution calorimetry studies on lasartan polymorphs. Pharm Res 1993; 10:1793–1795.
Yalkowsky SH, Banerjee S. Aqueous Solubility. Methods for Estimation for Organic Compounds. New York: Marcel Dekker Inc., 1992.
Yang, S. Determination of acid-base amphoteric dissociation constant of 2-methyl-5(4)-nitroimidazole by UV-spectrophotometry. Hua Xi Yi Ke Da Xue Xue Bao. 1999 Sep; 30(3): 345–346.
Yasushi Ishihama, Masahiro Nakamura, Toshinobu Miwa, Takashi Kajima, Naoki Asakawa. A rapid method for pKa determination of drugs using pressure-assisted capillary electrophoresis with photodiode array detection in drug discovery. J Pharm Sci 2002; 91:933–942.
Yoshioka S, Cartensen JL. Nonlinear estimation of kinetic parameters for solid-state hydrolysis of water-soluble drugs. Part 2. Rational presentation mode below the critical moisture content. J Pharm Sci 1990; 79:799–801.
Zimmerman I. Determination of overlapping pKa values from solubility data. Int J Pharm 1986b; 31:69–74.
Zimmerman I. Determination of pKa values from solubility data. Int J Pharm 1986a; 13:57–65.

Appendix 1: Some Representative Studies on pK_a and Log P Determination Reported in the Literature

Reported study	References
Capillary electrophoresis (CE) and diode array: Use of a rapid screening method for determination of pK_a of candidate drugs by pressure-assisted CE coupled with a photodiode array detector is described. Application of pressure during CE analysis allowed completion of one CE run in less than one minute, and the obtained pH-metric mobility shifts as well as the pH-metric UV spectrum were analyzed by a nonlinear regression fitting software to determine pK_a values. The difference between pK_a values by this method and by other conventional methods is within 0.25 units for 82 ionic functional groups of 77 drugs. The pK_a values of 96 compounds in dimethylsulfoxide (DMSO) solution on a 96-well microplate could be measured in one day. Our method provides rapid and accurate determination of pK_a values.	Yasushi Ishihama et al. (2002)
Antimicrobial activity, correlation with: Relationships between the antimicrobial activities of sulfonamides and the physicochemical properties including the acid dissociation constant (pK_a) and the hydrophobicity constant (pi) were determined. The minimal inhibitory concentrations (MIC) of sulfonamides against *Actinobacillus pleuropneumoniae*, a gram-negative veterinary pathogen, were used. High-performance liquid chromatography was applied for the determination of the electronic and hydrophobic parameters. Empirically determined relationships pointed out the dominant role of the degree of ionization on the antimicrobial activity. These data indicate that hydrophobic properties of sulfonamides, characterized by pi, are of minor importance for the in vitro antibacterial activity. Because of the restricted pK_a range (4.9–7.7), it could not be established whether the relationship between pK_a and activity was linear or bilinear. Whenever *o*,*m*-disubstituted sulfonamides were included, correlations decreased substantially. Relationships based on multicompartment equilibrium models were derived and indicated a bilinear relation between pK_a and MIC. Model-based equations showed that the antibacterial activity was governed by the extracellular ionic concentration of the sulfonamides whenever different intra- and extracellular pH values were assumed in the equilibrium model. The antimicrobial activities of the sulfonamides against gram-positive organisms were also related to the degree of ionization of sulfonamides in the agar medium.	Mengelers et al. (1997)
CE and potentiometry: The dissociation constants of new 2-amino-2-oxazolines were determined by CE based on a linear model. A series of eight 2-amino-2-oxazolines were investigated to determine their ionization constants. The K_a values were obtained from the plots of reciprocal effective mobility against inverse concentrations of protons. The potentiometric method (PM) was performed as a comparative method. No significant differences were observed between the determined dissociation constants using both methods.	Matoga et al. (2003)

(Continued)

Appendix 1: Some Representative Studies on pK_a and Log P Determination Reported in the Literature (Continued)

Reported study	References
Capillary zone electrophoresis (CZE), herbs: Knowledge of dissociation constants is important for predicting and understanding the migration behavior of analytes in CE. Icariin is the active component of the Chinese herb medicine *Epimedium grandiflorum* Morr. In order to determine the dissociation constant of icariin and to show many important pharmacology activities, a CZE method has been used to determine the ion mobility (mu A^-) and the pK_a value of icariin based on the nonlinear relation between mu_{eff} and [H^+], and the linear relation between the reciprocal of effective mobility of the solute ($1/mu_{eff}$) and the [H^+] of the buffer solution. In addition, the change of pK_a of icariin with the increase of ethanol concentration in the buffer was also investigated. Under the buffer condition of 24 mmol/L phosphate + 30% ethanol, the content of the active component icariin in the Chinese herb *E. grandiflorum* Morr. was quantitatively determined.	Wang et al. (2001)
Complexation with chloride ion: The biological activity of the anticancer drug cisplatin is supposed to be mediated by its reactive hydrolysis product cis-diammineaquachloroplatinum(II) ion (monoaqua). The monohydrated complex (monoaqua and its deprotonated form monohydroxy) was isolated from an equilibrium mixture of cisplatin in distilled water using a strong cationic exchanger. The structure of the monoaqua complex was established by 252Cf time-of-flight mass spectrometry (MS). The acid dissociation constant determined at 37°C by studying the influence of pH on the reaction between the monoaqua complex and the chloride ion (0.1 M). The concentration of the monohydrated compound was determined by liquid chromatography with postcolumn derivatization using sodium diethyldithiocarbamate as a reagent. The pK_a was determined to be 6.56 ± 0.01 (SEM). Thus, at physiological pH, the monoaqua complex is present mostly in its less reactive monohydroxy form.	Andersson et al. (1994)
D_2O and H_2O comparison: A linear correlation between pH-meter readings in equivalent D_2O and H_2O solutions, determined experimentally, leads to a novel equation, which allows for a direct re-calculation of pK_a values measured in D_2O into a H_2O equivalent: $pKH = 0.929pKH^* + 0.42$. The comparison of this equation with the previously used approach is discussed.	Krezel and Bal (2004)
Extended principal component analysis (EPCA): The pK_a of 3',3'',5',5''-tetrabromo-*m*-cresolsulfonaphthalein (Bromocresol Green) and *o*-cresolsulfonaphthalein (Cresol Red) was spectrophotometrically measured in a water/AOT/iso-octane microemulsion in the presence of a series of buffers carrying different charges at different water/surfactant ratios. Extended principal component analysis was used for a precise determination of the apparent pK_a and of the spectra of the acid and base forms of the dye. The apparent pK_a of dyes in water-in-oil microemulsions depends on the charge of the acid and base forms of the buffers present in the water pool. Combination with multiple linear regressions (MLR) increases the precision. Results are discussed taking into account the profile of the electrostatic potential in the water pool and the possible partition of the indicator between the aqueous core and the surfactant. The pK_a corrected for these effects are independent of w_0 and are close to the value of the pK_a in bulk water. On the basis of a tentative hypothesis, it is possible to calculate the true pK_a of the buffer in the pool.	Caselli et al. (2002)

Flow ratiometry, feedback-based:
Acid dissociation Constants (K_a) were determined through the rapid detection of the half equivalence point (EP1/2) based on a feedback-based flow ratiometry. A titrand, delivered at a constant flow rate, was merged with a titrant, whose flow rate was varied in response to a control voltage (V_c) from a controller. Downstream, the pH of the mixed solution was monitored. Initially, V_c was increased linearly. At the instance that the detector sensed EP1/2, the ramp direction of V_c changed downward. When EP1/2 was sensed again, V_c was increased again. This series of process was repeated automatically. The pH at EP1/2 was considered to be the pK_a of the analyte after an activity correction. Satisfactory results were obtained for different acids in various matrices with good precision (RSD approximately 3%) at a throughput rate of 56 sec/determination.
Tanaka and Tachibana (2004)

High-pressure liquid chromatography, reversed phase (HPLC, RP):
An RP-HPLC study for the pK_a determination of a series of basic compounds related to caproctamine, a dibenzylaminediamide reversible inhibitor of acetylcholinesterase, is reported. The 2-substituted analogs, bearing substituents with different electronegativity, were analyzed by RP-HPLC by using C18 C4 stationary phases with a mobile phase consisting of a mixture of acetonitrile (ACN) and triethylamine phosphate buffer (pH range comprised between 4 and 10). Typical sigmoidal curves were obtained, showing the dependence of the capacity factors upon pH. In general, the retention of the investigated basic analytes increased with increasing pH. The inflection point of the pH sigmoidal dependence was used for the dissociation constant determination at a fixed ACN percentage. When plotting pK_a versus percent of ACN in the mobile phase for two representative compounds, linear regression was obtained: the y intercept gave the aqueous $pK_a(w)$. The pK_a estimation by HPLC method was found to be useful to underline the difference of benzylamine basicity produced by the ortho aromatic substituents. The variation of pK_a values (6.15–7.80) within the series of compounds was correlated with the electronic properties of the ortho substituents through the Hammett sigma parameter, whereas the ability of substituents to accept H-bond was found to play a role in determining the conformational behavior of the molecules.
Bartolini et al. (2002)

HPLC, simulation:
An ab initio simulation of reversed-phase liquid chromatography for phenolic compounds was achieved based on molecular interaction energy values calculated using molecular mechanics calculations (MM2) of the CAChe program. The precision of the predicted retention factors from the molecular interaction energy values was equivalent to the predicted retention factors based on octanol-water partition coefficients (log P) calculated using the molecular orbital package (MOPAC). The prediction of retention factors of phenolic compounds in reversed-phase liquid chromatography in a given pH eluent was possible using the predicted dissociation constant (pK_a) from the atomic partial charge without a chemical experiment if the organic modifier effect was known.
Hanai (2004)

HPLC, structure–property relationship (SPR):
SPR investigations have been carried out for a series of N-bridged compounds synthetized recently as potential antithrombotic agents. The octanol/water partition coefficient and RP-HPLC retention (log K') were determined
Takacs-Novak et al. (1992)

(Continued)

Appendix 1: Some Representative Studies on pK_a and Log P Determination Reported in the Literature (*Continued*)

Reported study	References
experimentally for 10 imidazoquinazolone derivatives. Calculations including total geometry optimization and surface area have been carried out. The isotropic surface area (ISA) of supermolecules was obtained. The good correlations were found between log P, log k', and ISA values. This provides a simple and reliable means by using a single structural parameter for the log P prediction of polycyclic *N*-bridged compounds.	Stylli and Theobald (1987)
HPLC:	
The uptake of some ionogenic radiopharmaceuticals may be explained on the basis of a pH shift theory. An HPLC method for pK_a measurement has been developed for radiotracers. It has been tested with several amines and used to estimate the pK_{sa} (pK_a in the mixed solvent, 30% of ACN-water) of some 99mTc-PnAO complexes by observing the change in capacity factor k' with change in mobile phase pH. The nonionogenic complexes Tc-PnAO and Tc-DE-PnAO showed constant k' values while the ionogenic complexes Tc-MNE-PnAO and Tc-IP-PnAO had pK_{sa} values of about 6.6 and 8.3, respectively. The method is reproducible and is applicable to nearly all ionogenic radiopharmaceuticals.	Chen et al. (1999)
Nonaqueous capillary electrophoresis (NACE) indirect determination:	
NACE with indirect detection has been applied to the determination of fatty acids (FAs) and ascorbic acid (AA). C2-C18 FAs have been separated in less than 12 min using 8-hydroxy-7-iodoquinoline sulfonic acid as chromophores in NACE with indirect absorbance. The dissociation constant (pK_a) values of C8–C18 FAs obtained from the slope of the linear plot $-\log[(\text{mu } 0/\text{mu}) - 1]$ versus pH, using 20% isopropanol and 40% ACN as the organic modifier in NACE, are all above about two units than those obtained in aqueous solution. NACE with indirect laser-induced fluorescence, using merocyanine 540 (MC540) as fluorophores, has been performed for the analysis of AA and its stereoisomer, isoascorbic acid (IAA), and the limits of detection of AA and IAA are 0.30 and 0.17 μM, respectively. This method has been applied to the determination of AA in a lemon juice spiked with IAA as the internal standard in less than three minutes, and its concentration is 76.7 ± 0.4 mM.	Miller et al. (2000)
NACE:	
NACE is a chemical separation technique that has grown in popularity over the past few years. In this report, we focus on the combination of heteroconjugation and deprotonation in the NACE separation of phenols using ACN as the buffer solvent. By preparing various dilute buffers consisting of carboxylic acids and tetrabutylammonium hydroxide in ACN, selectivity may be manipulated based on a solute's dissociation constant and its ability to form heteroconjugation ions with the buffer components. Acetonitrile's low viscosity, coupled with its ability to allow for heteroconjugation, often leads to rapid and efficient separations that are not possible in aqueous media. In this report, equations are derived showing the dependence of mobility on various factors, including the pK_a of the analyte, the pH and concentration of the buffer, and	

the analyte–buffer heteroconjugation constant (K_f). The validity of these equations is tested as several nitrophenols are separated at different pH values and concentrations. Using nonlinear regression, the K_f values for the heteroconjugate formation between the nitrophenols and several carboxylate anions are calculated.

Partition coefficient, quantitative structure–property relationships (QSAR):

Correct QSAR analysis requires reliably measured or calculated log P values, log P being the most frequently utilized and most important physico-chemical parameter in such studies. Since the publication of theoretical fundamentals of log P prediction, many commercial software solutions are available. These programs are all based on experimental data of huge databases; therefore, the predicted log P values are mostly acceptable—especially for known structures and their derivatives. In this study, we critically reviewed the published methods and compared the predictive power of the commercial softwares (CLOGP, KOWWIN, SciLogP/ULTRA) with each other and with our recently developed automatic QS(P)AR program. We have selected a very diverse set of 625 known drugs (98%) and drug-like molecules with experimentally validated log P values. We have collected 78 reported "outliers" as well, which could not be predicted by the "traditional" methods. We used these data in the model building and validation. Finally, we used an external validation set of compounds missing from public databases. We emphasized the importance of data quality, descriptor calculation and selection, and presented a general, reliable descriptor selection and validation technique for such kind of studies. Our method is based on the strictest mathematical and statistical rules, fully automatic and after the initial settings, there is no option for user intervention. Three approaches were applied: MLR partial least squares analysis, and artificial neural network. Log P predictions with a MLR model showed acceptable accuracy for new compounds; therefore, it can be used for "in-silico-screening" and/or planning virtual/combinatorial libraries.

Eros et al. (2002)

Partition coefficient, software use:

Ten pairs of pyridazinone regioisomers were prepared, and their lipophilicity was described by the logarithm of the octanol/water partition coefficient (log P) determined experimentally and calculated with prediction methods. The 4- and 5-(substituted amino)-3(2H)-pyridazinone regioisomers were synthesized by nucleophilic substitution of one of the chloro atoms of 4,5-dichloro-2-methyl-3(2H)-pyridazinone or its 6-nitro derivative. Structures of new compounds were proven by spectroscopic methods. The experimental log P values were obtained by a shake-flask method in octanol and a Sorensen buffer (pH 7.4) solvent system. A consequent difference was found in the lipophilicity of regioisomers. For each isomer pair, the log P value of the 4-isomer was significantly (average by 0.75 log unit) higher than that of the 5-isomer. Some quantum chemical calculations as well as X-ray analysis of two pairs of regioisomers were also carried out to gain insight into the structural differences of regioisomers. The log P values were calculated by the fragmental approach KOWWIN and a QSPR analysis (3DNET). The a priori KOWWIN gave poor agreement, but with the programs KOWWIN with EVA (experimental value adjusted) and 3DNET, the results were generally in agreement with the experiment.

Anwair et al. (2003)

(Continued)

Appendix 1: Some Representative Studies on pK_a and Log P Determination Reported in the Literature (*Continued*)

Reported study	References
Partitioning, survey of methods: Experimental methods for octanol/water partition coefficient (log P) determination were surveyed. The terminology used in the literature, the lipophilicity/pH profile and the most important factors influencing the log P values have been discussed. Several new, recently developed direct log P determination methods are introduced including their advantages and limits of application. Some aspects of good laboratory practice of the shake-flask method are described and results of a validation study of pH-metric log P determination technique using PCA 101 pK_a and log P analyzer (Sirius, U.K.) were also shown. Questions to be answered in method selection for log P measurement are summarized in the flow chart. The author, based on her own experiences in lipophilicity measurements over decades, suggests the shake-flask method and the automated dual-phase potentiometric technique, as approaches fulfilling the GLP rules.	Takacsne Novak (1997)
Peptides, in purification stage: The method described here enables the proton dissociation constants of several amino acid residues of a peptide to be determined simultaneously in aqueous solution without prior knowledge of the exact concentration of the peptide. The method employs a nonlinear fitting program, the BEST program, or a linear least-squares method in prior knowledge of the exact concentration of the peptide. The method employs a nonlinear fitting program, the BEST program, or a linear least-squares method in combination with the BEST program. The possibility of integrating these procedures into a preparative chromatographic system for the "on-line" assessment of the pK_a values of peptides during the purification stage is an attractive and novel feature of this method.	Fang et al. (1995)
Potentiometric and HPLC, octanol system: Computerized method provides convenient and accurate determination of the ionization constant in aqueous solution and of the apparent ionization constant in the presence of octanol. From these parameters, partition coefficients and apparent partition coefficients are easily calculated and agree with data reported using the shaker technique or HPLC. The curve-fitting method has been applied to the differential titration technique in which the solvent curve is subtracted from the solution curve before calculations are begun. This method has been applied to the potentiometric titration of aqueous solutions of the salts of bases with very low solubility in water.	Clarke (1984)
Potentiometric titration, automated: The acid–base equilibrium constants of the beta-blockers atenolol, oxprenolol, timolol, and labetalol were determined by automated potentiometric titrations. The pK_a values were obtained in water-rich or water methanol medium (20% MeOH) to obviate the solubility problems associated with the compounds. The initial estimates of pK_a values were obtained from Gran's method and then, were refined by the NYTIT and ZETA versions of the LETAGROP computer program. The resultant values were 9.4 (I = 0.1 M KCl, 20% methanol) for atenolol, 9.6 (I = 0.1 M KCl) for oxprenolol, 9.4 (I = 0.1 M KCl, 20% methanol) for timolol and 7.4 and 9.4 (I = 0.1 M KCl) for labetalol. The potentiometric method was found to be accurate and easily applicable. The operational criteria for applying the methodology are indicated.	Martinez et al. (2000)

Potentiometric titration, sparingly soluble compounds:

Among aqueous organic mixtures, aqueous Me$_2$SO (80% w/w) presents properties particularly suitable for acid–base studies, and thermodynamically meaningful acidity constants can be obtained by a potentiometric technique, provided that the glass electrode is properly calibrated. Thermodynamic acidity constants of more than 100 acids have been potentiometrically determined at 25°C in this mixed solvent, and the selected series of acids has been divided into four classes according to the nature of the acidic group (COOH, OH, SH) and the structure of the acid (aliphatic, aromatic, heterocyclic). A linear relationship between the experimental pK_a values in water and in aqueous Me$_2$SO (80% w/w) has been the single classes and a group of equations has been derived (the asterisk denotes pK_a values in which infinite dilution in the mixed solvent is taken as a standard state). For the carboxylic acid class, the following "common" equation has been found: pK_a (H$_2$O) = $-0.80 + 0.67$ pK_a^* (Me$_2$SO, 80% w/w). As an application, pK_a values in water are reported for a representative number of sparingly soluble acids. These values have been calculated by means of the "common" equation, using pK_a values experimentally determined in aqueous Me$_2$SO (80% w/w). The calculated values are in good agreement with those expected from the acid structures.

Fini et al. (1987)

Potentiometric titration, uncertainties:

A procedure is presented for estimation of uncertainty in measurement of the p$K_{(a)}$ of a weak acid by potentiometric titration. The procedure is based on the ISO GUM. The core of the procedure is a mathematical model that involves 40 input parameters. A novel approach is used for taking into account the purity of the acid, the impurities are not treated as inert compounds only, and their possible acidic dissociation is also taken into account. Application to an example of practical p$K_{(a)}$ determination is presented. Altogether, 67 different sources of uncertainty are identified and quantified within the example. The relative importance of different uncertainty sources is discussed. The most important source of uncertainty (with the experimental set-up of the example) is the uncertainty of the pH measurement followed by the accuracy of the burette and the uncertainty of weighing. The procedure gives uncertainty separately for each point of the titration curve. The uncertainty depends on the amount of the titrant added, being lowest in the central part of the titration curve. The possibilities of reducing the uncertainty and interpreting the drift of the p$K_{(a)}$ values obtained from the same curve are discussed.

Koort et al. (2004)

Potentiometric, comparisons:

The pK_a and log P values of 23 structurally diverse compounds, including well-known drugs and two pharmacons under development, were determined by potentiometry. Also, the log P data were measured by the shake-flask method. Many of the samples were investigated at both of the participating laboratories in order to evaluate the reproducibility of the pH-metric log P technique. The interlaboratory evaluation of pK_a and log P data obtained by potentiometry showed excellent agreement (average delta pK_a = ±0.02 and delta log P = ±0.07). The log P values obtained by the two different methods, ranging from -1.84 to 5.80 (nearly eight orders of magnitude), were in very good concordance, as shown by the linear regression analysis: log PpH-metric = 0.9794 log P shake-flask $- 0.0397$ ($r = 0.9987$, $s = \pm 0.091$, $F = 8153$). The advantages of potentiometric log P determination are discussed.

Takacs-Novak and Avdeef (1996)

(Continued)

Appendix 1: Some Representative Studies on pK_a and Log P Determination Reported in the Literature (*Continued*)

Reported study	References
Potentiometric, organic solvent–water mixtures: The apparent acid dissociation constants $(p(s)K_a)$ of two water-insoluble drugs, ibuprofen and quinine, were determined pH-metrically in ACN water, dimethyl-formamide water, DMSO water, 1,4-dioxane-water, ethanol water, ethylene glycol-water, methanol water, and tetrahydrofuran water mixtures. A glass electrode calibration procedure based on a four-parameter equation (pH = alpha + SpcH + jH[H$^+$] + jOH[OH$^-$]) was used to obtain pH readings based on the concentration scale (pcH). We have called this four-parameter method the Four-Plus technique. The Yasuda Shedlovsky extrapolation $(p(s)K_a + \log [H_2O] = A/\epsilon + B)$ was used to derive acid dissociation constants in aqueous solution (pK_a). It has been demonstrated that the pK_a values extrapolated from such solvent–water mixtures are consistent with each other and with previously reported measurements. The suggested method has also been applied with success to determine the pK_a values of two pyridine derivatives of pharmaceutical interest.	Avdeef et al. (1999)
Spectrometric, ultraviolet (UV): The 2-methyl-5(4)-nitromidazole (BH) is an ampholyte and an intermediate of metronidazole. The amphoteric dissociation constant of BH is determined by equimolar UV-spectrophotometry. The acidic dissociation constant pK_a is 9.64 ± 0.04, and the basic dissociation constant pK_b is 12.98 ± 0.02.	Hua Xi et al. (1999)
Spectrophotometric and potentiometric methods: In order to establish the dissociation constants of organic compounds, the spectrophotometric and potentiometric methods are the most precise and useful ones. Comparing the results, we used both methods for three derivatives of 2-mercapto-5-R-amino-1,3,4-thiadiazole. For the pK_a determination by the spectrophotometric method, we measured the ratio between the concentration of dissociated and undissociated forms. The ratio was calculated from spectral data. The potentiometric method for pK_a consists of measuring the pH values within a potentiometric titration with 0.1 M NaOH. The results proved that the derivatives had a very low acidic character. The pK_a values were influenced by the properties of the amino group substituents, and the results confirm the theoretical considerations. The study confirms the thion-thiolic tautomery of the 2-mercapto-5-R-amino-1,3,4-thiadiazole derivates and their property of being ligands for the coordination of the cations of some representative and transitional metals, with applications in the preconcentration, detection, and the quantitative determination of polluting and toxic cations in environmental analysis.	Roman et al. (1998)
Spectrophotometric, multiwavelength (WApH): A WApH titration method was used for the determination of acid dissociation constants (pK_a values) of ionizable compounds. Microspeciation was investigated by three approaches: (*i*) selective monitoring of ionizable group by spectrophotometry, (*ii*) deductive method, and (*iii*) $k(z)$ method for determination of tautomeric ratio from co-solvent mixtures. It has been shown that the WApH technique, for such types of ampholytes, is able to deduce the microconstants and tautomeric ratios, which are in good agreement with the literature data.	Takacs-Novak and Tam (2000)

WApH:

pK_a values determined by the WApH technique for six ionizable substances, namely, benzoic acid, phenol, phthalic acid, nicotinic acid, p-aminosalicylic acid, and phenolphthalein are in excellent agreement with those measured pH-metrically. We have demonstrated that the first derivative spectrophotometry procedure provides a relatively simple way to visualize the pK_a values, which are consistent with those determined using the target factor analysis method. However, for ionization systems with insufficient spectral data obtained around the sought pK_a values or with closely overlapping pK_a values, the target factor analysis method outperforms the first derivative procedure in terms of obtaining the results. Using the target factor analysis method, it has been shown that the two-step ionization of phenolphthalein involves a colorless anion intermediate and a red colored di-anion.

Tam and Takacs-Novak (1999)

Structure–activity analysis:

The pK_a and therefore the physicochemical properties of a series of aminosulfonate-based compounds of phenol is correlated with the irreversible inhibition of the enzyme estrone sulfatase (ES). A strong correlation exists between the observed pK_a and inhibitory activity. The stability of the phenoxide ion, as indicated by the acid dissociation constant, is an important factor in the irreversible inhibition of this enzyme.

Ahmed et al. (2001)

Titrimetric and spectrophotometric methods, microscale:

This study describes the adaptation of conventional titrimetric and spectrophotometric techniques to a microscale for the determination of drug ionization constants (pK_a) and partition coefficients (log P). The apparatus for determining pK_a and compound purity (or equivalent weight) consists of a three-port conical glass microvial maintained at 25°C, a pH microelectrode, and a microinjection pump equipped with a 10 μL gastight syringe for titrant delivery. Sample mixing and protection from atmospheric CO_2, which is particularly important at the microscale, is accomplished using a fine stream of water-saturated N_2 bubbles. Simple titrimetric procedures combined with ionic equilibria models, which allow the accurate determination of pK_a and purity (or equivalent weight) using sample sizes in the microgram range and solution volumes of 10–100 μL were developed and validated using acetic acid and trometamine. Simultaneous determinations of pK_a, purity or equivalent weight, and octanol/water partition coefficient were shown to be possible from a single sample of a test solute by adapting the pH-metric technique to a microscale. Using benzoic acid as a model compound, a pK_a of 4.24 and octanol/water partition coefficient of 64 were obtained, in close agreement with the literature values. The principles employed in titrimetric analysis were also applied to demonstrate the spectrophotometric determination of benzoic acid's pK_a and partition coefficient using only 6 μg of the compound. The microscale titration method was then used to determine the two pK_a values of an "unknown" diprotic acid containing a carboxyl and an aromatic SH group. The phenyl thiol pK_a was confirmed using the microscale spectrophotometric procedure.

Morgan et al. (1998)

(Continued)

Appendix 1: Some Representative Studies on pK_a and Log P Determination Reported in the Literature (*Continued*)

Reported study	References
Titrimetry, multiprotic compounds: A generalized, weighted, nonlinear least squares procedure is developed, based on pH titration data, for the refinement of octanol-water partition coefficients (log P) and ionization constants (pK_a) of multiprotic substances. Ion-pair partition reactions, self-association reactions forming oligomers, and formations of mixed-substance complexes can be treated with this procedure. The procedure allows for CO_2 corrections in instances where the base titrant may have CO_2 as an impurity. Optionally, the substance purity and the titrant strength may be treated as adjustable parameters. The partial differentiation in the Gauss-Newton refinement procedure is based on newly derived analytical expressions. The new procedure was experimentally demonstrated with benzoic acid, 1-benzylimidazole, ($\pm$)-propranolol, and mellitic acid (benzenehexacarboxylic acid, AH6).	Avdeef (1993)
Yasuda-Shedlovsky equation: The aqueous pK values [(w/w)$pK_{(a)}$] of several sparingly water-soluble drugs with amino groups have been calculated from pK values determined in several methanol/water mixtures [(s/s)$pK_{(a)}$] by means of the Yasuda-Shedlovsky equation and by a linear equation that relates [(w/w)$pK_{(a)}$] of amino compounds with [(s/s)$pK_{(a)}$] obtained in any particular methanol/water mixture. Parameters of this last equation for amino compounds can be easily calculated from solely the methanol content of the solvent. Results from both approaches are consistent. However, the Yasuda-Shedlovsky equation requires several [(s/s)$pK_{(a)}$] determinations in solutions with different methanol contents and can be used only from measurements in solutions with a maximum methanol content of about 65% in weight. In contrast, the linear proposed equation allows a very good estimation of [(w/w)$pK_{(a)}$] from only one experimental [(s/s)$pK_{(a)}$] value, and it permits this estimation from [(s/s)$pK_{(a)}$] determined in a solution very rich in methanol. Therefore, it is suitable for highly insoluble compounds. The examined amino compounds cover a wide range [(w/w)$pK_{(a)}$] from 6.7 to 10.6, which have been estimated from experimental [(s/s)$pK_{(a)}$] values in the whole composition range.	Ruiz et al. (2003)

5 Release, Dissolution, and Permeation

INTRODUCTION

Newly discovered lead compounds that are ultimately formulated into drug delivery systems should be capable of existing either in a molecular dispersion, such as solutions or in an aggregate state, such as tablets, capsules, suspensions, and so on that are readily rendered into finer state of dispersion and dissolution. Regardless of the stage of aggregation in the final formulation, the active pharmaceutical ingredient (API) must be released from the drug delivery system and as the first step, should be dissolved in an aqueous environment; this will then be followed possibly by one or more transfers across nonaqueous barriers. The scope of preformulation studies has changed significantly over the past couple of decades with the availability of techniques to study the release and permeation characteristics of lead compounds. Whereas the design of drug delivery systems can alter release characteristics to some extent, the basic permeation characteristics remain an innate property based on the physico-chemical nature of the drug.

In this chapter, the preformulation studies that follow the basic studies of pK_a, log P, and solubility are discussed in chapter 4. In the recent years, several new approaches to study permeation have been developed that can be of great use in assessing the lead compound potential. The need to study release and permeation patterns has become a pivotal part of preformulation studies since the United States Food and Drug Administration (U.S. FDA) adopted a system of classification of drugs: the Biopharmaceutics Classification System (BCS), which was suggested by Gordon Amidon and his group in 1995, allows prediction of in vivo pharmacokinetic performance of drug products from measurements of permeability (determined as the extent of oral absorption) and solubility, and thus forms the scientific basis to allow waiver of in vivo bioavailability and bioequivalence testing of immediate-release solid dosage forms for high-permeability drugs that also exhibit rapid dissolution (1). Later in the chapter, a detailed discussion will be presented regarding this classification system.

RELEASE

Drug absorption depends on the release of the drug substance from the drug product (dissolution), the solubility, and the permeability across the gastrointestinal tract. The release characteristics of a drug delivery system are often determined by the manufacture of the product and highly affected by drug solubility, which also affects dissolution rates. The release step is followed by dissolution of the active ingredient. Dissolution of a pure substance follows the classic Noyes Whitney equation:

$$dc/dt = kS(Cs - Ct) \tag{1}$$

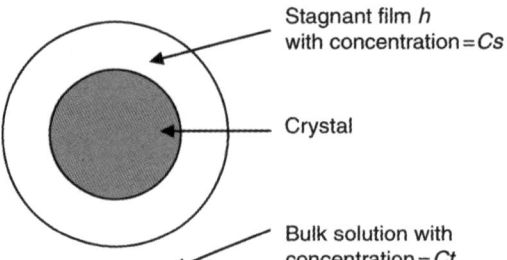

Figure 1 Diffusion layer model of dissolution.

where dc/dt is the rate of dissolution, k is the dissolution rate constant, S is the surface area of the dissolving solid, Cs is the saturation concentration of drug in the diffusion layer, and Ct is the concentration of drug in dissolution media (or the bulk).

This equation is of great value in the formulation studies wherein increase in the surface area of aggregates is the most powerful tool to optimize dissolution. The innate property in the equation that is subject to much of preformulation work refers to the solubility of the compound. In dissolution theory, it is assumed that an aqueous diffusion layer or stagnant liquid film of thickness h exists at the surface of a solid undergoing dissolution, as observed in Figure 1. This thickness h represents a stationary layer of solvent in which the solute molecules exist in concentrations from Cs to C. Beyond the static diffusion layer, at x greater than h, mixing occurs in the solution, and the drug is found at a uniform concentration, C, throughout the bulk phase (Fig. 1).

The diffusion layer model of dissolution assumes that the dissolution of drug at the solid/liquid interface into a concentrated layer surrounding the solid particle is more rapid than the diffusion of dissolved drug from that layer into the bulk solution. This diffusion is therefore rate-limiting in observed dissolution. As diffusion involves kinetic energy, it is highly dependent on the temperature. For an ideal solution, no heat is absorbed or given off upon dissolution; however for a real solution, the heat of the solution (ΔH) can be either negative (heat is given off) or positive (heat is absorbed). The mathematical relationship of solubility (Cs) to temperature is:

$$\log Cs = (-\Delta H/2.303\, RT) + \text{constant} \qquad (2)$$

where R is the gas constant and T is the absolute temperature. A plot of log Cs versus $1/T$ gives the value of the constant. A heat effect depends on whether the material absorbs heat (an endothermic process) or gives off heat (an exothermic process) when it dissolves. Most materials absorb heat as they dissolve. According to the La Chatelier's Principle, a system at equilibrium will adjust in such a manner as to reduce external stress. Therefore, if a substance absorbs heat when it dissolves and heat is added to the system, equilibrium can be restored, that is, the external stress can be reduced, by the absorption of heat. This can only be done in such a system by the dissolution of more of the substance, that is, an increase in solubility at the higher temperature until the equilibrium is restored.

The thermodynamic driving force for dissolution is therefore the heat of solution of the substance. For a crystalline solid, this represents the difference between the heat of sublimation of the compound and the heat of hydration of the ions. The heat of sublimation is the heat required to bring ions from the solid state to the gaseous state and is a measure of the energy required to pull apart the crystalline lattice. The heat of hydration is the heat given off by the hydration of those ions. For dissolution to be an endothermic process, the heat of sublimation is greater than the heat of hydration and ΔH is positive, that is, the heat is absorbed upon dissolution; therefore, solubility increases with an increase in temperature. If heat of sublimation is equal to the heat of hydration, solubility is independent of temperature.

As discussed in the previous chapter, the solubility of ionizable compounds is pH-dependent. For weak acids, as pH decreases, the solubility decreases. At equilibrium:

$$[HA]\,\text{solid} \longleftrightarrow [HA]\,\text{solution} \tag{3}$$

while the molar solubility (So) remains unchanged as a function of pH. The equilibrium dissociation constant is:

$$K_a = [H_3O^+] * [A^-]/[HA] \tag{4}$$

or

$$[A^-] = K_a * [HA]/[H_3O^+] = K_a * So/[H_3O^+] \tag{5}$$

The total solubility (S) is expressed as:

$$S = [A^-] + [HA] \tag{6}$$

or

$$S = K_a * So/[H_3O^+] + So \tag{7}$$

or

$$S - So = K_a * So/[H_3O^+] \tag{8}$$
$$\log(S - So) = \log K_a + \log So - \log[H_3O^+] \tag{9}$$
$$\log[(S - So)/So] = -pK_a + pH \tag{10}$$

Plotting $\log[(S - So)/S]$ versus pH (Fig. 2) allows experimental determination of S and So.

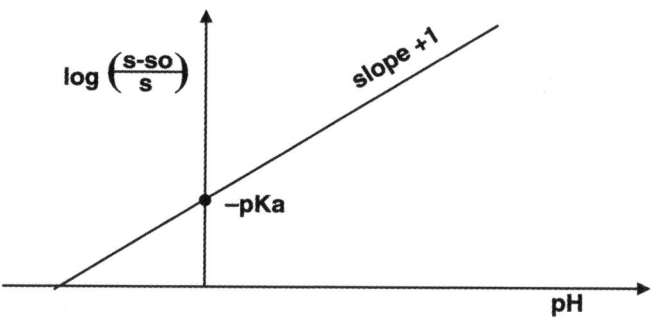

Figure 2 Calculation of solubility as a function of pH.

Therefore, because So is assumed to remain constant and pK_a is a constant, if pH decreases, the value of S must also decrease. In a similar manner, the solubility of a weak base decreases as pH increases.

There are several new ways emerging to measure the solubility of ionizable drugs, such as using the method called CheqSol® (chasing equilibrium solubility). Not only does CheqSol measure equilibrium and kinetic solubility rapidly and accurately, it also provides insights into compound behavior that will be of value for the better understanding of drug bioavailability, modeling of precipitation processes, and for investigating changes of crystalline form in suspensions.

The solubility of drug is an important physico-chemical property because it affects the bioavailability of the drug, the rate of drug release into the dissolution medium, and consequently, the therapeutic efficacy of the pharmaceutical product. The solubility of a molecule in various solvents is determined as a first step. This information is valuable in developing a formulation. Solubility is usually determined in a variety of commonly used solvents and some oils if the molecule is lipophilic.

The solubility of a material is usually determined by the equilibrium solubility method, which employs a saturated solution of the material, obtained by stirring an excess of the material in the solvent for a prolonged period until equilibrium is achieved. Common solvents used for solubility determination are:

- water
- polyethylene glycols
- propylene glycol
- glycerin
- sorbitol
- ethyl alcohol
- methanol
- benzyl alcohol
- isopropyl alcohol
- tweens
- polysorbates
- castor oil
- peanut oil
- sesame oil
- buffers at various pHs

Solubility Modulation

Poorly soluble compounds represent an estimated 60% of compounds in development and many major marketed drugs. It is important to measure and predict solubility and permeability accurately at an early stage, and interpret these data to help assess the potential for the development of the candidates. This requires developing an effective strategy to select the most appropriate tools to examine and improve solubility in each phase of development, optimization of solid-state approaches to enhance solubility including the use of polymorphs, co-crystals, and amorphous solids. All of these would affect the dissolution rates and bioavailability that can be studied with nanocrystal technology.

With this trend of increasingly insoluble drugs stretching resources, many companies are now re-evaluating their strategy. They know that there are many available technologies to measure and predict and finally improve solubility, and several new techniques are emerging. Studies that encompass this scope would include how membrane permeation of drugs can be enhanced by means of solubilizing agents, how the solid state is characterized and modified to improve solubility and drug performance, how salt screening and selection can impact on dissolution rate and oral absorption, applying nanocrystal technology to increase dissolution rate, and analyzing the use of pharmaceutical co-crystals in enhancing drug properties.

Many different approaches have been developed to overcome the solubility problem of poorly soluble drugs, for example, solubilization, inclusion compounds, and complexation. A basic disadvantage in these formulation approaches is that these can only be applied to a certain number of drugs exhibiting special features required for implementing the formulation principle (e.g., molecule fits into the cavity of the cyclodextrin ring). The use of solvent mixtures is also very limited due to toxicological considerations. In addition, more and more newly developed drugs are poorly soluble in aqueous media and simultaneously in organic media, thus excluding the use of solvent mixtures. Ideally, the formulation principle should be able to be applied to all or at least most of the poorly soluble drugs.

Solubilizers (e.g., organic solvents, detergents, and Pluronics) are often used to solubilize drugs in aqueous solution without considering their effects on biological systems, such as lipid membranes and multidrug resistance (MDR) efflux transporters (e.g., P-glycoprotein or MDR1). Liposomal solubilization is an effective approach for the delivery of potent, insoluble drug candidates. An alternative to other methods developed is the production of drug nanoparticles by high-pressure homogenization either as pearl milling or as the continuous high-pressure homogenization. Of importance is the consideration of metallic contamination during fast speed milling processes to keep it less than 1 ppm. Drug nanoparticles are produced by the dispersion of drug powder in an aqueous surfactant solution; the obtained presuspension is passed through a high-pressure piston-gap homogenizer, for example, 5–20 homogenization cycles at typically 1000–1500 bars and works on the principle that cavitation occurs in the aqueous phase. The particle suspension has a very high flow velocity when passing the tiny gap of the homogenizer, the static pressure on the water decreases below the vapor pressure of water, the water starts boiling at room temperature leading to the formation of gas bubbles, and at the exit of the gap the gas bubbles implode. The implosion shock waves disintegrate the drug particles to drug nanoparticles. Further improvement on nanoparticle production includes homogenization in nonaqueous phases or with reduced water content to produce more pronounced cavitation at higher temperatures. The chemical stability of the drugs is less impaired when homogenizing in nonaqueous or water-reduced media at low temperatures. The drug powder is dispersed in a nonaqueous medium (e.g., PEG 600, Miglyol 812) or a water-reduced mixture (e.g., water-ethanol), and the presuspension is homogenized in a piston-gap homogenizer. A suitable machine for lab scale is the Micron Lab 40 (APV Deutschland GmbH, Lübeck, Germany). Ostwald ripening occurs because of different saturation solubilities in the vicinity of very small and large particles. The particles produced are relatively homogeneous. The differences in the size in combination with the generally poor solubility of the drug nanoparticles are sufficiently low to avoid Ostwald ripening. Aqueous drug nanoparticle suspensions generally prove to be physically stable for several years.

The application of micronization and nanonization increases the surface area leading to an increased dissolution rate according to the Noyes-Whitney equation. However, this is only one aspect. The dissolution pressure is a function of the curvature of the surface that is much stronger for a curved surface of nanoparticles. For a size less than approximately 1–2 μm, the dissolution pressure increases distinctly leading to an increase in saturation solubility. In addition, the diffusional distance h on the surface of drug nanoparticles is decreased, thus leading to an increased concentration gradient $(Cs - Cx)/h$. The increase in surface area and concentration gradient leads to a greater increase in the dissolution velocity compared with a micronized product. In addition, the saturation solubility is increased as well, even though it is a thermodynamic parameter; the increase in solubility occurs as the supersaturation stage is reached. Saturation solubility and dissolution velocity are important parameters affecting the bioavailability of orally administered drugs. From this, nanoparticles have the potential to overcome these limiting steps.

Nanoparticle-based products are likely to have some unique characteristics: general adhesiveness of nanoparticles to the gut wall, adhesion to the gut wall being a reproducible process thus minimizing variation in drug absorption, increase in dissolution velocity overcoming this rate-limiting step, and an additional increase in the saturation solubility leading to an increased concentration gradient between gut and blood. Orally administered drug nanoparticles can increase the bioavailability and can be the only tool available to achieve a sufficient bioavailability with poorly soluble drugs. However, the possibility of faster absorption may have its own drawbacks, both from pharmacology and stability in the gut. For intravenous administration, the drug nanoparticles should possess a bulk population in the nanometer range by simultaneously having a low microparticle content, that is, especially particles larger than 5 μm, which can cause capillary blockade. The homogenization process yields a product with a minimized content of particles larger than 1 μm. Intravenous administration of drug nanoparticles allows achievement of sufficient blood levels and finds good application in the evaluation of new compounds. In addition, toxicologically critical excipients, such as cremophor EL used in taxol formulations can be avoided when stabilizing the drug nanoparticles with accepted emulsifiers, for example, lecithin or Tween 80. It is interesting to note that when taxol is administered with cremophor EL, the pharmacokinetics of the drug turns out to be nonlinear. For intravenous administration, a small particle size of less than 150 nm is only desirable in cases where one wants to pass fenestrated endothelia (e.g., treatment of tumors); however, this is a very limited case. More realistic and short-term achievable goals are passive targeting of drugs to treat mononuclear phagocytic system (MPS) infections (i.e., targeting the macrophages, e.g., treatment of *Mycobacterium tuberculosis* and *Mycobacterium avium* infections, especially in HIV patients). Here, it is more desirable to have larger particles to ensure fast and efficient removal from the blood streams by the macrophages. Another therapeutic goal is the creation of stealth drug nanoparticles circulating in the blood, minimizing free drug concentration but simultaneously prolonging the drug release by slow dissolution. For this purpose, very small particles are not suitable because they will dissolve too fast. Another therapeutic goal is targeting the non-MPS targets, for example, the brain and the bone marrow.

The particle size should be customized depending on the therapeutic requirements and purpose. The nanoparticle suspensions are physically stable for a long period of time if they are stabilized by emulsifiers/polymers in optimized composition. However, aqueous suspensions might not be the most convenient dosage form for the patient. The nanoparticle suspension can be used as granulation fluid to produce tablets or as wetting liquid for pellet production. The dispersions can also be spray-dried to be filled into hard gelatin capsules or sachets. Drug nanoparticles produced in Peg 600 or Miglyol can directly be filled into soft gelatine capsules. Lyophilization of drug nanoparticles produced in water-reduced media can be used to produce fast dissolving delivery systems. For parenteral application, nanoparticles can be lyophilized and reconstituted prior to injection with isotonic media (e.g., water with glycerol). There are also other areas of application, for example, ocular delivery (prolonged retention time) or topical application (increased saturation solubility leading to increased diffusion pressure into skin).

ASSAY SYSTEMS

As the U.S. FDA has begun accepting recommendations for waiver of bioequivalence requirement, protocols proving extremely expensive in the drug development cycle, there is a greater need to develop surrogate models that might prove useful sometime in securing waivers for all classes of drugs. Generally, the methods available currently show that the complexity of assay is directly proportional to its correlation with the absorption of drugs in humans (Fig. 3). Studies that correlated log P with human absorption profile and the suitability or lead candidates are elaborated in chapter 4. In this chapter, we will examine more complex assay systems.

Drug transport across epithelial cell barriers, especially the human small intestine, is difficult to predict. The intestinal epithelial cell barrier is a sophisticated organ that has evolved over hundreds of millions of years to become a "smart," effective, and selective xenobiotic screen. Nevertheless, there is large interindividual variability in the intestinal transport of drugs. Genetic variability in key proteins is believed to be causal. There is a pressing need to better understand the key processes and how the system components interact at the molecular, cellular, and tissue levels to control drug transport and determine drug absorption in small intestines.

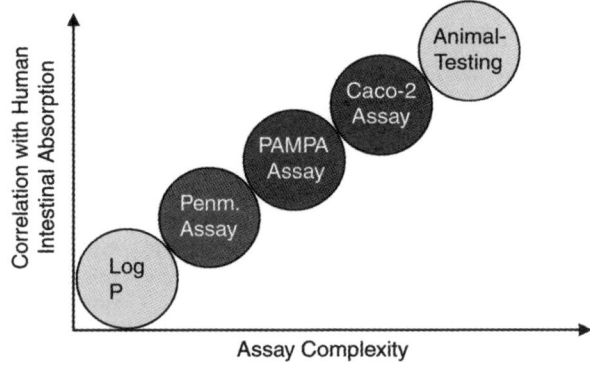

Figure 3 Assay complexity versus correlation with human absorption. Data from both complex biological and artificial permeation assays can provide valuable information regarding the absorption of a drug. *Source*: Courtesy of Millipore Corporation, Billerica, Massachusetts, U.S.A.

Is it feasible to construct an in silico framework to represent the drug absorption in small intestines at the cellular level with internal dynamic property and concert with the update molecular biochemical mechanism? This new generation of models and computational tools might integrate the available and emerging information at different levels to better account for and predict observed experimental results. Predicting aqueous solubility with in silico tools solubility is a key drug property. It is, however, difficult to measure accurately, especially for poorly soluble compounds, and thus numerous in silico models have been developed for its prediction. Some in silico models can predict aqueous solubility of simple, uncharged organic chemicals reasonably well; however, solubility prediction for charged species and drug-like chemicals is not very accurate. However, extrapolating solubility data to intestinal absorption from pharmacokinetic and physico-chemical data, and elucidating crucial parameters for absorption and the potential for improvement of bioavailability, are important at the preformulation stages.

The poor oral bioavailability of drugs is generally assumed to be due to physio-chemical problems, which result in poor solubility in the gastrointestinal (GI) tract or difficulty in diffusion through small intestine epithelial membrane. Furthermore, the biochemical process also contributes to oral bioavailability. The in vitro cell culture models of the intestinal epithelial cell barrier have evolved to become widely used experimental devices.

In the previous chapter, the log P factor was discussed in detail; in this chapter, we will examine other methods of testing transport across membranes.

Permeability Assays

The permeability assay uses an artificial membrane composed of hexadecane. Appendix 1 lists examples of how permeability values are assessed and correlated with solubility and the overall potential for a drug candidate to prove as a lead compound.

The automated systems comprise of multiwell systems is shown in Figure 4. The protocol for permeability assay is outlined below.

1. Into each well, add 15 µL of a 5% solution of hexadecane in hexane.
2. Dry for 45 minutes—one hour to ensure complete evaporation of hexane.
3. Add 300 µL of buffer with 5% dimethylsulfoxide (DMSO) at the desired pH to the acceptor plate.

(A) Lid

(B) Donor plate, 96 well filter plate

(C) Acceptor plate, 96-well plate insert in single well tray (supplied with plate) or PTFE acceptor plate

Figure 4 The 96-well permeability testing method. The support membrane is a 3-µm track etched polycarbonate of 10-µm thick; the artificial membrane in hexadecane and the recommended incubation time is four to six hours. Source: Courtesy of Millipore Corporation, Billerica, Massachusetts, U.S.A.

4. Place donor plate into the acceptor plate making sure that the underside of the membrane is in contact with the buffer.
5. Dissolve drugs of interest to the desired concentration. Add 150 µL of the drug at the desired concentration in 5% DMSO/phosphate buffer saline (PBS) at the desired pH to each well in the donor plate.
6. Cover and incubate at room temperature for four to six hours.
7. Transfer 100 µL/well from the donor plate and 250 µL/well from the acceptor plate to separate ultraviolet (UV)/visible (Vis) compatible plates and measure the UV/Vis absorption from 250 to 500 nm (SPECTRAmax® plate reader, Molecular Devices) for both plates.
8. Prepare drug solutions at the theoretical equilibrium (i.e., the resulting concentration if the donor and acceptor solutions were simply combined) and measure UV/Vis absorption from 250 to 500 nm for 250 µL/well of each.
9. Calculate log P_e and membrane retention using Eq. 11:

$$\log P_e = \log\left\{C* - \ln\left\{1 - \frac{[\text{drug}]_{\text{acceptor}}}{[\text{drug}]_{\text{equlibrium}}}\right\}\right\} \text{ where}$$

$$C = \left\{\frac{V_D * V_A}{(V_D + V_A) \text{ area} * \text{time}}\right\} \tag{11}$$

Parallel Artificial Membrane Permeability Analysis

Early drug discovery (ADME) (absorption and distribution in the body, metabolism and elimination from the body) assays, such as fast Caco-2 screens, can help in rejecting test compounds that lack good pharmaceutical profiles. A cost-effective, high-throughput method—parallel artificial membrane permeability analysis (PAMPA)—that uses a phospholipid artificial membrane that models the passive transport of epithelial cells, is becoming increasingly popular. The PAMPA assay utilizes a range of lipid components that model a variety of different plasma membranes. The support membrane is 0.45 µm hydrophobic polyvinylidene fluoride with a thickness of 130 µm and artificial membrane is lecithin in dodecane; recommended incubation time is 16–24 hours. The protocol is as follows.

1. Dissolve drugs of interest to the desired concentration.
2. Add 300 µL of the buffer with 5% DMSO at the desired pH to the acceptor plate.
3. Into each well add 5 µL of lipids in organic solvent (e.g., 2% lecithin in dodecane).
4. Add 150 µL of the drug at the desired pH and concentration in 5% DMSO/PBS to each well in the donor plate.
5. Place donor plate into the acceptor plate making sure that the underside of the membrane is in contact with the buffer. Steps three to five should be completed quickly, within 10 minutes.
6. Cover and incubate at room temperature for 16–24 hours.
7. Transfer 100 µL/well from the donor plate and 250 µL/well from the acceptor plate to separate UV/Vis compatible plates and measure the UV/Vis absorption from 250 to 500 nm (SPECTRAmax plate reader, Molecular Devices) for both plates.

8. Prepare drug solutions at the theoretical equilibrium (i.e., the resulting concentration if the donor and acceptor solutions were simply combined) and measure UV/Vis absorption from 250 to 500 nm for 250 µL/well of each.
9. Calculate log P_e and membrane retention using Eq. 10, described earlier.

The permeability and PAMPA assays as described are robust and reproducible assays for determining passive, transcellular compound permeability. Permeability and PAMPA are automation compatible, relatively fast (4–16 hours), inexpensive, straightforward, and their results correlate with human drug absorption values from published methods. The PAMPA assay provides the benefits of a more biologically relevant system. It is also possible to tailor the lipophilic constituents so that they mimic specific membranes, such as the blood–brain barrier (BBB). Optimization of incubation time, lipid mixture, and lipid concentration will also enhance the assay's ability to predict compound permeability.

Modifications of permeability and PAMPA systems have been reported, for example, using the pION PAMPA Evolution 96 System with double-sink and gut-box (2) as a new surrogate assay that predicts the GI tract absorption of candidate drug molecules at different pH conditions. Using Beckman Coulter's Biomek FX Single-Bridge Laboratory Automation Workstation PAMPA Assay System that features a 30-minute incubation time and an on-deck integrated gut-box and a SpectraMax microplate spectrophotometer, the permeability coefficients of drug standards with diverse physiochemical properties can be compared from both PAMPA and Caco-2 assays. It is automated using the Biomek FX Workstation.

These automated assays can be used for high-throughput ADME screening in early drug discovery. The double-sink PAMPA permeability assay mimics in vivo conditions by the use of a chemical sink in the acceptor wells and pH gradient in the donor wells. The use of the pION gut-box integrated on the deck has shortened the PAMPA assay incubation time to 30 minutes. The permeability coefficient and rank order correlate well with data obtained using the in vitro Caco-2 assay and in vivo permeability properties measured in rat intestinal perfusions.

Caco-2 Drug Transport Assays

Drug absorption generally occurs either through passive transcellular or paracellular diffusion, active carrier transport, or active efflux mechanisms. Several methods have been developed to aid in the understanding of the absorption of new lead compounds. The most common ones use an immortalized cell line (e.g., Caco-2, Madin-Darby canine kidney, and the like) to mimic the intestinal epithelium. These in vitro models provide more predictive permeability information than the artificial membrane systems (i.e., PAMPA and permeability assays, described previously) based on the cells' ability to promote (active transport) or resist (efflux) transport. Various in vitro methods are listed in the U.S. FDA guidelines. These are acceptable to evaluate the permeability of a drug substance, and includes a monolayer of suitable epithelial cells, and one such epithelial cell line that has been widely used as a model system of intestinal permeability is the Caco-2 cell line.

The kinetics of intestinal drug absorption, permeation enhancement, chemical moiety structure–permeability relationships, dissolution testing, in vitro/in vivo correlation, bioequivalence, and the development of novel polymeric materials are closely associated with the concept of Caco-2. As most drugs are known to absorb via intestines without using cellular pumps, passive permeability models came in the limelight. In a typical Caco-2 experiment, a monolayer of cells is grown on a filter separating two stacked micro well plates. The permeability of drugs through the cells is determined after the introduction of a drug on one side of the filter. The entire process is automated, and when used in conjunction with chromatography and/or mass spectroscopy (MS) detection, it enables any drug's permeability to be determined. The method requires careful sample analysis to calculate permeability correctly. Limitations of Caco-2 experiments are 21 days for preparing a stable monolayer and stringent storage conditions; however, tight-junction formation prior to use is the better choice. The villus in the small intestine contains more than one cell type; the Caco-2 cell line does not produce the mucus as observed in the small intestine and no P-450 metabolizing enzyme activity has been found in the Caco-2 cell line. Test compound solubility may pose a problem in Caco-2 assays because of the assay conditions. Finally, Caco-2 cells also contain endogenous transporter and efflux systems, the latter of which works against the permeability process and can complicate data interpretation for some drugs.

The Caco-2 cell line is heterogeneous and was derived from a human colorectal adenocarcinoma. Caco-2 cells are used as in vitro permeability models to predict human intestinal absorption because they exhibit many features of absorptive intestinal cells. This includes their ability to spontaneously differentiate into polarized enterocytes that express high levels of brush border hydrolases and form well-developed junctional complexes. Consequently, it becomes possible to determine whether passage is transcellular or paracellular based on a compound's transport rate. Caco-2 cells also express a variety of transport systems including di-peptide transporters and P-glycoproteins (Pgp). Due to these features, drug permeability in Caco-2 cells correlates well with human oral absorption, making Caco-2 an ideal in vitro permeability model. Additional information can be gained on metabolism and potential drug–drug interactions as the drug undergoes transcellular diffusion through the Caco-2 transport model.

Although accurate and well researched, the Caco-2 cell model requires a high investment of time and resources. Depending on a number of factors, including initial seeding density, culturing conditions, and passage number, Caco-2 cells can take as much as 20 days to reach confluence and achieve full differentiation. During this 20-day period, they require manual or automated exchange of media as frequently as every other day. The transport assays consume valuable drug compounds and normally require expensive, post-transport sample analyses (e.g., liquid chromatography (LC)/MS). Therefore, the use of the Caco-2 transport model in a high-throughput laboratory setting is only possible if the platform is robust, automation compatible, reproducible, and provides high-quality data that correlate well with established methodologies.

The Millipore MultiScreen Caco-2 assay system is a reliable 96-well platform for predicting human oral absorption of drug compounds (using Caco-2 cells or other cell lines whose drug transport properties have been well characterized). The MultiScreen system format is automation compatible and is designed to offer more cost-effective, high-throughput screening (HTS) of drugs than a 24-well system. The MultiScreen Caco-2 assay system exhibits good uniformity of cell growth and drug permeability across all 96 wells and low variability between production lots.

The plate design supports the use of lower volumes of expensive media and reduced amounts of test compounds. Using the MultiScreen Caco-2 assay system,

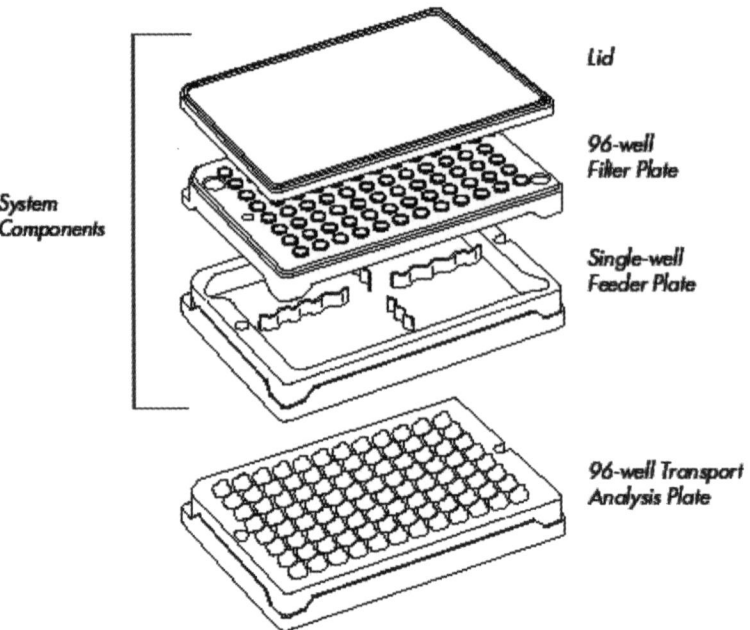

Figure 5 MultiScreen Caco-2 assay system. Components with single-well feeder plate and 96-well transport analysis plate. *Source*: Courtesy of Millipore Corporation, Billerica, Massachusetts, U.S.A.

standard drug compounds are successfully categorized as either "high" or "low" permeable, as defined by the FDA, and the permeability data correlate well with established human absorption values. The components of the Caco-2 assay system are shown in Figure 5.

The apparent permeability (P_{app}), in units of centimeters per second, can be calculated for Caco-2 drug transport assays using the following equation:

$$P_{app} = \left(\frac{V_A}{area * time}\right) * \left(\frac{[drug]_{acceptor}}{[drug]_{initial,donor}}\right) \tag{12}$$

where V_A is the volume (in mL) in the acceptor well, area is the surface area of the membrane (0.11 cm^2 for MultiScreen Caco-2 plate and 0.3 cm^2 for the 24-well plate), and time is the total transport time in seconds. For radio-labeled drug transport experiments, the counter per minute units obtained from the Trilux Multiwell Plate Scintillation Counter are used directly for the drug acceptor and the initial concentrations such that the formula becomes:

$$P_{app} = \left(\frac{V_A}{area * time}\right) * \left(\frac{CPM_{acceptor}}{CPM_{initial,donor}}\right) \tag{13}$$

Historically, it has been shown that a sigmoidal relationship exists between drug absorption rates as measured with the in vitro Caco-2 model and human absorption:

$$\% \text{ human absorption} = 100 \times \exp(a + b \times P_{app}) / [1 + \exp(a + b \times P_{app})] \quad (14)$$

Caco-2 cells are heterogeneous and their properties in final culture may differ based on the selection pressures of a particular laboratory. Direct comparison of compound permeability rates between laboratories is not possible unless the same Caco-2 cells and conditions are used. Therefore, transport rates and permeability classification ranges of specific drugs are expected to vary between reported studies. Most important is the ability to successfully classify compounds as low, medium, or high permeable drugs, and produce transport results that correlate to established human absorption values.

Several modifications of Caco-2 cell model have been tested; for example, CYP3A4-transfected Caco-2 cells are also used to define the biochemical absorption barriers. Oral bioavailability and intestinal drug absorption can be significantly limited by metabolizing enzymes and efflux transporters in the gut. The most prevalent oxidative drug-metabolizing enzyme present in the intestine is cytochrome P4503A4 (CYP3A4). Currently, more than 50% of the drugs in the market metabolized by P450 enzymes are metabolized by CYP3A4. Oral absorption of CYP3A4 substrates can also be limited by the MDR transporter Pgp, as there is extensive substrate overlap between these two proteins. Pgp is an ATP-dependent transporter on the apical plasma membrane of enterocytes that functions to limit the entry of drugs into the cell. There is a significant interaction between CYP3A4 and Pgp in the intestine. Although Caco-2 cells express a variety of uptake and efflux transporters found in the human intestine, a major drawback to the use of Caco-2 cells is that they lack CYP3A4. As such, no data regarding the importance of intestinal metabolism on limiting drug absorption can be obtained from normal Caco-2 cells. Caco-2 cells pretreated with 1,25-dihydroxyvitamin-D_3 (vitamin D_3) express higher levels of CYP3A4 compared with Caco-2, but still underestimate the amount of CYP3A4 in the human intestine. In the CYP3A4-transfected Caco-2 cells, P-gp can enhance the metabolism of orally dosed drugs by repeated cycling of the drug at the apical membrane.

Animal Model Testing

Whereas the quantity of substance available at the preformulation stages is generally small, in some instances, early animal testing for absorption potential are needed, particularly if the solid form of the new drug offers many options, such as amorphous forms, solvates, and so on. The absorption models used in animals are well described and will not be discussed here. Establishing good in vitro–in vivo correlation (IVIVC) at this stage proves useful because of limited access to sufficient compounds to run the entire absorption profiles. The IVIVC analysis can be made extensive or general conclusions can be drawn from limited studies; the choice depends on the amount of compound available and the nature or robustness of correlation observed.

In Vitro–In Vivo Correlation

The selection of a drug candidate marks the most crucial stage in the life cycle of drug development. Such selection is primarily based on the drug "developability" criteria, which include physico-chemical properties of the drug and the results obtained from preliminary studies involving several in vitro systems and in vivo animal models, which address efficacy and toxicity issues. During this stage, exploring the relationship between in vitro and in vivo properties of the drug in animal models provide an idea about the feasibility of the drug delivery system for a given drug. In such correlations, study designs including study of more than one formulation of the modified-release dosage forms and a rank order of release (fast/slow) of the formulations should be incorporated. Even though the formulations and methods used at this stage are not optimal, they prompt better design and development efforts in the future.

There are four levels of IVIVC that have been described in the FDA guidance, which include levels A, B, C, and multiple C.

- *Level A correlation*: This correlation represents a point-to-point relationship between in vitro dissolution and in vivo dissolution (input/absorption rate). Level A IVIVC is also viewed as a predictive model for the relationship between the entire in vitro release time course and entire in vivo response time course. In general, correlations are linear at this level. Although a concern of acceptable nonlinear correlation has been addressed, no formal guidance on the nonlinear IVIVC has been established. Level A correlation is the most informative and very useful from a regulatory perspective.
- *Level B correlation*: In Level B correlation, the mean in vivo dissolution or mean residence time is compared with the mean in vitro dissolution time by using statistical moment analytical methods. This type of correlation uses all of the in vitro and in vivo data; thus, it is not considered as a point-to-point correlation. This is of limited interest and use because more than one kind of plasma curve produces similar mean residence time.
- *Level C correlation*: This correlation describes a relationship between the amount of drug dissolved (e.g., percent dissolved in one hour) at one time point and one pharmacokinetic parameter [e.g., either area under the curve (AUC) or C_{max}]. Level C correlation is considered the lowest correlation level as it does not reflect the complete shape of the plasma concentration time curve. Similarly, a multiple Level C correlation relates one or more pharmacokinetic parameters to the percent drug dissolved at several time points of the dissolution profile and thus may be more useful. Levels B and C correlations can be useful in early formulation development, including the selection of the appropriate excipients, optimization of manufacturing processes, for quality control purposes, and characterization of the release patterns of newly formulated immediate-release and modified-release products relative to the reference.

The most basic IVIVC models are expressed as a simple linear equation between the in vivo drug absorption and in vitro drug dissolved (released):

$$Y(\text{in vivo absorbed}) = mX(\text{in vitro drug dissolved}) + C \qquad (15)$$

In this equation, m is the slope of the relationship, and C is the intercept. Ideally, $m = 1$ and $C = 0$, indicating a linear relationship. However, depending on the nature of the modified-release system, some data are better fitted using nonlinear models, such as Sigmoid, Weibull, Higuchi, or Hixson-Crowell.

In vivo release rate (X'_{vivo}) can also be expressed as a function of in vitro release rate ($X'_{rel,vitro}$) with parameters (a, b), which may be empirically selected and refined using appropriate mathematical processes:

$$X'_{vivo}(t) = X'_{rel,vitro}(a + bt) \tag{16}$$

An iterative process may be used to compute the time-scaling and time-shifting parameters. Integral to the model development exercise is model validation, which can be accomplished using data from the formulations used to build the model (internal validation) or using data obtained from a different (new) formulation (external validation). While internal validation serves the purpose of providing a basis for the acceptability of the model, external validation is superior and affords greater "confidence" in the model.

Generally, a plot of the fraction of drug absorbed (F_a) against the fraction drug dissolved (F_d) is made wherein the fraction absorbed is obtained by deconvoluting the plasma profile. Often the goal is to develop a profile that need not a priori be a linear or even a predefined function. For example,

$$F_a = \frac{1}{f_a}\left(1 - \frac{\alpha}{\alpha - 1}(1 - F_d) + \frac{1}{\alpha - 1}(1 - F_d)^\alpha\right) \tag{17}$$

where F_a is the fraction of the total amount of drug absorbed at time t, f_a is the fraction of the dose absorbed at $t = \#$, α is the ratio of the apparent first-order permeation rate constant (k_{paap}) to the first-order dissolution rate constant (k_d), and F_d is the fraction of the drug dose dissolved at time t.

Internal validation: Using the IVIVC model, for each formulation, the relevant exposure parameters (C_{max} and AUC) are predicted and compared with the actual (observed) values. The prediction errors are calculated using:

$$\text{Prediction error (\% PE)} = [(C_{max,observed} - C_{max,predicted})/$$
$$C_{max,observed}] \times 100 \tag{18}$$

The C_{max} can be replaced with corresponding AUC. The criteria set in the FDA guidance on IVIVC are as follows: for C_{max} and AUC, the mean absolute percent prediction error (% PE) should not exceed 10%, and the PE for individual formulations should not exceed 15%.

For establishing external predictability, the exposure parameters for a new formulation are predicted using its in vitro dissolution profile and the IVIVC model, and the predicted parameters are compared with the observed parameters. The PEs are computed as for the internal validation. For C_{max} and AUC, the PE for the external validation formulation should not exceed 10%. A PE of 10–20% indicates inconclusive predictability and illustrates the need for further study using additional data sets. For drugs with a narrow therapeutic index, external validation is required despite acceptable internal validation, whereas internal validation is usually sufficient with non-narrow therapeutic index drugs.

Several commercial software programs are available to study IVIVC; for example, PDx-IVIVC (3), which is a comprehensive IVIVC software program that performs deconvolution, calculating the fraction or percentage of drug absorbed and correlating it with in vitro fraction or percentage of dissolved data. It also allows Level C correlations (single or multiple), wherein a single-point relationship between a dissolution parameter, for example, percent dissoved in four hours and a pharmacokinetic parameter (e.g., AUC, C_{max}, T_{max}) is determined. A successful IVIVC model can be developed if in vitro dissolution is the rate-limiting step in the sequence of events leading to the appearance of the drug in the systemic circulation following oral or other routes of administration. Thus, the dissolution test can be utilized as a surrogate for bioequivalence studies (involving human subjects) if the developed IVIVC is predictive of in vivo performance of the product.

THE BIOPHARMACEUTICS DRUG CLASSIFICATION SYSTEMS

A biopharmaceutics drug classification scheme (BDCS) for correlating in vitro drug product dissolution and in vivo bioavailability is based on recognizing that drug dissolution and GI permeability are the fundamental parameters controlling the rate and extent of drug absorption. The BDCS is defined as:

Case 1. High solubility–high-permeability drugs,
Case 2. Low solubility–high-permeability drugs,
Case 3. High solubility–low-permeability drugs, and
Case 4. Low solubility–low-permeability drugs.

The testing methodology must be based on the physiological and physical–chemical properties controlling drug absorption. Generally, no IVIVC may be expected for rapidly dissolving (85% dissolved in less than 15 minutes) low-permeability drugs; where permeability is not a problem, even a simple one-point dissolution test is all that is needed to insure bioavailability. For slowly dissolving drugs, a dissolution profile is required with multiple time points in systems, which would include low pH, physiological pH and surfactants, and the in vitro conditions should mimic the in vivo processes (Table 1).

The BDCS should serve the needs of the earliest stages of discovery research where it can be useful in predicting routes of elimination, effects of efflux, and absorptive transporters on oral absorption, when transporter–enzyme interplay will yield clinically significant effects, such as low bioavailability and drug–drug interactions, the direction and importance of food effects, and transportor effects on postabsorption systemic levels following oral and intravenous doses.

Leads obtained through HTS tend to have higher molecular weights (MW) and greater lipophilicity than leads in the pre-HTS era. Poor absorption or permeation is more likely when there are more than five H-bond donors, ten H-bond acceptors, the MW is greater than 500, and the calculated log P is greater than five. This is also often referred to as Rule 5 of Lipinsky. However, Lipinsky specifically states that the Rule 5 only hold for compounds that are *not* substrates for active transporters. As almost all drugs are substrates for some transporter or the other, much remains to be studied about the Lipinsky's rule. In addition, unless a drug molecule can passively gain intracellular access, it is possible to simply investigate whether the molecule is a substrate for efflux transporters.

Several generalizations can be made about the interplay of transporters and the BDCS classification.

1. *Transporter effects are minimal for Class 1 compounds.* The high permeability/high solubility of such compounds allows high concentrations in the gut to saturate any transporter, both efflux and absorptive. Class 1 compounds may be substrates for both uptake and efflux transporters in vitro in cellular systems under the right conditions (e.g., midazolam and nifedipine are

Table 1 The Biopharmaceutics Drug Classification System as Defined by the Food and Drug Administration and Modified by Recent Findings

High solubility (e.g., when the highest dose strength is soluble in 250 mL or less of aqueous media over a pH range of 1–7.5 at 37°C) **High permeability** (e.g., absorption >90% compared with intravenous dose (drug + metabolite)	Low solubility Class 1 (generally about 8% of new leads): • High solubility • High permeability • Rapid dissolution for biowaiver • Route of elimination: metabolism, extensive • Transporter effects: minimal *Examples*: Abacavir; Acetaminophen; *Acyclovir*[b]; *Amiloride*[S,I]; Amitryptyline[S,I]; antipyrine; *Atropine*; **Buspirone**[c]; Caffeine; *Captopril*; Chloroquine[S,I]; **Chlorpheniramine**; Cyclophosphamide; Desipramine; **Diazepam**; **Diltiazem**[S,I]; **dihenhydramine**; Disopyramide; **Doxepin**; oxycycline; Enalapril; Ephedrine; Ergonovine; Ethambutol; Ethinyl estradiol; Fluoxetine[I]; Glucose; Imipramine[I]; Ketoprofen; **Ketorolac**; Labetolol; Levodopa[S]; Levofloxacin[S]; **Lidocaine**[I]; Lomefloxacin: **Meperidine**; Metoprolol; Metronidazole; **Midazolam**[S,I]; **Minocycline**; Misoprostol; **Nifedipine**[S]; Phenobarbital; Phenylalanine; Prednisolone; **Primaquine**[S]; Promazine; Propranolol; **Quinidine**[S,I]; **Rosiglitazone**; Salicylic acid; Theophylline; Valproic acid; **Verapamil**[I]; Zidovudine Class 2: • Low solubility • High permeability • Route of elimination: metabolism, extensive • Transporter: efflux transporter effects predominant *Examples*: **Amiodarone**[I]; **Atorvastatin**[S,I]; **Azithromycin**[S,I]; **Carbamazepine**[S,I]; **Carvedilol**; Chlorpromazine[I]; Ciprofloxacin[S]; **Cisapride**[S]; **Cyclosporine**[S,I]; **Danazol**; **Dapsone**; Diclofenac; Diflunisal; Digoxin[S]; *Erythromycin*[S,I]; Flurbiprofen; **Glipizide**; Glyburide[S,I]; Griseofulvin; Ibuprofen; **Indinavir**[S]; **Indomethacin**; **Itraconazole**[S,I]; **Ketoconazole**[I]; **Lansoprazole**[I]; **Lovastatin**[S,I]; *Mebendazole*; Naproxen; Nelfinavir[S]; Ofloxacin[I]; Oxaprozin; Phenazopyridine; Phenytoin[S]; Piroxicam; Raloxifene[S]; **Ritonavir**[S,I]; **Saquinavir**[S,I]; **Sirolimus**[S]; Sirolimus[S]; Spironolactone[I]; Spironolactone[I]; Talinolol[S]; **Tacrolimus**[S,I]; **Tamoxifen**[I]; Tacrolimus[S,I]; **Tamoxifen**[I]; **Terfenadine**[I]; Terfenadine[I]; Warfarin

(Continued)

Table 1 The Biopharmaceutics Drug Classification System as Defined by the Food and Drug Administration and Modified by Recent Findings (*Continued*)

	Class 3:	Class 4:
Low permeability	• High solubility • Low permeability • Route of elimination: renal and/or biliary elimination of unchanged drugs; metabolism is poor • Transporter: absorptive effects predominant Examples: Acyclovir; *Amiloride*[S,I]; Amoxicillin[S,I]; Atenolol; *Atropine*; Bidisomide; Bisphosphonates; Captopril; Cefazolin; Cetirizine; Cimetidine[S]; **Ciprofloxacin**[S]; Cloxacillin; Dicloxacillin[S]; **Erythromycin**[S,I]; Famotidine; Fexofenadine[S]; Folinic acid; *Furosemide*; Ganciclovir; *Hydrochlorothiazide*; Lisinopril; Metformin; *Methotrexate*; Nadolol; Penicillins; Pravastatin[S]; Ranitidine[S]; Tetracycline; Trimethoprim[S]; Valsartan; Zalcitabine	• Low solubility • Low permeability • Route of elimination: renal and/or biliary elimination of unchanged drug; metabolism is poor • Transporter: absorptive and efflux transporters can be predominant Examples: Amphotericin B; Chlorothiazide; Chlorthalidone; *Ciprofloxacin*[S]; Colistin; *Furosemide*; *Hydrochlorothiazide*; *Mebendazole*; *Methotrexate*; Neomycin

Notes: The compounds listed in *italic* are those that fall in more than one category according to different authors, which could be a result of the definition of the experimental conditions. The compounds listed in **bold** are primarily cytochrome P450 (CYP3A) substrates, where metabolism accounts for more than 70% of the elimination; superscript I and/or S indicate P-glycoprotein inhibitors and/or substrates, respectively. Classes 1 and 2 compounds are eliminated primarily via metabolism, whereas Classes 3 and 4 compounds are primarily eliminated unchanged into the urine and bile.

substrates for Ppg), but transporter effects will not be important clinically. It is thereofore possible that some compounds that should be considered Class 1 in terms of drug absorption and disposition are not Class 1 in BDCS due to the requirement of good solubility and rapid dissolution at low pH values. Such pH effects would not be limiting in vivo where absorption takes place from the intestine. Examples of this include the nonsteroidal anti-inflammatory drugs (NSAIDs) diclofenac, diflunisal, flurbiprofen, indomethacin, naproxen, and piroxicam; warfarin is almost completely bioavailable. In contrast, ofloxacin is listed as Class 2 because of its low solubility at pH 7.5.

2. *Efflux transporter effects will predominate for Class 2 compounds*. The high permeability of these compounds will allow ready access into the gut membranes and uptake transporters will have no effect on absorption, but the low solubility will limit the concentrations coming into the enterocytes, thereby preventing saturation of the efflux transporters. Consequently, efflux transporters will affect the extent of oral bioavailability and the rate of absorption of Class 2 compounds.

3. *Transporter-enzyme interplay in the intestines will be important primarily for Class 2 compounds that are substrates for CYP3A and Phase 2 conjugation enzymes*. For such compounds, intestinal uptake transporters will generally be unimportant due to the rapid permeation of the drug molecule into the enterocytes as a function of their high lipid solubility. That is, absorption of Class 2 compounds is primarily passive and a function of lipophilicity. However, due to the low solubility of these compounds, there will be little opportunity to saturate apical efflux transporters and intestinal enzymes, such as CYP3A4 and UDP-glucuronosyltransferases (UGTs). Thus, changes in transporter expression, and inhibition or induction of efflux transporters will cause changes in intestinal metabolism of drugs that are substrates for the intestinal metabolic enzymes. Note the large number of Class 2 compounds in Table 1 that are primarily substrates for CYP3A (compounds listed in bold) as well as substrates or inhibitors of the efflux transporter Pgp (indicated by superscripts S and I, respectively). Work in our laboratory has characterized this interplay in the absorptive process for the investigational cysteine protease inhibitor K77 and sirolimus, substrates for CYP3A and Pgp, and more recently for raloxifene, a substrate for UGTs and Pgp.

4. *Absorptive transporter effects will predominate for Class 3 compounds*. For Class 3 compounds, sufficient drug will be available in the gut lumen due to good solubility, but an absorptive transporter will be necessary to overcome the poor permeability characteristics of these compounds. However, intestinal apical efflux transporters may also be important for the absorption of such compounds when sufficient enterocyte penetration is achieved via an uptake transporter.

REFERENCES
1. http://www.fda.gov/cder/guidance/index.htm.
2. http://www.pion-inc.com/products.htm.
3. http://www.globomaxservice.com/pdxivivc.htm.

BIBILIOGRAPHY
Alsenz J, Haenel E. Development of a 7-day, 96-well Caco-2 permeability assay with high-throughput direct UV compound analysis. Pharm Res 2003 Dec; 20(12):1961–1969.

Ambudkar SV, Dey S, Hrycyna CA, Ramachandra M, Pastan I, Gottesman MM. Biochemical, cellular and pharmacological aspects of the multidrug transporter. Annu Rev Pharmacol Toxicol 1999; 39:361–398.

Amidon GL, Lennernas H, Shah VP, Crison JR. A theoretical basis for a biopharmaceutics drug classification: the correlation of in vitro drug product dissolution and in vivo bioavailability. Pharm Res 1995; 12:413–420.

Ano R, Kimura Y, Shima M, Matsuno R, Ueno T, Akamatsu M. Relationships between structure and high-throughput screening permeability of peptide derivatives and related compounds with artificial membranes: application to prediction of Caco-2 cell permeability. Bioorg Med Chem 2004, 2 Jan; 12(1):257–264.

Artursson P. Epithelial transport of drugs in cell culture. I: a model for studying the passive diffusion of drugs over intestinal absorptive (Caco-2) cells. J Pharm Sci 1990; 79:476–482.

Artursson P, Borchardt RT. Intestinal drug absorption and metabolism in cell cultures: Caco-2 and beyond. Pharm Res (NY) 1997; 14:1655–1658.

Artursson P, Karlsson J. Correlation between oral drug absorption in humans and apparent drug permeability coefficients in human intestinal epithelial (Caco-2) cells. Biochem Biophys Res Comm 1991; 175:880–885.

Artursson P, Palm K, Luthman K. Caco-2 monolayers in experimental and theoretical predictions of drug transport. Adv Drug Deliv Rev 2001; 46:27–43.

Avdeef A. Physicochemical profiling (solubility, permeability, and charge state). Curr Top Med Chem 2001 Sep; 1(4):277–351.

Avdeef A, Artursson P, Neuhoff S, Lazorova L, Grasjo J, Tavelin S. Caco-2 permeability of weakly basic drugs predicted with the double-sink PAMPA pK flux a method. Eur J Pharm Sci 2005 Mar; 24(4):333–349. Epub 2005, 20 Jan.

Avdeef A, Nielsen PE, Tsinman O. PAMPA—a drug absorption in vitro model 11. Matching the in vivo unstirred water layer thickness by individual-well stirring in microtitre plates. Eur J Pharm Sci 2004 Aug; 22(5):365–374.

Avdeef A, Strafford M, Block E, Balogh MP, Chambliss W, Khan I. Drug absorption in vitro model: filter-immobilized artificial membranes. 2. Studies of the permeability properties of lactones in Piper methysticum Forst. Eur J Pharm Sci 2001 Dec; 14(4):271–280.

Balimane PV, Pace E, Chong S, Zhu M, Jemal M, Pelt CK. A novel high-throughput automated chip-based nanoelectrospray tandem mass spectrometric method for PAMPA sample analysis. J Pharm Biomed Anal 2005, 21 Jun [Epub].

Benet LZ, Cummins CL. The drug efflux-metabolism alliance: biochemical aspects. Adv Drug Deliv Rev 2001; 50:S3–S11.

Benet LZ, Wu CY, Hebert MF, Wacher VJ. Intestinal drug metabolism and antitransport processes: a potential paradigm shift in oral drug delivery. J Control Release 1996; 39: 139–143.

Benet LZ, Cummins CL, and Wu CY. Transporter–enzyme interactions: implications for predicting drug–drug interactions from in vitro data. Curr Drug Metab 2003; 4:393–398.

Bergstrom CA. In silico predictions of drug solubility and permeability: two rate-limiting barriers to oral drug absorption. Basic Clin Pharmacol Toxicol 2005 Mar; 96(3):156–161.

Bergstrom CA, Strafford M, Artusson P. Absorption classification of oral drugs based on molecular surface properties. J Med Chem 2003; 46:558–570.

Bermejo M, Avdeef A, Ruiz A, et al. PAMPA—a drug absorption in vitro model 7. Comparing rat in situ, Caco-2, and PAMPA permeability of fluoroquinolones. Eur J Pharm Sci 2004 Mar; 21(4):429–441.

Bjornsson TD et al. The conduct of in vitro and in vivo drug–drug interaction studies: a PhRMA perspective. J Clin Pharmacol 2003; 43:443–469.

Blume HH Blume, Schut BS. The biopharmaceutics classification system (BCS): class III drugs—better candidates for BA/BE waiver? Eur J Pharm Sci 1999; 9:117–121.

Cestelli A, Catania C, D'Agostino S, et al. Functional feature of a novel model of blood–brain barrier: studies on permeation of test compounds. J Control Release 2001, 11 Sep; 76(1–2):139–147.

Chan EC, Tan WL, Ho PC, Fang LJ. Modeling Caco-2 permeability of drugs using immobilized artificial membrane chromatography and physicochemical descriptors. J Chromatogr A 2005, 29 Apr; 1072(2):159–168.

Chi-Yuan Wu, Leslie Z. Benet. Predicting drug disposition via application of BCS: transport/absorption/elimination interplay and development of a Biopharmaceutics Drug Disposition Classification System. Pharmaceut Res 2005 Jan; 22(1):11–23.
Chu I, Liu F, Soares A, Kumari P, Nomeir AA. Generic fast gradient liquid chromatography/tandem mass spectrometry techniques for the assessment of the in vitro permeability across the blood–brain barrier in drug discovery. Rapid Commun Mass Spectrom 2002; 16(15):1501–1505.
Corrigan OI. The biopharmaceutic drug classification and drugs administered in extended release (ER) formulations. In: Young D, DeVane J, Butler J, eds. In vitro–In vivo Correlations. Vol. 423. New York, NY: Plenum Press, 1997:111–128.
Crespi CL, Penman BW, Hu M. Development of caco-2 cells expressing high levels of cDNA-derived cytochrome P4503A4. Pharmaceut Res 1996; 13:1635–1641.
Crowe A, Bruelisauer A, Duerr L, Guntz P, Lemaire M. Absorption and intestinal metabolism of SDZ-RAD and rapamycin in rats. Drug Metab Dispos 1999; 27:627–632.
Crowe A, Lemaire M. In vitro and in situ absorption of SDZ-RAD using a human intestinal cell line (Caco-2) and a single pass perfusion model in rats: comparison with rapamycin. Pharm Res (NY) 1998; 15:1666–1672.
Cummins CL, Jacobsen W, Benet LZ. Unmasking the dynamic interplay between intestinal P-glycoprotein and CYP3A4. J Pharmacol Exp Ther 2002; 300:1036–1045.
Cummins CL, Mangravite LM, Benet LZ. Characterizing the expression of CYP3A4 and efflux transporters (P-gp, MRP1 and MRP2) in CYP3A4-transfected Caco-2 cells after induction with sodium butyrate and the phorbol ester 12-Otetradecanoylphorbol-13-acetate. Pharm Res (NY) 2001; 18:1102–1109.
Dash AK, Elmquist WF. Separation methods that are capable of revealing blood–brain barrier permeability. J Chromatogr B Analyt Technol Biomed Life Sci 2003, 25 Nov; 797(1–2):241–254.
Di L, Kerns EH. Profiling drug-like properties in discovery research. Curr Opin Chem Biol 2003 Jun; 7(3):402–408.
Di L, Kerns EH, Fan K, McConnell OJ, Carter GT. High throughput artificial membrane permeability assay for blood–brain barrier. Eur J Med Chem 2003 Mar; 38(3):223–232.
Dias VC, Yatscoff RW. Investigation of rapamycin transport and uptake across absorptive human intestinal cell monolayers. Clin Biochem 1994; 27:31–36.
Diez-Sales O, Guzman D, Cano D, Martin A, Sanchez E, Herraez M. A comparative "in vitro" study of permeability with different synthetic and biological membranes. Eur J Drug Metab Pharmacokinet 1991; Spec No 3:441–446.
Doppenschmitt S, Spahn-Langguth H, Regardh CG, Langguth P. Role of P-glycoprotein-mediated secretion in absorptive drug permeability: an approach using passive membrane permeability and affinity to P-glycoprotein. J Pharm Sci 1999 Oct; 88(10):1067–1072.
Dressman JB, Amidon GL, Fleisher D. Absorption potential: estimating the fraction absorbed for orally administered compounds. J Pharm Sci 1985; 74:588–589.
Dressman JB, Amidon GL, Reppas C, Shah VP. Dissolution testing as a prognostic tool for oral drug absorption: immediate release dosage forms. Pharm Res 1999; 15(1):11–22.
Drewe J, Guitard P. In vitro–in vivo correlation for modified-release formulations. J Pharm Sci 1993; 82(2):132–137.
Ecker GF, Noe CR. In silico prediction models for blood–brain barrier permeation. Curr Med Chem 2004 Jun; 11(12):1617–1628.
Faassen F, Vogel G, Spanings H, Vromans H. Caco-2 permeability, P-glycoprotein transport ratios and brain penetration of heterocyclic drugs. Int J Pharm 2003, 16 Sep; 263(1–2):113–122.
Fisher JM, Wrighton SA, Watkins PB, et al. First-pass midazolam metabolism catalyzed by 1 alpha, 25-dihydroxy vitamin D3-modified Caco-2 cell monolayers. J Pharmacol Exp Ther 1999; 289:1134–1142.
Food and Drug Administration. Guidance for Industry: Drug Metabolism/Drug Interaction Studies in the Drug Development Process: Studies In Vitro. Food and Drug Administration, Rockville, MD, 1997. Available at http://www.fda.gov/cder/guidance/index.htm.
Food and Drug Administration. Guidance for Industry: Waiver of In vivo Bioavailability and Bioequivalence Studies for Immediate-Release Solid Oral Dosage Forms Based on a

Biopharmaceutics Classification System. Food and Drug Administration, Rockville, MD, 2000. Available at http://www.fda.gov/cder/guidance/index.htm.

Fujikawa M, Ano R, Nakao K, Shimizu R, Akamatsu M. Relationships between structure and high-throughput screening permeability of diverse drugs with artificial membranes: application to prediction of Caco-2 cell permeability. Bioorg Med Chem 2005 1 Aug; 13(15):4721–4732.

Fung EN, Chu I, Li C, et al. Higher-throughput screening for Caco-2 permeability utilizing a multiple sprayer liquid chromatography/tandem mass spectrometry system. Rapid Commun Mass Spectrom 2003; 17(18):2147–2152.

Gaillard PJ, de Boer AG. Relationship between permeability status of the blood–brain barrier and in vitro permeability coefficient of a drug. Eur J Pharm Sci 2000 Dec; 12(2):95–102.

Galia E, Nicolaides E, Dressman JB. Evaluation of various dissolution media for predicting in vivo performance of class I and II drugs. Pharm Res 1998; 15:698–705.

Gan LS, Hsyu PH, Pritchard JF, Thakker D. Mechanism of intestinal absorption of ranitidine and ondansetron transport across Caco-2 cell monolayers. Pharm Res 1993; 10:1722–1725.

Ginski MJ, Polli JE. Prediction of dissolution-absorption relationships from a dissolution/Caco-2 system. Int J Pharm 1998; 177:117–125.

Ginski MJ, Taneja R, Polli JE. Prediction of dissolution-absorption relationships from a continuous dissolution/Caco-2 system. AAPS PharmSci 1999; 1(2):article 3. DOI: 10.1208/ps010203.

Hansen DK, Scott DO, Otis KW, Lunte SM. Comparison of in vitro BBMEC permeability and in vivo CNS uptake by microdialysis sampling. J Pharm Biomed Anal 2002, 1 Mar; 27(6):945–958.

Hatanaka T, Inuma M, Sugibayashi K, Morimoto Y. Prediction of skin permeability of drugs. I. Comparison with artificial membrane. Chem Pharm Bull (Tokyo) 1990 Dec; 38(12):3452–3459.

Hendriksen BA, Felix MV, Bolger MB. The composite solubility versus pH profile and its role in intestinal absorption prediction. AAPS PharmSci 2003; 5(1):E4.

Hidalgo IJ. Assessing the absorption of new pharmaceuticals. Curr Top Med Chem 2001 Nov; 1(5):385–401.

Hidalgo IJ, Raub TJ, Borchardt RT. Characterization of the human colon carcinoma cell line (Caco-2) as a model system for intestinal epithelial permeability, Gastroenterology 1989; 96:736–749.

Hilgars AR, Conradi RA, Burton PS. Caco-2 cell monolayers as a model for drug transport across the intestinal mucosa. Pharm Res 1990; 7:902–910.

Hovgaard L, Brondstad H. Drug delivery studies in Caco-2 monolayers. IV. Absorption enhancer effects of cyclodextrins. Pharm Res 1995; 12:1328–1332.

Humbert H, Cabiac MD, Bosshardt H. In vitro–in vivo correlation of a modified-release oral form of ketotifen: in vitro dissolution rate specification. J Pharm Sci 1994; 83:131–136.

Hwang KK, Martin NE, Jiang L, Zhu C. Permeation prediction of M100240 using the parallel artificial membrane permeability assay. J Pharm Sci 2003 Sep–Dec; 6(3):315–320.

Hwang SS, Bayne W, Theeuwes F. In vivo evaluation of controlled-release products. J Pharm Sci 1993; 82(11):1145–1150.

Hwang SS, Gorsline JJ, Louie J, Dye D, Guinta D, Hamel L. In vitro and in vivo evaluation of a once-daily controlled release pseudoephedrine product. J Clin Pharmacol 1995; 35:259–267.

Jacobsen W, Serkova N, Hausen B, Morris RE, Benet LZ, Christians U. Comparison of the in vitro metabolism of the macrolide immunosuppressants sirolimus and RAD. Transplant Proc 2001; 33:514–515.

Kalampokis A, Argyrakis P, Macheras P. A heterogeneous tube model of intestinal drug absorption based on probabilistic concepts. Pharm Res 1999 Nov; 16(11):1764–1769.

Kanfer I. Report on the International Workshop on the Biopharmaceutics Classification System (BCS): scientific and regulatory aspects in practice. J Pharm Sci 2002; 5:1–4.

Kansy M, Senner F, Gubernator K. Physicochemical high throughput screening: parallel artificial membrane permeation assay in the description of passive absorption processes. J Med Chem 1998; 41:1007–1010.

Kaplan B, Meier-Kriesche HU, Napoli KL, Kahan BD. The effects of relative timing of sirolimus and cyclosporine microemulsion formulation coadministration on the pharmacokinetics of each agent. Clin Pharmacol Ther 1998; 63:48–53.

Kariv I, Rourick RA, Kassel DB, Chung TD. Improvement of "hit-to-lead" optimization by integration of in vitro HTS experimental models for early determination of pharmacokinetic properties. Comb Chem High Throughput Screen 2002 Sep; 5(6):459–472.

Kasim NA, Whitehouse M, Ramachandran C, et al. Molecular properties of WHO essential drugs and provisional biopharmaceutical classification. Mol Pharmaceut 2004; 1:85–96.

Kerns EH, Di L, Petusky S, Farris M, Ley R, Jupp P. Combined application of parallel artificial membrane permeability assay and Caco-2 permeability assays in drug discovery. J Pharm Sci 2004 Jun; 93(6):1440–1453.

Kim RB, Wandel C, Leake B, et al. Interrelationship between substrates and inhibitors of human CYP3A and P-glycoprotein. Pharm Res (NY) 1999; 16:408–414.

Komura H, Kawahara I, Shigemoto Y, et al. High throughput screening of pharmacokinetics and metabolism in drug discovery (I)—establishment of assessment system for absorption to compounds with a wide diversity of physical properties. Yakugaku Zasshi 2005 Jan; 125(1):121–130.

Komura H, Shigemoto Y, Kawahara I, et al. High throughput screening of pharmacokinetics and metabolism in drug discovery (III)—investigation on in-silico model for membrane permeability and CYP1A2 inhibition. Yakugaku Zasshi 2005 Jan; 125(1):141–147.

Kramer SD, Wunderli-Allensbach H. Physicochemical properties in pharmacokinetic lead optimization. Farmaco 2001 Jan–Feb; 56(1–2):145–148.

Krause KP, Kayser O, Mäder K, Gust R, Müller RH. Heavy metal contamination of nanosuspensions produced by high-pressure homogenization. Int J Pharm 2000; 196:169–172.

Lasic DD, Ceh B, Stuart MC, Guo L, Frederik PM, Barenholz Y. Transmembrane gradient driven phase transitions within vesicles: lessons for drug delivery. Biochim Biophys Acta 1995, 1 Nov; 1239(2):145–156.

Lau YY, Wu C-Y, Okochi H, Benet LZ. Ex situ inhibition of hepatic uptake and efflux significantly changes metabolism: hepatic enzyme-transporter interplay. J Pharmacol Exp Ther 2004; 308:1040–1045.

Lennernas H. Human intestinal permeability. J Pharm Sci 1998; 87:403–410.

Lennernas H. Human jejunal effective permeability and its correlation with preclinical drug absorption models. J Pharm Pharmacol 1997; 49:627–638.

Lennernas H, Knutson L, Knutson T, et al. The effect of amiloride on the in vivo effective permeability of amoxicillin in human jejunum: experience from a regional perfusion technique. Eur J Pharm Sci 2002; 15:271–277.

Lennernas H, Palm K, Fagerholm U, Artursson P. Comparison between active and passive flux transport in human intestinal epithelial (Caco-2) cells in vitro and human jejunum in vivo. Int J Pharm 1996; 127:103–107.

Lindenberg M, Kopp S, Dressman JB. Classification of orally administered drugs on the World Health Organization Model of Essential Medicines according to the Biopharmaceutics Classification System. Eur J Pharm Biopharm 2004; 58:265–278.

Lipinski CA. Chris Lipinski discusses life and chemistry after the Rule of Five. Drug Discov Today 2003; 8:12–16.

Lipinski CA. Drug-like properties and the causes of poor solubility and poor permeability. J Pharmacol Toxicol Methods 2000 Jul–Aug; 44(1):235–249.

Lipinski CA et al. Experimental and computational approaches to estimate solubility and permeability in drug discovery and development settings. Adv Drug Deliv Rev 2001; 46:3–26.

Liu H, Sabus C, Carter GT, Du C, Avdeef A, Tischler M. In vitro permeability of poorly aqueous soluble compounds using different solubilizers in the PAMPA assay with liquid chromatography/mass spectrometry detection. Pharm Res 2003 Nov; 20(11):1820–1826.

Liu X, Tu M, Kelly RS, Chen C, Smith BJ. Development of a computational approach to predict blood–brain barrier permeability. Drug Metab Dispos 2004 Jan; 32(1):132–139.

Lobenberg R, Amidon GL. Modern bioavailability, bio-equivalence and biopharmaceutics classification system. New scientific approaches to international regulatory standards. Eur J Pharm Biopharm 2000; 50:3–12.

Lohmann C, Huwel S, Galla HJ. Predicting blood–brain barrier permeability of drugs: evaluation of different in vitro assays. J Drug Target 2002 Jun; 10(4):263–276.

Loo JC, Riegelman S. New method for calculating the intrinsic absorption rate of drugs. J Pharm Sci 1968; 57(6):918–924.

Mandagere AK, Thompson TN, Hwang KK. Graphical model for estimating oral bioavailability of drugs in humans and other species from their Caco-2 permeability and in vitro liver enzyme metabolic stability rates. J Med Chem 2002; 45:304–311.
Markowska M, Oberle R, Juzwin S, Hsu CP, Gryszkiewicz M, Streeter AJ. Optimizing Caco-2 cell monolayers to increase throughput in drug intestinal absorption analysis. Pharmacol Toxicol Methods 2001 Jul–Aug; 46(1):51–55.
Martinez MN, Amidon GL. A mechanistic approach to understanding the factors affecting drug absorption: a review of fundamentals. J Clin Pharmacol 2002; 42:620–643.
Marttin E, Verhoef JC, Merkus FW. Efficacy, safety and mechanism of cyclodextrins as absorption enhancers in nasal delivery of peptide and protein drugs. J Drug Target 1998; 6:17–36.
Masson M, Loftsson T, Masson G, Stefansson E. Cyclodextrins as permeation enhancers: some theoretical evaluations and in vitro testing. J Control Release 1999, 1 May; 59(1):107–118.
Mauger DT, Chinchilli VM. In vitro–in vivo relationships for oral extended-release drug products. J Biopharm Statist 1997; 7(4):565–578.
Mendell-Harary J, Dowell J, Bigora S, et al. Nonlinear in vitro–in vivo correlation. In: Young D, DeVane J, Butler J, eds. In vitro–In vivo Correlations. New York, NY: Plenum Press, 1997; 423:199–206.
Merisko-Liversidge E, Sarpotdar P, Bruno J, et al. Formulation and antitumor activity evaluation of nanocrystalline suspensions of poorly soluble anticancer drugs. Pharm Res 1996; 13(2):272–278.
Miret S, Abrahamse L, de Groene EM. Comparison of in vitro models for the prediction of compound absorption across the human intestinal mucosa. J Biomol Screen 2004 Oct; 9(7):598–606.
Modi NB, Lam A, Lindemulder E, Wang B, Gupta SK. Application of in vitro–in vivo correlation (IVIVC) in setting formulation release specifications. Biopharm Drug Dispos 2000; 21:321–326.
Müller RH, Böhm BHL. Nanosuspensions. In: Müller RH, Benita S, Böhm BHL, eds. Emulsions and Nanosuspensions for the Formulation of Poorly Soluble Drug. Stuttgar: Medpharm Scientific Publishers, 1998:149–174.
Nagahara N, Tavelin S, Artursson P. Contribution of the paracellular route to the pH-dependent epithelial permeability to cationic drugs. J Pharm Sci 2004 Dec; 93(12): 2972–2984.
Nielsen PE, Avdeef A. PAMPA—a drug absorption in vitro model 8. Apparent filter porosity and the unstirred water layer. Eur J Pharm Sci 2004 May; 22(1):33–41.
Noyes A, Whitney W. The rate of solution of solid substances in their own solutions. J Am Chem Soc 1897; 19:930–934.
Obata K, Sugano K, Machida M, Aso Y. Biopharmaceutics classification by high throughput solubility assay and PAMPA. Drug Dev Ind Pharm 2004 Feb; 30(2):181–185.
Ohwaki T, Ishii M, Aoki S, Tatsuishi K, Kayano M. Effect of dose, pH, and osmolarity on nasal absorption of secretin in rats. III. In vitro membrane permeation test and determination of apparent partition coefficient of secretion. Chem Pharm Bull (Tokyo) 1989 Dec; 37(12):3359–3362.
Pade V, Stavchansky S. Link between drug absorption solubility and permeability measurements in Caco-2 cells. J Pharm Sci 1998 Dec; 87(12):1604–1607.
Paine MF, Leung LY, Lim HK, et al. Identification of a novel route of extraction of sirolimus in human small intestine: roles of metabolism and secretion. J Pharmacol Exp Ther 2002; 301:174–186.
Park SB, You JO, Park HY, Haam SJ, Kim WS. A novel pH-sensitive membrane from chitosan–TEOS IPN; preparation and its drug permeation characteristics. Biomaterials 2001 Feb; 22(4):323–30.
Perez-Buendia MD, Gomez-Perez B, Pla-Delfina JM. Permeation mechanisms through artificial lipoidal membranes and effects of synthetic surfactants on xenobiotic permeability. Arzneimittelforschung 1993 Jul; 43(7):789–794.
Phillips J, Arena A. Optimization of Caco-2 Cell Growth and Differentiation for Drug Transport Studies. Millipore Corporation Protocol Note PC1060EN00, 2003.

Pinto M, Robin LS, Appay MD, et al. Enterocyte-like differentiation and polarization of the human colon carcinoma cell line Caco-2 in culture. Biol Cell 1983; 47:323–330.
Polli JE, Crison JR, Amidon GL. Novel approach to the analysis of in vitro–in vivo relationships. J Pharm Sci 1996; 85:753–760.
Polli JE, Ginski MJ. Human drug absorption kinetics and comparison to Caco-2 monolayer permeabilities. Pharm Res 1998; 15:47–52.
Polli JW, Wring SA, Humphreys JE, et al. Rational use of in vitro P-glycoprotein assays in drug discovery. J Pharmacol Exp Ther 2001; 299:620–628.
Polli L, JW Sanvordeker DR, Taneja R, et al. Summary workshop report: biopharmaceutics classification system—implementation challenges and extension opportunities. J Pharm Sci 2004; 3:1375–1381.
Popa-Burke IG, Issakova O, Arroway JD, et al. Streamlined system for purifying and quantifying a diverse library of compounds and the effect of compound concentration measurements on the accurate interpretation of biological assay results. Anal Chem 2004, 15 Dec; 76(24):7278–7287.
Radovanovic J, Duric Z, Jovanovic M, Ibric S, Petrovic M. An attempt to establish an in vitro–in vivo correlation: case of paracetamol immediate-release tablets. Eur J Drug Metab Pharmacokinet 1998; 23(1):33–40.
Rekhi GS, Eddington NE, Fossler MJ, Schwartz P, Lesko LJ, Augsburger LL. Evaluation of in vitro release rate and in vivo absorption characteristics of four metoprolol tartrate immediate-release tablet formulations. Pharm Dev Technol 1997; 2:11–24.
Rinaki E, Valsami G, Macheras P. Quantitative biopharmaceutics classification system: the central role of dose/solubility ratio. Pharm Res 2003 Dec; 20(12):1917–1925.
Ruell JA, Tsinman KL, Avdeef A. PAMPA—a drug absorption in vitro model. 5. Unstirred water layer in iso-pH mapping assays and pKa(flux)—optimized design (pOD-PAMPA). Eur J Pharm Sci 2003 Dec; 20(4–5):393–402.
Saitoh R, Sugano K, Takata N, et al. Correction of permeability with pore radius of tight junctions in Caco-2 monolayers improves the prediction of the dose fraction of hydrophilic drugs absorbed by humans. Pharm Res 2004 May; 21(5):749–755.
Smith DA. Design of drugs through a consideration of drug metabolism and pharmacokinetics. Eur J Drug Metab Pharmacokinet 1994; 3:193–199.
Strickley RG. Parenteral formulations of small molecules therapeutics marketed in the United States (1999)—Part I, PDA. J Pharm Sci Tech 1999; 53:324–349.
Suen Y, Tsinman K, Zhu Z, Threadgill G. Automation of a double-sink PAMPA permeability assay on the Biomek® FX Laboratory Automation Workstation. Pharm Discov 2005a, 1 June; xx.
Sugano K, Hamada H, Machida M, Ushio H. High throughput prediction of oral absorption: improvement of the composition of the lipid solution used in parallel artificial membrane permeation assay. J Biomol Screen 2001a, Jun; 6(3):189–196.
Sugano K, Hamada H, Machida M, Ushio H, Saitoh K, Terada K. Optimized conditions of bio-mimetic artificial membrane permeation assay. Int J Pharm 2001b, 9 Oct; 228(1–2):181–188.
Sugano K, Nabuchi Y, Machida M, Asoh Y. Permeation characteristics of a hydrophilic basic compound across a bio-mimetic artificial membrane. Int J Pharm 2004, 4 May; 275(1–2):271–278.
Sugano K, Takata N, Machida M, Saitoh K, Terada K. Prediction of passive intestinal absorption using bio-mimetic artificial membrane permeation assay and the paracellular pathway model. Int J Pharm 2002, 25 Jul; 241(2):241–251.
Sugaya Y, Yoshiba T, Kajima T, Ishihama Y. Development of solubility screening methods in drug discovery. Yakugaku Zasshi 2002 Mar; 122(3):237–246.
Sun D, Yu LX, Hussain MA, Wall DA, Smith RL, Amidon GL. In vitro testing of drug absorption for drug "developability" assessment: forming an interface between in vitro preclinical data and clinical outcome. Curr Opin Drug Discov Devel 2004; 7:75–85.
Sun YM, Hsu SC, Lai JY. Transport properties of ionic drugs in the ammonio methacrylate copolymer membranes. Pharm Res 2001 Mar; 18(3):304–310.
Swarbrick J, Lee G, Brom J, Gensmantel NP. Drug permeation through human skin II: permeability of ionizable compounds. J Pharm Sci 1984 Oct; 73(10):1352–1355.

Takahashi K, Tamagawa S, Katagi T, Rytting JH, Nishihata T, Mizuno N. Percutaneous permeation of basic compounds through shed snake skin as a model membrane. J Pharm Pharmacol 1993 Oct; 45(10):882–886.

Takamatsu N, Welage LS, Idkaidak NM, et al. Human intestinal permeability of piroxicam, propanolol, phenylalanine, and PEG 400 determined by jejunal perfusion. Pharm Res 1997; 14:1127–1132.

Taub ME, Kristensen ME, Frokjaer L. Optimized conditions for MDCK permeability and turbidimetric solubility studies using compounds representative of BCS classes I–IV. Eur J Pharm Sci 2002; 15:331–340.

Tolle-Sander S, Rautio J, Wring S, Polli JW, Polli JE. Midazolam exhibits characteristics of a highly permeable P-glycoprotein substrate. Pharm Res (NY) 2003; 20:757–764.

Tran TT, Mittal A, Gales T, et al. Exact kinetic analysis of passive transport across a polarized confluent MDCK cell monolayer modeled as a single barrier. J Pharm Sci 2004 Aug; 93(8):2108–2123.

Twist JN, Zatz JL. Membrane-solvent-solute interaction in a model permeation system. J Pharm Sci 1988 Jun; 77(6):536–540.

Uppoor VRS. Regulatory perspectives on in vitro (dissolution)/in vivo (bioavailability) correlations. J Control Rel 2001; 72:127–132.

van de Waterbeemd, H. The fundamental variables of the bio-pharmaceutics classification system (BCS): a commentary. Eur J Pharm Sci 1998; 7:1–3.

Veber DF, Johnson SR, Cheng HY, et al. Molecular properties that influence the oral bioavailability of drug candidates. J Med Chem 2002; 45:2615–2623.

Venkatesh S, Lipper RA. Role of the development scientist in compound lead selection and optimisation. J Pharm Sci 2000; 89(2):145–154.

Wagner JG, Nelson E. Kinetic analysis of blood levels and urinary excretion in the absorptive phase after single doses of drug. J Pharm Sci 1968; 53(11):1392–1403.

Wexler DS, Gao L, Anderson F, et al. Linking solubility and permeability assays for maximum throughput and reproducibility. J Biomol Screen 2005 Jun; 10(4):383–390.

Willmann S, Schmitt W, Keldenich J, Lippert J, Dressman JB. A physiological model for the estimation of the fraction dose absorbed in humans. J Med Chem 2004, 29 Jul; 47(16):4022–4031.

Wilson G, Hassan IF, Dix CJ, et al. Transport and permeability properties of human Caco-2 cells: an in vitro model of the intestinal epithelial cell barrier. J Controlled Release 1990; 11:25–40.

Wohnsland F, Faller B. High-throughput permeability pH profile and high-throughput alkane/water log P with artificial membranes. J Med Chem 2001; 44:923–930.

Xia XR, Baynes RE, Monteiro-Riviere NA, Leidy RB, Shea D, Riviere JE. A novel in-vitro technique for studying percutaneous permeation with a membrane-coated fiber and gas chromatography/mass spectrometry: part I. Performances of the technique and determination of the permeation rates and partition coefficients of chemical mixtures. Pharm Res 2003 Feb; 20(2):275–282.

Xia XR, Baynes RE, Monteiro-Riviere NA, Riviere JE. A compartment model for the membrane-coated fiber technique used for determining the absorption parameters of chemicals into lipophilic membranes. Pharm Res 2004 Aug; 21(8):1345–1352.

Xiang TX, Anderson BD. Development of a combined NMR paramagnetic ion-induced linebroadening/dynamic light scattering method for permeability measurements across lipid bilayer membranes. J Pharm Sci 1995 Nov; 84(11):1308–1315.

Yamamoto T, Tahara K, Setaka M, Yano M, Kwan T. Permeability properties of a multilayer planar membrane. J Biochem (Tokyo) 1980 Dec; 88(6):1819–1827.

Yazdanian M, Briggs K, Jankovsky C, Hawi A. The "high solubility" definition of the current Food and Drug Administration guidance on biopharmaceutical classification system may be too strict for acidic drugs. Pharm Res 2004; 21:293–299.

Yee S. In vitro permeability across Caco-2 cells (colonic) can predict in vivo (small intestinal) absorption in man-fact or myth. Pharm Res (NY) 1997; 14:763–766.

Youdim KA, Avdeef A, Abbott NJ. In vitro trans-monolayer permeability calculations: often forgotten assumptions. Drug Discov Today 2003, 1 Nov; 8(21):997–1003.

Young D, Chilukuri D, Becker R, Bigora S, Farrell C, Shepard T. Approaches to developing a Level-A IVIVC for injectable dosage forms. AAPS PharmSci 2002; 4(4):M1357.

Yu K, Gebert M, Altaf SA, Wong D, Friend DR. Optimisation of sustained-release diltiazem formulations in man by use of an in vitro/in vivo correlation. J Pharm Pharmacol 1998; 50:845–850.

Yu L, Amidon G, Polli J, et al. Biopharmaceutics classification system: the scientific basis for biowaiver extensions. Pharm Res 2002; 19:921–925.

Yu S, Li S, Yang H, Lee F, Wu JT, Qian MG. A novel liquid chromatography/tandem mass spectrometry based depletion method for measuring red blood cell partitioning of pharmaceutical compounds in drug discovery. Rapid Commun Mass Spectrom 2005b; 19(2):250–254.

Zhu C, Jiang L, Chen TM, Hwang KK. A comparative study of artificial membrane permeability assay for high throughput profiling of drug absorption potential. Eur J Med Chem. 2002 May; 37(5):399–407.

Appendix 1: Recent Studies Reported in Literature

Absorption intestinal-computer simulation-tube model: The drug flow in the gastrointestinal tract (GIT) was simulated with a biased random walk model in the heterogeneous tube model, while probability concepts were used to describe the dissolution and absorption processes. A certain amount of the drug was placed into the input end of the tube and allowed to flow, dissolve, and absorb along the tube. Various drugs with a diversity in dissolution and permeability characteristics were considered. The fraction of dose absorbed (F_{abs}) was monitored as a function of time measured in Monte Carlo steps (MCS). The absorption number, A_n, was calculated from the mean intestinal transit time and the absorption rate constant adhering to each of the drugs was examined. A fully computerized approach, which describes the flow, dissolution, and absorption of drug in the GIT in terms of probability concepts was developed. This approach can be used to predict F_{abs} for drugs with various solubility and permeability characteristics provided that probability factors for dissolution and absorption are available. — Kalampokis et al. (1999)

Biopharmaceutics Classification System (BCS)-high throughput (HT)-parallel artificial membrane permeation assay (PAMPA)-permeability: Solubility and permeability were measured by high-throughput solubility assay (HTSA) and PAMPA, respectively. HTSA was performed using simulated gastric fluid (SGF, pH 1.2) and simulated intestinal fluid without bile acid (SIF, pH 6.8). We categorize 18 drugs based on the BCS using HTSA and PAMPA. Fourteen out of 18 drugs were correctly classified (78% success rate). The result of the present study showed that HTSA could predict BCS class with a high success rate, and PAMPA could also be useful to predict the permeation of drugs. — Obata et al. (2004)

BCS-profiling permeability-charge state: About 30% of the drug candidate molecules are rejected due to pharmacokinetic-related failures. When poor pharmaceutical properties are discovered in development, the costs of bringing a potent but poorly absorbable molecule to a product stage by "formulation" can become very high. Fast and reliable in vitro prediction strategies are needed to filter out problematic molecules at the earliest stages of discovery. This review will consider recent developments in physico-chemical profiling used to identify candidate molecules with physical properties related to good oral absorption. Poor solubility and poor permeability account for many pharmacokinetic failures. The Food and Drug Administration's (FDA) BCS is an attempt to rationalize the critical components related to oral absorption. The core idea in the BCS is in vitro transport model, centrally embracing permeability and solubility, with qualifications related to pH and dissolution. The objective of the BCS is to predict in vivo performance of drug products from in vitro measurements of permeability and solubility. The BCS can be rationalized by the framework of the BCS could serve the interests of the earliest stages of discovery research. The BCS can be rationalized by considering Fick's first law, applied to membranes. When molecules are introduced on one side of a lipid membrane barrier (e.g., epithelial cell wall) and no such molecules are on the other side, passive diffusion will drive the molecules across the membrane. When certain simplifying assumptions are made, the flux equation in Fick's law reduces simply to a product of permeability and solubility. Many other measurable properties are closely related to permeability and solubility. Permeability (P_e) is a kinetic parameter related to lipophilicity (as indicated by the partition and distribution coefficients, log P and log D). Retention (R) of lipophilic molecules by the membrane (which is related to lipophilicity and may predict pharmacokinetic — Avdeef (2001) volumes of distribution) influences the characterization of permeability. Furthermore, strong drug interactions with serum proteins can influence permeability. The unstirred water layer on both sides of the membrane barrier can impose limits on permeability. Solubility (S) is a thermodynamic parameter, and is closely related to dissolution, a kinetic parameter. The unstirred water layer on the surfaces of suspended solids imposes limits on dissolution. Bile acids effect both solubility and dissolution, by a micellization effect. For ionizable molecules, pH plays a crucial role. The charge state that a molecule exhibits at a particular pH is characterized by the ionization constant (pK_a) of the molecule. Buffers effect pH gradients in the unstirred water layers, which can dramatically affect both permeability and dissolution of ionizable molecules. In this review, we will focus on the emerging instrumental methods for the measurement of the physico-chemical parameters Pe, S, pK_a, R, log P, and log D (and their pH-profiles). These physico-chemical profiles can be valuable tools for the medicinal chemists, aiding in the prediction of in vivo oral absorption.

BCS-quantitative-permeability: A simple absorption model that considers transit flow, dissolution, and permeation processes stochastically was used to illustrate the primary importance of dose/solubility ratio and permeability on drug absorption. Simple mean time considerations for dissolution, uptake, and transit were used to identify relationships between the extent of absorption and a drug's dissolution and permeability characteristics. The QBCS (Quantitative Biopharmaceutics Classification) developed relies on a (permeability, dose/solubility ratio) plane with cut-off points $2 \times 10(-6) - 10(-5)$ cm/sec for the permeability and 0.5–1 (unitless) for the dose/solubility ratio axes. Permeability estimates, P_{app} are derived from Caco-2 studies, and a constant intestinal volume content of 250 mL is used to express the dose/solubility ratio as a dimensionless quantity, q. A physiologic range of 250–500 mL was used to account for variability in the intestinal volume. Drugs are classified into the four quadrants of the plane around the cut-off points according to their P_{app}, q values, establishing four drug categories, that is, I ($P_{app} > 10(-5)$ cm/sec, $q \leq 0.5$), II ($P_{app} > 10(-5)$ cm/sec, $q > 1$), III ($P_{app} < 2 \times 10(-6)$ cm/sec, $q \leq 0.5$), and IV ($P_{app} < 2 \times 10(-6)$ cm/sec, $q > 1$). A region for borderline drugs ($2 \times 10(-6) < P_{app} < 10(-5)$ cm/sec, $0.5 < q < 1$) was also defined. For category I, complete absorption is anticipated, whereas categories II and III exhibit dose/solubility ratio-limited and permeability-limited absorption, respectively. For category IV, both permeability and dose/solubility ratio control drug absorption. Semiquantitative predictions of the extent of absorption were pointed out on the basis of mean time considerations for dissolution, uptake, and transit in conjunction with drug's dose/solubility ratio and permeability characteristics. A set of 42 drugs were classified into the four categories, and the predictions of intestinal drug absorption were in accord with the experimental observations. *Conclusions*: The QBCS provides a basis for compound classification into four explicitly defined drug categories using the fundamental biopharmaceutical properties, permeability, and dose/solubility ratio. Semiquantitative predictions for the extent of absorption are essentially based on these drug properties, which either determine or are strongly related to the in vivo kinetics of drug dissolution and intestinal wall permeation.

Rinaki et al. (2003)

(Continued)

Appendix 1: Recent Studies Reported in Literature (*Continued*)

Biopharmaceutics Drug Classification System (BDCS)-extended release (ER): A BDCS for correlating the invitro drug product dissolution and invivo bioavailability for IR products was proposed by Amidon et al. (1995). The classification arose from drug dissolution and absorption models, which identified the key parameters controlling drug absorption as the dimensionless numbers; the absorption number (An), the dissolution number (Dn), and the dose number (Do). This led to a biopharmaceutic classification of drugs into four groups, the establishment of a basis for determining the conditions under which in vitro–in vivo (IVIV) correlations are expected and the use of the classification to set drug bioavailability standards for IR products. These developments raise the issue of whether the biopharmaceutic classification has relevance to ER products. In contrast to IR products, drugs selected for ER products should have good GI permeability and an extended site of absorption. However, their permeability (P_{app}) may change depending on the site. Solubility (Cs), effective fluid volume, and hence Do may also vary with site. Of particular relevance to both permeability and solubility is the degree of ionization of the drug. Residence time at each site, pH changes, and the potential for drug degradation at different sites, the latter resulting in a restricted absorption window, will influence the time frame over which an IVIV relationship is possible. Of the drugs available in ER dosage forms approximately 63% are bases, 15% acids, and the remainder are either unionizable or small inorganic ions. Acidic drugs will tend to have low solubility high up in the GIT, with solubility increasing down the GIT. In contrast, increased ionization permeability should fall. Thus with acids, as the dosage form moves to a more alkaline environment down the GIT, absorption may change from dissolution control to membrane control depending on the pK_a of the drug. In contrast, bases will lose solubility with transit down the GIT, but become more permeable; absorption becoming more dissolution-release-controlled or in extreme cases solubility-controlled in the latter stages of the absorption phase. In the light of these considerations, a modified biopharmaceutic classification is proposed for ER products.

Corrigan (1997)

BDCS-solubility definition: The solubility and permeability values of 20 (18 acidic and 2 nonacidic) nonsteroidal anti-inflammatory drugs (NSAID) were determined. The NSAIDs were grouped into three different sets having acetic acid, propionic acid, or other acidic moieties, such as fenamate, oxicam, and salicylate. Two nonacidic NSAIDs (celecoxib and rofecoxib) were also included for comparison purposes. Equilibrium solubility values were determined at pH 1.2, 5.0, 7.4, and in biorelevant media fed with intestinal fluid and fasted intestinal fluids. Permeability classification was established relative to that of reference drugs in the Caco-2 cell permeability model. Permeability coefficients for all drugs were measured at concentrations corresponding to the lowest and highest marketed dose strengths dissolved in 250 mL of volume, and their potential interaction with cellular efflux pumps was investigated. All NSAIDs with different acidic functional groups were classified as highly permeable based on their Caco-2 cell permeability. Only ketorolac appeared to have a potential for interaction with cellular efflux pumps. Solubility classification was based on comparison of equilibrium solubility at pH 1.2, 5.0, and 7.4 relative to marketed dose strengths in 250 mL. The pK_a values for the acidic NSAIDs studied were between 3.5 and 5.1, and, as expected, their solubility increased dramatically at pH 7.4 compared with pH 1.2. Only three NSAIDs, ketorolac, ketoprofen, and acetyl salicylic acid, meet the current criteria for high solubility over the entire pH range. However, with the exception of ibuprofen, oxaprozin, and mefenamic acid, the

Yazdanian et al. (2004)

remaining compounds can be classified as Class I drugs (high solubility–high permeability) relative to solubility at pH 7.4. The use of bio-relevant media simulating gastric and intestinal milieu for solubility measurements or increasing the dose volume to 500 mL did not provide for a better boundary for solubility classification. Based on the current definition of solubility, 15 of the 18 acidic NSAIDs in this study will be classified as Class II compounds as the solubility criteria applies to the entire pH range of 1.2–7.4, although the low solubility criteria does not hold true over the entire pH range. Whence, of the 18 acidic drugs, 15 can be classified as Class I based on the pH 7.4 solubility alone. This finding is intriguing because these drugs exhibit Class I behavior as their absorption does not seem to be dissolution- or solubility-limited. It could then be argued that for acidic drugs, the boundaries for solubility are too restrictive. Solubility at pH > 5 (pH in duodenum) may be more appropriate because most compounds are mainly absorbed in the intestinal region. Consideration for an intermediate solubility classification for highly permeable ionizable compounds that reflects physiological conditions seems warranted.

Biomimetic artificial membrane-paracellular pathways-Renkin function: The purpose of this study was to construct and examine the prediction model for total passive permeation through the intestinal membrane. The paracellular pathway prediction model based on Renkin function (PP-RF) was combined with a bio-mimetic artificial membrane permeation assay (BAMPA), which is an in vitro method to predict transcellular pathway permeation, to construct the prediction model (BAMPA-PP-RF model). The parameters of the BAMPA-PP-RF model, for example, apparent pore radius and potential drop of the paracellular pathway, were calculated from BAMPA permeability, the dissociation constant, the molecular radius, and the fraction of a dose absorbed in humans consisting of 80 structurally diverse compounds. The apparent pore radius and the apparent potential drop obtained in this study were 5.61–5.65 A and 75–86 mV, respectively, and these were in accordance with the previously reported values. The mean square root error of the BAMPA-PP-RF model was 13–14%. The BAMPA-PP-RF model was shown to be able to predict the total passive permeability more adequately than BAMPA alone.

Sugano et al. (2002)

Biomimetic artificial membranes-factors: Effects of pH and co-solvents on the BAMPA were investigated to determine the optimal conditions for the prediction of oral absorption. The permeability (P_{am}) of 33 structurally diverse drugs to the PC/PE/PS/PI/CHO/1,7-octadiene membrane system [bio-mimetic lipid (BML) membrane] was measured at pH 5.5, 6.5, and 7.4. The pH dependence of P_{am} was in accordance with the pH partition theory. The better prediction of oral absorption (fraction of a dose absorbed) was shown under the pH 5.5 condition for determining the permeability of poorly soluble compounds were examined. Dimethysulfoxide (DMSO), ethanol (EtOH) and polyoxyethyleneglycol 400 (PEG 400) were added up to 30% to the transport medium as solubilizers. DMSO, EtOH and PEG 400 decreased P_{am} of hydrocortisone and propranolol. For example, DMSO (30%) decreased P_{am} of hydrocortisone and propanol by 60 and 70%₀₂, respectively. DMSO and PEG 400 also decreased P_{am} of ketoprofen. In contrast, EtOH produced an opposite effect on permeability, that is, an increased P_{am} of ketoprofen. Therefore, the high concentration of these co-solvents could lead to the under- or overestimation of drug permeability.

Sugano et al. (2001b)

(Continued)

Appendix 1: Recent Studies Reported in Literature (*Continued*)

Blood–brain barrier (BBB)–computerization–permeability: The objectives of this study were to generate a data set of BBB permeability values for drug-like compounds and to develop a computational model to predict BBB permeability from structures. The BBB permeability, expressed as permeability-surface area product (PS, quantified as log PS), was determined for 28 structurally diverse drug-like compounds using the in situ rat brain perfusion technique. A linear model containing three descriptors, log D, van der Waals surface area of basic atoms, and polar surface area, was developed based on the 23 compounds in our data set, where the penetration across the BBB was assumed to occur primarily by passive diffusion. The correlation coefficient (R^2) and standard deviation (SD) of the model-predicted log PS against the observed are 0.74 and 0.50, respectively. If an outlier was removed from the training data set, the R^2 and SD were 0.80 and 0.44, respectively. This new model was tested in two literature data sets, resulting in an R^2 of 0.77–0.94 and an SD of 0.38–0.51. For comparison, four literature models, log P, log D, log [D, MW(−0.5)], and linear free-energy relationship, were tested using the set of 23 compounds primarily crossing the BBB by passive diffusion, resulting in an R^2 of 0.33–0.61 and an SD of 0.59–0.76. In summary, we have generated the largest PS data set and developed a robust three-descriptor model that can quantitatively predict BBB permeability. This model may be used in a drug discovery setting to predict the BBB permeability of new chemical entities.	Liu et al. (2004)
BBB-cerebrospinal fluid (CSF) sampling-chromatography: The objective of this review is to emphasize the application of separation science in evaluating the BBB permeability to drugs and bioactive agents. Several techniques have been utilized to quantitate the BBB permeability. These methods can be classified into two major categories: in vitro and in vivo. The in vivo methods used include brain homogenization, CSF sampling, voltametry, autoradiography, nuclear magnetic resonance (NMR) spectroscopy, positron emission tomography (PET), intracerebral microdialysis, and brain uptake index (BUI) determination. The in vitro methods include tissue culture and immobilized artificial membrane (IAM) technology. Separation methods have always played an important role as adjunct methods to the methods outlined here for the quantification of BBB permeability and have been utilized the most with brain homogenization, in situ brain perfusion, CSF sampling, intracerebral microdialysis, in vitro tissue culture, and IAM chromatography. However, the literature published to date indicates that the separation method has been used the most in conjunction with intracerebral microdialysis and CSF sampling methods. The major advantages of microdialysis sampling in BBB permeability studies is the possibility of online separation and quantitation as well as the need for only a small sample volume for such an analysis. Separation methods are preferred over nonseparation methods in BBB permeability evaluation for two main reasons. First, when the selectivity of a determination method is insufficient, interfering substances must be separated from the analyte of interest prior to determination. Second, when a large number of analytes are to be detected and quantified by a single analytical procedure, the mixture must be separated to each individual component prior to determination. Chiral separation in particular can be essential to evaluate the stereo-selective permeation and distribution of agents into the brain. In conclusion, the usefulness of separation methods during BBB permeability evaluation is immense and more application of these methods is foreseen in the future.	Dash et al. (2003)

BBB-in vitro correlation: The ability of a drug to penetrate the BBB is essential for its use in the pharmaceutical treatment of central nervous system (CNS) disorders. Five different in vitro methods to predict BBB permeability of drugs were compared and evaluated in the present study. All assays were performed with a consistent set of seven compounds and in the same physiological buffer to provide a basis for direct comparison of the results. Octanol-buffer and liposome-buffer partition coefficients were most conveniently obtained, but failed to predict BBB permeability for certain drugs. The incorporation of drugs into lipid monolayers at the air–buffer interface was found to be a poor predictor of BBB permeability and was furthermore not considered suitable for screening due to the demanding experimental requirements. Permeability studies using Caco-2 cell monolayers provided a good correlation to an in vitro model of the BBB, which was based on primary cultured porcine brain capillary endothelial cells (PBCEC). However, differences in drug permeability between the intestine and brain-derived cells were detected, limiting the advantages of the easy handling of the Caco-2 cell line compared with the more time-consuming primary culture of the BCEC.

Lohmann et al. (2002)

BBB-microdialysis-correlation: Studies presented in this report were designed to assess the correlation of the bovine brain microvessel endothelial cell (BBMEC) apparent permeability coefficient (P_{app}) and in vivo BBB extracellular fluid (ECF) concentration to free drug in plasma. The compounds studied have a broad range of physico-chemical characteristics and have widely varying in vitro and in vivo permeability across the BBB. BBMEC permeability coefficients vary in magnitude from a low value of $0.9 \times 10(-5)$ cm/sec to a high value of $7.5 \times 10(-5)$ cm/sec. Corresponding in vivo measurements of BBB permeability are represented by clearance (CL_{in}) into the brain ECF and range from a low value of 0.023 μmol/min/g to a high value of 12.9 μmol/min/g. While it is apparent that in vitro data from the BBMEC model can be predictive of the in vivo permeability of a compound across the BBB, there are numerous factors both prior to and following entry into the brain, which impact the ultimate uptake of a compound. Even in the presence of high BBB permeability, factors, such as high plasma protein binding, active efflux across the BBB, and metabolism within the CNS can greatly limit the ultimate concentrations achieved. In addition, concentrations in the intracellular space may not be the same as concentrations in the extracellular space. While these data show that the BBMEC permeability is predictive of the in vivo BBB permeability, the complexity of the living system makes prediction of brain concentrations difficult, based solely on the in vitro measurement.

Hansen et al. (2002)

BBB-novel model: Drug delivery to the CNS is subject to the permeability limitations imposed by the BBB. Several systems in vitro have been described to reproduce the physical and biochemical behavior of intact BBB, most of which lack the feature of the in vivo barrier. We developed a fully formed monolayer of RBE4 B-immortalized rat brain microvessel endothelial cells (ECs), grown on top of polycarbonate filter inserts with cortical neuronal cells grown on the outside. Neurons induce ECs to synthesize and sort occludin to the cell periphery. Occludin localization is regulated by both compositions of the substratum and soluble signals released by cortical co-cultured neurons. The observed effects do not require strict physical contact among cells and neurons. To assess the physiological function of the barrier, we examined the transendothelial transfer of three test compounds: dopamine, L-tryptophan, and L-DOPA. Polycarbonate filter inserts, where ECs were co-cultured with neurons,

Cestelli et al. (2001)

(Continued)

Appendix 1: Recent Studies Reported in Literature (*Continued*)

were assumed as open two-compartment vertical dynamic models. Permeation studies demonstrated that the ECs/neuron co-cultures possess permeability characteristics approaching those of a functional BBB: the system behaves as a selective interface that excludes dopamine permeation, yet permits L-tryptophan and L-DOPA to pass through. The movement of test compounds from the donor to the acceptor compartment was observed at a distinct time from the start of co-culture. Transfer was determined using standard kinetic equations. Different performance was observed after five and seven days of co-culture. After five days, dopamine, L-tryptophan, and L-DOPA passively permeated through the membrane as indicated by fittings with a first-order kinetic process equation. After seven days of co-culture, occludin localizes at EC periphery, dopamine does not cross the barrier to any further extent, while the transfer of L-tryptophan and L-DOPA fits well with a saturable Michaelis-Menten kinetic process, thus indicating the involvement of a specific carrier-mediated transport mechanism. Permeation studies confirmed that culture of ECs in the presence of neurons induces the characteristic permeability limitations of a functional BBB.

BBB-permeability-liquid chromatography/mass spectroscopy- (LC/MS): Rapid, generic gradient liquid chromatography/tandem mass spectroscopy (LC/MS/MS) assays, designed to accelerate sample analyses, have been developed to keep pace with the productivity of advanced synthetic procedures. In this study, LC/MS/MS was combined with an in vitro, cell-based, BBB model to evaluate the potential of new chemical entities (NCEs) to cross the BBB. This in vitro assay provides the permeability of discovery compounds across a monolayer of a primary culture of BBMEC in a fraction of the time that is required for in vivo studies (brain/plasma concentrations), using only 2 mg of the compound. The results are consistent with in vivo brain/plasma concentration ratio data.

Chu et al. (2002)

BBB-permeability-transendothelial resistance: An in vitro BBB model was used, comprising of BCEC and astrocytes co-cultured on semipermeable filter inserts. Experiments were performed under control and challenged experimental circumstances, induced to simulate drug effects. The apparent BBB permeability coefficient for two markers for paracellular drug transport, sodium fluorescein ($P_{app,FLU}$), MW = 376 Da) and FITC-labeled dextran ($P_{app,FD4}$, MW = 4 kDa), was determined. Transendothelial electrical resistance (TEER) was used to quantify basal and (simulated) drug-induced changes in permeability of the in vitro BBB. The relationship between P_{app} and TEER was determined. Drug effects were simulated by exposure to physiologically active endogenous and exogenous substances (i.e., histamine, deferroxamine mesylate, adrenaline, noradrenaline, bradykinin, vinblastine, sodium nitroprusside, and lipopolysaccharide). *Results:* $P_{app,FLU}$ and $P_{app,FD4}$ in control experiments varied from 1.6 up to 17.6 (10(−6) cm/sec) and 0.3 up to 7.3 (10(−6) cm/sec), respectively; while for individual filters $P_{app,FLU}$ was four times higher than $P_{app,FD4}$ ($R^2 = 0.97$). As long as TEER remained above 131 Ωcm² for FLU or 122 Ωcm² for FD4 during the transport assay, P_{app} remained independent from the basal permeability of the in vitro BBB. Below these TEER values, P_{app} increased exponentially. This nonlinear relationship between basal BBB permeability and P_{app} was described by a one-phase exponential decay model. From this model, the BBB permeability status-independent permeability coefficients for FLU and FD4 (P_{FLU} and P_{FD4}) were estimated to be 2.2 ± 0.1 and 0.48 ± 0.03 (10(−6) cm/sec), respectively. In

Gaillard and de Boer (2000)

the experimentally challenged experiments, a reliable indication for P_{FLU} and P_{FD4} could be estimated only after the (simulated) drug-induced change in BBB permeability was taken into account. The assessment of basal BBB permeability status during drug transport assays was essential for an accurate estimation of the in vitro permeability coefficient of a drug. To accurately extrapolate the in vitro permeability coefficient of a drug to the in vivo situation, it is essential that drug-induced changes in the in vitro BBB permeability during the drug transport assay are determined.

BBB-VolSurf: In silico prediction methods are important for CNS drugs. Starting with simple regression models based on the calculation of lipophilicity and polar surface area, the field developed via PLS methods to grid-based approaches (e.g., VolSurf). Additionally, the use of artificial neural networks gain increasing importance. However, permeation through the BBB is also influenced by active transport systems. For nutrients and endogenous compounds, such as amino acids, monocarboxylic acids, amines, hexoses, thyroid hormones, purine bases, and nucleosides, several transport systems regulating the entry of the respective compound classes into the brain have been identified. The other way round there is the striking evidence that expression of active efflux pumps like the multidrug transporter P-glycoprotein (Pgp) on the luminal membrane of the brain capillary endothelial cells accounts for poor BBB permeability of certain drugs. Undoubtedly, Pgp is an important impediment for the entry of hydrophobic drugs into the brain. Thus, proper prediction models should also take into account the active transport phenomena.

Ecker and Noe (2004)

Caco-2-absorption-permeability-link: Effective permeability coefficients of the model drugs (naproxen, phenytoin, propranolol, diltiazem, salicylic acid, ephedrine, cimetidine, chlorothiazide, and furosemide) at 37°C and pH 7.2 were estimated using the Caco-2 cell line. Saturation solubilities of the model drugs were estimated at pH 7.2 and at 37°C. Data obtained from the permeability and solubility experiments were employed in classifying the drugs into high- and low-permeability solubility groups. The permeability coefficients ranged from $1 \times 10(-7)$ to $4 \times 10(-5)$ cm/sec, and a good correlation was observed between the permeability coefficients in Caco-2 cells and percent absorbed in humans. Drugs in the high-permeability, high-solubility class are completely absorbed (90% or higher). The study results indicate that there is a strong link between permeability measured in Caco-2 cells, solubility, and the fraction of drug absorbed in humans.

Pade and Stavchansky (1998)

Caco-2-HT-LC/MS/MS-TEER-intestinal absorption: The TEER of cell monolayers was measured on 0.33, 1.0, and 4.7 cm² polycarbonate membranes using epithelial volt-ohmmeter, over a 25-day period. Absorptive transport was determined on all compounds tested using LC-MS/MS assays, or liquid scintillation spectrometry. The effect of multiple compounds in one well compared with single compounds was assessed with atenolol, nadolol, metoprolol, and propranolol for mixtures of four compounds and with RWJ-53308, atenolol, terbutaline, propranolol, naproxen, piroxicam, topiramate, and furosemide for mixtures of eight compounds. The apparent permeability (P_{app}) values correlated well between single analytes and mixtures of four and eight analytes in each well. Drug permeability decreased slightly with an increase in well size. The TEER value increased with the number of days in culture for each of the 6-, 12-, and 24-well sizes. It was demonstrated that the 24-well format system is ideal for high-throughput assessment. Furthermore, the approach of mixing four or eight analytes in each well to further increase throughput was also demonstrated to be valid.

Markowska et al. (2001)

(Continued)

Appendix 1: Recent Studies Reported in Literature (*Continued*)

Caco-2-pore radius tight junction (TJ)-permeability: The apparent permeability coefficient (Pa_{app}) of drugs was measured using the Caco-2 assay and the PAMPA, and values were corrected with the pore radius of TJs. *Results*: An equation for calculating the pore radius of TJs from the P_{app} of lucifer yellow was obtained. The optimal pore radius of TJs in Caco-2 monolayers for predicting human F_a was calculated to be 7 A. The correlation between the actual and predicted F_a was improved by using the P_{app} corrected with the pore radius of TJs. Permeability in the PAMPA, which was corrected using the pore radius and membrane potential, was well correlated with that in the Caco-2 assay. Most of the hydrophilic drugs tested in this study were absorbed mainly through the paracellular pathway. The results suggest the necessity of optimizing paracellular permeation for the prediction of F_a, and also the importance of the paracellular pathway to the absorption of hydrophilic drugs. This method might contribute to the setting of appropriate dosages and the development of hydrophilic drugs.	Saitoh et al. (2004)
Caco-2-ultraviolet (UV) absorption model: Caco-2 cells were seeded at high density in 96-well plates in novel cell culture boxes. After seven days, drug permeability studies were performed. Samples were analyzed by a new UV detection method. With increased cell seeding density, functional Caco-2 monolayers with polarized efflux transporters were established after seven days in 96-well polycarbonate filter plates in a standard medium. For faster feeding and to eliminate medium replacement in each individual well, plates were completely submerged in the medium in novel cell culture boxes, and only medium outside the plate was exchanged. For high-throughput sample analysis, a novel UV-transparent transport buffer was established that allowed direct quantification of permeated drug from its UV absorption. In vitro permeability studies analyzing 22 passively absorbed drugs in the new model correlated well with reported human permeability values ($r^2 = 0.8725$). The new seven-day, 96-well Caco-2 permeability model tied to UV analysis offers considerable time, cost, and resource savings compared with the traditional model. It has a potential for automation and makes it possible to determine the permeability of passively diffusing compounds and to classify them according to the BCS in a truly medium- to high-throughput mode.	Alsenz and Haenel (2003)
Copolymer membrane-ionic drug-transport-aspirin-ambroxol: *Purpose*: Ammonio methacrylate copolymer is a pharmaceutical excipient widely used as a coating material for encapsulation of pellet and tablet dosage forms. Because of the charged ammonio function groups within the polymer, ionic drugs may interact with the coating film while transporting through it. The kinetic swelling and drug permeation properties of the ammonio methacrylate copolymer membranes were studied to delineate the effect of ionic interaction between the ionic drugs and the membranes. *Methods*: The pH and ionic strength of the solutions and the charged properties of drugs were varied to study the effects on the transport properties through the membranes. Ambroxol was chosen as a model cationic drug and aspirin as a model anionic drug. The degree of membrane swelling in the drug-free solution decreases as the ionic strength increases, but it is irrelevant to the pH. With the presence of ionic drugs, the degree of membrane swelling is affected by the drug species and the pH of the solutions in addition to the effect of ionic strength. The degree of swelling for a membrane in a solution containing aspirin is higher at a lower pH and is lower for ambroxol at a lower pH. Aspirin experiences a three-stage permeation and ambroxol a two-stage one. The ion-exchange reaction between the anionic carboxylic groups in aspirin and the cationic ammonio groups in the membranes results in a slow permeation stage during the transient state. The pseudo steady-state permeability for each drug follows the trend as	Sun et al. (2001)

the degree of membrane swelling in the drug media at various pH and ionic strengths. However, it is much higher for aspirin than ambroxol although the degree of membrane swelling is higher in an ambroxol solution than that in an aspirin solution. The permeability of ambroxol through the membrane is largely reduced because of the Donnan exclusion effect. The interaction between ionic drugs with the cationic groups in the membranes affects the ionic strength of the solutions and results in a pH-dependent degree of swelling. The ionic interaction also determines the drug permeation rates and the transient permeation behaviors.

Pgp-Caco-2-BBB: The GI absorption and Pgp efflux transport of heterocyclic drugs was investigated with the Caco-2 cell model. Based on the calculation of the physico-chemical properties, a good oral absorption was predicted for all the drugs tested in this study, which corresponded well with the measured Caco-2 permeabilities (P_{app}). Generally a high permeability of the tested heterocyclic drugs was measured being in agreement with earlier published human in vivo absorption data. Based on the transport data of domperidone and verapamil, it was found that the Pgp efflux transporter was expressed in the Caco-2 cells. Many of the drugs tested were indicated to be potential Pgp efflux substrates. As Pgp is expressed at the BBB as well, it was expected that CNS penetration will be impaired if a drug is a Pgp substrate. However, no correlation could be found between brain penetration in rats and the Pgp efflux ratio as measured with the Caco-2 cells. From the data, it is concluded that Pgp efflux ratios as determined in in vitro high-throughput screening (HTS) tests, where the transport conditions are fixed (pH gradient, concentration, and the like), cannot be used routinely to predict a possible limited brain penetration.

Faassen et al. (2003)

Pgp-Caco-2-permeability: It has been shown in vivo and in vitro that Pgp may be able to influence the permeability of its substrates across biological membranes. However, the quantitative contribution of the secretion process mediated by Pgp on the overall permeability of membranes has not been determined yet. In particular, observations need to be clarified in which substrates showing high affinity to Pgp, for example, verapamil, apparently do not seem to be greatly influenced by Pgp in their permeability and consequently also with respect to their extent of GI absorption after oral administration, whereas weaker substrates of Pgp, for example, talinolol, have clearly shown Pgp-related absorption phenomena, such as nonlinear intestinal permeability and bioavailability. Experiments with Caco-2 cell monolayers and mathematical simulations based on a mechanistic permeation model aid in clarifying the underlying mechanism for these observations and quantify the influence of passive membrane permeability and affinity to Pgp to the overall transmembrane drug flux. In addition, the concentration range of the drug at which Pgp-mediated transport across the biological membrane is relevant should be examined. The permeability of various drugs in Caco-2 monolayers was determined experi-mentally and modeled using a combination of passive absorptive membrane permeability and a Michaelis-Menten type transport process in the secretory direction. The passive permeabilities were experimentally obtained for the apical and basolateral membranes by efflux experiments using Caco-2 monolayers in the presence of a Pgp inhibitor. The Michaelis-Menten parameters were determined by a newly developed radioligand-binding assay for the quantification of drug affinity to Pgp. The model was able to accurately simulate the permeability of Pgp substrates, with differing passive membrane permeabilities and Pgp affinities. Using the outlined approach, permeability versus donor-concentration profiles were calculated, and the relative contribution of passive and active transport processes to the overall membrane permeability was evaluated. A model is presented to quantitatively describe and

Doppenschmitt et al. (1999)

(Continued)

Appendix 1: Recent Studies Reported in Literature *(Continued)*

predict direction-dependent drug fluxes in Caco-2 monolayers by determining the affinity of a compound to the exsorptive transporter Pgp and its passive membrane permeability. It was shown that a combination of high Pgp affinity with good passive membrane permeability, for example, in the case of verapamil, will readily compensate for the Pgp-mediated reduction of intestinal permeability, resulting in a narrow range in which the permeability depends on the apical drug concentration. On the other hand, the permeability of compounds with low passive membrane permeability (e.g., talinolol) might be affected over a wide concentration range despite low affinity to Pgp. — Kariv et al. (2002)

HT-microtiter-radiometric-fluorometric-cytochrome P450 (CYP): Development of predictive in vitro surrogate methods for traditional approaches assessing bioavailability and pharmacokinetics of lead compounds must be made to both keep pace with HT lead identification and to mitigate the high costs associated with progression of compounds with poor chances of developmental success. Indeed, opportunities for improvement still exist in the lead optimization (LO) phase versus the lead identification phase, where HT methodologies have been nearly optimized. Review of examples, limitations, and development of HT microtiterplate-based assays for evaluating metabolic liabilities, such as in vitro radiometric and fluorometric assays for inhibition of CYP450 activity, determination of stability of a compound in liver microsomes, or cloned CYPs coupled to reconstituting systems are described. Parallel approaches to improve speed, resolution, sample preparation, as well as data analysis using LC/MS and LC/MS/MS approaches and technologies to assess compound integrity and biotransformation by automation and multiplexing are also discussed. Realization of the benefits in automation of cell-based models for determining drug permeability to predict drug absorption are still hampered by bottlenecks in analytical analysis of compounds. The implementation and limitations of surrogate physiochemical methods for passive adsorption, such as immobilized artificial membranes (IAM) and PAMPA, and compound solubility by laser nephelometry are reviewed as well. Additionally, data from a HT 96-well equilibrium dialysis device, showing good correlation to classical methods, is presented. Finally, the impact of improvements in these downstream bottlenecks in LO and preclinical drug discovery are discussed in this review.

HT-pharmacokinetics-CYP1A2 inhibition: Pharmacokinetic and metabolic screening plays an important role in the optimization of a lead compound in drug discovery. Since these screening methods are time-consuming and labor intensive, in silico models would be effective to select compounds and guide derivatization prior to the screening. We investigated in silico models for permeability in Caco-2 cells, brain distribution and CYP450 inhibition using molecular weight, lipophilicity [c log $D(7.4)$], polar surface area (PSA), and number of rotatable bonds (RB). A variety of test compounds was selected from different Caco-2 assay projects. The permeability determined exhibited a good correlation with a combination of PSA and c log $D(7.4)$ rather than with PSA alone. In the brain distribution, PSA, in addition to lipophilicity, was one of the determinant parameters, and compounds were significantly distributed to the brain in rats with the decrease in the PSA value. When this approach was adapted to CYP1A2 inhibition in the fluorometric assay, the inhibitory potential for two plane core structures was successfully predicted by utilizing the number of RB, PSA, and c log $D(7.4)$. In particular, an increase in the number of RB weakened the inhibitory potential due to a loss of the plane structures. These results suggest that PSA and RB are key parameters to design chemical structures in terms of the improvement of both membrane permeability in the brain and GI and CYP1A2 inhibitions, respectively. — Komura et al. (2005)

HT-Caco-2-structure–pharmacokinetic parameter relationship (SPR): The application of combinatorial chemistry and HTS to biological targets has led to efficient identification of lead compounds in wide therapeutic areas. However, the physico-chemical properties of some lead compounds are lipophilic with low water solubility. As these parameters determine in vivo absorption, we established robust screening methods for solubility and Caco-2 membrane permeability, which are applicable to our screening strategy based on the SPR. Of test compounds with different core structures, turbidimetric solubility and apparent solubility as determined by high-performance liquid chromatography (HPLC)-UV analysis after dilution of aqueous media from DMSO stock solution was overestimated in comparison with the corresponding thermodynamic solubility obtained using a traditional shake-flask method. A new powder-dissolution method providing thermodynamic solubility similar to that in the traditional method was developed using 96-well plates for equilibrium dialysis. The throughput of the method was almost the same as that using the apparent solubility method. In a conventional Caco-2 assay, membrane permeability (P_{app}) of some lipophilic compounds was underestimated due to low solubility in the apical site and adhesion to the device, resulting in a poor relationship between the in vivo absorption fraction and the P_{app} values. The addition of 0.1% Gelucire 44/14 into the apical site and 4% bovine serum albumin into the basolateral site improved the relationship. These newly developed methods are therefore useful to optimize lead compounds with less water solubility and high lipophilicity on the basis of SPR.

Komura et al. (2005)

HT-drug metabolism and pharmacokinetics (DMPK)-LC/MS/MS: In the current drug discovery environment, HT analytical assays have become essential to keep pace with the screening demands for DMPK attributes. This has been dictated by advances primarily in chemical procedures, notably combinatorial and parallel syntheses, which have resulted in many-fold increases in the number of compounds requiring DMPK evaluation. Because of its speed and specificity, LC/MS/MS has become the dominant technology for sample analysis in the DMPK screening assays. For HT assays, analytical speed and other factors, such as method development, data processing, quality control, and report generation, must be optimized. The four-way multiplexed electrospray interface (MUX), which allows for the analysis of four LC eluents simultaneously, has been adopted to maximize the rate of sample introduction into the mass spectrometer. Generic fast-gradient HPLC methods that are suitable for approximately 80% of the new chemical entities encountered have been developed. In-house-written software programs have been used to streamline information flow within the system, and for quality control by automatically identifying analytical anomalies. By integrating these components together with automated method development and data processing, a system capable of screening 100 compounds per week for Caco-2 permeability has been established.

Fung et al. (2003)

HT-LC/UV/chemiluminescent nitrogen detection (CLND)/evaporative light scattering detection (ELSD)/MS: As part of an overall systems approach to generating highly accurate screening data across large numbers of compounds and biological targets, we have developed and implemented streamlined methods for purifying and quantifying compounds at various stages of the screening process, coupled with automated "traditional" storage methods (DMSO, −20°C). Specifically, all of the compounds in our drug-like library are purified by LC/MS/UV and are then controlled for identity and concentration in their respective DMSO stock solutions by CLND/ELSD and MS/UV. In addition, the compound-buffer solutions used in the various biological assays are quantified by LC/UV/CLND to determine the concentration of the compound actually present during screening. Our results show that LC/UV/CLND/ELSD/MS is a widely applicable method that can be used to purify, quantify, and identify most small organic molecules from compound libraries. The LC/UV/CLND technique is a simple and sensitive

Popa-Burke et al. (2004)

(Continued)

Appendix 1: Recent Studies Reported in Literature (*Continued*)

method that can be easily and cost-effectively employed to rapidly determine the concentrations of even small amounts of any N-containing compound in aqueous solution. We present data to establish error limits for concentration determination that are well within the overall variability of the screening process. This study demonstrates that there is a significant difference between the predicted amount of soluble compound from stock DMSO solutions following dilution into assay buffer and the actual amount present in assay buffer solutions, even at the low concentrations employed for the assays. We also demonstrate that knowledge of the concentrations of compounds to which the biological target is exposed is critical for accurate potency determinations. Accurate potency values are in turn particularly important for drug discovery, for understanding structure–activity relationships, and for building useful empirical models of protein–ligand interactions. Our new understanding of relative solubility demonstrates that most, if not all, decisions that are made in early discovery are based upon missing or inaccurate information. Finally, we demonstrate that careful control of compound handling and concentration, coupled with accurate assay methods, allows the use of both positive and negative data in analyzing screening data sets for structure–activity relationships that determine potency and selectivity.

HT-models-absorption: Compounds with good biological activity may fail to become drugs due to insufficient oral absorption. Selection of drug development candidates with adequate absorption characteristics should increase the probability of success in the development phase. To assess the absorption potential of new chemical entities, numerous in vitro and in vivo model systems have been used. Many laboratories rely on cell culture models of intestinal permeability, such as Caco-2, HT-29, and MDCK. To attempt to increase the throughput of permeability measurements, several physico-chemical methods, such as IAM columns and PAMPA have been used. More recently, much attention has been given to the development of computational methods to predict drug absorption. However, it is clear that no single method is sufficient for studying drug absorption, but most likely a combination of systems will be needed. HT, less reliable methods could be used to discover "loser" compounds, whereas lower throughput, more accurate methods could be used to optimize the absorption properties of lead compounds. Finally, accurate methods are needed to understand absorption mechanisms (efflux-limited absorption, carrier-mediated, intestinal metabolism) that may limit intestinal drug absorption. This information could be extremely valuable to medicinal chemists in the selection of favorable chemo-types. This review describes different techniques used for evaluating drug absorption and indicates their advantages and disadvantages.

Hidalgo (2001)

HT-permeability-microtiter plate: This study reports on a novel, HT assay, designed to predict passive, transcellular permeability in early drug discovery. The assay is carried out in 96-well microtiter plates and measures the ability of compounds to diffuse from a donor to an acceptor compartment, which are separated by a 9–10-μm hexadecane liquid layer. A set of 32 well-characterized, chemically diverse drugs was used to validate the method. The permeability values derived from the flux factors between donor and acceptor compartments show a good correlation with GI absorption in humans. For comparison, correlations based on experimental or calculated octanol/water distribution coefficients [log D(o/w, 6.8)] were significantly lower. In addition, this simple and robust assay allows determination of pH permeability profiles, critical information to predict GI absorption of ionizable drugs, and is difficult to obtain from cell culture experiments. Correction for the unstirred water layer effect allows to differentiate between effective and intrinsic membrane permeability and opens up the dynamic range of the method. In addition, alkane/water partition coefficients can be derived from intrinsic membrane permeabilities, making this assay the first HT method able to measure alkane/water log P in the microtiter plate format.

Wohnsland and Faller (2001)

HT-profiling: Measurement and application of compound properties for candidate selection and optimization is an emerging trend. Property-based design supplements successful activity-based strategies to produce drug-like candidates. HTS hits are evaluated for integrity and aggregation to ensure quality leads. Solubility data assures accurate activity assays and predicts absorbance. Cellular and artificial membrane permeability assays indicate compound penetration through membranes in cells, intestines, and BBB. Lipophilicity and pK_a provide fundamental structure design elements. Stability in liver, plasma, and buffer evaluates compound's lifetime. Drug–drug interaction is predicted using CYP inhibition assays. Drug-like properties are vital to successful drug candidates and enhance drug discovery.

Di et al. (2003)

HT-solubility/permeability: First, solubility is determined at four pH values by comparing the concentration of a saturated compound solution with its dilute, known as the concentration. The filtered, saturated solution from the solubility assay is then used as input material for the membrane permeability determination. The permeability assay is a parallel artificial membrane technique whereby a membrane is created on a solid support, PAMPA. The two artificial membranes presented here model the GIT and the BBB. Data are presented for control compounds, which are well documented in the literature and exemplify a range of solubility and membrane permeability. The advantages of the combination method are: (*i*) reduction of sample usage and preparation time, (*ii*) elimination of interference from compound precipitation in membrane permeability determination, (*iii*) maximization of input concentration to permeability assay for improved reproducibility, and (*iv*) optimization of sample tracking by streamlining data entry and calculations. BBB permeability ranking of compounds correlates well with literature CNS activity.

Wexler et al. (2005)

IAM chromatography-HT-absorption: An IAM chromatographic method was developed and validated. Absorption profiles of 32 structurally diverse compounds (acidic, basic, neutral, and amphoteric) were then evaluated based on their IAM retention factor (log k_{IAM}), molecular weight (MW), calculated log P (C log P), PSA, hydrogen bonding capacity (HBD and HBA), and calculated Caco-2 permeability (QPCaco). Using regression and stepwise regression analysis, experimental Caco-2 permeability was correlated against log k_{IAM} and a combination of various physico-chemical variables for quantitative structural-permeability relationship (QSPR) study. For the 32 structurally diverse compounds, log K_{IAM} correlated poorly with Caco-2 permeability values ($R^2 = 0.227$). Stepwise regression analysis confirmed that C log, PSA, HBD, and HBA parameters are not statistically significant and can be eliminated. Correlation between Caco-2 cell uptake and log k_{IAM} was enhanced when molecular size factor (MW) was included ($R^2 = 0.555$). The exclusion of 11 compounds (paracellularly and actively transported, Pgp substrates and blocker, and molecules with MW lesser than 200 and greater than 800) improved the correlation between Caco-2 permeability, IAM, and MW factors to R^2 value of 0.84. The results showed that IAM chromatography can only profile the passive absorption of drug molecules. Finally, it was confirmed in this study that the IAM model can accurately identify the Caco-2 permeability of nontransported Pgp substrates, such as verapamil and ketoconazole, through passive permeation because of their high permeability. IAM chromatography, combined with molecular size factor (MW), is useful for elucidating biopartitioning mechanism of drugs.

Chan et al. (2005)

(Continued)

Appendix 1: Recent Studies Reported in Literature (*Continued*)

MDCK monolayer-passive transport-Pgp: The commonly used approximate formula for the passive permeability coefficient is based on the initial rate of permeation across cell monolayers, requires measurement during the linear phase of permeation, and is not applicable when there is significant back flux of compound or mass balance problem. To develop a rigorous equation that can be used at any time point, that is, valid outside of the linear phase, the mass action equations were integrated for a standard single-barrier model of passive permeability. The simple analytical solution found also allows correction for both loss of drug (e.g., due to binding and/or hydrolysis) and sampling volume loss for multiple time point experiments. To test this equation, we measured the passive permeation of three well-characterized drugs (amprenavir, quinidine, and loperamide) across confluent monolayers of MDCKII-hMDR1 cells. The potent Pgp inhibitor, GF120918, was used to inhibit Pgp activity, and so only passive permeability was determined. Dramatically different time-dependent behavior was observed for the three compounds, with loperamide showing significant loss of compound, and loperamide and quinidine causing plasma membrane modifications over time. The simple and exact equation for the permeability coefficient developed here works from start of transport to equilibrium, being valid when the commonly used approximate equation may not be. Thus, the exact equation is safer to use in any context, even for single time point estimates in HT permeability assays. — Tran et al. (2004)

Membrane-coated fiber (MCF)-lipophilic transport: A polymer membrane coated onto a section of inert fiber was used as a permeation membrane in the MCF technique. When MCFs were immersed into a donor solution, the compounds in the solution partitioned into the membrane. At a given permeation time, a fiber was removed from the solution and transferred into a gas chromatography injector for quantitative analysis. The permeation process of a given chemical from the donor phase into the membrane was described by a one-compartment model by assuming first-order kinetics. A mathematical model was obtained that describes the cumulative amount of a chemical permeated into the membrane as a function of the permeation time in an exponential equation. Two constants were introduced into the compartment model that were clearly defined by the physiochemical parameters of the system (a kinetic parameter and the equilibrium absorption amount), and were obtained by regression of the experimental data sampled over a limited time before equilibrium. This model adequately described the permeation kinetics of the MCF technique. All theoretical predictions were supported by the experimental results. The experimental data correlated well with the mathematical regression results. The partition coefficients, initial permeation rate, uptake, and elimination rate constants were calculated from the two constants. The compartment model can describe the absorption kinetics of the MCF technique. The regression method based on the model is a useful tool for the determination of the partition coefficients of lipophilic compounds when it takes too long for them to reach permeation equilibrium. The kinetic parameter and the initial permeation rate are unique parameters of the MCF technique that could be used in the development of quantitative structure-activity relationship (QSAR) models. — Xia et al. (2004)

Membrane lipid-polysiloxane-permeation: Through the use of permeation/lipophilicity correlations, the mechanisms of permeation of selected test compounds across artificial lipoidal membranes of the polysiloxane type, in the absence and in the presence of a nonionic surfactant (Polysorbate 80), are investigated, in order to design "in vitro" conditions and features suitable for reproducing "in vivo" intestinal absorption tests, as well as to validate some conclusions arising from "in situ" rat gut experiments about the effects of the synthetic surfactants on drug and xenobiotic absorption processes. Six 4-alkylanilines showing a perfect homology were used as test compounds. The reported results clearly show that the in situ biophysical absorption (diffusion) models are completely reproduced by in vitro tests, provided that perfect sink conditions are achieved. Further selection of artificial membrane polarity should be necessary, however, in order to exactly equalize in vitro and in situ permeation rates. As far as the synthetic surfactant action on permeability is concerned, our conclusions are similar to those drawn from in situ studies, except that the effect of the surfactant on membrane polarity is much smaller and the micelle-solubilizing effect is somewhat larger. The disruption of the aqueous stagnant diffusion layers adjacent to the membranes by the surfactant has been conclusively demonstrated. A clear first-element deviation for aniline, which prevents its inclusion as a term of the tested series, has been observed; this feature should be borne in mind whenever any in vivo/in vitro correlation has to be established.

Perez-Buendia et al. (1993)

Membrane-chitosan/tetra ethyl ortho silicate (TEOS)-permeation: A novel organic–inorganic composite membrane was prepared, using TEOS as an inorganic material and chitosan as an organic compound. Equilibrium and oscillatory swelling studies were conducted to investigate swelling behaviors of the membrane according to the pH of the swelling medium. Drug permeation experiments were also performed in phosphate buffer solution of the pH of 2.5 and 7.5, respectively. Lidocaine HCl, sodium salicylate, and 4-acetamidophenol were selected as model drugs to examine the effect of ionic property of drug on the permeation behavior. The effects of membrane composition and the external pH on the swelling and the drug permeation behavior of interpenetrating polymer networks (IPN) membrane could be summarized as follows, chitosan incorporated into TEOS IPN swelled at pH 2.5 while it shrunk at pH 7.5. This swelling behavior was completely reversible and the membrane responded rapidly to the change in environmental pH condition. According to the swelling behavior, an increase in pH from 2.5 to 7.5 yielded an increase in the rate of drug permeation because of the shrinking of the incorporated chitosan in TEOS IPN, while decrease in pH resulted in a low permeation rate. The optimal TEOS-chitosan ratio for maximum pH sensitivity existed and drug permeation was influenced not only with the external pH, but also with the ionic interactions between the drug and membrane.

Park et al. (2001)

Membrane-filter immobilized-permeability-botanical: The assessment of transport properties of 23 drugs and natural product molecules was made by using the in vitro model based on filter-IAM, assembled from phosphatidylcholine in dodecane, in buffer solutions at pH 7.4. Five of the compounds were lactones extracted from the roots of the kava-kava plant. Experiments were designed to test the effects of stirring (0–600 rpm) during assays and the effects of varying the assay times (2–15 hr).

Avdeef et al. (2001)

(Continued)

Appendix 1: Recent Studies Reported in Literature (*Continued*)

The highly mobile kava lactones permeated in the order: dihydromethisticin (40) > yangonin (37) > kavain (34) > methisticin (32) > desmethoxyyangonin (26), the numbers in parentheses being the measured effective permeabilities in units of 10(−6) cm/sec. By comparison, commercial drugs ranked: phenazopyridine (35) > testosterone (19) > propranolol (13) > ketoconazole (6.3) > piroxicam (2.2) > caffeine (1.7) > metoprolol (0.8) > terbutaline (0.01). In addition to permeability measurements, membrane retention of compounds was determined. More than 60% of yangonin, desmethoxyyangonin, ketoconazole, and phenazopyridine were retained by the artificial membranes containing phospholipids. Stirring during assay significantly increased the observed permeabilities for highly mobile molecules, but had minimal impact on the poorly permeable molecules. The influence of hydrogen bonding was explored by determining permeabilities using filters coated with dodecane, free of phospholipids. In the filter-IAM method, concentrations were determined by microtitre plate UV spectrophotometry and by LC-MS. HT was achieved with direct UV by the use of 96-well microtitre plate formats and with LC-MS by the use of cassette dosing (five-in-one).

Membranes-biologic-synthetic-comparison: Permeability coefficients of seven compounds belonging to a true homologous series (4-alkylanilines) through several different synthetic and biological membranes were assayed in a two-chamber diffusion cell. Permeability-lipophilicity relationships for the experimental data were established and compared in order to ascertain whether the behavior of these membranes was similar to that of the human skin. In all cases, the best fit for the permeation-lipophilicity correlation was provided by the bilinear model. It was demonstrated that this type of correlation, when a dimethylpolysiloxane membrane is used, is due to the existence of a supplementary stagnant aqueous layer adjacent to the membrane in the receptor compartment. This is clear from the fact that when Polysorbate 80 is added to the receptor solution, the effect of this layer is abolished. In these conditions, hyperbolic equation gives, consequently, the best fit for penetration-lipophilicity correlation. On the basis of the data obtained with rat skin and Polysorbate 80 in the receptor solution, it can be concluded that for biological membranes the bilinear model obtained is due to their heterogeneous nature. The optimum lipophilicity value for penetration according to the bilinear model was not the same for all the membranes assayed. Human and rat skin were qualitatively similar in behavior.

Diez-Sales et al. (1991)

Monolayer systems-comparison: In designing effective therapeutic strategies, novel drugs must exhibit favorable pharmacokinetic properties. The physico-chemical characteristics of a drug, such as pK_a, MW, solubility, and lipophilicity, will influence the way the drug partitions from the aqueous phase into membranes, and thus, will influence its ability to cross cellular barriers, such as the lining of the GIT and the BBB. Physico-chemical characteristics also influence the degree to which a drug is able to cross a barrier layer, and the route by which it does this; whether transcellular (across the cells)—by diffusion, carrier-mediated transport or transcytosis—or paracellular—by diffusing through the tight junctions between the cells. The in vitro model systems that are currently employed to screen the permeation characteristics of a drug often represent a compromise between HT with low predictive potential and low throughput with high predictive potential. Here, we will examine the way in which in vitro cellular permeability assays are often performed and the assumptions that are implied but sometimes forgotten, and we will make simple suggestions for improving the methodological techniques and mathematical equations used to determine drug permeability.

Youdim et al. (2003)

Multilayer membrane-permeability-cellulose lipid: The permeability properties of multilayer planar membranes of uniformly oriented lipids between a pair of cellulose sheets were investigated. The effect of the two cellulose sheets supporting the lipid membrane on the glucose or the Ca permeation was subtracted empirically, and values of $5 \times 10(-6)$ and $8 \times 10(-6)$ cm/sec were thus obtained for the permeability coefficients of an egg yolk lecithin (egg PC)–cholesterol membrane of about 200 bilayers to Ca^{2+} and glucose, respectively. These values are discussed as compared with the permeability coefficients of other model membranes. The membrane permeability was moderately affected by the addition of chemical substances to egg PC membranes. It was reduced by the presence of cholesterol, but enhanced by the presence of isopropanol, n-butanol, or thymol in the same solution above a critical concentration of each compound. These and previous observations suggest that a close correlation may exist between the permeability of the membrane and the orientation of the membrane lipids. The glucose permeation was drastically suppressed by the presence of Ca^{2+} (10 mM) in the sample solution with the membrane containing phosphatidylserine, but was not at all suppressed with the membrane of the egg PC–cholesterol mixture.

Yamamoto et al. (1980)

PAMPA modified-BBB-HT: The recent advances in HTS for biological activities and combinatorial chemistry have greatly expanded the number of drug candidates. Rapid screening for BBB penetration potential early in drug discovery programs provides important information for compound selection and guidance of synthesis for desirable CNS properties. In this paper, we discuss a modification of the PAMPA for the prediction of BBB penetration (PAMPA-BBB). The assay was developed with 30 structurally diverse commercial drugs and validated with 14 Wyeth Research compounds. The PAMPA-BBB assay has the advantages of: predicting passive BBB penetration with high success, HT, low cost, and reproducibility.

Di et al. (2003)

PAMPA-biomimetic-Caco-2-comparison: Several in vitro assays have been developed to evaluate the GI absorption of compounds. Our aim was to compare three of these methods: (*i*) the BAMPA method, which offers a HT, noncellular approach to the measurement of passive transport; (*ii*) the traditional Caco-2 cell assay, the use of which as a HT tool is limited by the long cell differentiation time (21 days); and (*iii*) The BioCoat HTS Caco-2 assay system, which reduces Caco-2 cell differentiation to three days. The transport of known compounds (such as cephalexin, propranolol, or chlorothiazide) was studied at pH 7.4 and 6.5 in BAMPA and both Caco-2 cell models. Permeability data obtained was correlated to known values of human absorption. Best correlations ($r = 0.9$) were obtained at pH 6.5 for BAMPA and at pH 7.4 for the Caco-2 cells grown for 21 days. The Caco-2 BioCoat HTS Caco-2 assay system does not seem to be adequate for the prediction of absorption. The overall results indicate that BAMPA and the 21-day Caco-2 system can be complementary for an accurate prediction of human intestinal absorption.

Miret et al. (2004)

PAMPA-Caco-2-permeability-fluoroquinolones: PAMPA was used to measure the effective permeability, P_e, as a function of pH from 4 to 10, of 17 fluoroquinolones, including three congeneric series with systematically varied alkyl chain length at the 4′N-position of the piperazine residue. The permeability values spanned over three orders of magnitude. The intrinsic permeability, P_o, and the membrane permeability, P_m, were determined from the pH dependence of the effective permeability. The pK_a values were determined potentiometrically. The PAMPA method employed stirring, adjusted such that the unstirred water layer (UWL) thickness matched the estimated 30–100-μm range in the human small intestine. The intrinsic permeability coefficients [10(−6) cm/sec], representing the permeability of the uncharged form of the drug, are for 4′N-R-norfloxacin: 0.7 (R=H), 49 (Me), 132 (n-Pr), 365 (n-Bu); 4′N-R-ciprofloxacin: 2.7 (H), 37 (Me), 137 (n-Pr), 302 (n-Bu); 4′N-R-3′-methylciprofloxacin: 3.8 (H), 20 (Me), 51 (Et), 160 (n-Pr), 418 (n-Bu). Increasing the alkyl chain length in the congeneric

Bermejo et al. (2004)

(*Continued*)

Appendix 1: Recent Studies Reported in Literature (*Continued*)

series resulted in increased permeability, averaging about 0.34 log units per methylene group, except that of the first (H-to-Me), which was about 1.2 log units. These results were compared with Caco-2 and rat in situ permeability measurements. The in situ closed loop technique used for obtaining permeability values in rat showed a water layer thickness effect quite consistent with in vivo expectations. The rat-PAMPA correlation ($r^2 = 0.87$) was better than that of rat-Caco-2 ($r^2 = 0.63$). Caco-2-PAMPA correlation indicated that $r^2 = 0.66$. The latter correlation improved significantly ($r^2 = 0.82$) when the Caco-2 data were corrected for the UWL effect. — Zhu et al. (2002)

PAMPA-Caco-2-log *D*, log *P*-PSA-correlation: Artificial membrane permeability measurement is a potentially HT and low-cost alternative for in vitro assessment of drug absorption potential. It will be an ideal screening/profiling tool in the lead generation program of drug discovery research if it is proven to be generally applicable for classifying drug absorption potential and is advantageous over other in vitro or in silico methods. This study provides an in-depth evaluation of the method in close comparison with Caco-2, log *D*, log *P*, PSA, and QSPR predictions using a large and diverse compound set. It showed that the accuracy of using artificial membrane permeability in assessing drug absorption is comparable with Caco-2, but significantly better than log *P*, log *D*, PSA, and QSPR predictions. This study also explored the artificial membrane composition by adopting a hydrophilic filter membrane for artificial membrane (lecithin-dodecane) support. The use of hydrophilic filter membrane increased the rate of permeation significantly and reduced the transport time to two hours or less as compared with over 10 hr when a hydrophobic filter membrane is used.

PAMPA-Caco-2-permeability-double sink: The aim of this study was to analyze pH-dependent permeability of cationic drugs in Caco-2 cell monolayers using the $pK_{a,flux}$ method and to correlate the results with those obtained in PAMPA. The pH-dependent permeability of verapamil and propranolol was studied in Caco-2 cell monolayers. The data were subsequently processed using software developed for the PAMPA $pK_{a,flux}$ method. Literature values for an additional nine cationic drugs were also analyzed. Double-sink PAMPA data were also obtained for the same cationic drugs, to compare with the Caco-2 data. The Algorithm Builder program was then used to develop a predictive model of Caco-2 permeability based on PAMPA permeability and calculated Abraham molecular descriptors. From the relationship between permeability and pH, it was shown that in PAMPA, only the uncharged form of the drugs permeated across the membrane barrier, while charged and ionized forms of the drugs were significantly permeable in Caco-2. The charged form of permeability, P_i, was therefore determined and subsequently subtracted from all permeability coefficients in Caco-2 to obtain those derived from the PAMPA model. In this study, we have shown that permeability coefficients obtained in PAMPA can predict the passive transcellular permeability in Caco-2. — Avdeef et al. (2005)

PAMPA-HT-oral absorption: Modified composition of lipid solution was used to make a lipid membrane on the filter support. First, they changed the chain length of organic solvent [PC/alkyldienes (C7–C10)]. A negative charge was then added to the membrane to mimic the intestinal membrane (PC/stearic acid/1,7-octadiene and PC/PE/PS/PI/cholesterol/1,7-octadiene). Finally, they examined the predictability of the PC/PE/PS/PI/CHO/1,7-octadiene membrane using structurally diverse compounds. Permeability coefficients of tested compounds were increased as the chain length of alkyldiene became shorter. — Sugano et al. (2001)

The addition of a negative charge to the membrane increased the permeability of the basic compounds. However, the negatively charged membrane with stearic acid showed different permeability profiles from PC/PE/PS/PI/CHO. The predictability of the PC/PE/PS/PI/CHO/1,7-octadiene membrane was adequate ($r = 0.858$, $n = 31$) for use during the early stages of the drug discovery/development process.

PAMPA-hydrophilic basic compound (HBC): The permeation characteristics of a HBC in a bio-mimetic PAMPA were investigated. The bio-mimetic PAMPA membrane was constructed on a hydrophobic filter by impregnating a lipid solution consisting of phosphatidylcholine (0.8%, w/w), phosphatidylethanolamine (0.8%, w/w), phosphatidylserine (0.2%, w/w), phosphatidylinositol (0.2%, w/w), cholesterol (1.0%, w/w), and 1,7-octadiene (97.0%, w/w). The pH–permeability curve (pH 3–10), the effect of lipid composition, concentration dependency (0.02–2.00 mM), and inhibition by other cationic compounds, were investigated for several HBCs. Ketoprofen and methylchlorpromazine were also employed as an acidic and a quaternary ammonium compound, respectively. At pH 3–6, the permeability of timolol, a HBC, was higher than expected from the pH-partition hypothesis, especially in the PI-containing membrane, whereas the pH–permeability curve of ketoprofen followed the pH-partition hypothesis. Permeation of HBC was saturable and inhibited by basic and quaternary ammonium compounds. Similar results were also found for methylchlorpromazine. The permeation characteristics of HBC observed in the present study are not usually expected in a passive permeation process across an artificial membrane. The participation of facilitated permeation of cationic species was suggested, in addition to a simple passive diffusion of un-dissociated species. Ion pair transport was suggested as a possible permeation mechanism of cationic species. However, further investigation is necessary to clarify the reason for the permeation characteristics of HBC.

PAMPA-LC/MS: The analytical techniques typically used for PAMPA sample analysis are HPLC-UV, LC/MS or more recently, the UV-plate reader. The LC techniques, though sturdy and accurate, are often labor- and time-intensive and are not ideal for HT. On the other hand, UV-plate reader technique is amenable to HT, but is not sensitive enough to detect the lower concentrations that are often encountered in early drug discovery work. This article investigates a novel analytical method, a chip-based automated nanoelectrospray mass spectrometric method for its ability to rapidly analyze PAMPA permeability samples. The utility and advantages of this novel analytical method is demonstrated by comparing PAMPA permeability values obtained from nanoelectrospray with those from conventional analytical methods. Ten marketed drugs having a broad range of structural space, physico-chemical properties, and extensive intestinal absorption were selected as test compounds for this investigation. PAMPA permeability and recovery experiments were conducted with model compounds followed by analysis by UV-plate reader, UV-HPLC, and the automated nanoelectrospray technique (nanoESI-MS/MS). There was a very good correlation ($r^2 > 0.9$) between the results obtained using nanoelectrospray and the other analytical techniques tested. Moreover, the nanoelectrospray approach presented several advantages over the standard techniques, such as higher sensitivity and ability to detect individual compounds in cassette studies, making it an attractive HT analytical technique. Thus, it has been demonstrated that nanoelectrospray analysis provides a highly efficient and accurate analytical methodology to analyze.

Sugano et al. (2004)

Balimane et al. (2005)

(Continued)

Appendix 1: Recent Studies Reported in Literature (*Continued*)

PAMPA-lipophilicity-factors: In PAMPA, if more lipid is used than needed to fill all the pores of a microfilter, the excess lipid layer on both sides of the lipophilic filter increases the "apparent" porosity, ε_a, of the filter. The specific resistance of the artificial membrane barrier is lowered with increasing lipid excess. If this effect is not recognized, and the uncorrected value of filter porosity, ε, is used, then the calculated intrinsic permeability and the UWL permeability coefficient of the permeating molecule is significantly overestimated, resulting in underestimates of the thickness of the UWL. UWL corrections are important in pharmaceutical research for in vitro–in vivo correlations (IVIVC) aimed at predicting oral absorption and BBB penetration characteristics of lead candidate compounds. The novel concept of the apparent porosity is introduced, described, and its utility is demonstrated with the drugs diclofenac, desipramine, caffeine, and piroxicam. The PAMPA data of Wohnsland and Faller is taken as an example, where the improved extraordinarily efficient stirring is thought to be better explained in terms of normal stirring when apparent porosity is taken into account in the calculation of the effective permeability coefficient.	Nielsen and Avdeef (2004)
PAMPA-microtiter-permeability-unstirred: Many plate-based in vitro assays of membrane permeability (e.g., Caco-2, MDCK, PAMPA) of sparingly soluble candidate molecules report permeability of water, and not of the intended membrane barrier. This is so because the UWL on both sides of the membrane barrier is rate limiting for these highly permeable molecules. The thickness of this water layer can be 1500–4000 μm in unstirred assays. Under in vivo conditions, however, the UWL is believed to be 30–100-μm thick. Lightly stirred in vitro assays, using plate shakers, cannot lower the thickness of the water layer to match that found in vivo. In this study, 55 lipophilic drugs were employed to characterize the effect of stirring in PAMPA. Highly efficient individual well magnetic stirring at speeds greater than 110 rpm has been demonstrated to lower the UWL thickness to the in vivo range. Stirring at 622 rpm has lowered the layer thickness to 13 μm in some cases, which had not been previously achieved for plate-based permeability assays. With diminished water layer contribution at 622 rpm, for example, the effective permeability of progesterone is $2754 \times 10(-6)$ cm/sec. The new stirring apparatus used in this study is not only suitable for PAMPA, but can also be used in Caco-2 assays. Because of the diminished resistance of the thinner water layer, the stirred PAMPA permeation time has decreased from the usual 15 hr to about 15 min for lipophilic compounds.	Avdeef et al. (2004)
PAMPA-monolayer model comparison: Data from permeability profiling using the PAMPA and cell monolayer (Caco-2 and MDR1-MDCKII) methods were compared for two published compound sets and one in-house set. A majority of compounds in each set correlated ($R^2 = 0.76-0.92$), indicating the predominance of passive diffusion in the permeation of these compounds. Compounds that did not correlate were grouped into two subsets. One subset had higher PAMPA permeability than cell monolayer permeability and consisted of compounds that are subject to secretory mechanisms: efflux or reduced passive diffusion of bases under Caco-2 when run under a pH gradient. The other subset had higher cell monolayer permeability than PAMPA permeability and consisted of compounds that are subject to absorptive mechanisms: paracellular, active transport, or increased passive diffusion of acids under Caco-2 when run under a pH gradient. Given the characteristics of the two methods, these studies suggest how PAMPA and Caco-2 can be synergistically applied for efficient and rapid investigation of permeation mechanisms in drug discovery. During early discovery, all compounds can be rapidly screened using PAMPA at low and neutral pHs to assess passive diffusion permeability to indicate potential for GI and cell assay permeation. During intermediate discovery, selected compounds can be additionally assayed by apical-to-basolateral Caco-2, which, in combination with PAMPA data, indicates susceptibility to additional permeation mechanisms (secretory and absorptive). During mid-to-late discovery, selected candidates can be examined in detail via multiple directional Caco-2 experiments and with transporter inhibitors for complete characterization of permeation mechanisms.	Kerns et al. (2004)

PAMPA-multimembrane model–pH dependence-Caco-2-hexadecane membrane (HDM)-2/4/A1: The investigation of the contribution of the paracellular route to the pH-dependent permeability to cationic drugs in three models expressing different drug permeabilities: HDMs, Caco-2, and 2/4/A1 cell monolayers was made. The high- and low-permeability drugs alfentanil and cimetidine were used as model drugs. The paracellular permeability was calculated: (i) from the assumption that the ionized form (P_{mi}) permeates a cell monolayer only by the paracellular route, and (ii) on the basis of the pore-restricted diffusion. For both drugs, sigmoidal relationships between membrane permeability and pH were observed in all models. The P_{mi} was in excellent agreement with the paracellular permeability of cimetidine in the two cell models, whereas no significant P_{mi} of the drugs could be observed in HDM. The results showed that the paracellular route has a significant role in the permeability of small basic hydrophilic drugs, such as cimetidine in leaky, small intestine-like epithelia, such as 2/4/A1. In contrast, in tighter epithelia such as Caco-2 and in artificial membranes such as HDM, the permeability of the ionized forms of the drugs and the paracellular permeability are lower or insignificant, respectively. These findings will have implications in the experimental design and data interpretation of pH-dependent drug transport experiments in cell culture models as well as in artificial membrane models, such as HDM and PAMPA. — Nagahara et al. (2004)

PAMPA-p$K_{a,flux}$ optimized design (pOD)-permeability: Iso-pH mapping unstirred PAMPA was used to measure the effective permeability, P_e, as a function of pH from 3 to 10, of five weak monoprotic acids (ibuprofen, naproxen, ketoprofen, salicylic acid, benzoic acid), an ampholyte (piroxicam), five monoprotic weak bases (imipramine, verapamil, propranolol, phenazopyridine, metoprolol), and a diprotic weak base (quinine). The intrinsic permeability, P_o, the UWL permeability, P_u, and the apparent pK_a (p$K_{a,flux}$) were determined from the pH dependence of log P_e. The underlying permeability–pH equations were derived for multiprotic weak acids, weak bases, and ampholytes. The average thickness of the UWL on each side of the membrane was estimated to be nearly 2000 μ, somewhat larger than that found in Caco-2 permeability assays (unstirred). As the UWL thickness in the human intestine is believed to be about forty times smaller, it is critical to correct the in vitro permeability data for the effect of the UWL. Without such correction, the in vitro permeability coefficient of lipophilic molecules would be indicative only of the property of water. In single-pH PAMPA (e.g., pH 7.4), the uncertainty of the UWL contribution can be minimized if a specially selected pH (possibly different from 7.4) were used in the assay. From the analysis of the shapes of the log P_e–pH plots, a method to improve the selection of the assay pH, called pOD-PAMPA, was described and tested. From an optimally selected assay pH, it is possible to estimate P_o, as well as the entire membrane permeability–pH profile. — Ruell et al. (2003)

PAMPA-prediction-parallale assays: Kansy et al. (1998) first introduced the PAMPA in 1998. In this system, the permeability through a membrane formed by a mixture of lecithin and an inert organic solvent on a filter support is assessed. PAMPA shows definite trends in the ability of molecules to permeate membranes by transcellular passive diffusion. Its simplicity, low cost, HT, and wide pH range make it very attractive in modern drug discovery. Based on this concept, Wohnsland et al., Sugano et al., and Zhu et al. modified the assay and used it to screen compound permeability. We used PAMPA for the permeation prediction of M100240, which was unable to be determined by cell-based assays due to compound instability. In this study, 92 commercially available agents provided the structural diversity used to generate a mathematical prediction model for human fraction absorbed, M100240—an acetate thioester of MDL 100, 173. Permeation of M100240 and MDL 100, 173 was evaluated using the PAMPA. The donor and recipient solutions consisted of 0.5 N HCl (pH 1.5) or phosphate-buffered saline (pH 5.5 or 7.4) with 2% DMSO. The donor solution also contained 200 mM M100240 or MDL 100, 173. It was predicted that M100240 is likely to be well absorbed via passive diffusion across the human GIT following oral administration. — Hwang et al. (2003)

(Continued)

Appendix 1: Recent Studies Reported in Literature (*Continued*)

PAMPA-QSAR-permeability: To evaluate the absorption of drugs with diverse structures across a membrane via the transcellular route, their permeability was measured using the PAMPA. The permeability coefficients obtained by PAMPA were analyzed using a classical QSAR approach with simple physico-chemical parameters and 3D-QSAR, VolSurf. Correlation equations were formulated for diverse drugs similar to the equation obtained for peptide-related compounds in the previous study. The hydrogen-bonding ability of molecules, not only the hydrogen-accepting ability but also the hydrogen-donating ability, in addition to hydrophobicity at a particular pH, was significant in determining variations in PAMPA permeability coefficients. Based on this result, an in silico good prediction model for the passive transcellular permeability of diverse structural compounds was obtained. The artificial lipid-membrane permeability coefficients of the drugs, except salicylic acid, were well correlated with the Caco-2 permeability in a previous report suggesting the importance of absorption by the transcellular mechanism for these drugs.	Fujikawa et al. (2005)
PAMPA-QSAR-VolSurf: To evaluate absorption of compounds across the membrane via a transcellular route, the permeability of peptide derivatives and related compounds was measured by the PAMPA. The permeability coefficients by PAMPA were analyzed quantitatively using classical QSAR and VolSurf approaches with the physico-chemical parameters. The results from both approaches showed that hydrogen bonding ability of molecules in addition to hydrophobicity at a particular pH were significant in determining variations in PAMPA permeability coefficients. The relationship between Caco-2 cell permeability and artificial lipid membrane permeability was then determined. The compounds were sorted according to their absorption pathway in the plot of the Caco-2 cell and PAMPA permeability coefficients.	Ano et al. (2004)
PAMPA-UV/VIS LC: This study compares the use of UV-VIS detection with LC/MS detection for the PAMPA permeability determination of compounds in the drug discovery stage. LC/MS detection offers a selective and sensitive method for the determination of the PAMPA permeability for compounds that do not contain a UV chromophore or possess a low UV extinction coefficient. To enhance the reliability of our permeability measurements for compounds with low aqueous solubility, we demonstrated the use of LC/MS detection as a means for facilitating the study of solubilizing agents to enhance aqueous solubility that normally would interfere with UV-VIS detection. In doing so, the PAMPA assay can be expanded to study the in vitro permeability of poorly water soluble compounds and evaluate the effects of solubilizers' on the membrane permeability of different compounds. This might be useful in selecting solubilizers for poorly water soluble compounds to be used for further in vivo studies. A diverse set of 20 drugs using UV-VIS detection were compared with data using LC/MS detection. A PAMPA screening method was designed, which used solubilizers (Brij 35, Cremophor EL, ethanol, and Tween 80) for compounds with low aqueous solubility. The stability of the artificial membrane was determined using various solubilizer concentrations (0.1–5% w/v) to ensure that the phospholipid membrane was not disrupted. Two compounds, amiodarone and miconazole, with low aqueous solubility yielding an undetected response in the PAMPA assay using UV-VIS detection were subjected to the different solubilizing agents and their PAMPA permeability was measured using LC/MS detection. Most of the compounds showed similar PAMPA permeability using the two detection systems. However, for compounds lacking a UV chromophore or with a low UV extinction coefficient, LC/MS was the detection method of choice for determination of PAMPA permeability values.	Liu et al. (2003)

LC/MS also gave reliable quantification data for compounds containing impurities, as well as compounds that were not stable during the assay. Although many solubilizers were found to interfere with UV-VIS detection, the LC/MS approach was applicable to determine the permeability values of compounds with normally low aqueous solubility. LC/MS detection offered greater sensitivity and selectivity as compared with UV-VIS detection for the PAMPA assay. With this added versatility in detection, PAMPA can be used in both discovery and preformulation applications, which has not been described earlier.

Permeability-molecular surface area-in vitro-in silico model: The permeability values obtained from the Caco-2 cell monolayers have been traditionally used to devise in silico models for the prediction of drug absorption. In this paper, the use of molecular surface areas as descriptors of permeability and solubility will be reviewed. Moreover, a virtual filter for the prediction of oral drug developability based on the successful combination of in vitro and in silico models of drug permeability and aqueous drug solubility will be discussed. — Bergstrom (2005)

Permeability-NMR-light scattering: A combined method using NMR line-broadening for permeant lifetime determination and dynamic light scattering for vesicle size determination has been developed for the measurement of permeability coefficients of ionizable permeants across large phospholipid:cholesterol unilamellar vesicles. The method has been validated by examining its reproducibility and the influence of various factors that might affect the permeability measurements. The vesicle hydrodynamic diameter was varied between 0.1 and 0.2 μ by extruding multilamellar vesicles through polycarbonate membranes with different pore sizes (0.03–0.2 μ). For these large unilamellar vesicles, the normalized size distributions analyzed by the CONTIN method had standard deviations <0.36, which led to errors in permeability coefficients <10% as predicted from a theoretical model developed here. The permeability coefficient for acetic acid is independent of its concentration, vesicle hydrodynamic diameter, the concentration of Pr^{3+}, and ionic strength over the ranges 0.01–0.2 M, 0.1–0.2 μ, 0.004–0.04 M, and 0.03–0.3 M, respectively. Membrane/water and decane/water partition coefficient measurements of acetic acid indicate that the effects of permeant binding onto the bilayer membrane and self-association are negligible within the permeant concentration range 0.01–0.2 M. The addition of Pr^{3+} ions induces vesicle fusion with rates increasing with temperature and decreasing with cholesterol concentration in the membranes. While the intravesicular resonance intensity for acetic acid decreases continuously with time due to vesicle fusion under certain conditions, the corresponding line width and chemical shift remain constant over the same period, highlighting an important advantage of this NMR method over those based on detecting net flux in response to a concentration gradient as there is no means in the latter experiments of discerning vesicle leakiness from passive diffusion rates. The effective chemical nature of a dimristoylphosphatidylcholine:cholesterol bilayer barrier microenvironment was explored by comparing the transport of two permeants, D-(−)-mandelic acid and phenylacetic acid, with their relative bulk solvent/water partition coefficients using three reference solvents (n-decane, 1,9-decadiene, and isoamyl alcohol). Using the NMR line-broadening method, the permeability coefficients for these two permeants were determined to be $(2.9 \pm 0.4) \times 10(-4)$ cm/sec and $(3.9 \pm 0.7) \times 10(-2)$ cm/sec, respectively, at 294 K and $X_{chol} = 0.3$. The incremental free energy of transport for the additional OH group in D-(−)-mandelic acid, $\Delta G_0 = +2.9$ kcal/mol, resembles most closely that for the transfer of this group from water to 1,9-decadiene, suggesting that the barrier domain resides in the acyl chain region and is slightly more polar/polarizable than a saturated hydrocarbon, possibly due to the presence of a double bond in cholesterol and/or the proximity of the barrier domain to the hydrophilic interface. — Xiang and Anderson et al. (1995)

(Continued)

Appendix 1: Recent Studies Reported in Literature (*Continued*)

Permeation enhacer-cyclodextrin-permeability: It is well known that cyclodextrins can enhance the permeation of poorly soluble drugs through biological membranes. However, the permeability will decrease if cyclodextrin is added in an excess of the concentration needed to solvate the drug. The mechanism of cyclodextrin effect on drug permeability has not been fully explained. The effect of cyclodextrins cannot be explained due to increased solubility of the drug in the aqueous donor phase nor can it be explained by assuming that cyclodextrins act as classical permeation enhancers, that is, by decreasing the barrier function of the lipophilic membrane. In the present work, we have modeled the effect of cyclodextrins in terms of mixed barrier consisting of both diffusion and membrane-controlled diffusion, where the diffusion of the drug in the aqueous diffusion layer is significantly slower than in the bulk of the donor. This diffusion model is described by a simple mathematical equation where the properties of the system are expressed in terms of two constants P_M/K_d and $M_{1/2}$. Data for the permeation of hydrocortisone through hairless mouse skin in the presence of various cyclodextrins, and cyclodextrin polymer mixtures, were fitted to obtain values for these two constants. The rise in flux with increased cyclodextrin complex concentration and fall with excess cyclodextrin was accurately predicted. Data for the permeation of drugs through semi-permeable cellophane membrane could also be fitted to the equation. It was concluded that cyclodextrins act as permeation enhancers carrying the drug through the aqueous barrier, from the bulk solution toward the lipophilic surface of biological membranes, where the drug molecules partition from the complex into the lipophilic membrane.
— Masson et al. (1999)

Permeation model-solute flux: Methyl- and propylparaben flux from various alcohol donors through polydimethylsiloxane membranes was investigated. Flux from saturated alcohol vehicles was markedly increased relative to water and glycol systems. The uptake of neat alcohol, a measure of solvent membrane interaction, gave a good rank order correlation to the flux data for a particular paraben. The major influence of the alcohols was an increase in membrane solubility of paraben, with a smaller effect on the diffusion coefficient. High paraben donor solubility indirectly reduced the solvent–membrane interaction leading to attenuated flux. Paraben membrane solubility was influenced by the amount of alcohol sorbed from saturated systems and the affinity of the paraben for the alcohol. This conforms to the concept of imbibed alcohol molecules being organized into clusters. The alteration in barrier properties of the membrane was found to require the presence of sorbed alcohol and was reversible upon removal of the solvent.
— Twist and Zatz (1988)

Permeation-skin-gas chromatography (GC)/MS: A silastic membrane was coated onto a fiber to be used as a permeation membrane. The MCF was immersed in the donor phase to partition the compounds into the membrane. At a given partition time, the MCF was transferred into a GC injector to evaporate the partitioned compounds for quantitative and qualitative analyses. This technique was developed and demonstrated to study the percutaneous permeation of a complex mixture consisting of 30 compounds. Each compound permeated into the membrane was identified and quantified with GC/MS. The standard deviation was less than 10% in 12 repeated permeation experiments. The partition coefficients and permeation rates in static and stirred donor solutions were obtained for each compound. The partition coefficients measured by this technique were well correlated ($R^2 = 0.93$) with the reported octanol/water partition coefficients. This technique can be used to study the percutaneous permeation of chemical mixtures. No expensive radiolabeled chemicals were required. Each compound permeated into the membrane can be identified and quantified. The initial permeation rate and equilibrium time can be obtained for each compound, which could serve as characteristic parameters regarding the skin permeability of the compound.
— Xia et al. (2003)

Permeation-solubility-intestinal absorption: Absorption simulations were carried out for virtual monobasic drugs having a range of pK_a, log D, and dose values as a function of presumed solubility and permeability. Results were normally expressed as the combination that resulted in 25% absorption. Absorption of basic drugs was found to be a function of the whole solubility/pH relationship rather than a single solubility value at pH 7. In addition, the parameter spaces of greatest sensitivity were identified. We compared three theoretical scenarios: the GIT pH range overlapping: (*i*) only the salt solubility curve, (*ii*) the salt and base solubility curves, or (*iii*) only the base curve. Experimental solubilities of 32 compounds were determined at pHs of 2.2 and 7.4, and they nearly all fitted into two of the postulated scenarios. Typically, base solubilities can be simulated in silico, but salt solubilities at low pH can only be measured. We concluded that quality absorption simulations of candidate drugs in most cases require experimental solubility determination at two pHs, to permit calculation of the whole solubility/pH profile. — Hendriksen et al. (2003)

Pharmacokinetic-physical property relationship review: The ADME (absorption and distribution in the body, metabolism and elimination from the body) profile of a drug determines its pharmacokinetic behavior. Modern drug design includes the modeling of pharmacokinetic parameters that are of concern are intestinal absorption, BBB passage, and metabolism. Traditionally, experimental parameters, such as partition coefficients and chromatographic capacity factors have been used for the estimation of intestinal absorption or BBB passage of newly synthesized compounds. Several studies have shown a sigmoidal relationship between intestinal absorption and lipophilicity. The latter is usually expressed by the apparent partition coefficient log D in a biphasic system at physiological pH or by the affinity to a lipophilic phase determined by chromatographic techniques. In contrast, structure-based descriptors need no experimental investigation of the compound studied. The most relevant descriptors give information on hydrogen-bonding characteristics and molecular volume. In recent years, attempts have been made to recognize substrates for MDR proteins by their structure characteristics without crucial success. There is evidence that MDR is not only driven by direct protein-substrate recognition, but also by the behavior of the compound in the lipid environment of the protein. — Kramer and Wunderli-Allenspach (2001)

Physiologic model-physiologically based pharmacokinetic model (PB/PK): A physiologically based model for GI transit and absorption in humans is presented. The model can be used to study the dependency of the fraction dose absorbed (F_{abs}) of both neutral and ionizable compounds on the two main physico-chemical input parameters [the intestinal permeability coefficient (P_{int}) and the solubility in the intestinal fluids (S_{int})] as well as the physiological parameters, such as the gastric emptying time and the intestinal transit time. For permeability-limited compounds, the model produces the established sigmoidal dependence between F_{abs} and P_{int}. In case of solubility-limited absorption, the model enables calculation of the critical mass-solubility ratio, which defines the onset of nonlinearity in the response of fraction absorbed to dose. In addition, an analytical equation to calculate the intestinal permeability coefficient based on the compound's membrane affinity and MW was used successfully in combination with the PB-PK model to predict the human fraction dose absorbed of compounds with permeability-limited absorption. Cross-validation demonstrated a root-mean-square prediction error of 7% for passively absorbed compounds. — Willmann et al. (2004)

(Continued)

Appendix 1: Recent Studies Reported in Literature (*Continued*)

Red blood cells (RBC) partitioning–LC/MS/MS: A novel LC/MS/MS-based depletion method for measuring compound partitioning between human plasma and RBCs in a drug discovery environment is presented. Conventionally, RBC partitioning is determined by separate measurements of drug concentrations in equilibrating plasma and whole blood or RBC using separate standards prepared in their respective matrices, that is, in plasma and whole blood or RBC lysates. The process is very tedious, labor-intensive, and difficult to automate. In addition, interferences from the heme and other highly abundant cellular composites make the measurement of the drug concentration in whole blood or RBC inevitably variable even with a highly specific LC/MS/MS method. Therefore, there is an imminent need to develop a straightforward and fast method to assess the partitioning of drug-like compounds in RBC. This work describes an LC/MS/MS-based depletion assay that measures the compound concentration in plasma that has been equilibrating with RBC. Compounds were spiked into fresh human whole blood and plasma, respectively to a final concentration of 500 nM. Both the spiked whole blood and plasma control were incubated at 37°C for up to 60 min. During the time course, aliquots of plasma and whole blood from both incubation mixtures were sampled at 10 and 60 min. The whole blood samples were centrifuged to yield the plasma. The plasma samples from both incubations were extracted using a protein precipitation method, and analyzed using LC/MS/MS under the multiple-reaction monitoring (MRM) mode. The RBC partitioning ratio was calculated using the analyte peak area responses of the plasma samples through an equation deduced in this work. The method was first tested using two commercial compounds, phenoprobamate and acetazolamide, to determine the optimal incubation conditions and the concentration dependency of the assay. The assay reproducibility was also assessed by three interday assays for phenoprobamate. This method was further evaluated using 20 commercial compounds of different classes with a wide range of RBC partitioning coefficients and the results were compared with those reported in the literature. Excellent correlation ($R^2 = 0.9396$) was found between the measured and literature values. In addition, several proprietary compounds were assayed using both the new and traditional methods, and the measured partitioning ratios from the two methods, are equivalent. The experiments in this work demonstrate that the LC/MS/MS-based depletion method can provide direct and accurate measurement of RBC partitioning for compounds in drug discovery.

Yu et al. (2005)

Skin permeation–artifical membrane comparison: In order to measure the contribution of lipid and pore (aqueous) pathways to the total skin permeation of drugs, and to establish a predictive method for the steady-state permeation rate of drugs, the relationship between permeability through excised hairless rat skin and some physico-chemical properties of several drugs were compared with those through polydimethylsiloxane (silicone) and poly(2-hydroxyethyl methacrylate) (pHEMA) membranes, as typical solution-diffusion and porous membranes, respectively. A linear relationship was found between the permeability coefficients of drugs for the silicone membrane and their octanol/water partition coefficients. For the pHEMA membrane, the permeability coefficients were almost constant independent of the partition coefficient. On the other hand, the skin permeation properties could be classified into two types: one involves the case of lipophilic drugs, where the permeability coefficient is correlated to the partition coefficient, similar to the silicone membrane; and the other involves hydrophilic drugs, where the permeability coefficients were almost constant, similar to pHEMA membrane. From these results, the stratum corneum, the main

Hatanaka et al. (1990)

Skin-snake-model percutaneous absorption: Relationships between the in vitro permeability of basic compounds through shed-snake skin as a suitable model membrane for human stratum corneum and their physio-chemical properties were investigated. Compounds with low pK_a values were selected to compare the permeabilities of the nonionized forms of the compounds. Steady-state penetration was achieved immediately without a lag time for all compounds. Flux rate and permeability coefficient were calculated from the steady-state penetration data and relationships between these parameters and the physico-chemical properties were investigated. The results showed that permeability may be controlled by the lipophilicity and the molecular size of the compounds. Equations were developed to predict the permeability from the MWs and the partition coefficients of basic compounds. — Takahashi et al. (1993)

Solubility-ADME-HT-disadvantages: There are currently about 10,000 drug-like compounds. These are sparsely, rather than uniformly, distributed through chemistry space. True diversity does not exist in experimental combinatorial chemistry screening libraries. ADME and chemical reactivity-related toxicity is low, while biological receptor activity is of higher dimension in chemistry space, and this is partly explainable by evolutionary pressures on ADME to deal with endobiotics and exobiotics. ADME is hard to predict for large data sets because current ADME experimental screens are multimechanisms, and predictions get worse as more data accumulates. Currently, screening for biological receptor activity precedes or is concurrent with screening for properties related to "drugability." In the future, "drugability" screening may precede biological receptor activity screening. The level of permeability or solubility needed for oral absorption is related to potency. The relative importance of poor solubility and poor permeability towards the problem of poor oral absorption depends on the research approach used for lead generation. Rational drug design approaches have advanced clinical candidate leads to time-dependent higher MW, higher H-bonding properties, unchanged lipophilicity, and, hence, poorer permeability. A HTS-based approach for early candidates leads to higher MW, unchanged H-bonding properties, higher lipophilicity, and, hence, poorer aqueous solubility. — Lipinski (2000)

Solution precipitation (SP)-powder dissolution-solubility (PD) screening: SP method, in which the sample solutions are prepared by adding the drug solution in DMSO to buffers followed by filtering off the precipitate using 96-well filterplate and a PD method, in which the solid samples are dissolved to the buffer in the HPLC vial equipped with the filter membrane in the HPLC autosampler were used. An HPLC equipped with a photodiode array detector is used to measure the concentration of the sample solutions in both methods. The SP method was used for HTS of the solvating process of the candidates in aqueous solutions with lower sample consumption, and the PD method was used for screening both intermolecular interaction in solid state and solvation in aqueous solution with more sample amount than that of the SP method. Therefore, the solubility screening from early to final stages of LO process would be successfully accomplished by using both methods complementarily. — Sugaya et al. (2002)

6 Solid-State Properties

INTRODUCTION
Solid-state characterization is one of the most important functions of the preformulation group, which is assigned the responsibility of making recommendations for further formulation work on a lead compound. Physical properties have a direct bearing on both physical and chemical stabilities of the lead compound. Much of the later work on formulation will depend on how well the solid state is characterized from the decisions to compress the drug into tablets to the selection of appropriate salt forms. The studies reported in this section, of course, apply to those drugs that are available in solid form, crystalline or amorphous, pure or amalgamated.

Physical properties affected by the solid-state properties can influence both the choice of the delivery system and the activity of the drug, as determined by the rate of delivery. Chemical stability, as affected by the physical properties, can be significant. Whereas, it is always desirable to enhance chemical stability (a pursuit of the synthetic chemist), modulation of physical properties, such as reducing the hygroscopicity by increasing the hydrophobicity of an acid, or by moving to carboxylic rather than sulfonic or mineral acid, or to use an acid of higher pK_a to raise the pH of a solution often provides more stable compounds. Stability is also improved by decreasing the solubility and increasing the crystallinity by increasing the melting. It is important to realize that factors that improve the chemical stability often impact adversely the physical properties. Therefore, a fine balance must be achieved when selecting between the physical properties of a chemical property modulation.

The stability of the salt could also be an important issue, and depending on the pK_a, many properties can change, including indirectly related physical characteristics, such as volatility (e.g., hydrochloride salts are often more volatile than sulfate salts). Discoloration of the salt form of drugs is also prominent for some specific forms, as the oxidation reactions (often accompanied by hydrolysis) are a result of factors, such as affinity for moisture, surface hydrophobicity, and so on. Hydrolysis of a salt back to the free base may also take place if the pK_a of the base is sufficiently weak.

CRYSTAL MORPHOLOGY
A crystalline species is defined as a solid that is composed of atoms, ions, or molecules arranged in a periodic, three-dimensional (3D) pattern. A 3D array is called a lattice, as shown in Figure 1. The requirement of a lattice is that each volume, which is called a unit cell, is surrounded by identical objects. Three vectors, a, b, and c, are defined in a right-handed sense for a unit cell. However, as three vectors are quite arbitrary, a unit cell is described by six scalars, a, b, c, α, β, and γ without directions (Fig. 2). Several kinds of unit cells are possible, for example, if $a = b = c$ and $\alpha = \beta = \gamma = 90°$, the unit cell is cubic. It turns out that only seven different kinds of unit cells are necessary to include all the possible lattices. These correspond to the seven crystal systems as shown in Table 1.

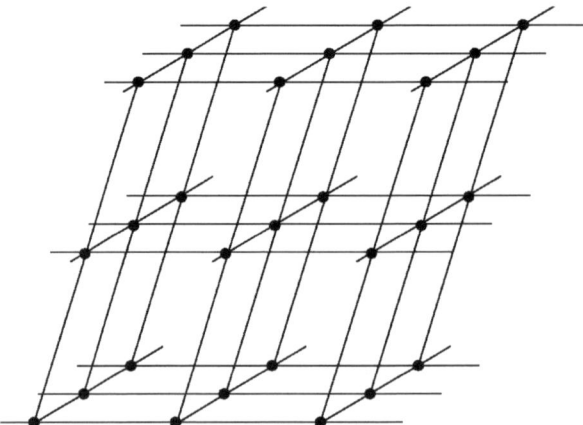

Figure 1 Crystal lattice.

The seven different point lattices can be obtained simply by putting points at the corners of the unit cells of the seven crystal systems. However, there are more possible arrangements of points, which do not violate the requirements of a lattice. The French crystallographer, Bravais, proposed 14 possible point lattices, as shown in Figures 3 and 4, as a result of combining the seven crystal systems and centered points.

Symmetry operations are divided into macroscopic and microscopic operations. Macroscopic operations can be deduced from the arrangement of well-developed crystal faces, without any knowledge of the atomic arrangement inside the crystals, whereas the microscopic operations depend on the atomic arrangement (Table 2) that cannot be inferred from the external growth of the crystal. Reflection, rotation, inversion, and rotation–inversion are included in macroscopic operations, whereas glide planes and screw axes belong to microscopic operations. The combination of macroscopic operations with the seven crystal systems leads to 32 possible groups, and they are called 32 point groups. The microscopic symmetry operations describe the way in which the atoms or molecules in crystals are combined to 32 point groups with 14 Bravais lattices, resulting in 230 combinations, called 230 space groups.

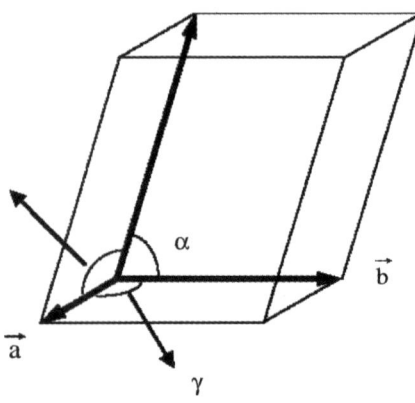

Figure 2 Scalars of lattice structure.

Table 1 Seven Crystal Systems

Crystal system	Axial lengths and angles
Cubic	$a = b = c$
	$\alpha = \beta = \gamma = 90°$
Tetragonal	$a = b \neq c$
	$\alpha = \beta = \gamma = 90°$
Orthorhombic	$a \neq b \neq c$
	$\alpha = \beta = \gamma = 90°$
Rhombohedral	$a = b = c$
Trigonal	$\alpha = \beta = \gamma \neq 90°$
Hexagonal	$a = b \neq c$
	$\alpha = \beta = 90°, \gamma = 120°$
Monoclinic	$a \neq b \neq c$
	$\alpha = \gamma = 90° \neq \beta$
Triclinic	$a \neq b \neq c$
	$\alpha \neq \beta \neq \gamma \neq 90°$

A crystalline particle is characterized by definite external and internal structures. Habit describes the external shape of a crystal, whereas polymorphic state refers to the definite arrangement of molecules inside the crystal lattice. Crystallization is invariably employed as the final step for the purification of a solid. The use of different solvents and processing conditions may alter the habit of recrystallized particles, besides modifying the polymorphic state of the solid. Subtle changes in crystal habit at this stage can lead to significant variation in raw-material characteristics. Furthermore, various indices of dosage form performance, such as particle orientation, flowability, packing, compaction, suspension stability, and dissolution can be altered even in the absence of significantly altered polymorphic state. These effects are a result of the physical effect of different crystal habits. In addition, changes in crystal habit either accompanied or not by polymorphic transformation during processing or storage, can lead to serious implications of physical stability in dosage forms. Therefore, in order to minimize the variations in raw-material characteristics, to ensure the reproducibility of results during preformulation, and to correctly judge the cause of instability and poor performance of a dosage form, it is essential to recognize the importance of changes in crystal surface appearance and habit of pharmaceutical powders.

The crystal habit is also affected by impurities present in the crystallizing solution; often these impurities provide the earliest nucleation of crystal growth, and become an integral part of the crystal. In some instances, the presence of impurities inhibit crystal growth, as shown, when certain dyes or heavy metals are mixed with solutions. If an impurity can adsorb at the growing face, it can significantly alter the course of crystal growth and geometry. The habits bound by plane faces are termed *euhedral* and those with irregularly shaped ones are called *anhedral*. The symmetry of a crystal is generally studied by using optical goniometer that allows the measurement of the angles between the crystal faces. This technique is of use only when good crystals of size >0.05 mm in each direction can be obtained, which is generally not the case.

Chemical crystallography provides accurate and precise measurements of molecular dimensions in a way that no other science can begin to approach. Historically, single crystal X-ray diffraction was used to determine the structure of what

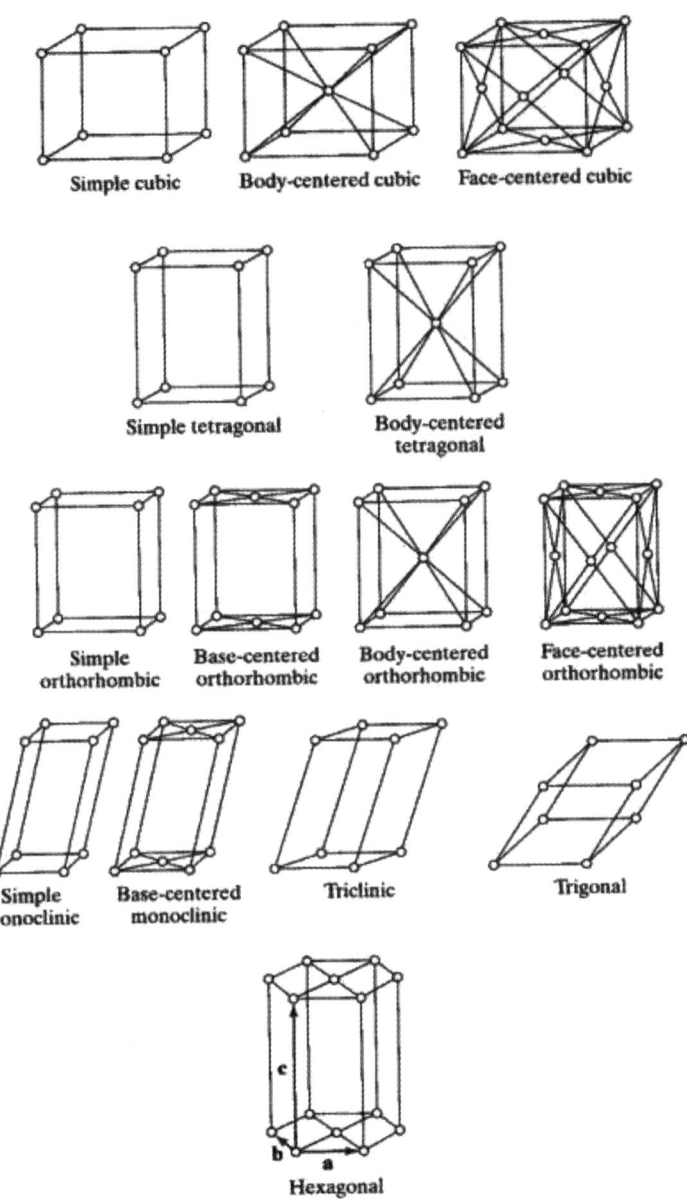

Figure 3 Bravais lattice system.

was thought of as "small molecules." Twenty years ago, it was possible to solve structures with an average of only 100 nonhydrogen atoms. However, with developments in hardware and software, the upper limit has risen to about 500 and recently, even a 1000-atom structure was solved. The APEX II line of Chemical Crystallography Solutions (1) allow single crystal structure determination. The APEX II detector is suitable for fast processing. The Brucker SHellXTL software system

Solid-State Properties

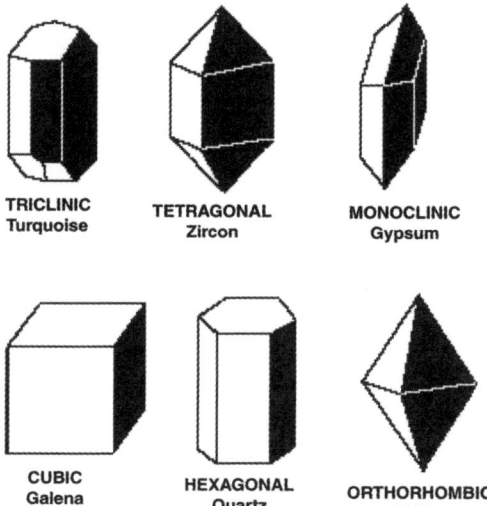

Figure 4 Common crystal habits.

works well with these systems and provides a complete characterization that is suitable for publication would include the following data:

1. Data collection
   - source of sample and conditions of crystallization;
   - habit, color, and dimensions of the crystal;
   - formula, formula weight;
   - unit cell parameters and volume with esds. The number of data and theta range of data used to determine the cell parameters;
   - crystal type and space group;
   - Z, density, and absorption coefficient;
   - instrument and temperature of data collection and cell parameter determination.
2. Structure solution
   - number of data collected, unique [R(int)];
   - method and program used for structure solution;
   - absorption correction details.
3. Structure refinement
   - method and program for refinement;
   - number of data refined, restraints, parameters;
   - weighting scheme;
   - R_1 (observed data), wR_2 (all data), S values;
   - final maximum shift/error;
   - final maximum and minimum of difference electron density map.
4. Tables and figures
   - positional parameters and isotropic or equivalent displacement parameters;
   - bond distances, angles, and torsion angles;
   - anisotropic displacement parameters;

Table 2 Common Crystal Habits

Habit	Description	Example
Acicular	Needle-like, slender, and/or tapered	Rutile in quartz
Amygdaloidal	Almond-shaped	Heulandite
Anhedral	Poorly formed, distorted	Olivine
Bladed	Blade-like, slender, and flattened	Kyanite
Botryoidal or globular	Grape-like, hemispherical masses	Smithsonite
Columnar	Similar to fibrous: long, slender prisms often with parallel growth	Calcite
Coxcomb	Aggregated flaky or tabular crystals closely spaced	Barite
Dendritic or arborescent	Tree-like, branching in one or more direction from central point	Magnesite in opal
Dodecahedral	Dodecahedron, 12-sided	Garnet
Drusy or encrustation	Aggregate of minute crystals coating a surface	Uvarovite
Enantiomorphic	Mirror-image habit and optical characteristics; right- and left-handed crystals	Quartz
Equant, stout, stubby, or blocky	Squashed, pinnacoids dominant over prisms	Zircon
Euhedral	Well formed, undistorted	Spinel
Fibrous or columnar	Extremely slender prisms	Tremolite
Filiform or capillary	Hair-like or thread-like, extremely fine	Natrolite
Foliated or micaceous	Layered structure, parting into thin sheets	Mica
Granular	Aggregates of anhedral crystals in matrix	Scheelite
Hemimorphic	Doubly terminated crystal with two differently shaped ends	Hemimorphite
Mamillary	Breast-like: intersecting large rounded contours	Malachite
Massive or compact	Shapeless, no distinctive external crystal shape	Serpentine
Nodular or tuberose	Deposit of roughly spherical form with irregular protruberances	Geodes
Octahedral	Octahedron, eight-sided (two pyramids base to base)	Magnetite
Plumose	Fine, feather-like scales	Mottramite
Prismatic	Elongate, prism-like: all crystal faces parallel to c-axis	Tourmaline
Pseudo-hexagonal	Ostensibly hexagonal because of cyclic twinning	Aragonite
Pseudomorphous	Occurring in the shape of another mineral through pseudomorphous replacement	Tiger's eye
Radiating or divergent	Radiating outward from a central point	Pyrite suns
Reniform or colloform	Similar to mamillary: intersecting kidney-shaped masses	Hematite
Reticulated	Acicular crystals forming net-like intergrowths	Cerussite
Rosette	Platy, radiating rose-like aggregate	Gypsum
Sphenoid	Wedge-shaped	Sphene
Stalactitic	Form as stalactites or stalagmites; cylindrical or cone-shaped	Rhodochrosite
Stellate	Star-like, radiating	Pyrophyllite
Striated/striations	Surface growth lines parallel or perpendicular to c-axis	Chrysoberyl
Tabular or lamellar	Flat, tablet-shaped, prominent pinnacoid	Ruby
Wheat sheaf	Aggregates resembling hand-reaped wheat sheaves	Zeolites

- structure factor tables (often required for review but discarded by the journal);
- torsion angles (optional);
- least-square planes (optional);
- hydrogen bond geometry (optional);
- a labeled figure showing the displacement ellipsoids;
- a packing diagram showing relevant intermolecular interactions.

The modeling of the habits of crystals is a subject of many sophisticated computer programs, such as CERIUS2 (2) that also provides the effect of additives. The Bravais, Friedel, Donny, and Harker (BFDH) model and the attachment energy model, in conjunction with force field methods, are used in habit prediction. The attachment energy approach gives the growth morphology of the crystal studied, but it is also possible to calculate the shape of a small particle in equilibrium with its growth environment by computing the surface energy of each relevant face.

Surface interactions between solvent molecules and growing faces can also be modeled. It is well known that the stronger the solvent binds to a particular face, the more it will inhibit the growth of that face, so as to affect the morphology. This can be simulated by the computer. The ability to predict the crystal morphology, that is, to identify the key growth faces, combined with the ability to analyze the surface chemistry of each of the faces in detail (including interactions with solvent molecules, excipients and impurities), enables rational control of morphology and crystal growth. For example, an undesirable morphology (a plate) can be transformed into a more isometrical shape.

In addition to morphological assessments of crystals, optical microscopy can be used to measure their refractive indices. To identify the crystal, it is not necessary to measure the principal refractive indices; simply measuring two that are unique and reproducible is sufficient. These are termed the key refractive indices that, according to these researches, are all that are needed to identify any particular compound.

POLYMORPHISM

Both organic and inorganic pharmaceutical compounds can crystallize into two or more solid forms that have the same chemical composition; this is called polymorphism. Polymorphs have different relative intermolecular and/or interatomic distances and unit cells, resulting in different physical and chemical properties, such as density, solubility, dissolution rate, bioavailability, and so on. Crystal structures containing solvents (or water) are often called psudopolymorphs, with distinct physical and chemical properties. It is possible for each pseudopolymorph to have many polymorphs. In polymorphism, the crystal lattice formation can take place through two mechanisms: packing polymorphism and conformational polymorphism. Packing polymorphism represents the formation of different crystal lattices of conformationally rigid molecules that can be rearranged stably into different 3D structures through different intermolecular mechanisms. When a nonconformationally rigid molecule can be folded into alternative crystal structures, the polymorphism is categorized as conformational polymorphism.

Polymorphs and pseudopolymorphs can also be classified as monotropes or enantiotropes, depending upon whether or not one form can transform reversibly to another. In a monotropic system, Form I does transform to Form II, because the transition temperature cannot appear before the melting temperature (Fig. 5,

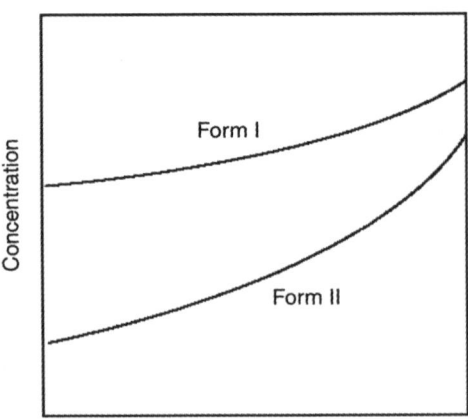

Figure 5 Monotropic system as a function of temperature (x-axis).

monotropy). In Figure 6 (enantiotropy), Form II is stable over a temperature range below the transition temperature, at which two solubility curves meet, and Form I is stable above the transition temperature. At the transition temperature, reversible transformation between two forms occurs. Figure 7 (enantiotropy with metastable phases) shows the kinetic effects on the thermodynamic property of solubility, which shows Ostwald ripening effect. An unstable system does not necessarily transform directly into the most stable state, but into one which most closely resembles its own, that is, into another transient state, whose formation from the original is accompanied by the smallest loss of free energy.

When a decision needs to be made on whether two polymorphs are enantiotropes or monotropes, it is very useful to use the thermodynamic rules developed by Burger and Ramberger, which are tabulated in Table 3.

The stability of polymorphs is thermodynamically related to their free energy. The more stable polymorph has the lower free energy at a given temperature. The aforementioned classification of polymorphic substances into monotropic and enantiotropic classes, from the lattice theory perspective is not always appropriate. There is a need to explore the way the crystal lattice structures of

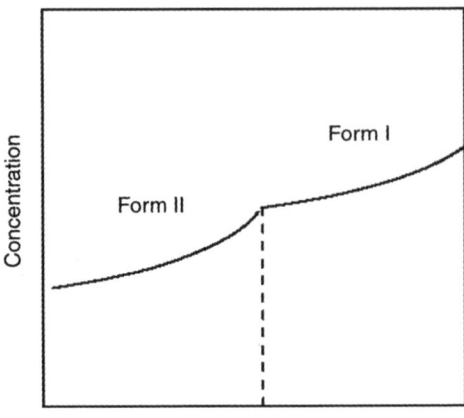

Figure 6 Enantiotropic system as a function of temperature (x-axis).

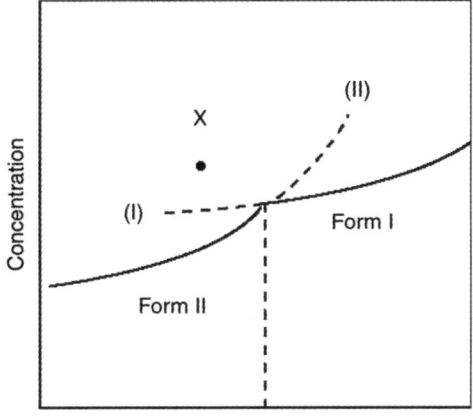

Figure 7 Enantiotropic system with metastable phases as a function of temperature (x-axis).

polymorphs are related. At a transition point, with the temperature and the pressure fixed, it is possible for interconversion to take place between two polymorphs only in the case where the structures of the polymorphs are related. If complete rearrangement is required by atoms or molecules during transformation, no point of contact for reversible interconversion exists. Therefore, the existence of enantiotropes or monotropes in thermodynamics and phase theory corresponds to related or unrelated lattice structures in structural theory. Transformation between polymorphs that have completely different lattice structures exhibits dramatic changes in properties. The difference in energy between polymorphs is not always considerable as shown with diamond/graphite. In most cases, polymorphs in this category are required to break bonds and rearrange atoms or molecules, and consequently, the polymorphs have a monotropic relation.

For the study of polymorphs that are structurally related, the structural relationships between the polymorphs should be established first; second, it should be explained why a particular substance is able to arrange its structural units in two closely related lattices, and finally, there should be a description of the manner and conditions under which rearrangement of the units from one lattice type to another can happen. It is the physical form of the drugs that is responsible for its degradation in the solid state. Selection of a polymorph that is chemically more stable is a solution in many cases. Different polymorphs also

Table 3 Thermodynamic Rules for Polymorphic Transitions

Enantiotropy	Monotropy
Transition < melting I	Transition > melting I
I stable > transition	I always stable
II stable < transition	—
Transition reversible	Transition irreversible
Solubility I higher < transition	Transition I always lower
Solubility I lower > transition	—
Transition II → I endothermic	Transition II → I exothermic
$\Delta H_f^I < \Delta H_f^{II}$	$\Delta H_f^I > \Delta H_f^{II}$
IR peak I before II	IR peak I after II
Density I < II	Density I > density II

lead to different morphology, tensile strength and density of powder bed, which collectively contribute to the compression characteristics of materials. Some investigation of polymorphism and crystal habit of a drug substance, as it relates to pharmaceutical processing, is desirable during its preformulation evaluation, especially when the active ingredient is expected to constitute the bulk of the tablet mass.

Various techniques are available for the investigation of the solid state. These include microscopy (including hot-stage microscopy, HSM), infrared spectrophotometry (IRS), single-crystal X-ray and X-ray powder diffraction (XRPD), thermal analysis, and dilatometry.

A preformulation study plan must challenge the crystal structure to determine if any polymorphs exist. It is possible for a new compound to show polymorphic forms only when subjected to stress—physical and chemical. Most organic compounds are capable of exhibiting polymorphism because of their complex flexible structure. The window of physico-chemical stress that a drug is generally subjected to during manufacturing is at times not able to adduce the differentiation of a drug into its possible polymorphic forms. For example, enantiotropic state is the state in which one polymorph can be reversibly changed into another by varying the temperature or pressure. One way of assessing whether the solid is a metastable form of the compound is to slurry the compound in a range of solvents. In this way, a solvent-mediated phase transformation may be detected using the usual techniques. The monotropic state exists when the transformation between the two forms is irreversible. As all polymorphs are deproviding the lowest energy polymorph, the most able polymorph is often needed to assure consistency in the physico-chemical properties; this is necessary for consistency in manufacturing procedures and in bioavailability. The right polymorph, at times, is not necessarily the most stable polymorph; unstable forms like amorphous forms (that are most constrained) are often used because of their higher solubility and often a better bioavailability profile.

The manufacturing factors that may be affected by the choice of a particular polymorphic form include granulation, milling and compression, stability (particularly for semisolid forms), amount of dose delivered in metered inhalers, crystallization from different solvents at different speeds and temperature, precipitation, concentration or evaporation, crystallization from the melt, grinding and compression, lyophilization, and spray drying. In the manufacturing processing, crystallization is a major problem and it can be avoided by a careful study of polymorphic transition, particularly in supercritical fluids.

Polymorphism is frequently a function of the type of salt, because the presence of counter ions can make the crystals form differently, leading to widely variable physico-chemical properties, as described earlier under the description of polymorphism. Generally, salts exhibiting polymorphism should be avoided.

An interesting example of polymorphic structure differentiation is that of human immunodeficiency virus (HIV) protease inhibitors. The HIV protease inhibitors pose a serious problem in their bioavailability. Invirase showed only modest market performance, and it was soon superseded by drugs, such as ritonavir (Norvir) and indinavir sulfate (Crixivan®) that had better bioavailability. Three years after initial approval, saquinavir was reintroduced in a formulation with sixfold higher oral bioavailability relative to the original product. Ritonavir was originally launched as a semisolid dosage form, in which the waxy matrix contained the dispersed drug in order to achieve acceptable oral bioavailability. Two years after its introduction, ritonavir exhibited latent crystal polymorphism, which caused the semisolid capsule formulation of Norvir to be removed from the market.

The acceptance criterion for polymorphic forms in a drug substance is generally based on the considerations given in Scheme 1.

HIGH-THROUGHPUT CRYSTAL SCREENING

The search for crystal forms and salts of compounds emerging from discovery must rely on automation and miniaturization of crystallization trials. Currently, development chemists may experiment with 1–10 mg per trial on a total budget of tens to hundreds of milligram. Although, the material is usually recoverable at a cost in time and effort, the traditional experimentation remains linear in nature. The search for crystal forms in such a linear fashion is time consuming, and the pressure keeps mounting to test a compound in toxicology and the clinic. The technical solution provided by high-throughput (HT) crystallization is the possibility of parallel, miniaturized trials of a larger experimental space (solvents, combinations, processing parameters, and so on). In order to increase significantly the productivity of crystallization efforts, one must be able to conduct parallel experiments at the level of micrograms per trial. In this way, valuable time and material can be saved, while generating useful physico-chemical information to support development decisions. For instance, if crystalline forms are found, the program can confidently move forward to assessing their utility. Even when a crystal form remains elusive, the information from crystallization trials on the compound and some of its congeners may help the medicinal chemists design the optimal compound to advance the program. High-throughput crystallization screening provides a way to address polymorphism issues much earlier, and can help to avoid late discoveries of polymorphism in pharmaceutical systems; the use of such technologies as CrystalMax® (4) enables parallel, miniaturized crystallization of compounds in cycles of one to two weeks. With a capacity of up to 10,000 crystallization experiments per week, this technology enables the discovery and characterization of diverse solid forms of active pharmaceutical ingredients, including leads to enhance solubility, selection of salts, co-crystals, and the like. The technology allows design, execution and analysis of thousands of crystallization trials on hundreds of micrograms of crystalline material per well in microliter volumes within a 96-well array format. The FAST® (4) HT technology from the same vendor allows the discovery of novel solution formulations of poorly soluble compounds, either for intravenous or oral use. The technology uses a 96-well format to conduct parallel screening of thousands of combinations of semi-aqueous formulations. Other tools, like HT crystallization tools with miniaturization come from Symyx (5), Aventium (6), and Solvias (7).

In the recent years, sophisticated modeling tools have become available, such as the Cerius^2 (8), where various modules allow the analysis of crystallization, crystal growth, and material form characterization. In brief, this technique uses a simulated annealing and a rigid-body Rietveld refinement procedure, whereby the calculated and measured XRPD patterns are compared; if they agree sufficiently, the structure is deemed to be solved. Other modules offered by Cerus include:

- C^2. HP Morphology is an advanced method for predicting crystal morphology for salts and solvates.

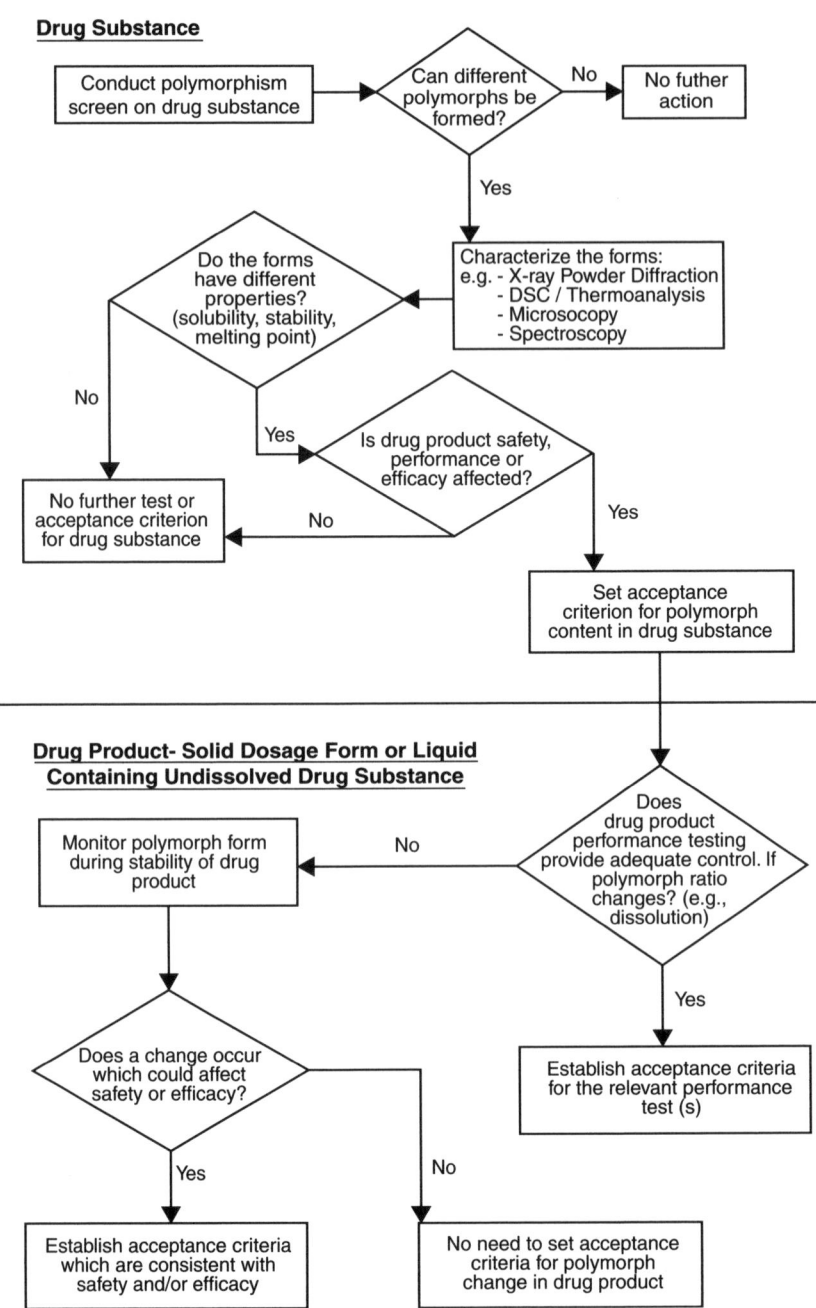

Scheme 1 Some acceptance criteria for polymorphism in drug substances and drug products. *Source*: Courtesy of Ref. 3.

- C^2. Morphology predicts and analyzes the morphology of crystals from their internal crystal structure, which helps relate morphological features to structure and understand the likely effects of solvents and growth modifying additives.
- C^2. Polymorph predicts the polymorphs of organic crystalline materials, such as drugs, pigments, or fine chemicals, from their molecular structure.
- C^2. Polymorph is used to predict unsynthesized polymorphs or to determine structures in combination with techniques, such as Rietveld refinement. The Cereus Polymorph Predictor (see subsequently) is based on a potential energy function program and a search program to locate potential minima of that potential function. It does have limitations, for example, by neglecting polarization effects, the results are less accurate for molecules, such as salts that are common in the pharmaceutical industry. Moreover, atoms, such as fluorine and divalent sulfur are not optimally parameterized. The molecules need to be rigid; however, a few successes for flexible molecules have been reported.
- C^2. Powder Fit provides crystal structure determination capabilities by helping to determine the parameters required to generate a simulated powder pattern by analyzing the peak positions, profiles, and background of an experimental pattern.
- C^2. Powder Indexing completes a comprehensive package of software modules for crystal structure determination from powder data. It is possible to establish unit cell and symmetry information, and use this to assist Rietveld refinement or crystal structure predictions.
- C^2. Powder Solve provides crystal structure determination capabilities by sampling a vast number of trial structures, subsequently proposing a structure for which the simulated pattern most closely matches the experimental one.
- C^2. Diffraction-Amorphous simulates noncrystalline diffraction, including small angle scattering. Comparison with experimental data helps determine amorphous structure, polymer chain conformation, copolymer sequence structure, and orientation.
- C^2. Diffraction-Crystal simulates powder, fiber, and single-crystal diffraction from crystalline models, which helps interpret the experimental data from molecular, inorganic, and polymeric crystalline materials.
- C^2. Diffraction-Faulted simulates powder diffraction from faulted or layered structures, which helps characterize structures, such as zeolites and clays.
- C^2. EXAFS (extended X-ray absorption fine structure) jointly developed with the CCLRC Daresbury Laboratory, U.K. (9), integrates the EXAFS analysis and refinement techniques of Daresbury's EXCURV92 software with Cerius^{2}s' modeling tools, radically improving the ability to interpret the EXAFS data.
- C^2. HRTEM simulates high-resolution transmission electron microscope images from crystals, interfaces, and defect structures. One can set up and interpret EM experiments that investigate technologically important materials.
- C^2. IR/Raman is a computational instrument that predicts IR/Raman spectra.
- C^2. LEED/RHEED helps interpret low-energy electron diffraction patterns and reflection high-energy electron diffraction from surfaces.
- C^2. Rietveld performs crystal structure refinement and quantitative phase analysis using powder diffraction data and the Rietveld method. Combining refinement programs and advanced modeling tools allows a faster route to determining the structures of both inorganic and molecular crystals. An effective way to know the atomic structure is by means of diffraction techniques using neutrons from nuclear reactors and particle accelerators or X-rays from

X-ray tubes and synchrotrons. The single-crystal diffraction technique, using relatively large crystals of the material, gives a set of separate data from which the structure can be obtained. However, the *powder* diffraction technique is used in conditions where it is not possible to grow large crystals. The drawback of this conventional *powder* method is that the data grossly overlap, thereby preventing proper determination of the structure. The "Rietveld method" creates an effective separation of these overlapping data, thereby allowing an accurate determination of the structure. An even more widely used application of the method is the determination of the components of chemical mixtures.

Crystalline index of refraction: As different polymorphs have different internal structures, they belong to different crystal systems; therefore, polymorphs can be distinguished using polarized light and a microscope. The crystals can be either isotropic or anisotropic. In isotropic crystals, the velocity of light is the same in all directions, whereas anisotropic crystals have two or three different light velocities or refractive indices. In terms of crystal systems, only the cubic system is isotropic and the other six are anisotropic.

SOLVATES

In addition to polymorphs, solvates (inclusion of the solvent of crystallization) are also often formed during the crystallization process. These forms are also called pseudopolymorphs. The solvent molecules fill the spaces in the crystal lattice, and generally reduce the solubility and dissolution rates. This phenomenon is thermodynamically driven. If the solvate contains an organic solvent, this would not be admitted by the regulatory authorities. According to the International Conference on Harmonization (ICH) guidelines, the Class I solvents, such as benzene, carbon tetrachloride and 1,2-dichloromethane must be avoided, as these are carcinogenic. The Class II solvents should be limited and include nongenotoxic animal carcinogens, such as cyclohexane and acetonitrile. The Class III solvents, including acetic acid, alcohol, acetone, which have low toxicity potential, are allowed as long as the daily permissible dose does not exceed 50 mg. Generally, an allowed solvate would likely be removed during the manufacturing process, but in some instances, the presence of the solvate is desired, as in the case of the beclomethasone dipropionate product of Glaxo that includes trichlorofluoromethane solvate. This solvate prevents crystal growth in sprays containing trichlorofluoromethane as propellant. A U.S. Patent issued to Glaxo 5270305 demonstrates the use of trichlorfluoromethane.

HYDRATES

When the solvate happens to be water, these are called hydrates, wherein water is entrapped through hydrogen bonding inside the crystal, and strengthens the crystal structure, thereby invariably reducing the dissolution rate (Table 4). The water molecules can reside in the crystal either as isolate lattice, where they are not in contact with each other; as lattice channel water, where they fill space; and metal coordinated water in salts of weak acids, where the metal ion coordinates with the water molecule. Metal ion coordinates may also fill channels, such as in the case of nedocromil sodium trihydrate. Crystalline hydrates have been classified by structural aspects into three classes: isolated lattice sites, lattice channels, and metal-ion coordinated water. There are three classes, which are discernible by the commonly available analytical techniques.

Table 4 Drug Substance Hydrate Forms as Reported in the Pharmacopeia

Compound	Water of hydration
Aminophylline	2
Ampicillin	3
Beclomethasone dipropionate	0 or 1
Caffeine	1
Calcium citrate	4
Calcium gluceptate	0 or variable; effloresces
Calcium gluconate	0 or 1
Dextrose	1
Diatrizoic acid	2
Dibasic sodium phosphate	0, 1, or 2
Ephedrine	1/2
Fluocinolone acetonide	2
Hydrocortisone hemisuccinate	1
Magnesium citrate oral solution	1
Magnesium gluconate	2
Magnesium sulfate	0, loses gradually
Monosodium sodium phosphate	0, 1, or 2
Naloxone hydrochloride	2
Nitrofurantoin	0 or 1
Potassium gluconate	0 or 1
Prednisolone	0 or 1
Saccharin sodium	1/3; effloresces
Sodium acetate	3
Sodium citrate	0 or 2
Sodium sulfate	0 or 1; effloresces
Succinyl chloride	2
Theophylline	0 or 1
Thioguanine	0 or 1/2
Thiothixene hydrochloride	0 or 2
Zinc sulfate	1 or 7

Source: United States Pharmacopeia (USP) 24.

1. Class I includes isolated lattice sites, and represents the structures with water molecules that are isolated and kept from contacting other water molecules directly in the lattice structure. Therefore, water molecules exposed to the surface of crystals may be easily lost. However, the creation of holes that were occupied by the water molecules on the surface of the crystals does not provide access for water molecules inside the crystal lattice. The thermogravimetric analysis (TGA) and differential scanning calorimetry (DSC) for the hydrates in this class show sharp endotherms. Cephradine dihydrate is an example of this class of hydrates.
2. Class II includes hydrates that have water molecules in channels. The water molecules in this class lie continuously next to the other water molecules, forming channels through the crystal. The TGA and DSC data show interesting characteristics of channel hydrate dehydration. Early onset temperature of dehydration is expected, and broad dehydration is also characteristic of the channel hydrates. This is because the dehydration begins from the ends of channels that are open to the surface of crystals. Subsequently, dehydration keeps happening until all the water molecules are removed through the

channels. Ampicillin trihydrate belongs to this class. Some hydrates have water molecules in two-dimensional (2D) space, and they are called planar hydrates.
3. Class III includes ion-associated hydrates. Hydrates contain metal-ion coordinated water and the interaction between the metal ions and water molecules is the major force in the structure of crystalline hydrates. The metal–water interactions may be quite strong relative to the other nonbonded interactions, and therefore, dehydration occurs at very high temperatures. In TGA and DSC thermograms, very sharp peaks corresponding to dehydration of water bonded with metal ions are expected at high temperatures.

Hydrates can also exist in various polymorphs, such as in the case of amiloride hydrochloride. A myriad of methods are available to study hydrates and their polymorphs, including DTA, DSC, XRPD, and moisture-uptake studies.

AMORPHOUS FORMS

Solid powders, wherein no particular order of molecules are technically noncrystalline, are called amorphous forms. The amorphous forms are formed by vapor condensation, supercooling of a melt, precipitation from solution, and milling and compaction of crystals. These are more like liquids, where the molecular interaction has weakened; in most instances, there would be some crystalline forms among the amorphous forms as well. This two-state model is described in the USP. The amorphous forms are thermodynamically unstable, as they have high energy (that went into breaking intermolecular bonds). As a result, they might turn into a crystalline form, particularly in suspension dosage forms, and even in solid dosage forms, wherein the atmospheric moisture might serve as the nucleation points.

Discovery programs frequently yield amorphous compounds as a result of time pressures, the methods used to isolate them on small scales, and the increasing complexity of newly discovered molecules. Amorphous compounds carry inherent risks because of their physicochemical nature, and as a result, very few Food and Drug Administration (FDA)-approved drugs appear in amorphous forms; examples include accupril/accuretic, intraconazole, accolate (zafirlukast), viracept (nelfinavir mesylate), paroxetine. Other drugs that are available in amorphous forms include: celecoxib, amifostine, cefuroxime axetil, cefpodoxime proxetil, and novobiocin. In addition to being a physically metastable physical form, amorphous forms are generally less stable chemically. They also tend to have very low bulk densities, making the materials difficult to isolate and handle. The irregular shape of the powder of amorphous forms creates high surface area, which attracts water molecules making them inherently more hygroscopic.

Although, all these problems can be resolved, generally, the amorphous forms are to be avoided unless the differences in solubility make a significant impact on the bioavailability.

HYGROSCOPICITY

Water molecules have polar ends, and readily form hydrogen bonding. As a result, several compounds interact with water molecules by surface adsorption, condensation in capillaries, bulk retention, and chemical interaction, and are called hygroscopic. At times, the interaction between the compounds and water is so strong that the interacting water vapors result in dissolving the compound. This process is called deliquescence, wherein a saturated layer of solution is formed around the particles. Most of these interactions are dependent on critical water vapor pressure or relative humidity (RH). Moisture also induces hydrolysis and other degradation reactions. In addition, its presence affects the physical properties, such as powder flow, dissolution, and even crystal structure. The impact of moisture on the physical or chemical properties of compounds depends on the strength of bonding between the water molecules and the surrounding space where the water molecules are contained. In a tightly bound state, the water molecules are generally not available to induce chemical reactions. Free water molecules can participate in the creation of a liquid environment around the crystal lattice, where the pH may be altered as a result of the dissolution process. Similarly, water molecules held as crystal hydrates or trapped in an amorphous form are not available to modify the milieu interior of solid powders. It is noteworthy that some hydrates upon taking up moisture convert into hydrates (discussed earlier). This transition can be useful in formulation studies, and this property should be tested for hygroscopic compounds.

The classification of compounds into different hygroscopic categories is based on two types of models: one where the RH and temperature are kept constant and gain in the weight of compound is recorded as per the definitions of the European Pharmacopoeia, the compound tested is stored at 25°C for 24 hours at 80% RH. A slightly hygroscopic compound would show less than 2% m/m mass gain, hygroscopic compounds show less than 15%, and very hygroscopic compounds show more than 15% m/m mass gain; the deliquescent compounds simply liquefy. The dynamic model tests hygroscopic nature at various humidities; compounds showing no mass gain at 90% are called nonhygroscopic, those that do not gain at 90% are slightly hygroscopic, but those that gain 5% over a week's period are called moderately hygroscopic. Where mass increases at 40–50% humidity, these compounds are called very hygroscopic.

Generally, a compound that is very hygroscopic would be less desirable, but if studies show that despite moisture uptake the compound stays stable and workable in the formulation studies; this is an important consideration.

High hygroscopicity is undesirable for many reasons, including handling problems, requirement of special storage conditions, and chemical and physical stability problems. It is difficult to develop acceptance criteria for the amount of moisture, and large batch to batch variations are inevitable. Even if it were possible to define reasonable acceptance criteria, if the compound shows changes in crystal structure as a function of moisture content, this leads to problems in solubility and dissolution profiles that may not be acceptable. Stability of salts at accelerated temperature is complicated when there is significant sorption of moisture, because the properties related to the removal of moisture will be highly dependent on the choice and amount of excipients, the manufacturing process of the final dosage form, and even the impurity profile, both in the lead compound and in the excipients. As a rule, any property, such as hygroscopicity, which makes it difficult to create acceptance criterion should be minimized. Solid-state stability, as a result of hygroscopicity often plays a significant role in determining the dissolution rate, for example, napsylate salts often provide a more stable physical form and thus allow better dissolution.

SOLUBILITY

Solubility is a function of hygroscopicity, polymorphism, and chemical nature or pK_a of the salt. If the pK_a is at least two units lower than the pH of the medium, complete dissolution can be achieved; the opposite holds true for basic compounds. Even though in the early phase of the study the quantity of the compound might be limited, solubility studies need to be carried out as a function of pH, leaving sufficient quantity even after the formation of salt. The solids formed (both wet and dry) should then be studied using the usual techniques, such as DSC, TGA, and XRPD. The method of determining solubility can often provide variable results. The in situ technique often proves more useful to screen out poor solubility compounds; the traditional methods are always preferred. Solubility increase leads to improved bioavailability and liquid formulation, and can be achieved by increasing the melting point or the hydrophobicity of the conjugate anion. Reduced solubility is desired for suspension and controlled release dosage forms, and can be achieved by decreasing the pK_a and increasing the solubility of the conjugate acid.

The choice of salt is greatly determined by solubility considerations; the pH of the resultant solution is important, because the salts of the stronger acids produce liquids with a lower pH to promote the dissolution of the basic compounds. However, in places, where a common ion effect can operate, such as the use of hydrochloride in gastric fluid, the useful solubility window might be limited, and this modification might not work well. Similarly, when determining the solubility, there can often be significant differences in results obtained depending on whether it is determined in water, saline, or buffer, as a function of the nature of salt. It is not as straightforward as in the case of hydrochloride, where a common ion effect is clearly the most important observation. There is complex effect of pH, common ion effect, and dielectric properties of media.

The dissolution of solid particles of salts can be inhibited if the parent acid or base precipitates at the surface of the particles undergoing dissolution. For example, stearate salts show reduced dissolution if stearic acid layer precipitates on the surface in an acidic pH environment.

The choice of salt is often determined by taste consideration, such as the use of benzathine salts of penicillin V; low solubility salts have lesser taste, but also dissolve slowly, and are often used for preparing depot preparations, such as benzathine salts of penicillin G and V. Similarly, the napsylate salt provides better organoleptic properties as a result of its low solubility when compared with hydrochloride forms.

The stoichiometry of the salts is established by a detailed study of the physical structure using XRPD techniques, and it can at times be not what the chemical structures would generally indicate.

Salt Form

Combinatorial chemistry offers many advantages, including the synthesis of larger molecular weight drugs, which are mostly lipophilic. The bioavailability considerations require converting them into salt forms. This trend is apparent from recent regulatory approvals by FDA, where more than 50% of the new drugs approved have been in salt forms. The common methods used for the characterization and the screening of salt forms include:

- XRPD analysis
- thermal analysis (DSC, TGA, thermomechanical analysis)
- dynamic vapor sorption (DVS)
- nuclear magnetic resonance (NMR)

- dissolution (including intrinsic)
- microscopy (light and polarized)
- density (intrinsic and bulk)
- particle size (optical, laser light, and light obscuration)

A process flow for the selection of the best salt form is given in Scheme 2.

The choice of using a cation or an anion form is always based on all the factors described earlier. There are fewer salt-forming species for weak acids than there are for weak bases, and the available information suggests that, in general, alkali metal salts exhibit greater solubility than the corresponding alkaline earth salts. Among cations, the most frequently found ion is sodium (62%), followed by potassium and calcium (10%); this is followed by zinc and meglumine (3%), lithium, magnesium diethanolamine, benzathine, ethyldiamine, aluminum, chloroprocaine, and choline (in decreasing order of frequency). Among anions, the most frequently used counter ion is hydrochloride (almost 50%), followed by sulfate (8%), bromide and chloride (5%), diphosphate, citrate, maleate (3%), iodine mesylate, hydrobromide (2%), acetate, pamoate (1%), isothionate, methylsulfate, salicylate, lactate, methylbromide, nitrate, bitartrate, benzoate, dihydrochloride, gluconate, carbonate, edisylate, mandelate, methylnitrate, subacetate, succinate, benzenesulfonate, calcium edentate, camsylate, edentate, fumarate, glutamate, hydrobromine, napsylate, pantothenate, stearate, gluceptate, bicarbonate, estolate, esylate, glycollylarsinate, hexylresorcinate, lactobionate, maleate, mucate, polygalactoronate, teoclate, and triethiodide (in decreasing order of frequency). The choice of counter ions is a function of the pK_a of the weak acid involved in the formation of salt. Table 5 lists the pK_a values of weak acids that are most frequently used in salt formation.

To form a salt of a basic compound, the pK_a of the salt-forming acid has to be less than or equal to the pK_a of the basic center of the compound. As a result, very weak basic compounds have a pK_a value around 2. Bases with higher pK_a have a greater range of possibilities for salt formation. As most drugs are weak bases, it is not surprising that hydrochloride, sulfuric, and toluenesulfonic salts are very common.

The chances are high that a newly discovered drug substance would have polarizable groups that make the compound capable of interacting with the receptor sites. The most common polarizable ends are present in acidic or basic compounds. Neutral compounds are inert and mostly inactive, for example, perfluorodecalin is a balanced cyclohexane, wherein the ring electronegativity is neutralized by the strong fluorine molecules. This molecule is inert, and is not degraded by the body. As the availability of the ionizing center leads to salt formation, the salt formation studies are one of the most important studies at the prenomination stage, because this can affect the solubility, dissolution hygroscopity, taste, physical and chemical stability, or polymorphism properties of the newly discovered drug substance. It is most likely for these studies to be conducted after establishing the basic physico-chemical properties, in order to allow the comparison with properties in various salt forms. In the order of importance, the selection process follows the following theme.

Melting Point
Solubility is increased when the melting point of the salt is lower, or where there is improved hydrogen bonding (with water), and as a result the hydroxyl groups in

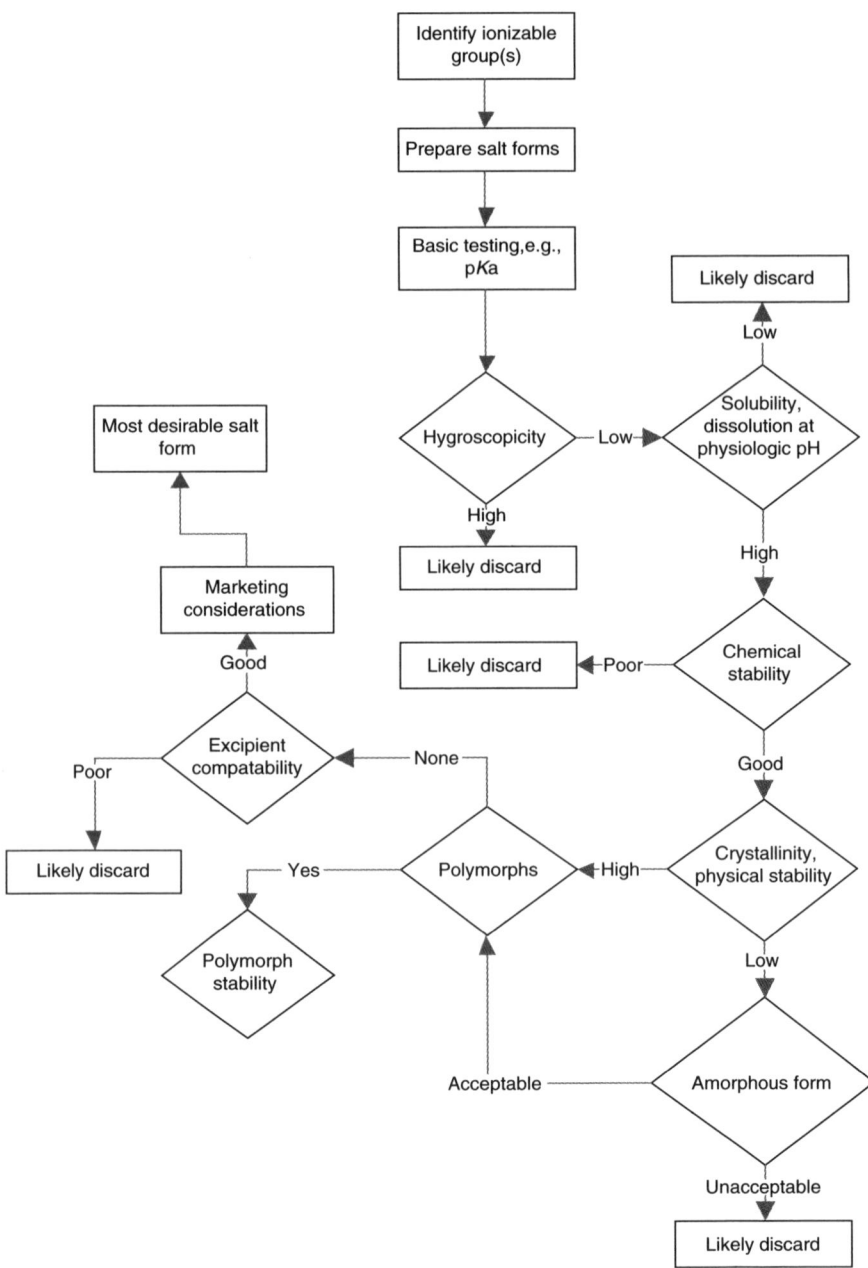

Scheme 2 Decision flow diagram to select the best salt form.

Table 5 The pK_a of Common Weak Acids Used in Salt Formation

Acid	pK_a
Acetate	4.76
Ascorbate	4.21
Benzoate	4.20
Besylate	2.54
Citrate	3.13
Fumarate	3.0, 4.4
Gluconate	3.60
Hydrobromide	−8.0
Hydrochloride	−6.1
Malate	3.5, 5.1
Mesylate	1.92
Napsylate	0.17
Oleate	∼4.0
Phosphate	2.15, 7.20, 12.38
Succinate	4.2, 5.6
Sulfate	−3.0
Tartrate	3.0, 4.3
Tosylate	−0.51

the conjugate acid improve the solubility and the hydrophobic groups reduce the solubility. Often, it is desired to prepare salts with a lower solubility to mask taste, provide slower dissolution, and increase its chemical stability. Melting point increase results in process problem and reduced solubility; this can be achieved by the use of more flexible aliphatic acids with aromatic bases. Move to more highly substituted acids that destroy the crystal symmetry. A decrease in the melting point generally improves the solubility and allows the formation of oil, and can be achieved by the use of small counter-ions, for example, chloride or bromide, or by the use of aromatic conjugate anions if aromatic base, or by using small hydroxyl acids if the drug has good hydrogen bonding potential. The melting point is generally decreased by increasing the hydroxylation of the conjugate acid, and in the cases of common ion dependence, by moving small organic acids. In the case of the sodium salt of drugs, the logarithm of aqueous solubility is often inversely related to melting point.

Dissolution

The factors described earlier that affect the solubility of a lead compound when choosing a particular salt form—a polymorphic form—a specific crystalline form directly affects the most critical parameter that determines the drug activity, which is the dissolution rate in the biological milieu. The first step in the commencement of dissolution is the wettability of solid particles—there is a direct correlation between wettability and bioavailability. As the milieu of drug administration sites is mostly aqueous in nature, low wettability makes the particles less hygroscopic.

The dissolution of the salts leads to a change in the pH of the dissolution media because of the buffering effect. A base dissolved in acidic media increases the pH, because the acidic counter-ions are trapped into salt forms. Similarly, as the salts dissolve, the pH shift depends on whether it is the acid or the basic component, which is weaker. The final balance is always dependent on the relative pK_a of the acidic and alkaline components. This is an important consideration as it explains the difference in the results obtained when the studies are conducted in water or buffer. When enteric protection is desired, the dissolution rates should be determined in 0.1-N HCl, wherein many differences in the dissolution rates between water and buffer are obviated.

STUDY METHODS

At the preformulation stage, the limitations of the quantity of the sample available determines to a great degree the type of study to be conducted. Some physical properties are fundamental in nature, whereas others are a manifestation of these basic properties. For example, melting point determination reveals much about the internal structure of crystals, the solubility and dissolution characteristics; the latter properties are the derived properties. As a result, techniques available to study the aforementioned properties are categorized by the U.S. FDA in a decreasing order of importance (FDA, The Gold Sheet, 1985). The following is an expanded list of these methodologies available for evaluation:

- Melting point (HSM)
- IRS
- XRPD
- thermal analytical techniques [e.g., DSC, differential thermal analysis (DTA), TGA, and the like]
- solid-state Raman spectroscopy
- crystalline index of refraction
- phase solubility analysis
- solution pH profile determination
- solution calorimetry
- comparative intrinsic dissolution rates
- cross-polarization/magic angle spinning (CP/MAS) solid-state NMR
- hygroscopicity measurement (particularly for salts)

Thermal Analysis

There are a number of interrelated thermal analytical techniques that can be used to characterize the salts and the polymorphs of candidate drugs. The melting point of a salt can be manipulated to produce compounds with desirable physicochemical properties for specific formulation types. Of the thermal methods available for investigating polymorphism and related phenomena, DSC, TGA, and HSM are the most widely used methods.

Differential Scanning Calorimetry

This is one of the most frequently used methods to study solid-state properties. The flux type DSC involves heating the sample and reference samples at a constant rate using thermocouples, to determine how much heat is flowing into each sample and thus finding the differences between the two. Examples of such DSC instrumentation are those provided by Mettler and duPont. The power compensation DSC (e.g., Perkin-Elmer), an exothermic or endothermic event, occurs when a sample is heated, and the power added or subtracted to one or both of the furnaces to compensate for the energy change occurring in the sample is measured. Thus, the system is maintained in a thermally neutral position at all times, and the amount of power required to maintain the system at equilibrium is directly proportional to the energy changes that are occurring in the sample. In both types of DSC measurements only a few milligrams of the compound suffices. The sample can be heated in an open pan or in hermetically sealed chambers, where there may or may not be vents to release moisture or solvents; the compound may be subjected to pyrolysis in the testing phase.

Whereas the instrumentation available in the recent years has become very sophisticated, making such analysis possible with great consistency, the interpretation of the results is highly dependent on a keen understanding of the factors that affect the results. For example, such subtle factors as the type of pan, the heating rate used, the nature and mass of the compound, the particle size distribution, packing and porosity, pretreatment and dilution of the sample, and the use of the nitrogen cover can significantly alter the DSC profile obtained, and should be controlled to secure consistency in the repeat results.

A well designed and properly replicated DSC profile would yield such physical properties as melting (endothermic), solid-state transitions (endothermic), glass transitions, crystallization (endothermic), decomposition (exothermic) and dehydration or desolvation (endothermic), purity (of high purity compounds; though much less reliable than high-performance liquid chromatography, HPLC).

A heating rate of 10°C/min is a useful compromise between the speed of analysis and detecting any heating rate-dependent phenomena. If any heating rate-dependent phenomena are evident, the experiments should be repeated by varying the heating rate, in order to identify the nature of the transition that might be the result of polymorphism or particle size. It is noteworthy that milling the powder size may alter the profile significantly, and can be confused with polymorphic changes. Using different heating rates often resolves this problem.

A number of parameters can be measured from the various thermal events detected by DSC. For example, for a melting endotherm, the onset, peak temperatures, and enthalpy of fusion can be derived. The onset temperature is obtained by extrapolation from the leading edge of the endotherm to the baseline. The peak temperature is the temperature corresponding to the maximum of the endotherm, and the enthalpy of fusion is derived from the area of the thermogram. It is an accepted custom that the extrapolated onset temperature is taken as the melting point; however, some users report the peak temperature in this respect. We tend to report both for completeness.

Recycling experiments can also be conducted, whereby a sample is heated and then cooled. The thermogram might show a crystallization exotherm for the sample, which on subsequent reheating might show a melting point different from the first run. In a similar way, amorphous forms can be produced by cooling the molten sample to form a glass.

The calibration of a DSC employs the use of standards; the most common ones are listed in Table 6. These standards must meet a certain criterion of purity. A two-point calibration is often needed, for example, using indium and lead.

A variation of DSC is the MDSC (modulated DSC), wherein heat is applied sinusoidally, such that any thermal events are resolved into reversing and nonreversing components to allow complex and even overlapping processes to be deconvoluted. The heat flow signal in conventional DSC is a combination of

Table 6 Standards for Thermal Analysis in the Order of Increasing Melting Point

Temperature (°C)	Substance
0	Water
26.87	Phenoxybenzene
114.2	Acetanilide
151.4	Adipic acid
156.6	Indium
229	Tin
232	Caffeine
327.5	Lead
419.6	Zinc

"kinetic" and heat capacity responses, and Fourier transform (FT) techniques are used to separate the heat flow component from the underlying heat flow signal. The cyclic heat flow part of the signal (heat capacity, C_p × heating rate) is termed the reversing heat flow component. The nonreversing part is obtained by subtracting this value from the total heat flow curve. It is important to note that all the noise appear in the nonreversing signal. The limitations of MDSC studies include the requirement of a sufficient number of cycles to cover thermal events. In cases where the samples do not follow the signal or where there is fluctuation in temperature during the sinusoidal ramp, these compounds may not be suitable for this study.

Hot-Stage Microscopy

Hot-stage microscopy is a thermal analytical technique, whereby a few milligrams of the material is spread on a microscope slide, which is then placed in the hot stage and heated at various rates and under different atmospheric environments, including very low temperatures. The events can be recorded using video systems. Hot-stage microscopy is routinely used in conjunction with other methods. Although many newer automated methods to observe the melting behavior of crystals are available, to a trained eye, this classic method remains one of the most powerful tools.

Thermogravimetric Analysis

Thermogravimetric analysis is used to detect the amount of weight lost on heating a sample. It is based on a sensitive balance that records the weight of the sample (generally 5–10 mg) as it is heated under nitrogen. Thermogravimetric analysis experiments can detect the presence of water or solvent in different locations in the crystal lattice. This technique has an advantage over a Karl Fischer titration or a loss on drying experiment that can only detect the total amount of moisture present. In addition, TGA requires smaller quantities of the compounds than the other two techniques. However, the use of very little sample in TGA can yield erroneous results because of buoyancy and convection current effects. The total amount of moisture lost in TGA experiments is not affected by the heating rate; however, the temperature at which it occurs may vary. It is noteworthy that the dehydration mechanism and activation of the reaction may be dependent on the practice size and sample weight. The TGA is calibrated using magnetic standards.

Solution Calorimetry

Solution calorimetry involves the measurement of heat flow when a compound dissolves into a solvent. There are two types of solution calorimeters, that is, isoperibol and isothermal. In the isoperibol technique, the heat change caused by the dissolution of the solute gives rise to a change in the temperature of the solution. This results in a temperature–time plot from which the heat of the solution is calculated. In contrast, in isothermal solution calorimetry (where, by definition, the temperature is maintained constant), any heat change is compensated by an equal, but opposite, energy change, which is then the heat of solution. The latest microsolution calorimeter can be used with 3–5 mg of compound. Experimentally, the sample is introduced into the equilibrated solvent system, and the heat flow is measured using a heat conduction calorimeter.

Dissolution of a solute involves several thermal events, such as heat associated with wetting, breakage of lattice bonds, and salvation energy. The peak can be integrated directly to give an enthalpy of dissolution. The relative stability of polymorphs can be investigated in this way by the magnitude and sign (endothermic/exothermic) of the enthalpy of dissolution. A more endothermic (or less exothermic) response indicates that the energy of solvation of the solute does not compensate for the breaking of lattice bonds, and it is therefore the more stable solid (polymorphs).

Solution calorimetry can also be used to evaluate amorphous/crystalline content in a binary mixture. The enthalpy of solution for the amorphous compound is an exothermic event, whereas that of the crystalline hydrate is endothermic. Enthalpy of solution is a sum of several thermal events, that is, heat of wetting (incorporating sorption process, such as surface sorption and complexation), disruption of the crystal lattice, and solvation. The order of magnitude of solution enthalpy for the crystalline compound suggests that the disruption of the crystal lattice predominates over the heat of solvation. In addition, the ready solubility of the compound in aqueous media is probably governed by entropy considerations.

Solution calorimeters are calibrated using KCl in water (for endothermic processes) and Tris-HCl in 0.01-M HCl (for exothermic processes) standards. For example, the heat of solution ΔH_s) of KCl at 25°C (298.15 K) is 235.86 ± 0.23 J/g. Similarly, the ΔH_s for tris HCl at 25°C is −29.80 kJ/mol.

Isothermal Microcalorimetry

Isothermal microcalorimetry can also be used to determine, among other things, the hygroscopicity of substances. In the ramp mode, this technique can be used, like DVS, to examine milligram quantities of compound. This instrument utilizes a perfusion attachment with a precision flow switching valve. The moist gas is pumped into a reaction ampoule through two inlets, one that delivers dry nitrogen at 0% RH and the other that delivers nitrogen that has been saturated by passing it through two humidifier chambers maintained at 100% RH. The required RH is then achieved by the switching valve, which varies the proportion of dry to saturated gas. The RH can then be increased or decreased to determine the effect of moisture on the physico-chemical properties of the compound.

It is probably more popular to perform microcalorimetry in the static mode. In the so-called internal hygrostat method, the compound under investigation is sealed into a vial with a sealed pipette tip containing the saturated salt solution chosen to give the required RH.

Infrared Spectroscopy

Infrared spectroscopy differentiates solid-state structures of compounds just as well as it differentiates and identifies the chemical structures and peculiarities. This is because the different arrangements of atoms in the solid state lead to different molecular environments, which in turn induce variability in stretching frequencies. These differences are used to distinguish the polymorphic forms of a compound. The presence of solvent or water can be detected using this technique as a result of the broad —OH stretch associated with water.

The IRS is applied to studies in a number of ways: by Nujol mull, KBr disc, or the diffuse reflectance (DR) technique. In the KBr disc technique, the compound is mixed with KBr and compressed into a disc using a press and die. This compression can be a disadvantage if the compound undergoes a polymorphic transformation under pressure.

Nowadays, most instruments use a FT-infrared (FT-IR) system, a mathematical operation used to translate a complex curve into its component curves. In an FT-IR instrument, the complex curve is an interferogram, or the sum of the constructive and destructive interferences generated by overlapping light waves, and the component curves are the IR spectrum. The standard IR spectrum is calculated from the Fourier-transformed interferogram, giving a spectrum in percent transmittance ($\%T$) versus light frequency (cm^{-1}).

An interferogram is generated because of the unique optics of an FT-IR instrument. The key components are a moveable mirror and a beam splitter. The moveable mirror is responsible for the quality of the interferogram, and it is very important to move the mirror at constant speed. For this reason, the moveable mirror is often the most expensive component of an FT-IR spectrometer. The beam splitter is just a piece of semireflective material, usually Mylar film sandwiched between two pieces of IR-transparent material. The beam splitter splits the IR beam 50/50 to the fixed and moveable mirrors, and then recombines the beams after being reflected at each mirror. The Fourier transform is named after its inventor, the French geometrician and physicist Baron Jean Baptiste Joseph Fourier, born in 1830.

The FT-IR spectra of amorphous forms are often less well defined and can be used to characterize various polymorphic forms. Heating experiments are also possible using IRS, where the variable temperature IRS is conducted to confirm that a solid–solid transition takes place on heating various forms of the compounds.

The disadvantages of conventional IRS, like the need to compress the samples, is overcome when the *diffuse reflectance Fourier transform* (*DRIFT*) technique is used, whereby a few milligrams of the compound is dispersed in approximately 250 mg of KBr, and the spectrum is obtained by reflection from the surface.

Many substances in their natural states (e.g., powders and rough surface solids) exhibit DR, that is, the incident light is scattered in all directions as opposed to specular (mirror-like) reflection, where the angle of incidence equals the angle of reflection. In practice, the DR spectra are complex, and are strongly dependent upon the conditions under which they are obtained. These spectra can exhibit both absorbance and reflectance features as a result of the contributions from transmission, internal and specular reflectance components and scattering phenomena in the collected radiation. The DR spectra are further complicated by sample preparation, particle size, sample concentration, and optical geometry effects, to name a few. Specular reflection, whether it occurs from a glossy sample surface or from a crystal surface, produces inverted bands ("Restsrahlen bands") in the DR spectrum, which reduces the usefulness of traditional transmission reference spectra. For highly absorbing samples, these Restsrahlen bands are strong. Grinding and diluting the sample with nonabsorbing powder, such as KBr, KCl, Ge, or Si can minimize or eliminate these effects. Grinding reduces the contribution of reflection from large particle faces. Diluting ensures deeper penetration of the incident beam, thus increasing the contribution to the spectrum of the transmission and internal reflection component. The resulting spectra have an appearance more similar to that of the transmittance spectra than bulk reflectance spectra. If sample dilution is not feasible, the spectra may still be improved by using an optical geometry that employs a low incident angle and an offline collection angle.

The DR spectrum of a dilute sample of "infinite depth" (i.e., up to 3 mm) is usually calculated with reference to the diffuse reflectance of the pure diluent to yield the reflectance, $R_{i\ddot{A}}$. $R_{i\ddot{A}}$ is related to the concentration of the sample, c, by the Kubelka–Munk (K–M) equation:

$$f(R_{i\ddot{A}}) = (1 - R_{i\ddot{A}})^2 / 2R_{i\ddot{A}} = 2.303 ac/s \tag{1}$$

where a is the absorptivity and s is the scattering coefficient. The scattering coefficient depends on both particle size and degree of sample packing. Thus, the K–M function can be used for accurate quantitative analysis, provided the particle size and packing method are strictly controlled. For good diffuse reflectors, plots of the K–M function, $f(R_{i\ddot{A}})$, are analogous to absorbance plots for transmission spectra. Care must be taken in applying the K–M equation when $R_{i\ddot{A}}$ is much less than about 30%, because deviations from linearity can occur when the sample concentration is high.

A modification of the aforementioned DR model is the *praying mantis model*, where the preferred offline type incorporates two 6:1 90 degree off-axis ellipsoidal mirrors. One of the ellipsoids focuses the incident beam on the sample, whereas the second collects the radiation diffusely reflected by the sample. Both ellipsoidal mirrors are tilted forward; therefore, the specular component is deflected behind the collecting ellipsoid and permits the collection of primarily the diffusely reflected component. Another advantage of the "praying mantis" design is the ability to expand the available sampling area indefinitely by rotating the ellipsoids and positioning the sampling point above the optical plane. This accessory may also be used for specular reflectance at a 41.50 angle of incidence. This is achieved by tilting the sample angle as the alignment mirror. Specular sample holders are available for this purpose. Although diffuse reflection spectroscopy primarily measures the spectrum of the bulk, it can be very sensitive to the nature of the sample, for example, powders with a high surface area. Thus, it is valuable for catalysis and oxidation studies. In this application, it is important to measure the spectrum under controlled atmospheres and at high or low temperatures. The "praying mantis" model has a large sampling space between the ellipsoids for additional accessories, such as vacuum chambers. This cell is specially designed to conduct diffuse reflection spectroscopy studies in controlled atmospheres at high (up to 750°C) or low (liquid nitrogen) temperatures and under vacuum or high pressure (e.g., up to 1500 psi).

X-Ray Powder Diffraction

X-rays are part of the electromagnetic spectrum lying between ultraviolet and gamma rays, and they are expressed in angstrom units (Å). Diffraction is a scattered phenomena, and when X-rays are incident on crystalline solids, they are scattered in all directions. Scattering occurs as a result of the radiation wavelength being in the same order of magnitude as the interatomic distances within the crystal structure. X-rays are extensively used to characterize a crystal. In Figure 8, the relationship between the interplanar spacing and the angle of an incident beam is described by Bragg's equation.

The interference is constructive when the phase shift is proportional to 2π; this condition can be expressed by Bragg's law:

$$n\lambda = 2d \sin(\theta) \qquad (2)$$

where n is an integer, λ is the wavelength of X-rays, and moving electrons, protons, and neutrons, d is the spacing between the planes in the atomic lattice, and θ is the angle between the incident ray and the scattering planes.

Bragg's equation gives an easy way to understand XRPD. Powder X-ray diffraction data collected on crystalline samples gives information about peak intensities and peak positions. Peak intensities are determined by the contents of unit cells, and peak positions are closely related to the cell constants. Interplanar spacing is a function of Miller indices and cell constants. Therefore, if the cell constants are known for a crystalline compound, peak positions corresponding to Miller indices can be obtained from the Bragg's equation: the wavelength, λ, is machine-specific. The determination of cell parameters in structure determination of XRPD pattern is a reverse process to find cell constants from peak positions. Here, note that the cell constants for a unit cell are not affected by the contents in the unit cell. The contents in the unit cell have effects on the peak intensities.

The X-ray diffraction experiment requires an X-ray source, the sample under investigation, and a detector to pick up the diffracted X-rays. Figure 9 is a schematic diagram of a powder X-ray diffractometer.

The X-ray radiation most commonly used is that emitted by copper, whose characteristic wavelength for the K radiation is 1.5418 Å. When the incident beam strikes a powder sample, diffraction occurs in every possible orientation of 2θ. The diffracted beam may be detected by using a moveable detector, such as a

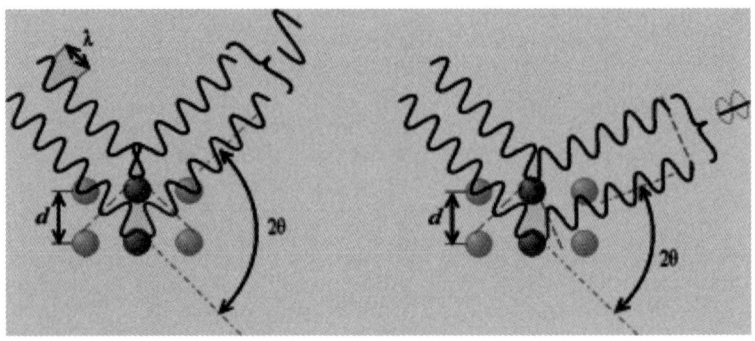

Figure 8 Braggs diffraction.

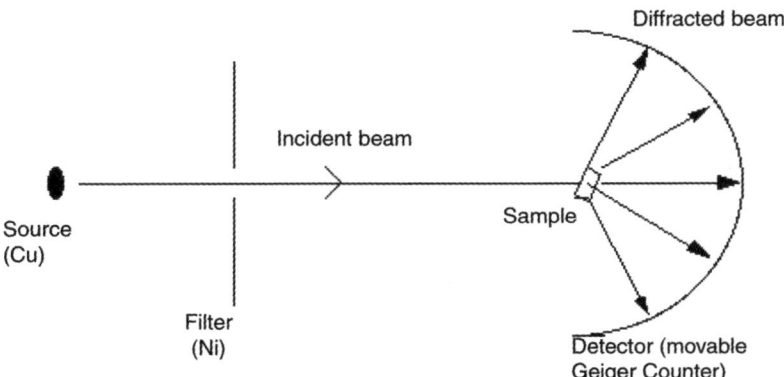

Figure 9 Design plan of a typical X-ray powder diffractometer.

Geiger counter, which is connected to a chart recorder. In normal use, the counter is set to scan over a range of 2θ values at a constant angular velocity. Routinely, a 2θ range of 5–70° is sufficient to cover the most useful part of the powder pattern. The scanning speed of the counter is usually 2θ of 2°/min and therefore, about 30 minutes are needed to obtain a trace.

An X-ray diffractometer is made up to an X-ray tube generating X-rays from, for example, Cu, Ka, or Co source and a detector. The most common arrangement in pharmaceutical powder studies is the Bragg–Brentano θ–θ configuration. In this arrangement, the X-ray tube is moved through angle θ, and the detector is moved through angle θ. The sample is fixed between the detector and the X-ray shown in Figure 3.

The powder pattern consists of a series of peaks that have been collected at various scattering angles, which are related to d-spacing, so that the unit cell dimensions can be determined. In most cases, the measurement of the d-spacing will suffice to positively identify a crystalline material. If the sample is of amorphous nature, that is, does not show long-range order, the X-rays are not coherently scattered, and no peaks will be observed.

Although XRPD analysis is a relatively straightforward technique for the identification of solid-phase structures, there are sources of error, including the following:

- *Variations in particle size.* Large particle sizes can lead to nonrandom orientation, and hence particles <10 µm should be used, that is, the sample should be carefully ground. However, if the size is too small, for example, 1 µm, it leads to the broadening of the diffraction peaks. Indeed, if the crystal sizes are too small, then the sample may appear to be amorphous.
- *Preferred orientation.* If a powder consists of needle- or plate-shaped particles, these tend to become aligned parallel to the specimen axis, and thus certain planes have a greater chance of reflecting the X-rays. To reduce the errors caused by this source, the sample is usually rotated. Alternatively, the sample can be packed into a capillary.
- *Statistical errors.* The magnitude of statistical errors depends on the number of photons counted. To keep this number small, scanning should be carried out at an appropriately slow speed.

- *Sample height.* The sample should be at the same level as the top of the holder. If the sample height is too low, the pattern shifts down the 2θ scale, and if it is too high, it moves up the 2θ scale.
- *Sample preparation procedures.* The greatest potential source of problems is grinding, which can introduce strain, amorphism, and polymorphic changes. Even the contamination from the process of grinding (e.g., in a mortar) can significantly affect the diffraction pattern. Furthermore, the atmosphere surrounding the sample can create problems as a result of the loss or gain of moisture or carbon dioxide. This is particularly true if a heating stage is used; particularly when a compound undergoes a solid-state transition from a low-melting form to a high-melting form, this can be detected by a change in the diffraction pattern. Using the Anton Parr TTK-450 temperature attachment, the compound can be investigated between subambient temperature and several hundred degrees. In cases where desolvation occurs upon micronization, heating the sample makes the peaks shaper and stronger, indicating an increase in crystallinity. This is analogous to the annealing exotherm observed in the DSC thermogram. In a similar way, the sample can be exposed to varying degrees of humidity in situ, and the diffraction pattern determined.
- *Irradiation effects.* Sample exposure can result in solid-state reactions, cabonization, polymorphic changes, and the like, as a result of high energy exposure of the sample.
- *Size of sample.* The limited amount of compound available can be problematic. However, modern diffractometers can use the so-called zero-background holders (ZBH). These are made from a single crystal of silicon that has been cut along a nondiffracting plane, and then polished to an optically flat finish. Thus, X-rays incident on this surface will be negated by Bragg extinction. In this technique, a thin layer of grease is placed on the ZBH surface, and the sample of ground compound is placed on the surface. The excess is removed such that only a monolayer is examined. The total thickness of the sample and grease should be of the order of a few microns. It is important that the sample is deagglomerated so that the monolayer condition is met. Using this technique, the diffraction pattern of approximately 10 mg of the compound can be obtained. One disadvantage of the ZBH is that weak reflections may not be readily detectable because of the small sample size used.
- *Calibration.* The XRPD should be properly calibrated using the standards available from reliable sources, such as the Laboratory of the Government Chemist (LGC) in the United Kingdom or the National Institute of Standards and Technology (NIST) in the United States. Analyzing one or two peaks of LaB6 (line broadening calibrator), at least weekly, should give confidence in the diffractometer performance and alert the user to any problems that may develop. The common external standards are: silicon, α-quartz, gold, and silicon (SRM 640b). The primary standards for internal d-spacing include silicon (SRM 640b), fluorophlogopite (SRM 675), and the secondary standards for internal d-spacing are: tungsten, silver, quartz, and diamond. The internal quantitative intensity standards are: Al_2O_3 (SRM 676), α- and β-silicon nitride (SRM 656), oxides of Al, Ce, Cr, Ti, and Zn (SRM 674a), α-silicon dioxide (SRM 1878a), and cristobalite (SRM 1879a). A typical external sensitivity standard is Al_2O_3 (SRM 1976).

Phase Solubility Analysis
Solubility is generally estimated by visual observation. The solubility of a compound is initially determined by weighing out 10 mg (or other suitable amount) of the compound. To this is added 10 µL of the solvent of interest. If the compound does not dissolve, a further 40 µL of the solvent is added, and its effect is noted. Successive amounts of the solvent are then added until the compound is observed to dissolve. This procedure should give an approximate value of the solubility. This method does not take into account the kinetic aspects of the dissolution processes involved in solubility measurements. To determine more accurately the concentration of a saturated solution of a compound, the following procedure can be used. A known volume of the solvent, water or buffer is taken into a scintillation vial, and the compound is added until saturation is observed. The solution is then stirred or shaken and the experiment restarted. It is recommended that the experiment be conducted at least overnight or longer, for low solubility compounds. Depending on the amount of the compound available, replicate experiments should be carried out. After stirring or shaking, the solvent should be separated from the suspension by centrifugation or by filtration using polytetrafluoroethylene (PTFE) filters. The filtrate is then assayed preferably by HPLC; however, ultraviolet (UV)-visible spectroscopy can also be used to determine the solubility, if compound stability or impurities are not an issue. This is termed as the thermodynamic solubility. It is also useful to measure the pH of the filtrate, and to characterize any undissolved material by DSC to detect any phase changes that might have occurred.

For high-throughput screening (HTS) of solubility, where the amount of the compound might be severely restricted, reporting kinetic solubility might be adequate. This can be accomplished by using techniques, such as a 96-well microtiter technique with an integral nephelometer, where aliquots of the aqueous solution are placed in the microtiter wells, to which are added 1 µL of the compounds in dimethylsulfoxide (DMSO), and the plate is shaken. The turbidity of the solutions is then measured using the nephelometer. The process is repeated up to 10 times. If turbidity is detected in a cell, the experiment is terminated, that is, solution additions are stopped, and the solutions are ranked in terms of the number of additions that caused turbidity. This method is suitable to rank the compounds in terms of their solubility and not to measure solubility accurately. A transformation of the amorphous form to a crystalline form would decrease the dissolution rate in most instances. Once the presence of polymorphs is established, the solubility of the polymorphs and the thermodynamic quantities involved in the transition from a metastable to stable polymorph can be calculated. Experimentally, the solubility of the polymorphs are determined at various temperatures, and subsequently, the log of the solubility is plotted against the reciprocal of the temperature (the van't Hoff method), from which the enthalpy of solution can be calculated from the slope. If the lines intersect, it is known as the transition temperature, and one consequence of this is that there may be a transition from one polymorph to another, depending on the storage conditions.

Dynamic Vapor Sorption
Measurement of the hygroscopic properties of a compound can be conveniently carried out on small quantities of compound using a DVS system (10) from

Surface Measurement Systems (UK) that allows highly accurate measurements under different conditions and materials. The IGA$_{sorp}$ is designed to measure accurately the magnitude and kinetics of moisture sorption onto materials. It is fully automated and combines an ultrasensitive microbalance with precise measurement and control of both humidity and temperature. The IGA® (11) series of instruments uniquely utilize the IGA method to intelligently determine the equilibrium uptakes and kinetics, and the fully automated system is capable of isothermal, isochoric, and temperature programmed determinations. The RH is generated by bubbling nitrogen through a water reservoir, where it is saturated with moisture. Using a mixing chamber, the moist nitrogen is mixed with dry nitrogen in a fixed ratio, thus producing the required RH. The moist nitrogen is then passed over the sample, and the instrument is programmed such that the increase in weight caused by moisture is monitored with time using an ultrasensitive microbalance. The compound takes up moisture and reaches equilibrium, at which point the next RH stage is programmed to start. The adsorption and desorption of moisture can be studied using this instrument, and the effect of temperature can be investigated as well. Using this technique, a quantity as small as 1 mg can be assessed (12).

The SGA-100 Symmetrical Gravimetric Analyzer is a continuous vapor flow sorption instrument for obtaining water and organic vapor isotherms at temperatures ranging from 0°C to 80°C at ambient pressure. As a result of its symmetrical design (both the sample and the reference side of the microbalance are subjected to identical temperature, relative humidity, and flow rate), this instrument achieves great accuracy and stability. Another benefit of this design is the ability to perform absolute or differential adsorption experiments. In addition to isotherms, "isohumes™", maintaining constant humidity and equilibrating the sample to a series of temperatures, heats and kinetics of adsorption, hydrate formation can also be studied. The core of the instrument is an isothermal aluminum block containing the sample chamber, which permits very tight control of temperature and RH at the sample. The temperature within the block is kept stable by a constant temperature bath capable of temperature control within 0.01°C. Because of the easy access to the sample and the absence of glass hang-down tubes, this instrument is very easy to operate and highly reliable. The SGA-100 is also very compact, with a footprint of only 18″ × 20″ (13). The MB-300G-HP also from VTI Corp is specifically designed to obtain adsorption/desorption isotherms at values above the atmospheric pressures. A stainless steel microbalance head is integrated with a constant temperature bath or cryostat for optimal temperature control, a 5000 torr or higher pressure transducer, and our high-quality hardware and Microsoft™ Windows™-based software. The design of the MB-300G-HP provides very easy access to the sample. When a cryostat is used, the experimental temperature range is between −190°C and 600°C. The user can also select a constant temperature bath for experimental temperatures of 0–80°C and integrate a furnace for temperatures up to 500°C. The standard configuration is for up to six atmospheres of pressure; however, higher pressures can be achieved by integrating optional equipment. Isobars can also be obtained effortlessly.

The Rubotherm system provides gravimetric sorption measurements carried out in a closed measuring cell completely thermostated up to 700 K or down to 77 K without temperature differences or corrosive parts inside the whole measuring cell with pressures up to 100 MPa and a mass sensitivity down to 1 μg using a magnetic suspension balance to measure mass transfer under controlled environments (14).

Normally, the moisture sorption–desorption profile of the compound is investigated. This can reveal a range of phenomena associated with the solid. For example, on reducing the RH from a high level, hysteresis (separation of the sorption–desorption curves) may be observed. There are two types of hysteresis loops: an open hyteresis loop, where the final moisture content is higher than the starting moisture content due to so-called ink-bottle pores, where condensed moisture is trapped in pores with a narrow neck, and the closed hysteresis loop may be closed due to compounds having capillary pore sizes.

When there is a large uptake of moisture, this often indicates a phase change. In this case, the desorption phase is characterized by only small decreases in the moisture content (depending on the stability of the hydrate formed) until at low RH, when the moisture is lost. In some cases, hydrated amorphous forms are formed on desorption of the hydrate formed on the sorption phase. In this case, the sorption of moisture caused the sample to crystallize as a hydrate, which at higher RH crystallizes into a higher hydrate. The higher hydrates are generally more stable to decreasing RH until the humidity level reaches to less than 10% when most of the sorbed moisture can be lost to regenerate the amorphous forms.

In terms of salt selection procedure, the critical relative humidity (CRH) of each salt should be identified. This is defined as the point at which the compound starts to sorb moisture. Clearly, compounds or salts that exhibit excessive moisture uptake should be rejected. The level of this uptake is debatable, but those exhibiting deliquescence (where the sample dissolves in the moisture that has been sorbed) should be automatically excluded from further consideration.

The automation of moisture sorption measurements is a relatively recent innovation. Prior to this advance, moisture sorption of compounds ($\sim$10 mg) was determined by exposing weighed amounts of compound in dishes placed in sealed desiccators containing saturated salt solutions. Saturated solutions of salts that give defined RH (as a function of temperature) have long been in use. The relative humidity of a saturated solution at 25°C ranges between 0% for silica gel and 100% for water, potassium acetate (20%), calcium chloride (32%), sodium bromide (58%), potassium bromide (84%), and dipotassium hydrogen phosphate (92%). The test samples are placed in chambers containing these salts and then after saturation sorption, analyzed using methods, such as TGA or HPLC, and so on, to ascertain if there had been any phase change due to sorption in the solid state; this may require additional testing using scanning electron microscopy (SEM), DSC or XRPD.

Dissolution Testing
During the preformulation stage, an understanding of the dissolution rate of the drug candidate is necessary, as this property of the compound is recognized as a significant factor involved in drug bioavailability. Dissolution of a solid usually takes place in two stages: salvation of the solute molecules by the solvent molecules followed by transport of these molecules from the interface into the bulk medium by convection or diffusion. The major factor that determines the dissolution rate is the aqueous solubility of the compound; however, other factors, such as particle size, crystalline state (polymorphs, hydrates), pH, and buffer concentration can affect the rate. Moreover, physical properties, such as viscosity and wettability can also influence the dissolution process.

Ideally, dissolution should simulate in vivo conditions. To do this, it should be carried out in a large volume of dissolution medium, or there must be some mechanism whereby the dissolution medium is constantly replenished by fresh solvent. Provided this condition is met, the dissolution testing is defined as taking place under sink conditions. Conversely, if there is a concentration increase during dissolution testing, such that the dissolution is retarded by a concentration gradient, the dissolution is said to be nonsink. While the use of the USP paddle dissolution apparatus is mandatory when developing a tablet, the rotating disc method has great utility with regard to preformulation studies. The intrinsic dissolution rate is the dissolution rate of the compound under the condition of constant surface area. The rationale for the use of a compressed disc of pure material is that the intrinsic tendency of the test material to dissolve can be evaluated without formulation excipients.

Intrinsic dissolution rates of compounds obtained from rotating discs can be theoretically determined. Under hydrodynamic conditions, the intrinsic dissolution rate is usually proportional to the solubility of the solid. However, the dissolution rate obtained will depend on the rotation speed. Several modifications of rotating disc apparatus have been introduced to force zero intercepts. A disc is generally prepared by compressing about 200 mg of the candidate drug in a hydraulic press; the IR press often proves useful as it gives a disc with a diameter of 1.3 cm. It should be noted that some compounds do not compress well and may exhibit elastic compression properties; that is, the disc may be very weak, rendering the experiment impossible. In addition to poor compression properties, another complication is that some compounds can undergo polymorphic transformations because of the application of pressure. This should therefore be borne in mind if there is insufficient compound to perform, for example, XRPD postcompaction.

If the disc has reasonable compression properties, it is then attached to a holder and set in motion in the dissolution medium (water, buffer, or simulated gastric fluid): we use a rotation speed of 100 rpm. A number of analytical techniques can be used to follow the dissolution process; however, UV-visible spectrophotometry and HPLC with fixed or variable wavelength detectors (or diode array) appear to be the most common. The UV system employs a flow through system and does not require much attention; however, if HPLC is used, then any aliquot taken should be replaced by an equal amount of solvent. The intrinsic dissolution rate is given by the slope of the linear portion of the concentration versus time curve divided by the area of the disc and has the units of $mg/min\ cm^2$.

High-Performance Liquid Chromatography

High-performance liquid chromatography is used to assess the degradation compounds in testing the stability of new drugs since in these studies the identification of degradation products is very important. Combined with mass spectrometer and the newer instrumentation, liquid chromatography/tandem mass spectroscopy (LC/MS/MS), and so on, offer powerful tools for the elucidation of degradation mechanism.

Isocratic elution is often the most desirable method as it does not require postequilibration phase for the next analysis; this can be an important consideration if a matrix of factors and excipients are studied for interaction. Gradient elution offers the advantage of sharper peaks, increased sensitivity, greater peak capacity, and selectivity (increased resolving power).

The type of detector to be used is usually dictated by the chemical structure of the compound under investigations. As most compounds of pharmaceutical interest contain aromatic rings, UV detection is the most common detection method.

When using this technique, the most appropriate wavelength is selected from the UV spectrum of the pure compound and that of the system suitability sample. Usually, the λ_{max} is chosen; however, in order to remove unwanted interference, it may be necessary to move away from this value. Where possible, the use of wavelength <250 nm should be avoided because of the high level of background interference and solvent adsorption. In practical terms, this requires the use of far-UV grade solvents and the avoidance of organic buffers.

Other types of detection include refractive index, fluorescence, or mass selective detectors. The use of other types of detectors, such as those based on fluorescence, may be used for assaying compounds that can be specifically detected at low concentrations in the present of nonfluorescent species. However, as few compounds are naturally fluorescent, they require chemical modification, assuming they have a suitable reactive group, to give a fluorescent derivative.

During the early stage of development, the amount of method validation carried out is likely to be limited because of compound availability. However, a calibration curve should be obtained using either an internal standard or external standard procedure. The latter procedure is commonly employed by injecting a fixed volume of standard samples containing a range of known concentrations of the compound of interest. Plots of peak height and/or area versus concentration are checked for linearity by subjecting the data to linear regression analysis. Other tests may be carried out, such as the limit of detection, precision of the detector response, accuracy, reproducibility, specificity, and ruggedness, if more extensive validation is required.

REFERENCES

1. http://www.brukeraxs.de/index.php?id=home0.
2. http://www.accelrys.com/products/cerius2/.
3. http://pharmquest.com.
4. http://www.transformpharma.com/te_platforms.htm
5. http://www.symyx.com.
6. http://www.avantium.nl.
7. http://www.solvias.com/.
8. http://www.accelrys.com/products/cerius2/.
9. http://www.cclrc.ac.uk/Activity/DL.
10. http://www.smsuk.co.uk/index.php.
11. http://www.metpowerin.com/moisure_sorption_analysis.htm.
12. http://www.hidenisochema.com/products/IGAsorp.html.
13. http://www.vticorp.com/products/sga100.htm.
14. http://www.rubotherm.de/ENGL/MAINFR031.HTM.

BIBLIOGRAPHY

Achanta AS, Adusumil P, James K, Rhodes C. Hot-melt coating: water sorption behaviour of excipient films. Drug Dev Ind Pharm 2001; 27:241.

Ahlneck C, Lundgren P. Methods for the evaluation of solid state stability and compatibility between drug and excipient. Acta Pharm Suec 1985; 22:305–314.

Ahmed H, Buckton G, Rawlins DA. Use of isothermal microcalorimetry in the study of small degress of amorphous content of a hydrophobic powder. Int LJ Pharm 1996; 130:195–201.

Aker MJ, Fites AL, Robinson RL. Formulation design and development of parenteral suspensions. J Parenteral Sci Tech 1987; 41:88–96.

Akers MJ. Preformulation screening of antioxidant efficiency in parenteral solutions. J Parenteral Drug Assoc 1979; 33:346–356.

Amidon GE. Physical test methods for powder flow characterization of pharmaceutical materials. A review of methods. Pharm Forum 1999; 25:8298–8305.

Ammar HO, Ibrahim SA, El–Mohsen A. Effect of chelating agents on the stability of injectable isoniazid solutions. Pharmazie 1982; 37:270–271.

Asahara K, Yamada H, Yoshida S, Hirose S. Stability prediction of nafamostat mesylate in an intravenous admixture containing sodium bisulfite. Chem Pharm Bull 1990; 38:492–497.

Atkinson TW, Greenway MJ, Holland SJ, Merrifield DR, Scott HP. The use of laser diffraction particle size analysis to predict the dispersibility of a medicament in a paraffin based ointment. Spec Publ R Soc Chem 1992; 102:139–152.

Atkinson TW, White S. Hydrophobic drug substances: the use of laser diffraction particle size analysis and dissolution to characterize surfactant stabilized suspensions. Spec Publ R Soc Chem 1992; 102:133–142.

Aulton M, Wells JA. In: Aulton ME, ed. Pharmaceutics. The Science of Dosage Form Design. Edinburgh: Churchill Livingstone, 1988.

Barra J, Somma R. Influence of the physicochemical variability of magnesium state on its lubricant properties: possible solutions. Drug Dev Ind Pharm 1996; 22:1105–1120.

Biro EJ, Racz I. The role of zeta potential in the stability of albendazole suspensions. S T P Pharma Sci 1998; 8:311–315.

Blackett PM, Buckton G. A microcalorimetric investigation of the interaction of surfactants with crystalline and partially crystalline salbutamol sulphate in a model inhalaltion aerosol system. Pharma Res 1995; 12:1689–1693.

Blagden N, Davey RJ, Rowe R, Roberts R. Disappearing polymorphs and the role of reaction by products: the case of sulphathiazole. Int J Pharm 1998; 172:169–177.

Boerefijin R, Ning Z, Ghadiri M. Disintegration of weak lactose agglomerates for inhalation applications. Int J Pharm 1998; 172:199–209.

Bommireddi A, Li L, Stephens D, Robinson D, Ginsberg E. Particle size determination of a flocculated suspension using a light scattering particle analyzer. Drug Dev Ind Pharm 1998; 24:1089–1093.

Briggner L-E, Buckton G, Bystrom K, Darcy P. Use of isothermal microcalorimetry in the study of changes in crystallinity induced during the processing or powders. Int J Pharm 1994; 105:125–135.

Brittian HG, ed. Physical characterization of pharmaceutical solid. New York: Marcel Dekker, 1995.

Brittain HG, Bogdanowich SJ, Bugey DE, DeVincetis J, Lwen G, Newman AW. Physical characterization of pharmaceutical solids. Pharm Res 1991; 8:963–973.

Buckton G. Characterisation of small changes in the physical properties of powders of significance for dry powder inhaler formulations. Adv Drug Dev Rev 1997; 26:17–27.

Buckton G. Surface characterization: understanding sources of variability in the production and and use of pharmaceuticals. J Pharm Pharmacol 1995; 47:265–275.

Buckton G, Darcy P. The use of gravimetric studies to assess the degree of crystallinity of predominantly crystalline powders. Int J Pharm 1995; 123:265.

Buckton G, Yonemochi E, Moffat AC. Water sorption and near IR spectroscopy to study the differences between microcrystalline cellulose and silicified microcrystalline cellulose before and after wet granulation. Int J Pharm 1999; 20:181–1861.

Bugay DE, Newman AW, Findlay WP. Quantitation of cefepime–2HCl dehydrate in cefepime–2HCl monohydrate by diffuse reflectance IR and powder X-ray diffraction techniques. J Pharm Biomed Anal 1996; 15:49–61.

Burger A, Ramberger R. On the polymorphism of pharmaceuticals and other molecular crystals. I. Theory of Thermodynamic Rules. Mikrochim Acta II 1979a; 259–271.

Burger A, Ramberger R. On the polymorhism of pharmaceuticals and other molecular crystals. II. Applicability of Rhermodynamic Rules. Mikrochim Acta II 1976; 273–316.

Burnett D, Thielmann F, Booth. Determining the critical relative humidity for moisture induced phase transitions. Int J Pharm 2004; 287:123.

Byron PR, Naini V, Phillips EM. Drug carrier selction–important physicochemical characteristic. Respir Drug Deliv 1996; V:103–113.

Cartensen JT, Ertell C, Geoffroy J-E. Physcio-chemical properties of particulate matter. Drug Dev Ind Pharm 1993; 19:195–219.

Carter PA, Rowley G, Fletcher EJ, Hill EA. An experimental investigation of triboelectrification in cohesive and non-cohesive pharmaceutical powders. Drug Dev Ind Pharm 1992; 18:1505–1526.

Carter PA, Rowley G, Fletcher EJ, Stylianopouplos V. Measurement of electrostatic charge decay in pharmaceutical powders and polymer materials used in dry powder inhaler devices. Drug Dev Ind Pharm 1998; 24:1083–1088.

Cavatur RK, Suryanarayanan R. Characterization of frozen aqueous solutions by low temperature X-ray powder diffractometry. Pharm Res 1998; 15:194–199.

Chang H-K, Whitworth CW. Aspirin degradation in mixed polar solvents. Drug Dev Ind Pharm 1984; 10:515–526.

Chansiri G, Lyons RT, Patel MV, Hem ST. Effect of surface charge on the stability of oil/water emulsions during steam sterilization. J Pharm Sci 1999; 88:454–458.

Chaumeil JC. Micronization: a method of improving the bioavailability of poorly soluble drugs. Meth Find Exp Clin Pharm 1998; 20:211–215.

Chrzanowski FA, Ulissi LA, Fegely BJ, Newman AC. Preformulation excipient compatibility testing, application of a differential scanning calorimetry method versus a wet granulation simulating isothermal stress method. Drug Dev Ind Pharm 1986; 12:783–800.

Clark AR. The use of laser diffraction for the evaluation of the aerosol clouds generated by medical nebulizers. Int J Pharm 1995; 115:69–78.

Collins-Gold LC, Lyons RT, Bartlow LC. Parenteral emulsions for drug delivery. Adv Drug Del Rev 1990; 5:189–208.

Craig DQM, Royall PG, Kett VL, Hopton ML. The relevance of the amorphous state to pharmaceutical dosage forms: glassy drugs and freeze dried systems. Int J Pharm 1999; 179:179–207.

Curzons AD, Merrifield DR, Warr JP. The assessment of crystal growth of organic pharmaceuticl material by specific surface measurement. J Phys D: Appl Phys 1993; 26:B181–B187.

Dahlback M. Behaviour of nebulizing solutions and suspensions. J Aerosol Med 1994; 7(suppl):S13–S18.

Dalby RN, Phillips EM, Byron PR. Determination of drug solubility in aerosol propellants. Pharm Res 1991; 8:1206–1209.

Decamp WH. Regulatory considerations in crystallization processes for bulk pharmaceutical chemicals: a reviews perspective. In: Myerson et al., eds. Proceedings of the 1995 Conference on Crystal Growth Org Mater. Washington DC: American Chemical Society, 1996:65–71.

Deicke A, Suverkrup R. Dose uniformity and redispersibility of pharmaceutical suspensions I: quantification and mechanical modeling of human shaking behaviour. Eur J Pharm Biopharm 1999; 48:225–232.

Dobetti L, Grassano A, Forster R. Physicochemical characterization of ITF 296, a novel antiischemic drug. Bull Chem Pharm 1995; 134:384–389.

Duddu SP, Weller K. Importance of glass transition temperature in accelerated stability testing of amorphous solids: case study using a lyophilized aspirin formulation. J Pharm Sci 1996; 85:345–347.

Duncan-Hewitt WC, Grant DJW. True density and thermal expansivity of pharmaceutical solids: comparison of methods and assessment of crystallinity. Int J Pharm 1986; 28:75–84.

Dunitz JD, Bernsteim J. Disappearing polymorphs. Acc Chem Res 1995; 28:193–200.

Durig T, Fassihi AR. Identification of stabilizing and destabilizing effects of excipient drug interactions in solid dosage form design. Int J Pharm 1993; 97:161–170.

Duro R, Alverez C, Martinez-Pachecon R, Gomez-Amoza JL, Concheiro A, Souto C. The adsorption of cellulose ethers in aqueous suspensions of pyrantel pamoate: effects of zeta potential and stability. Eur J Pharm Biopharm 1998; 45:181–188.

Feeley JC, York P, Sumby BS, Dicks H. Determination of surface properties and flow characteristics of salbutamol sulphate, before and after micronisation. Int J Pharm 1998; 172:89–96.

Flynn GL. Isotonicity–collagative properties and dosage form behaviour. J Parenteral Drug Assoc 1979; 33:292–315.

Folger M, Muller-Goymann CC. Investigations on the long-term stability of an O/W cream containing either bufexamac or betamethasone-17-valerate. Eur J Pharm Biopharm 1994; 40:58–63.

Forbes RT, Davis KG. Water vapor sorption studies on the physical stability of a series of spray-dried protein/sugar powders for inhalation. J Pharm Sci 1998; 87:1316.

Fu RC-C, Lidgate DM, Whately JL, McCullough T. The biocompatibility of parenteral vehicles—in vitro/in vivo screening comparison and the effect of excipients on hemolysis. J Parenteral Sci Tech 1987; 41:164–168.

Further C. Interparticulate attraction mechanisms. Drug Pharm Sci 1996; 71:1–15.

Girona V, Pacareu C, Riera A, Pouplana R, Castillo M, Bolos J. Spectrophotometric determination of the stability of an ampicillin-dicloxacillin suspension. J Pharm Biomed Anal 1988; 6:23–28.

Gould PL, Goodman M, Hanson PA. Investigation of the solubility relationship of polar, semi-polar and non-polar drugs in mixed co-solvent systems. Int J Pharm 1984; 19:149–159.

Hacche L, Dickason MA, Oda RM, Firestone BA. Particle size analysis for ofloxacin, prednisolone acetate, and an ophthalmic suspension containing the ofloxacin/steroid combination. Part Sci Technol 1992; 10:37–47.

Hammerlund ER. Sodium chloride equivalents, cryoscopic properties, and hemolytic effects of certain medicinals in aqueous solution IV: supplemental values. J Pharm Sci 1981; 70:1161–1163.

Hartauer KJ, Guillory JK. A comparison of diffuse reflectance FT-IR spectroscopy and DSC in the characterization of a drug excipient interaction. Drug Dev Ind Pharm 1991; 17:617–630.

Hatley RHM, Franks S, Brown, Sandhu G, Gray M. Stabilization of a pharmaceutical drug substance by freeze-drying: a case study. Drug Stabil 1996; 1:73–85.

Her LH, Nail SL. Measurement of the glass transition temperatures of freeze concentrated solutes by differential scanning calorimetry. Pharm Res 1994; 11:54–59.

Hindle M, Byron PR. Size distribution control of raw materials for dry powder inhalers using the aerosizer with aero-disperser. Pharm Tech 1995; 19:64–78.

Hoelgaard A, Moller N. Hydrate formation of metronidazole benzoate in aqueous suspensions. Int J Pharm 1983; 15:213–221.

Holgado MA, Fernandez-Arevalo M, Gines JM, Caraballo I, Rabasco AM. Compatibility study between carteolol hydrochloride and tablet excipients using differential scanning calorimetry and hot stage microscopy. Pharmazie 1995; 50:195–198.

Jashnani RN, Byron PR. Dry powder aerosol generation in different environments: performance comparisons of albuterol, albuterol sulfate, adipate and albuterol stearate. Int J Pharm 1996; 130:13–24.

Johansson D, Abrahamsson B. In vitro evaluation of two different dissolution enhancement principles for a sparingly soluble drug administered as extended-release (ER) tablet. Proc 24th Int Symp Controlled Release Bioact Mater, 1997:363–364.

Jonkman-De Vries JD, Rosing H, Henrar REC, Bult A, Beijen JH. The influence of formulation excipients on the stability of the novel antitumor agent carzelesin (U-80, 224): PDA J Pharm Sci Technol 1995; 49:283–288.

Jutta JC. Compatibility of nebulizer solubility admixture. Ann Pharmacother 1997; 31:487–489.

Kanerva H, Kiesvaara J, Muttonen E, Yliruusi J. Use of laser light diffraction in determining the size distribution of different shaped particles. Pharm Ind 1993; 55:849–853.

Kawashima Y, Sergano T, Hino T, Yamamoto H, Takeughi H. Effect of surface morphology of carrier lactose on dry powder inhalation property of pranlukast. Int J Pharm 1998; 172:179–188.

Kenley RA, Lee MO, Sakumar L, Powell MF. Temperature and pH dependence of fluocinolone acetonide degradation in a topical cream formulation. Pharm Res 1987; 4:342–347.

King S, Miyame A, Koda S, Morimoto Y. Effect of grinding on the solid-state stability of cefixime trihydrate. Int J Pharm 1989; 56:125–134.

Kriwet K, Muller-Goymann CC. Binary diclofenac diethylamine-water systems: Micelles, vesicles, and lyotropic liquid crystals. Eur J Pharm Biopharm 1993; 39:234–238.

Krzyzaniak JF, Yalkowsky SH. Lysis of human red blood cells3: Effect of contact time on surfactant–induced haemolysis. PDA J Pharm Sci Tech 1998; 52:66–69.

Kulvanich P, Stewart PJ. The effect of particle size and concentration on the adhesive characterisitics of a model drug-carrier interactive system. J Pharm Pharmacol 1987; 39:673–678.

Leung SS, Padden BE, Munson EJ, Grant DJW. Solid state characterization of two polymorphs of aspartame hemihydrate. J Pharm Sci 1998; 87:501–507.

Levy MY, Benita S. Design and characterization of a submicronized o/w emulsion of diazepam for parenteral use. Int J Pharm 1989; 54:103–112.

Lidgate DM, Fu RC, Fletiman JS. Using a microfluidizer to manufacture parenteral emulsions. Pharm Tech Int 1990:30–33.
Lipp R, Muller-Fahrnow A. Use of X-ray crystallography for the characterization of single crystals grown in steroid containing transdermal drug delivery systems. Eur J Pharm Biopharm 1999; 47:133–138.
Loudon GC. Mechanistic interpretation of pH-rate profiles. J Chem Ed 1991; 68:973–984.
MacCallion ONM, Taylor KMG, Bridges PA, Thomas M, Taylor AJ. Jet nebulisers for pulmonary drug delivery. Int J Pharm 1996; 130:1–11.
Mackin L, Zanon R, Demote M. Quantification of low levels of amorphous content in micronised active batches using DVS and isothermal microcalorimetry. Int J Pharm 2002; 231:847.
Martens-Lobenhoffer J, Jens RM, Losche D, Gollnick H. Long term stability of 8-methoxypsoralen in ointments for topical PUVA therapy ("Cream-PUVA"). Skin Pharmacol Appl Skin Physiol 1999; 12:266–270.
Mazzenga GC, Berner B, Jordan F. The transdermal delivery of zwitterionic drugs II: The flux of zwitterions salt. J Controlled Release 1992; 20:163–170.
Mendenhall DW. Stability of parenterals. Drug Dev Ind Pharm 1984; 10:1297–1342.
Merrifield DR, Carter PL, Clapham D, Sanderson FD. Addressing the problem of light instability during formulation development. In: Tønneson HH, ed. Photostab Drugs: Drug Formulations [Int Meet Photostable Drugs, 1995]. London: Taylor and Francis, 1996:141–154.
Midoux N, Hosek P, Pailleres L, Authelin JR. Micronization of pharmaceutical substances in a spiral jet mill. Powder Technol 1999; 104:113–120.
Moldenhauer JE. Determining whether a product is steam sterilizable. Pharm PDA J Sci Technol 1998; 52:28–32.
Monkhouse DC. Excipient compatibility possibilities and limitations in stability prediction. In stability testing in the EC, Japan and the USA. In: Crim W, Krummen K, eds. Scientific and Regulatory Requirements, Paperback APV 32. Stuttgar: Wissenschaftlidie Verlassgellschaft mbH, 1993:67–74.
Monkhouse DC, Maderich A. Whither compatibility testing. Drug Dev Ind Pharm 1989; 15(13):2115–2130.
Morefield EM, Feldkamp JR, Peck GE, White JL, Hem SL. Preformulation information for suspensions. Int J Pharm 1987; 34:263–265.
Mullin JW. Crystal size and size distribution: the role of test sieving. Anal Proc 1993; 30:455–456.
Munson JW, Hussain A, Bilous R. Precautionary note for use bisulphate in pharmaceutical formulations. J Pharm Sci 1977; 66:1775–1776.
Myrdal PB, Simamora P, Surakitbanharn Y, Yalkowsky SH. Studies in phlebitis. VII: In vitro and in vivo evaluation of pH-solubilized levemopamil. J Pharm Sci 1995; 84:849–852.
Na GC, Jr, Stevens HS. Yuan BO, Ragagoplan N. Physical stability of ethyl diatrizoate nanocrystalline suspension in steam sterilization. Pharm Res 1999; 16:569–574.
Nayak AS, Cutie AJ, Jochsberger T, Kay AI. The effect of various additives on the stability of isoproternol hydrochloride solutions. Drug Dev Ind Pharm 1986; 12:589–601.
Niemi L, Laine E. Effect of water content on the microstructure of an o/w cream. Int J Pharm 1991; 68:205–214.
Niemi L, Turakka L, Kahela P. Effect of water content and type of emulgator on the release of hydrocortisone from o/w creams. Acta Pharm Nord 1989; 1:23–30.
Nikander, K. Some technical, physicochemical and physiological aspects of nebulization of drugs. Eur Respir Rev 1997; 7:168–172.
Niven RW. Aerodynamic particle-size testing using a time-of-flight aerosol beam spectrometer. Pharm Technol 1993; 17:64–78.
Notari RE. On the merits of a complete kinetic stability study. Drug Stabil 1996; 1:1–2.
Nyqvist H, Wadsten T. Preformulation of solid dosage forms: light stability testing of polymorphs as part of a preformulation program. Acta Pharm Technol 1986; 32:130–132.
Nyqvist R, Lundgren P, Jansson I. Studies on the physical properties of tablets and tablet excipients. Part 2. Testing of light stability of tablets. Acta Pharm Suec 1980; 17:148–156.
Otsuka M, Hasegawa H, Matsuda Y. Effect of polymorphic transformation during the extrusion–granulation process on the pharmaceutical properties of carbmazepine granules. Chem Pharm Bull 1997; 45:894–898.

Otsuka M, Matsuda Y. Physicochemical characterization of Phenobarbital polymorphs and their pharmaceutical properties. Drug Dev Ind Pharm 1993; 19:2241–2269.
Ostwald W. Studdien über die Bildung und Umwandlung fester Körper. Zeitschrift für Physikalische Chemie 1897; 22:289–330.
Parasrampuria J, Li LC, Dudleston A, Zhang H. The impact of an autoclave cycle on the chemical stability of parenteral products. J Parenteral Sci Technol 1993; 47:177–179.
Parrot EL. Milling of pharmaceutical solids. J Pharm Sci 1974; 63:813–829.
Payne RS, Roberts RJ, Rowe RC, McPartlin M, Bashal A. The mechanical properties of two forms of primidone predicted from their crystal structures. Int J Pharm 1996; 145:165–173.
Pedersen S, Kristenen HG. Change in crystal density of acetylsalicylic acid during compaction. S T P Pharma Sci 1994; 4:201–206.
Pedersen S, Kristensen HG. Cornett C. Solid-state interactions between trimethoprim and parabens. S T P Pharma Sci 1994; 4:292–297.
Peramal VL, Tabburie S, Craig DQM. Characterisation of the variation in the physical properties of commercial creams using thermogravimetric analysis and rheology. Int J Pharm 1997; 155:91–98.
Phadke DS, Collier JL. Effect of degassing temperature on the specific surface area and other physical properties of magnesium stearate. Drug Dev Ind Pharm 1994; 20:853–858.
Phillies, GDJ. Quasielastic light scattering. Anal Chem 1990; 62:1049A–1057A.
Phillips EM, Byron PR. Surfactant promoted crystal growth of micronized methylprednisolone in trichloromonofluoromethane. Int J Pharm 1994; 110:9–19.
Phillips EM, Byron PR, Dalby RN. Axial ratio measurements for early detection of crystal growth in suspension-type metered dose inhalers. Pharm Res 1993; 10:454–456.
Phipps MA, Winnike RA. Long ST, Viscomi F. Excipient compatibility as assessed by isothermal microcalorimetry. J Pharm Pharmacol 1998; 50(suppl):9.
Pikal MJ, Shah S. The collapse temperature in freeze–drying: dependence on measurement methodology and rate of water removal from the glassy phase. Int J Pharm 1990; 62:165–186.
Podczeck F, Newton JM. James MB. Assessment of adhesion and autoadhesion forces between particles and surfaces. I. The investigation of autoadhesion phenomena of salmeterol xinafoate and lactose monohydrate particles using compacted powder surfaces. J Adhesion Sci Techn 1994; 8:1459–1472.
Podczeck F, Newton JM, James MB. Assessment of adhesion and autoadhesion forces between particles and surfaces. Part II. The investigation of adhesion phenomena of salmeterol xinafoate and lactose monohydrate particles in particle-on particle and particle-on-surface contact. J Adhesion Sci Technol 1995a; 9:475–486.
Podczeck F, Newton JM, James MB. Adhesion and autoadhesion measurements of micronized particles of pharmaceutical powders to compacted powder surfaces. Chem Pharm Bull 1995b; 43:1953–1957.
Podczeck F, Newton JM, James MB. The adhesion force of micronized salmeterol xinafoate to pharmaceutically relevant materials. J Phys D: Appl Phys 1996a: 29:1878–1884.
Podczeck F, Newton JM, James MB. The influence of physical properties of the materials in contact on the adhesion strength of particles of salmeterol base and salmeterol salts to various substrate materials. J Adhesion Sci Technol 1996b; 10:257–268.
Poirier J, Donaldson J, Barbeam A. The specific vulnerability of the substantia nigra to MPTP is related to the presence of transition metals. Biochim Biophys Res Commun 1985; 128:25–33.
Provent B, Chulia D, Carey J. Particle size and caking tendency of a powder. Eur J Pharm Biopharm 1993; 39:202–207.
Roberts RJ, Rowe RC, York P. The relationship between indentation hardness of organic solids and their molecular structure. J Mater Sci 1994; 29:2289–2296.
Rowe RC, Wakerly MJ, Roberts RJ, Grundy RU, Upjohn NG. Expert systems for parenteral development. PDA J Pharm Sci Technol 1995; 49:257–261.
Rubino JT. The effects of cosolvents on the action of pharmaceutical buffers. J Parenteral Sci Technol 1987; 41:45–49.
Rubino JT, Yalkowsky SH. Solubilization by cosolvents. Part 3. Diazepam and benzocaine in binary solvents. Bull Parenteral Drug Assoc 1985; 39:106–111.

Rubino JT, Blanchard J, Yalkowsky SH. Solubilization by cosolvents. Part 2. Phenytoin in binary and ternary solvents. Bull Parenteral Drug Assoc 1984; 38:215–221.
Rubino JT, Yalkowsky SH. Cosolvency and cosolvent polarity. Pharm Res 1987b; 4: 220–230.
Rubino JT, Yalkowsky SH. Cosolvency and deviations from log-linear solubilization. Pharm Res 1987a; 4:231–235.
Sallam E, Saleem H, Zaru R. Polymorphism and crystal growth of phenylbutazone in semi-solid preparations. Part one: characterisation of isolated crystals from commercial creams of phenylbutazone. Drug Dev Ind Pharm 1986; 12:1967–1994.
Samir RD. Preformulation aspects of transdermal drug delivery systems. In: Ghosh TK, Pfister WR, eds. Transdermal Topical Drug Delivery Systems. Buffalo Grove, Ill, USA: Interpharm Press, 1997:139–166.
Schoeni MH, Kraemer R. Osmolality changes in nebulizer solutions. Agents Actions 1989; 31:225–228.
Scott G. Antioxidants. Bull Chem Soc Jpn 1988; 61:165–170.
Selzer T, Radau M, Kreuter J. Use of isothermal heat condition microcalorimetry to evaluate stability and excipient compatibility of a solid drug. Int J Pharm 1998; 171:227–241.
Serrajuddin ATM, Mufson D. pH solubility profiles of organic bases and their hydrochloride salts. Pharm Res 1985; 2:65–68.
Serrajuddin ATM, Thakur AB, Ghoshal, et al. Selection of solid dosage form composition through drug-excipient compatibility testing. J Pharm Sci 1999; 88:696–704.
Shalaev EY, Zografi G. How does residual water affect the solid-state degradation of drugs in the amorphous state? J Pharm Sci 1996; 85:1137–1141.
Shamblin SL, Hancock BC, Zografi G. Water vapor sorption by peptides, proteins and their formulations. Eur J Pharm Biopharm 1998; 45:239.
Sing KSW. Adsorption methods for surface area determination. In: Stanley-Wood NG, Lines RW, eds. Particle size analysis. Proceedings of the 25th Anniversary Conference (1991), The Royal Society of Chemistry, 1992:13–32.
Smith A. Use of thermal analysis in predicting drug-excipient interactions. Anal Proc 1982; 19:554–556.
Spencer R, Dalder B. Sizing up grinding mills. Chem Eng 1997; 104(4)84–87.
Staniforth JN. Pre-formulation aspects of dry powder aerosols. Respir Drug Delivery V 1996; 65–73.
Stark G, Fawcett JP, Tucker IG, Weatherall IL. Instrumental evaluation of color of solid dosage forms during stability testing. Int J Pharm 1996; 143:93–100.
Stock N. Direct non-destructive color measurement of pharmaceuticals. Anal Proc 1993; 30:41–43.
Streng WH, Yu DH-S, Zhu C. Determination of solution aggregation using solubility, conductivity, calorimetry, and pH measurements. Int J Pharm 1996; 135:43–52.
Sweetana S, Akers MJ. Solubility principles and practices for parenteral drug dosage form development. PDA J Parenteral Sci Technol 1996; 50:330–342.
Taylor KMG, Venthoye G, Chawla A. Pentamide isoethionate delivery from jet; nebulisers. Int J Pharm 1992; 85:203–208.
Thakur AJ, Morris K, Grosso JA, et al. Mechanism and kinetics of metal ion mediated degradation of fosinopril sodium. Pharm Res 1993; 10:800–809.
Thoma K, Holzmann C. Photostability of dithranol. Eur J Pharm Biopharm 1998; 46:201–208.
Ticehurst MD, Rowe RC, York P. Determination of the surface properties of two batches of salbutamol sulphate by inverse gas chromatography. Int J Pharm 1994; 111:241–249.
Ticehurst MD, York P, Rowe RC, Dwivedi SK. Characterisation of the surface properties of α-lactose monohydrate with inverse gas chromatography used to detect batch variation. Int J Pharm 1996; 141:93–99.
Timsina MP, Martin GP, Marriot C, Ganderton D, Yianneskis M. Drug delivery to the respiratory tract using dry powder inhalers. Int J Pharm 1994; 101:1–13.
Tiwary AK, Panpalia GM. Influence of crystal habit on trimethoprim suspension formulation. Pharm Res 1999; 16:261–265.
Tiwari D, Goldman D, Malick WA, Madan PL. Formulation and evaluation of albuterol metered dose inhalers containing tetrafluoroethane (P134a), a non-CFC propellant. Pharm Dev Technol 1998; 3:163–174.

Tu Y-H, Wang D-P, Allen LV. Stability of a nonaqueous trimethoprim preparation. Am J Hosp Pharm 1989; 46:301–304.

Turner JL. The regulatory perspective to particle size specification. Anal Proc 1987; 24:80–81.

Tzou T-S, Pachta RR, Coy RB, Schultz RK. Drug form selection in albuterol-containing metered dose inhaler formulations and its impact on chemical and physical. J Pharm Sci 1997; 86:1352–1357.

Ugwu SO, Apte SP. Systematic screening of antioxidants for maximum protection against oxidation: an oxygen polarograph study. Pharm PDA J Sci Technol 1999; 53: 252–259.

van der Houwen OAGJ, de Loos MR, Beijnen JH, Bult A, Underberg WJM. Systematic interpretation of pH–degradation profiles. A critical review. Int J Pharm 1997; 155:137–152.

Vemuri S, Taracatac C, Skluzacek R. Color stability of ascorbic acid tablets measured by a tristimulus colorimeter. Drug Dev Ind Pharm 1985; 11:207–222.

Venkataram S, Khohlokwane M, Wallis SH. Differential scanning calorimetry as a quick scanning technique for solid state stability studies. Drug Dev Ind Pharm 1995; 21:847–855.

Vernon B. Using a ball mill for high–purity milling applications. Powder Bulk Eng 1994; 8(6):53–62.

Vervaet C, Byron PR. Drug-surfactant-propellant interactions in HFA-formulations. Int J Pharm 1999; 186:13–30.

Villiers MM, Tiedt LR. An analysis of fine grinding and aggregation of poorly solubile drug powders in a vibrating ball mill. Pharmazie 1996; 51:564–567.

Wada Y, Matsubara T. Pseudo-polymorphism and crystalline transition of magnesium stearate. Thermochim Acta 1992; 196:63–84.

Waltersson JO. Factorial designs in pharmaceutical preformulation studies. Part 1. Evaluation of the application of factorial designs to a stability study of drugs in suspension form. Acta Pharm Suec 1986; 23:129–138.

Wang Y-CJ, Kowal RR. Review of excipients and pH's for parenteral products used in the United States. J Parenteral Drug Assoc 1980; 34:452–462.

Ward GH, Schultz RK. Process-induced crystallinity changes in albuterol sulfate and its effect on powder physical stability. Pharm Res 1995; 12:773–779.

Ward GH, Yalkowsky SH. Studies in phlebitis. VI: dilution-induced precipitation of amiodarone HCl. J Parenteral Sci Technol 1993; 47:161–165.

Washington C. Particle size analysis in the pharmaceutics and other industries. Chichester, UK: Ellis Horwood, 1992.

Wells JI. Pharmaceutical preformulation. The physicochemical properties of drug substances. Chichester, UK: Ellis Horwood, 1988.

Whateley TL, Steele G, Urwin J, Smail GA. Particle size stability of intralipid and mixed total parenteral nutrition mixtures. J Clin Hosp Pharm 1984; 9:113–126.

Williams NA, Guglielmo J. Thermal mechanical analysis of frozen solutions of mannitol and some related stereoisomers: evidence of expansion during warming and correlation with vial breakage during lyophilization. J Parenteral Sci Technol 1993; 47:119–123.

Williams NA, Lee Y, Polli GP, Jennings TA. The effects of cooling rate on solid phase transitions and associated vial breakage occurring in frozen mannitol solutions. J Parenteral Sci Technol 1986; 40:135–141.

Williams NA, Polli GP. The lyophilization of pharmaceuticals: a literature review. J Parenteral Sci Technol 1984; 38:48–56.

Williams NA, Schwinke DL. Low temperature properties of lyophilized solutions and their influence on lyophilization cycle design: pentamide isothionate. J Pharm Sci Technol 1994; 48:135–139.

Williams RO III, Brown J, Liu J. Influence of micronization method on the performance of a suspension triamcinolone acetonide pressurized metered-dose inhaler formulation. Pharm Dev Technol 1999; 4:167–179.

Williams RO III, Repka M, Liu J. Influence of propellant composition on drug delivery from a pressurized metered dose inhaler. Drug Dev Ind Pharm 1998; 24:763–770.

William RO III, Rodgers TL, Liu J. Study of solubility of steroids in hydrofluoroalkane propellants. Drug Dev Ind Pharm 1999; 25:1227–1234.

Wirth DD, Stephenson GA. Purification of dirithromycin. Impurity reduction and polymorph manipulation. Org Proc Res Dev 1997; 1:55–60.

Wirth DD, Baertschi SW, Johnson RA, et al. Maillard reaction of lactose and fluoxetine hydrochloride, a secondary amine. J Pharm Sci 1998; 87:31–39.
Wong MWY, Mitchell AG. Phsyicochemical characterization of a phase change produced during the wet granulation of chlorophromazine hydrochloride and its effects on tableting. Int J Pharm 1992; 88:261–273.
Yalabik-Kas HS. Stability assessment of emulsion systems. PST, Pharma 1985; 1: 978–984.
Yalkowsky SH, Rubino JT. Solubilization by cosolvents. Part 2. Organic solvents in propylene glycol-water mixtures. J Pharm Sci 1985; 74:416–421.
Yalkowsky SH, Valvani SC, Johnson BW. In vitro method for detecting precipitation of parenteral formulations after injection. J Pharm Sci 1983; 72:1014–1017.
Yamaoka T, Nakamachi H, Miyata K. Studies on the characteristics of carbochromen hydrochloride crystals. II. Polymorphism and cracking in the tablets. Chem Pharm Bull 1982; 30:3695–3700.
Yang SK. Acid catalyzed ethanolysis of temazepam in anhydrous and aqueous ethanol solution. J Pharm Sci 1994; 83:898–902.
York P. Crystal engineering and particle design for the powder compaction process. Drug Dev Ind Pharm 1992; 18:677–721.
York P. Powdered raw materials: characterizing batch uniformity. Respir Drug Deliv 1994; IV:83–91.
York P, Ticehurst MD, Osborn JC, Roberts RJ, Rowe RC. Characterisation of the surface energetics of milled dl-propranolol hydrochloride using inverse gas chromatography and molecular modeling. Int J Pharm 1998; 174:179–186.
Zhang H-J, Xu G-D. The effect of particle refractive index on size measurement. Powder Technol 1992; 70:189–192.
Zhang Y, Johnson KC. Effect of drug particle size on content uniformity of low-dose solid dosage forms. Int J Pharm 1997; 154:179–183.
Ziller KH, Rupprecht H. Control of crystal growth in drug suspensions. Part 1. Design of a control unit and application to acetaminophen suspensions. Drug Dev Ind Pharm 1988a; 14:2341–2370.
Ziller KH, Rupprecht HH. Control of crystal growth in drug suspensions. Part 2. Influence of polymers on dissolution and crystallization during temperature cycling. Pharm Ind 1988b; 52:1017–1022.

7 Dosage Form Considerations in Preformulation

INTRODUCTION

Preformulation studies inevitably extend beyond the basic characterization of the lead compound, because what is considered as an acceptable characteristic of a lead compound will largely depend on the intended or anticipated dosage form. For example, the solubility issues will largely determine the route of administration; conversely, if a particular route of administration is the only desired route, then preformulation studies should attempt to find out the structural changes necessary for the candidate molecule.

In most instances the choice of a prospective dosage form will depend on a variety of factors:

1. rate of entry to body tissues desired
2. onset of action desired
3. aqueous and nonaqueous solubility
4. irritability of solution of drug
5. stability of drug at the site of administration
6. storage and handling requirements for the dosage form
7. shelf life desired
8. patient acceptance vis-à-vis the customary routes for the defined class

Scheme 1 lists some of the pivotal factors that go into the selection of an appropriate drug delivery system, to particularly with reference to the first and foremost requirement: dissolution of drug.

It is noteworthy that the dissolution rate considerations play a pivotal role in the selection of a dosage form and hence the time spent on studying this at the preformulation stages. Although it may be too early to set dissolution rate criteria for the dosage forms, preformulation studies can be very useful where a definite dosage form is envisioned. Schemes 2–4 show the decision-making process for setting the acceptance criteria for dissolution rates. There are three questions that need to be addressed at this stage:

1. What types of drug release acceptance criteria are appropriate (Scheme 2)?
2. What specific test conditions and acceptance criteria are appropriate for immediate release products (Scheme 3)?
3. What are the appropriate acceptable ranges for extended release dosage forms (Scheme 4)?

SOLID DOSAGE FORM CONSIDERATIONS

Most pharmaceutical companies would rather have their new molecule enter the market as a tablet or capsule for a variety of safety, cost, and marketing considerations. As a result, almost 70% of all drugs administered today are in solid

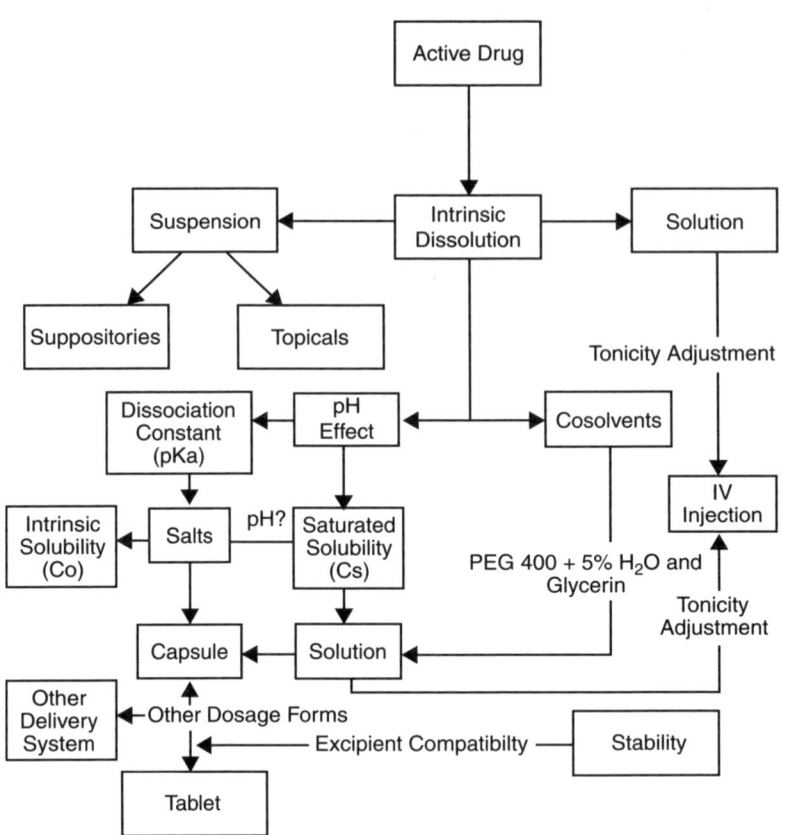

Scheme 1 Selection criteria for dosage form selection. *Source*: Ref. 20. (Courtesy of Pharmquest Corporation, Mountain View, California, U.S.A.)

dosage forms. When so intended, the default form should be a solid dosage form (unless it is predetermined in the case of therapeutic proteins or other drugs that must be administered by parenteral route or other specific routes for specificity of the desired activity). The typical parameter studies for solid dosage forms relate to the ability of a powder mix to flow well in manufacturing machines, and to the intrinsic characteristics that make it compressible. Some examples of properties studied include: crystal structures (polymorphs), external shapes (habits), compression properties, cohesion, powder flow, micromeretics, crystallization, yield strengths and effects of moisture and hygroscopicity, particle size, true bulk and tapped density, and surface area.

Particle Size Studies

The particle size of a new drug substance is a critical parameter, as it affects every phase of formulation and its effectiveness. Appropriate particle size is required to achieve optimal dissolution rate in solid dosage forms, and to control sedimentation and flocculation in suspensions. Small particle size (2–5 μm) is required

Dosage Form Considerations in Preformulation 243

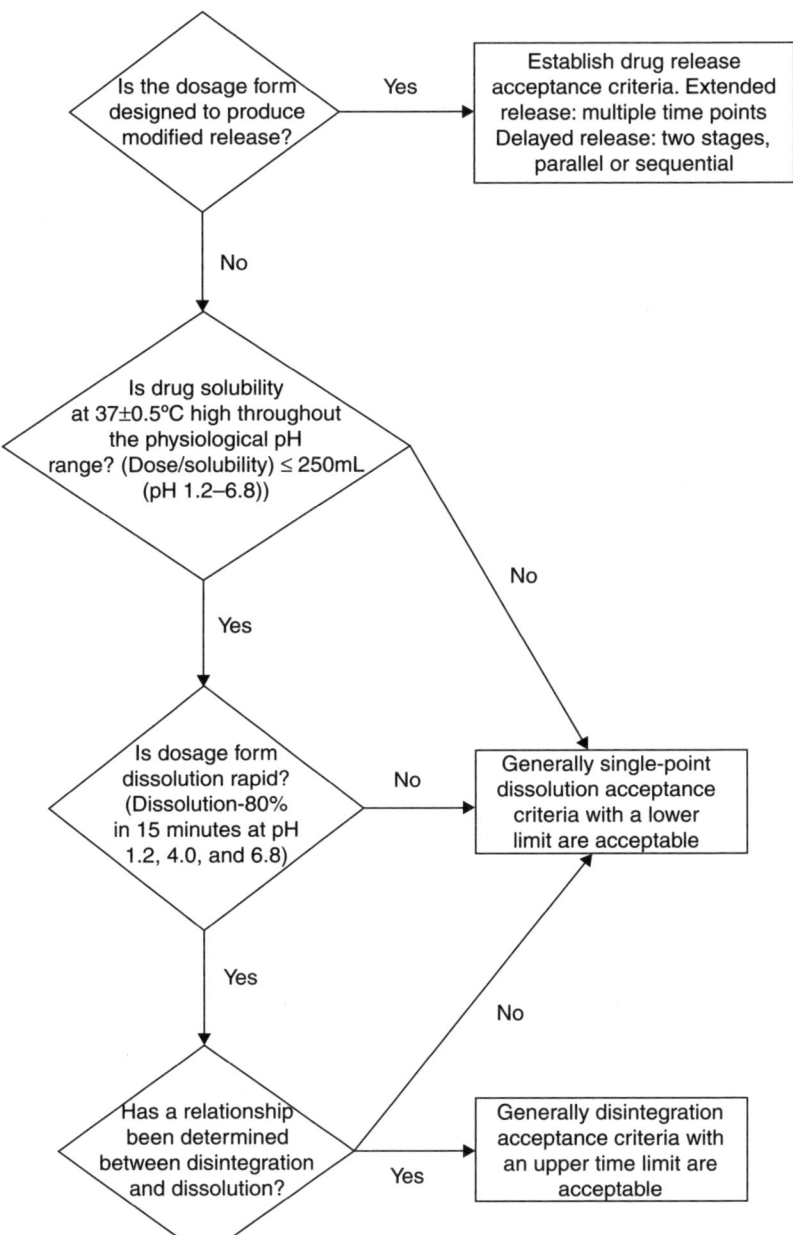

Scheme 2 Setting acceptance criteria for drug product dissolution, that is, what types of drug release acceptance criteria are acceptable? *Source*: Courtesy of Pharmquest Corporation, Mountain View, California, U.S.A.

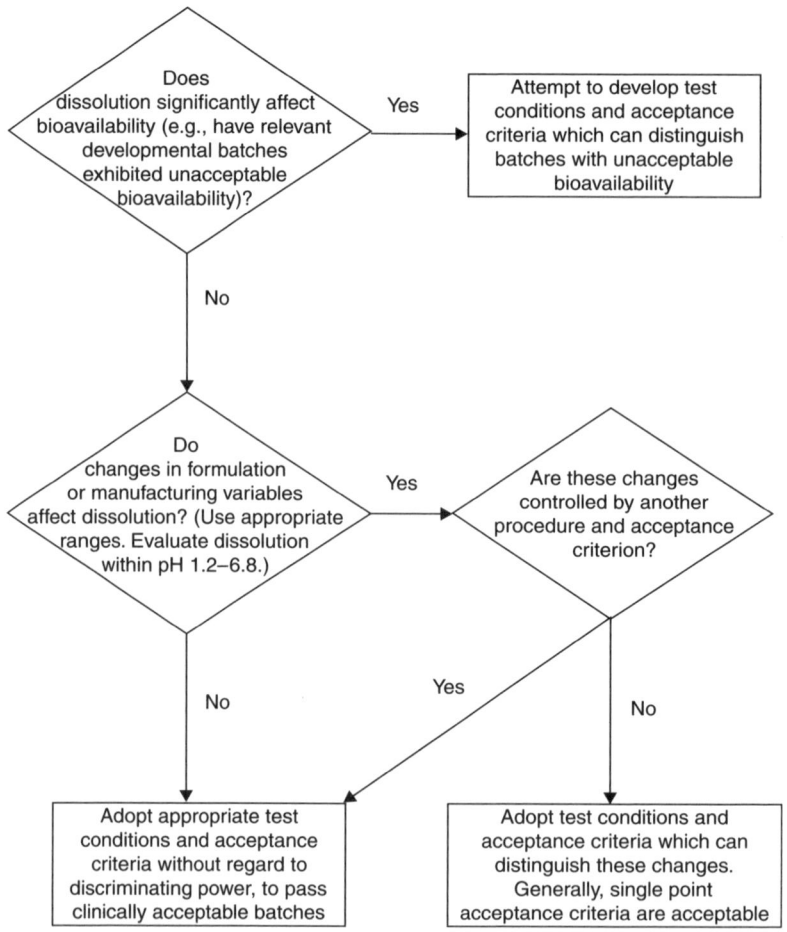

Scheme 3 Setting acceptance criteria for drug product dissolution, that is, what specific test conditions and acceptance criteria are needed for the immediate release of dosage forms? *Source*: Courtesy of Pharmquest Corporation, Mountain View, California, U.S.A.

for inhalation therapy. The content uniformity and compressibility are governed by the particle size. As a result, the preformulation studies must develop a specification of particle size as early as possible in the course of the studies and develop specifications that need to be adhered to throughout the studies.

Conventional methods of grinding in mortar or ball milling (where sample quantity is sufficient; generally it is not and limited to about 25–100 mg) or micronization techniques are used to reduce the particle size. The method used can have significant effect on the crystallinity, polymorphic structures (often to amorphous forms), and drug substance stability that can range from discoloration to significant chemical degradation. Changes in polymorphic forms can be determined by performing X-ray powder diffraction (XRPD) before and after milling.

Micronization, where possible, allows an increase in the surface area to the maximum, which can make an impact on the solubility, dissolution, and as a

Dosage Form Considerations in Preformulation

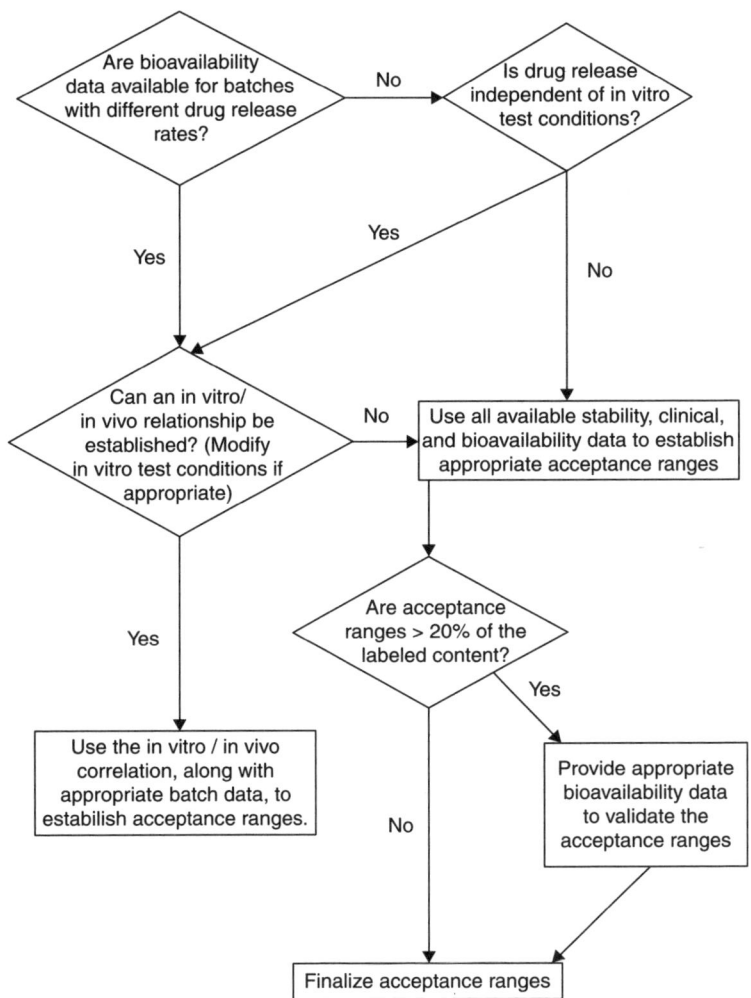

Scheme 4 Setting acceptance criteria for drug product dissolution, that is, the appropriate acceptance ranges for extended release dosage forms. *Source*: Courtesy of Pharmquest Corporation, Mountain View, California, U.S.A.

result, bioavailability. As the aim of most preformulation studies is to determine if a solid dosage form can be administered, knowing that the reduction of the particle size, where it changes the dissolution rates, can be pivotal in decision-making for the selection of dosage forms. In the process of micronization, the drug substance is fed into a confined circular chamber, where it is suspended in a high velocity stream of air. Interparticulate collisions result in a size reduction. Smaller particles are removed from the chamber by the escaping air stream toward the center of the mill, where they are discharged and collected. Larger particles recirculate until their particle size is reduced. Micronized particles are typically less than 10 μm in diameter. In some instances, micronization can prove counterproductive, where it results in increased aggregation (leading to reduced surface area) or alteration of crystallinity, which must be studied using methods like microcalorimetry, dynamic vapor sorption, or inverse gas chromatography (IGC).

The introduction of dynamic vapor sorption (DVS) in 1994 revolutionized the world of gravimetric moisture sorption measurement, bringing the use of outdated, time- and labor-intensive desiccator into the modern world of cutting-edge instrumentation and overnight vapor sorption isotherms. With a resolution down to 0.1 μg, a 1% change in the mass of a 10-mg sample on exposure to the humidity-controlled gas flow is both easily discernable and reproducible. DVS is a valued tool for studies related to polymorphism, compound stability, bulk and surface adsorption effects of water and organic vapors. The dynamic vapor sorption studies would typically show percent mass increases, but often a hysteresis loop relationship is observed, where there is crystallization of compound that results in the expelling of excess moisture. This effect can be important in some formulations, such as dry powder inhaler devices, as it can cause agglomeration of the powders and variable flow properties. The DVS is a useful study when amorphous forms are involved upon size reduction. In many cases, a low level of amorphous character cannot be detected by techniques, such as XRPD; microcalorimetry can detect <10% amorphous content (the limit of detection is 1% or less). The amorphous content of a micronized drug can be determined by measuring the heat output caused by the water vapor inducing the crystallization of the amorphous regions.

Figure 1 shows a typical DVS chart for microcrystalline cellulose and Figure 2 the chart for lactose. The reaction to moisture is dramatically represented in this study.

Excellent instrumentation support and advice is available through Surface Measurement Systems (SMS) (1), manufacturer of DVS-Advantage and DVS-1000 and 2000 series of equipment for dynamic vapor interaction studies. The DVS-HT

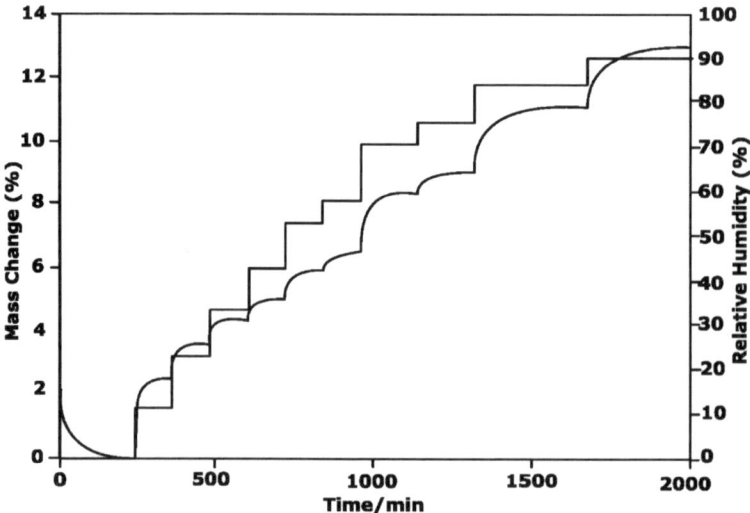

Figure 1 The dynamic vapor sorption chart for microcrystalline cellulose. The percentage mass change is based on a 10-mg sample of microcrystalline cellulose reference material. Steps refer to relative humidity changes. *Source*: Courtesy of Surface Measurement Systems (1).

Dosage Form Considerations in Preformulation 247

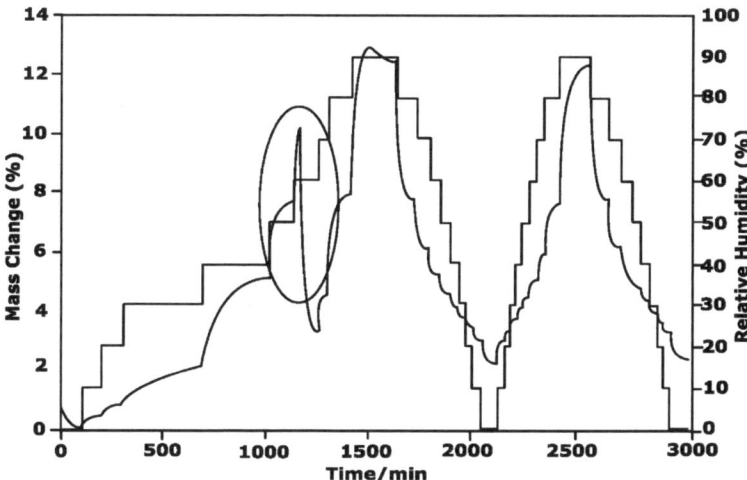

Figure 2 The dynamic vapor sorption chart for lactose. A humidity-induced recrystallization event of spray-dried lactose is marked. Steps refer to relative humidity changes. *Source*: Courtesy of Surface Measurement Systems (1).

represents the first new generation in gravimetric vapor sorption analyzers for more than a decade by Surface Measurement Systems (5 Wharfside, Rosemont Road, Alperton, Middlesex, HA0 4PE, U.K.). The DVS-HT is recommended for stability studies to select the optimal drug and excipient combinations. The DVS-HT features:

- rapid screens of salts, solvates, hydrates, polymorphs, and co-crystals;
- large-scale preformulation and formulation studies;
- characterization of polymers, food ingredients, and fine particles;
- process optimization monitoring of surface and bulk chemistry;
- quality control of incoming raw materials;
- investigation of batch-to-batch variations in material formulations;
- at-Line process analytical technology (PAT) support of production performance to specifications.

Although microcalorimetery remains the workhorse of studies, the use of IGC is becoming more popular to determine the changes to drug substances upon micronization. The IGC differs from traditional gas chromatography insofar as the stationary phase is the powder under investigation. The behavior of pharmaceutical solids, during either processing or use, can be noticeably affected by the surface energetics of the constituent particles. Several techniques exist to measure the surface energy, for example, sessile drop, and dynamic contact angle measurements. IGC is an alternative technique where the powder surface is characterized by the retention behavior of minute quantities of well-characterized vapors that are injected into a column containing the material of interest. Recently published articles using IGC on pharmaceutical powders have ranged from linking surface energetic data with triboelectric charging to studying the effect of surface moisture on surface energetics. Molecular modeling has also recently been used to explore the links between IGC data and the structural and chemical factors that influence the surface properties, thereby achieving predictive knowledge regarding powder behavior during processing. In this type of study, a range of nonpolar and polar adsorbates (probes) are used, for example, alkanes, from hexane to decane, acetone, diethyl ether, or ethyl acetate. The retention volume, that is, the net volume of carrier gas (nitrogen) required to elute the probe, is then measured.

IGC is a gas phase technique for characterizing surface and bulk properties of solid materials. The principles of IGC are very simple, being the reverse of a conventional gas chromatographic (GC) experiment. A cylindrical column is uniformly packed with the solid material of interest, typically a powder, fiber, or film. A pulse or constant concentration of gas is then injected down the column at a fixed carrier gas flow rate, and the time taken for the pulse or concentration front to elute down the column is measured by a detector. A series of IGC measurements with different gas phase probe molecules then allows access to a wide range of physicochemical properties of the solid sample. The flow and retention of gas is shown in Figure 3.

The injected gas molecules passing over the material adsorb on the surface with a partition coefficient K_s:

$$K_s = V_N/W_s \qquad (1)$$

where V_N is the net retention volume—the volume of carrier gas required for eluting the injection through the column, and W_s is the mass of the sample. V_N is a measure of extent of the interaction of the probe gas with the solid sample, and is the fundamental data obtained from an IGC experiment. A wide range of surface and bulk properties can be calculated from it. The surface partition coefficient (K_s) of the probes between carrier gas and surfaces of test powder particles can then be calculated. From this, a free energy can be calculated which can show that one batch may favorably adsorb the probes when compared with another, implying a difference in the surface energetics. The experimental parameter measured in IGC experiments is the net retention volume, V_N. This parameter is related to the surface partition coefficient, K_s, which is the ratio between the concentration of the probe molecule in the stationary and the mobile

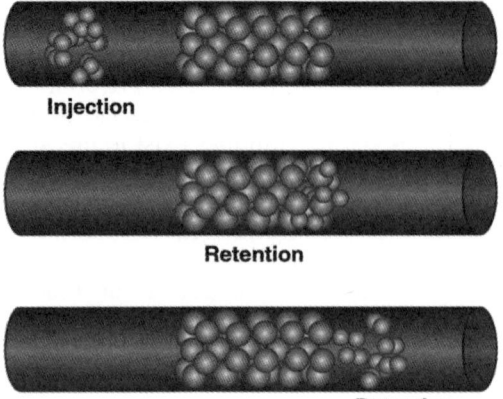

Figure 3 Inverse gas chromatography principles.

phases shown by

$$K_s = \frac{V_s}{m} \times A_{sp} \tag{2}$$

where m is the weight of the sample in the column, and A_{sp} is the specific surface of the sample in the column.

From K_s, the free energy of adsorption $(-\Delta G_A)$ is defined by

$$\Delta G_A = RT \ln\left(K_s \times \frac{P_{sg}}{P}\right) a^{(\gamma_s^{LD})^{1/2}} \tag{3}$$

where P_{sg} is the standard vapor state (101 KN/m^2) and P is the standard surface pressure, which has a value of 0.338 mN/m.

In a typical experiment, the samples are micronized to various particle sizes and γ_s^{LD} is measured and plotted against their median particle size. This will show that as the particle size decreases, the surface of the particles become more energetic. Depending on which functional groups are exposed more upon micronization, there can be an increase or decrease in electron donation, as the particle size decreases. Therefore, using moisture sorption, microcalorimetry, IGC, molecular modeling, and other techniques, the consequences of the particle size reduction process can be assessed. Moreover, surface energetics can be measured directly and predictions made about the nature of the surface, which ultimately could affect properties, such as the flow of powders or adhesion of particles. The IGC is a useful tool, and the newer IGC chromatographs are advanced instruments for the characterization of particulates, fibers, and thin films. It opens up a whole new world of sorption solutions. Some of the applications reported include surface energetics (as described earlier), heat of sorption, sorption isotherms, phase transitions, diffusion kinetics, and so on. The new revolutionary IGC from SMS is the world's first commercial IGC. The unique SMS flow control technology provides accurate and reproducible humidity control. The standard instrument configuration comes with a thermal conductivity detector (TCD) and a flame ionization detector (FID). This combination allows the differentiation between the moisture and the organic solvent elutants. Further detectors (e.g., mass spectrometer) might be added according to customer requirements. For further information, the reader is referred to. Another supplier of DVS is VTI Technologies (2). The SGA-100 symmetrical gravimetric analyzer of VTI is a continuous vapor flow sorption instrument for obtaining water and organic vapor isotherms at temperatures ranging from 0°C to 80°C at ambient pressure. As a result of its symmetrical design (both the sample and the reference side of the microbalance are subjected to identical temperature, relative humidity (RH), and flow rate). Another benefit of this design is the ability to perform absolute or differential adsorption experiments. In addition to isotherms, "isohumes^{TM}," maintaining constant humidity and equilibrating the sample to a series of temperatures, heats, and kinetics of adsorption, hydrate formation can also be studied. The core of the instrument is an isothermal aluminum block containing the sample chamber, which permits very tight control of temperature and RH at the sample. The temperature within the block is kept stable by a constant temperature bath capable of temperature control within 0.01°C.

Particle Size Distribution

Particle size reduction particularly mandates the study of particle size distribution studies using techniques, such as sieving, optical microscopy in conjunction with image analysis, electron microscopy, the Coulter counter and laser diffractometers, depending on the anticipated size of the particles. Although, the size characterization is simple for spherical particles, the study of irregular particles requires specialized methods. The Malvern Mastersizer series (3) is an example of an instrument that measures particle size by laser diffraction. The use of this technique is based on light scattered through various angles, which is directly related to the diameter of the particle. Thus, by measuring the angles and intensity of scattered light from the particles, a particle size distribution can be deduced. It should be noted that the particle diameters reported are the same as those produced by spherical particles under similar conditions. In the former, each particle is treated as spherical and essentially opaque to the impinging laser light. Figure 4 shows different methods of detection and the size of the particles.

Two different light scattering (DLS) methodologies can be used to characterize particles. The classical, also known as "static" or "Rayleigh" scattering or multiple angle laser light scattering, provides a direct measure of mass.

The DLS, which is also known as "photon correlation spectroscopy" (PCS) or "quasi-elastic light scattering" (QELS), uses the scattered light to measure the rate of diffusion of the particles. This motion data is conventionally processed to derive a size distribution for the sample, where the size is given by the "Stokes radius" or "hydrodynamic radius" of the protein particle. This hydrodynamic size depends on both mass and shape (conformation). Dynamic scattering is particularly good at sensing the presence of very small amounts of aggregated particles and studying samples containing a very large range of masses. It can be quite valuable for comparing the stability of different formulations, including real-time monitoring of changes at elevated temperatures. For submicron materials,

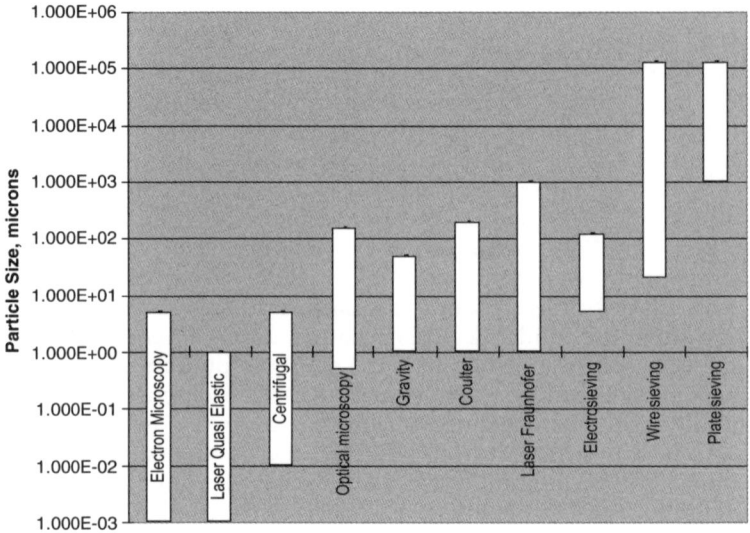

Figure 4 Techniques of particle size detection and their limits.

particularly colloidal particles, QELS is the preferred technique. Two theories dominate the theory of light scattering—the Fraunhofer and Mie. According to the Fraunhofer theory, the particles are spherical, nonporous, and opaque; diameter is greater than wavelength, particles are distant enough from each other, have random motion, and all the particles diffract the light with the same efficiency, regardless of size and shape. The Mie theory takes into account the differences in refractive indices between the particles and the suspending medium. If the diameter of the particles is above 10 µm, then the size produced by utilizing each theory is essentially the same. However, discrepancies may occur when the diameter of the particles approaches that of the wavelength of the laser source.

The following are the values reported from diffraction experiments:

- $D(v, 0.1)$ is the size of particles for which 10% of the sample is below this size.
- $D(v, 0.5)$ is the volume (v) median diameter, of which 50% of the sample is below and above this size.
- $D(v, 0.9)$ is the size of the particle for which 90% of the sample is below this size.
- $D[4, 4]$ is the equivalent volume mean diameter calculated using:

$$D[4, 3] = \frac{\sum d4}{\sum d3} \quad (4)$$

- $D[3, 2]$ is the surface area mean diameter; also known as the Sauter mean, where d = diameter of each unit.
- Log difference represents the difference between the observed light energy data and the calculated light energy data for the derived distribution.
- Span is the measurement of the width of the distribution, and is calculated using

$$\text{Span} = \frac{D(v, 0.9) - D(v, 0.1)}{D(v, 0.5)} \quad (5)$$

The dispersion of the powder is important in achieving reproducible results. Ideally, the dispersion medium should have the following characteristics:

- have a suitable absorbancy
- not swell the particles
- disperse a wide range of particles
- slow sedimentation of particles
- allow homogeneous dispersion of the particles
- be safe and easy to use

In terms of sample preparation, it is necessary to deggregate the samples, so that the primary particles are measured. To achieve this, the sample may be sonicated, although there is a potential problem of the sample being disrupted by the ultrasonic vibration. To check for this, it is recommended that the particle dispersion be examined by optical microscopy.

Although, laser light diffraction is a rapid and highly repeatable method in determining the particle size distributions of pharmaceutical powders, the results obtained can be affected by the particle shape. The laser light scattering generally reports broader size distribution compared with image analysis. In addition, the refractive index of the particles can introduce an error of 10% under most circumstances, and should be accounted for. Another laser-based instrument, relying on light scattering, is the aerosizer. The aerosizer measures particles one at a time in the range of 0.20–700 μm. The particles may be in the form of a dry powder or may be sprayed from a liquid suspension as an aerosol. The particles are blown through the system and dispersed in air to a preset count rate. The aerosizer operates on the principle of aerodynamic time of flight. The particles are accelerated by a constant, known force caused by the airflow, and are forced through a nozzle at nearly sonic velocity. Smaller particles are accelerated at a greater rate than large particles as a result of a greater force-to-mass ratio. Two laser beams measure the time of flight through the measurement region by detecting the light scattered by the particles. Statistical methods are used to correlate the start and stop times of each particle in a particular size range (channel) through the measurement zone. The time of flight is used in conjunction with the density of the particles, and calibration curves are established to determine the size distribution of the sample.

Surface Area

As the surface area exposed to the site of administration determines the speed with which a particle dissolves in accordance with the Noyes–Whitney equation, these determinations are important. In addition, in those instances where the particle size is difficult to measure, a gross estimation of the surface area is the second best parameter to characterize the drug. The most common methods of surface area measurement, including gas adsorption (nitrogen or krypton), based on what is most commonly described as the Braunauer, Emmet, and Teller (BET) method, is applied either as a multipoint or single-point determination.

Adsorption is defined as the concentration of gas molecules near the surface of a solid material. The adsorbed gas is called *adsorbate*, and the solid where adsorption takes place is known as the *adsorbent*. Adsorption is a physical phenomenon (usually called physisorption) that occurs at any environmental condition (pressure and temperature), but it becomes measurable only at very low temperatures. Thus physisorption experiments are performed at very low temperatures, usually at the boiling temperature of liquid nitrogen at atmospheric pressure. Adsorption takes place because of the presence of an intrinsic surface energy. When a material is exposed to a gas, an attractive force acts between the exposed surface of the solid and the gas molecules. The result of these forces is characterized as physical (or van der Waals) adsorption, in contrast to the stronger chemical attractions associated with chemisorption. The surface area of a solid includes both the external surface and the internal surface of the pores.

Because of the weak bonds involved between the gas molecules and the surface (<15 KJ/mol), adsorption is a reversible phenomenon. Gas physisorption is considered nonselective, thus filling the surface step-by-step (or layer by layer) depending on the available solid surface and the relative pressure. Filling the first layer enables the measurement of the surface area of the material, because the amount of gas adsorbed when the monolayer is saturated is proportional to the entire surface area of the sample. The complete adsorption/desorption analysis is called an adsorption isotherm. The six IUPAC (International Union for Physical and Applied Chemistry) standard adsorption isotherms are shown in Figure 5; they differ because the systems demonstrate different gas/solid interactions (4).

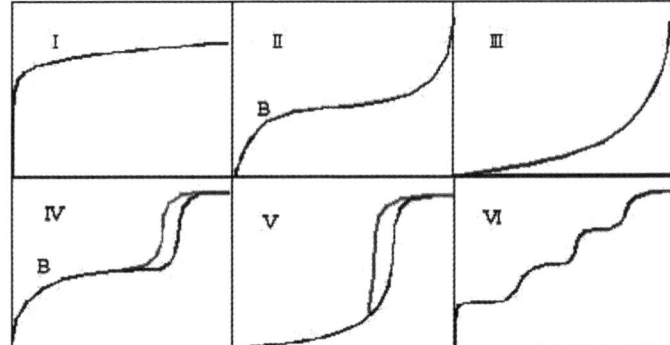

Figure 5 The six types of International Union for Physical and Applied Chemistry isotherms. The type I isotherm is typical of microporous solids and chemisorption isotherms. Type II is shown by finely divided nonporous solids. Types III and V are typical of vapor adsorption (i.e., water vapor on hydrophobic materials). Types V and VI feature a hysteresis loop generated by the capillary condensation of the adsorbate in the mesopores of the solid. The rare type VI, the step-like isotherm, is shown by nitrogen adsorbed on special carbon.

Once the isotherm is obtained, a number of calculation models can be applied to different regions of the adsorption isotherm to evaluate the specific surface area (i.e., BET, Dubinin, Langmuir, and the like) or the micro- and mesopore volume and size distributions (i.e., Barett-Joyner-Halenda, Dubinin-Radushkevich, Horvath and Kawazoe, Saito and Foley, and the like).

The surface area of a solid material is the total surface of the sample that is in contact with the external environment. It is expressed as square meters per gram of dry sample. This parameter is strongly related to the pore size and the pore volume, that is, the larger the pore volume, the larger the surface area, and the smaller the pore size, the higher the surface area. The surface area results from the contribution of the internal surface area of the pores along with the external surface area of the solid or the particles (in case of powders). Whenever a significant porosity is present, the fraction of the external surface area to the total surface area is small.

The BET isotherm for type II adsorption processes (typical for pharmaceutical powders) is given by:

$$\frac{P}{V(P_o - P)} = \frac{1}{cV_{mon}} + \left\{\frac{c-1}{cV_{mon}}\right\}\left\{\frac{P}{P_o}\right\} \tag{6}$$

where P is the partial pressure of the adsorbate, V is the volume of gas adsorbed at pressure p, V_{mon} is the volume of the gas at monolayer coverage, P_o is the saturation pressure and c is related to the intercept. Thus, by plotting $P/V (P_o - P)$ versus P/P_o, a straight line of slope $c - 1/cV_{mon}$ and intercept $1/cV_{mon}$ will be obtained. The total surface area is thus:

$$S_t = \frac{V_{mon} N A_{CS}}{M} \tag{7}$$

where N is the Avogadro's number, A_{CS} is the cross-sectional area of the adsorbate, and M is the molecular weight (MW) of the adsorbate. It follows that the specific surface area is given by S_t/m, where m is the mass of the sample. According to the U.S. Pharmacopeia (USP), the data are considered to be acceptable, if on linear regression, the correlation coefficient is not less than 0.9975, that is, r^2 is not less than 0.995.

It should be noted that, experimentally, it is necessary to remove gases and vapors that may be present on the surface of the powder. This is usually achieved by drawing a vacuum or purging the sample in a flowing stream of nitrogen. Raising the temperature may not always be advantageous, as often it is found that the specific surface area of the sample decreases with an increase in temperature. Thermal degassing can therefore affect the results. From other measurements using differential scanning calorimetry (DSC) and thermogravimetric analysis (TGA), it was found that raising the temperature changed the nature of the samples. Hence, it was recommended that magnesium stearate should not be degassed at elevated temperature.

Porosity

Most solid powders contain a certain void volume of empty space. This is distributed within the solid mass in the form of pores, cavities, and cracks of various shapes and sizes. The total sum of the void volume is called the porosity. Porosity strongly determines the important physical properties of materials, such as durability, mechanical strength, permeability, adsorption properties, and so on. The knowledge of pore structure is an important step in characterizing materials and predicting their behavior.

There are two main and important typologies of pores: closed and open pores. Closed pores are completely isolated from the external surface, not allowing the access of external fluids in neither liquid nor gaseous phase. Closed pores influence parameters like density and the mechanical and thermal properties. Open pores are connected to the external surface, and are therefore accessible to fluids, depending on the pore nature/size and the nature of fluid. Open pores can be further divided into dead-end or interconnected pores. Further classification is related to the pore shape, whenever is possible to determine it. The characterization of solids in terms of porosity consists in determining the following parameters:

- *Pore size*: Pore dimensions cover a very wide range. Pores are classified according to three main groups depending on the access size:
  - Micropores: less than 2-nm diameter;
  - Mesopores: between 2 and 50-nm diameter;
  - Macropores: larger than 50-nm diameter.
- *Specific pore volume and porosity*: The internal void space in a porous material can be measured. It is generally expressed as a void volume (in cc or mL) divided by a mass unit (g).
- *Pore size distribution*: It is generally represented as the relative abundance of the pore volume (as a percentage or a derivative) as a function of the pore size.
- *Bulk density*: Bulk density (or envelope density) is calculated by the ratio between the dry sample mass and the external sample volume.
- *Percentage porosity*: The percentage porosity is represented by the ratio between the total pore volume and the external (envelope) sample volume multiplied by 100.
- *Surface area*: See discussion on page 252.

Instrumentation for Particle Size, Surface Area, and Porosity

The following discussion lists some of the most widely used instruments; this by no means represents a comprehensive list or constitutes an endorsement of one instrument over another.

The *Sorptomatic 1990* (5) is a completely computerized instrument based on a static volumetric principle to characterize solid samples by the technique of gas adsorption. It is designed to perform physisorption measurements to determine the specific surface area and mesopore size distribution of porous materials by using inert gases, such as nitrogen, argon, carbon dioxide, and so on. This instrument can also perform chemisorption measurements on activated solids like catalysts or acid/basic materials (as zeolites), using reactive gases (i.e., hydrogen, carbon monoxide, or oxygen) or corrosive gases (i.e., gaseous dry ammonia). It has a specially designed pretreatment unit, and hence it is possible to connect up to six gases to the special flow gas burette to properly activate the catalyst before the experiment. The software enables the determination of metal-specific surface area and dispersion in supported catalysts from chemisorption isotherms.

Static volumetric gas adsorption requires a high vacuum pumping system to be able to generate a good vacuum over the sample of at least 10^{-4} torr. The system features stainless steel plumbing with high vacuum fittings to ensure precise results, as the experiment is carried out starting from high vacuum, and increasing the pressure step by step up to the *adsorbate* saturation pressure. A schematic of the instrument is shown in Figure 6.

The principle behind this method consists of introducing consecutive known amounts of *adsorbate* to the sample holder, which is kept at liquid nitrogen temperature (77 K). Adsorption of the injected gas onto the sample causes the pressure to slowly decrease until an equilibrium pressure is established in the manifold. The injection system of the Sorptomatic 1990 consists of a calibrated piston, where both the pressure and the injection volume can be automatically varied by the system according to the adsorption rate and the required resolution. The piston method is advantageous over other methods as it does not increase the manifold dead volume, even as the system waits for pressure equilibration. A small dead volume over the sample makes the instrument very sensitive to the

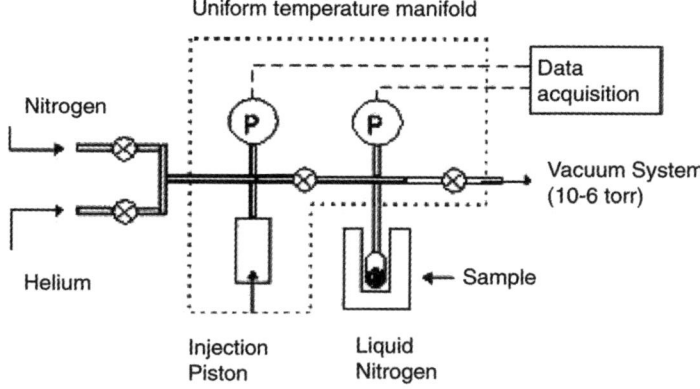

Figure 6 Static volumetric gas adsorption system.

amount of gas adsorbed. The equilibrium pressure is measured by a transducer chosen according to the pressure range where adsorption is established during the experiment. The raw experimental data are the equilibrium pressures and the amount of gas adsorbed for each step. The gas uptake is calculated directly from the equilibrium pressure values, but a dead volume calibration has to be performed before or after the measurement by a "blank run" (i.e., an analysis using an inert gas, typically helium, not adsorbed on the sample in the analytical conditions).

The static volumetric method is very precise, and is considered a very accurate technique to evaluate the surface area and the pore size in the region of micro- and mesopores. However, it is not advisable whenever a fast measurement of surface area is required, because this method involves a long analysis time to produce highly accurate and reliable results.

The *QSurf quick surface area analyzers* (5) offer a simple, straightforward analytical layout enabling easy and rapid operations, and is tailored to increase lab productivity. QSurf is a fast surface area analyzer designed for specific surface area and total pore volume determinations in porous and nonporous materials by dynamic gas adsorption and desorption. Available in four different configurations, single- and multipoint BET analysis and one to three analytical ports, QSurf is capable of accommodating the vast diversity in analytical and throughput requirements of modern laboratories. Based on flow technique, these new instruments are fully automated, thus allowing unattended operations.

The QSurf analyzers are based on the dynamic method for specific surface area and total pore volume determinations in porous and nonporous materials. The gas adsorbed and desorbed by the samples under test is measured by a TCD. The sample under test is submitted to a flow (at atmospheric pressure) in a gas mixture between helium and nitrogen in different percentages. Helium is generally not adsorbed, and it is called the carrier gas. The carrier gas first flows through the reference channel of a TCD, then through the sample holder, and finally through the analytical channel of the TCD. At the start of the experiment, the sample is kept at room temperature; thus, adsorption does not take place. In this situation, the thermal conductivity of the carrier in the two TCD channels is the same, as the gas composition is the same before and after the sample. Subsequently, the system immerses the sample in a liquid nitrogen bath, cooling the sample at a very low temperature. In this condition adsorption begins, but only nitrogen is adsorbed. Therefore, the gas composition after the sample holder is changed and the carrier thermal conductivity in the reference and analysis channels is different. The TCD generates a signal that is proportional to the amount of nitrogen adsorbed. When the signal returns to the starting baseline, the sample is saturated by the adsorbate at a certain partial pressure. The next step consists in removing the sample holder from the coolant. When the sample temperature rises to room temperature, the phenomenon of desorption takes place, and nitrogen is released from the sample surface. Additionally, in this case, the composition of the carrier in the two TCD channels is different because the sample is releasing the adsorbate. The TCD generates again a signal (of opposite polarity with respect to adsorption) that is proportional to the amount of gas desorbed. The Qsurf integrates the aforementioned peaks, and compares the resulting integrals to a calibration peak previously determined by injecting a known dose of pure adsorbate (calibration step is performed by an automatic loop valve). The desorption peak integration thus provides the amount of gas adsorbed by the sample and the gas mix percentage permits to calculate the partial pressure of nitrogen over the sample during the adsorption/desorption stages.

Micromeritics offers a comprehensive line of instrumentation for particle size, surface, and porosity analysis (6). The *TriStar 3000* gas adsorption analyzer is a fully automated, three-station, surface area, and porosimetry analyzer that delivers high-quality surface area data (and more) at an affordable price. It can increase the speed and efficiency of routine QC analyses, yet has the accuracy, resolution, and data reduction capability to meet most research requirements. Designed with the user in mind, the TriStar 3000 provides versatility in analysis methods and data reduction, allowing one to optimize analyses for a wide range of applications. The *ASAP 2020* accelerated surface area and porosimetry analyzer uses gas sorption techniques for research and quality control applications. Also available is the chemical adsorption ("Chemisorption") option, which uses the static volumetric technique to determine the percent metal dispersion, active metal surface area, size of active particles, and surface acidity of catalyst materials. The Micromeritics Gemini V Series of surface area analyzers rapidly and reliably produces accurate and repeatable surface area and porosity determinations of the sample material. Its simplicity of use and ruggedness has earned the Gemini its place in laboratories worldwide as an essential tool in both research and quality control environments. The *FlowSorb III* is an entry-level single-point and multipoint BET surface area instrument. The FlowSorb measures the surface area using the flowing gas method, which involves the continuous flow of an adsorptive and inert gas mixture over the sample at atmospheric pressure. It is ruggedly constructed and ideal for demanding analytical environments.

True Density

Density is the ratio of the mass of an object to its volume and for solids this term describes the arrangement of molecules. The study of compaction of powders is described by the Heckel equation:

$$\ln\left[\frac{1}{1-D}\right] = KP + A \tag{8}$$

where D is the relative density, which is the ratio of the apparent density to the true density, K is determined from the linear portion of the Heckel plot, and P is the pressure. The densities of molecular crystals can be increased by compression. Information about the true density of a powder can be used to predict whether a compound will cream or sediment in a suspension, such as metered dose inhaler (MDI) formulation. Therefore, suspensions of compounds that have a true density less than these figures will cream (rise to the surface), and those that are denser will sediment. It should be noted, however, that the physical stability of a suspension is not merely a function of the true density of the material. The true density is thus a property of the material, and is independent of the method of determination. In this respect, the true density can be determined using three methods: displacement of a liquid, displacement of a gas (pycnometry), or floatation in a liquid. The liquid displacement is tedious and tends to underestimate the true density. Displacement of a gas is more accurate, but needs relatively expensive instrumentation. As an alternative, the floatation method is simple to use and inexpensive.

Gas pycnometry is probably the most commonly used method in the pharmaceutical industry for measuring true density. Gas pycnometers rely on the measurement of pressure changes, as a reference volume of gas, typically helium, added to, or deleted from, the test cell.

Flow and Compaction of Powders

The flow properties of a powder will determine the nature and quantity of excipients needed to prepare a compressed or powder dosage form. This refers mainly to factors such as the ability to process the powder through machines. To make a quick evaluation, the compound is compressed using an infrared (IR) press and die under 10 torr of pressure with variable dwell times, and the resulting tablets are tested with regard to their crushing strength after storing them for about 24 hours. If longer dwell times result in higher crushing strength, then the material is likely to be plastic; elastic material will show capping at low dwell times; and the brittle material will not show any effect of dwell times. It is recommended that the compressed tablets be subject to XPRD to record any changes in the polymorphic forms.

There appears to be a relationship between indentation hardness and the molecular structure of organic materials. However, a prerequisite for predicting indentation hardness is the knowledge of the crystal structure. As a result, highly sophisticated computational methods and extensive crystallography libraries have recently become available to study the. For example, Pfizer Research relies on the Cambridge Structural Database (CSD) (7), which is one of the largest repository of small molecule crystal structures. The CSD is the principal product of the Cambridge Crystallographic Data Centre (CCDC). It is the central focus of the CSD system, which also comprises software for database access, structure visualization and data analysis, and structural knowledge bases derived from the CSD. CSD records bibliographic, chemical, and crystallographic information for organic molecules and metal-organic compounds, whose 3D structures have been determined using X-ray diffraction (XRD) or neutron diffraction. The CSD records results of single crystal studies and powder diffraction studies, which yield 3D atomic coordinate data for at least all non-H atoms. In some cases, the CCDC is unable to obtain coordinates, and incomplete entries are archived to the CSD. The CSD is distributed as part of the CSD system, which includes software for search and information retrieval (ConQuest), structure visualization (Mercury), numerical analysis (Vista), and database creation (PreQuest). The CSD system also incorporates IsoStar, a knowledge base of intermolecular interactions containing data derived from both the CSD and the Protein Data Bank. Some software listed here are available for free use.

X-ray microtomography, such as the one available from Skyscan (8) is used to analyze the effect of compaction on powder particles. It allows for the noninvasive three-dimensional (3D) analysis of resulting structures, and has shown that the structure may be controlled by the choice of porogen and the method of solvent removal. Simple seeding of the substrate surface with drug crystals can be used initially with a view to incorporating more sophisticated substrate polymorph approaches. The Skyscan-1172 represents a new generation in desktop X-ray micro-CT systems. A novel architecture in which both the sample stage and the X-ray camera are moveable allows an unprecedented combination of image resolution, sample size accommodation, scan speed, and sample throughput. This innovative flexible scanner geometry of the Skyscan-1172 is particularly advantageous over intermediate resolution levels, where scans are around ten times faster (to obtain the same or better image quality) compared with previous scanners with a fixed source-detector design. The Skyscan-1172 features two X-ray camera options: the high-performance 10-megapixel option, and the economy 1.3-megapixel option. The former, 10-megapixel camera allows the maximum scanning versatility, with an image field width of 68 mm (in dual image camera shift mode) or 3 mm (in standard single camera image mode). A nominal resolution (pixel size) of lower than 1 µm is attainable. A scannable height of around 70 mm allows for either large samples or automatic batch scanning of a column of smaller samples. The system obtains multiple X-ray "shadow" transmission images of the object from different angular views, as the object rotates on a high-precision stage. From these shadow images, cross-section images of the object are reconstructed by a modified Feldkamp cone-beam algorithm, creating a complete 3D representation of internal microstructure and density over a selected range of heights in the transmission images. The best micro-CT images are obtained from objects in which microstructure coincides with contrast in X-ray absorption of the sample's constituent materials.

Color

The color of a powder sample is used to indicate the presence of solvents, distribution of particle size, and other possible differences in the different lots of a new lead compound. In some instances, degradation of drug can be correlated with color changes to such a degree that accurate color measurements can be used as a tool to provide product specification. The compendia often describe the color of the substances, but mostly in subjective terms. Historically, the color evaluation has been a subjective measurement; however, newer quantitative measurement systems make this a more objective process. There are two basic methods for measuring the colors of surfaces:

- The first is to imitate the analysis made by the eye in terms of responses to three stimuli. This technique, known as "tristimulus colorimetry," sets out to measure X, Y, and Z directly.
- The second method is to determine the reflectance (R) for each wavelength band across the range of the spectrum to which the eye is sensitive, and then to calculate the visual responses by summing the products of R and the standard values for the distribution of the sensitivity of the three-color responses.

The tristimulus method has theoretical advantages, where the materials to be measured are fluorescent; but, there are serious practical problems in assuming that a tristimulus colorimeter exactly matches human vision, that is, in eliminating color blindness from the instrument.

Two commonly used types of color measurement equipment are a colorimeter and a spectrophotometer. A tristimulus colorimeter has three main components:

- a source of illumination (usually a lamp functioning at a constant voltage)
- a combination of filters used to modify the energy distribution of the incident/reflected light
- a photoelectric detector that converts the reflected light into an electrical output

Each color has a fingerprint reflectance pattern in the spectrum. The colorimeter measures color through three wide-band filters corresponding to the spectral sensitivity curves. Measurements made on a tristimulus colorimeter are normally comparative, the instrument being standardized on glass or ceramic standards. To achieve the most accurate measurements, it is necessary to use calibrated standards of similar colors to the materials to be measured. This "hitching post" technique enables reasonably accurate tristimulus values to be obtained even when the colorimeter is demonstrably color blind. Tristimulus colorimeters are most useful for quick comparison of near-matching colors. They are not very accurate. Large differences are evident between the various instrument manufacturers. However, colorimeters are less expensive than spectrophotometers.

To get a precise measurement of color, it is advisable to use a spectrophotometer. A spectrophotometer measures the reflectance for each wavelength, and allows to calculate the tristimulus values. The advantage over tristimulus colorimetry is that adequate information is obtained for the calculation of color values for any illuminant, and that metamerism is automatically detected. Metamerism is a psychophysical phenomenon commonly defined incompletely as "two samples which match when illuminated by a particular light source and then do not match when illuminated by a different light source." In actuality, there are several types of metamerism, of which the sample and illuminant metamerism are the most common. In sample metamerism, two color samples appear to match under a particular light source, and then do not match under a different light source. Illuminant metamerism appears when different light sources illuminate the same sample and the differences are revealed. The observer metamerism refers to the spot where each individual perceives color slightly differently. The geometric metamerism arises when identical colors appear different when viewed at different angles, distances, light positions, and so on.

In a spectrophotometer, the light is usually split into a spectrum by a prism or a diffraction grating before each wavelength band is selected for measurement. Instruments have also been developed in which narrow bands are selected by interference filters. The spectral resolution of the instrument depends on the narrowness of the bands utilized for each successive measurement. In theory, a spectrophotometer could be set up to compare reflected light directly with incident light, but it is more usual to calibrate against an opal glass standard that has been calibrated by an internationally recognized laboratory. Checks must also be made on the optical zero, for example, by measurements with a black light trap, because dust or other problems can give rise to stray light in an instrument (which would give false readings). Spectrophotometers contain monochromators and photodiodes that measure the reflectance curve of color every 10 nm or less. The analysis generates typically 30 or more data-points, with which a precise color composition can be calculated.

A large number of suppliers provide colorimeters, including a large array of equipment from Hunter Lab's Labscan XE with special adapter for small quantity of powders offering an excellent choice in preformulation work. The instrument has a 3-mm port and requires 0.4 cc powder to perform the testing (9).

Electrostaticity
When subjected to attrition, powders can acquire an electrostatic charge, the intensity of which is often proportional to physical force applied, as static electrification of two dissimilar materials occurs by the making and breaking of surface contacts

(tribo-electrification or friction electrification). Electrostatic charges are often used to induce adhesive character to bind drugs to carrier systems, for example, glass beads coated with hydroxypropylmethyl cellulose–containing drugs. The net charge on a powder may be either electropositive or electronegative depending on the direction of electron transfer. The mass charge density can vary from 10^{-5} to 100 µC/kg, depending on the stress, ranging from gentle sieving to micronization process. This can be done using electric detectors to determine polarity and the electrostatic field. The electrostaticity results in significant changes in the powder flow properties.

Studies on tribo-electrification and potential charge buildup on equipment and particle surfaces, and subsequent adhesion caused by static charge often overlook the fact that all materials (whether they have a net surface charge or not) exhibit surface energy forces that are very short range, but come into play once the surfaces are "touching." These van der Waals forces are caused by the dispersive and polar surface energies inherent at material boundaries. Dry powders with mass-median particle sizes larger than around 100 to 200 µm, seldom exhibit strong "cohesive" powder behavior, and such powders are usually described as "free flowing." As the particle size decreases, however, the amount of surface area per unit mass increases, and surface-energy forces have a greater influence on bulk powder flow characteristics. For contacting particles that are smaller than 2–20 µm, such forces can be strong enough to cause small amounts of plastic deformation on the particle surfaces near the points of contact—even with no applied external loads. The bulk behavior of such fine powders can be dominated by their "cohesivity." It is well known that powders comprised of finer particles are more cohesive, and, when very cohesive powders are placed in a rotating drum, they do not usually flow easily, nor do they form a smooth top surface. Instead, cohesive powders build up large overhanging "chunks" that can break off and collapse or cascade in random avalanches onto the material further down the slope. Placing the rotating drum in a centrifuge at an elevated g-level can cause a "nonflowable" cohesive powder to flow.

Caking

Powders cake as a result of agglomeration owing to factors such as static electricity, hygroscopicity, particle size, impurities of the powder and storage conditions, stress temperature, RH and storage time, and so on. The mechanisms involved in caking are based on the formation of five types of interparticle bonds, such as bonding resulting from mechanical tangling: bonding resulting from steric effects; bonds via static electricity; bonds as a result of free liquid; and bonds caused by solid bridges. During the process of micronization, the formation of localized amorphous zones can lead to caking, as these zones are more reactive to factors described earlier, especially when exposed to moisture. The mechanisms involve moisture sorption as a result of surface sintering and recrystallization at well below the critical RH. In most instances, the increase in RH begins to show some impact at values above 20%, resulting in most dramatic effects above 75–80% RH for powders that are subject to humidity effects.

Polymorphism

Because polymorphism can have an effect on so many aspects of drug development, it is important to fix the polymorph (usually the stable form) as early as possible in the development cycle. Although it is not necessary to create additional solid-state forms by techniques or conditions unrelated to the synthetic process for the purpose of clinical trials, regulatory submission of a thorough study of the effects of solvent, temperature, and possibly pressure on the stability of the solid-state forms is advised. A conclusion that polymorphism does not occur with a compound must be substantiated by crystallization experiments from a range of solvents. This should also include solvents that may be involved in the manufacture of the drug product, for example, during granulation.

As it is hoped that the issue of polymorphism is resolved during prenomination and early development, it can remain a concern when the synthesis of the drug is scaled up into a larger reactor or transferred to another production site. It is not unlikely that a metastable form identified in prenomination may not be reproduced in later batch products because of some unrecorded conditions in the early phases of development. Related substances, whether identified or not, can significantly alter the predominance of a specific polymorph. To develop a reliable commercial recrystallization process, the following scheme should be followed in the production of candidate drugs:

1. selection of solvent system
2. characterization of the polymorphic forms
3. optimization of process times, temperature, solvent compositions, and the like
4. examination of the chemical stability of the drug during processing
5. manipulation of the polymorphic form, if necessary

Many analytical techniques have been used to quantitate mixtures of polymorphs, for example, XRPD has been used to quantify the various polymorphs. Assay development requires the creation of calibration curves and validation, which can be a difficult task where mixed polymorphs are present, and requires a study to prove that there is no polymorphic transformation during analysis or change in the hydration of crystals, if that is also a concomitant problem. Although at the preformulation stage the dosage form considerations are still developing, there is a need to answer questions like: how would a polymorph change and should this be subjected to manufacturing equipment stress, such as granulation or drying of granules, wet or dry granulation and compression? In addition to the polymorphism of active drugs, the excipients, like magnesium stearate, can be present in various polymorphic forms that can significantly alter the behavior of active drug in the formulation stages. Studies using XRPD, IR, or scanning electron microscopy (SEM) should be made for excipients and the active drug.

SOLUTION FORMULATIONS

Solution dosage forms offer several advantages, particularly the resolution of bioavailability problems, instant administration as injectable forms (though nonsolution forms are also given parenterally). At the preformulation stage, more important factors are the solubility (and any pH dependence) and stability of the new compound.

Solubility

In case a solution form is desired, and the compound has low solubility, there are several techniques, some very simple to some very complex, to achieve the desirable property of the lead drug, including pH manipulation, use of co-solvents, surfactants, emulsion formation, and adding complexing agents. In a more complex stage, the liposomes or similar drug delivery systems can be used.

As many compounds are weak acids or bases, their solubilities become a function of pH. However, the ionic strength of the medium plays a significant role, and as a result most parenteral formulations are buffered to prevent the crystallization of drugs.

The use of co-solvents improves the solubility as a result of the polarity of the co-solvent mixture being closer to the drug than it is in water:

$$\log S_m = f \log S_c + (1 - f) \log S_w \tag{9}$$

where S_m is the solubility of the compound in the solvent mix, S_w is the solubility in water, S_c is the solubility of the compound in pure co-solvent, f is the volume fraction of the co-solvent, and σ is the slope of the plot of $\log(S_m/S_w)$ versus f. There is a definite correlation between the s value to indices of the co-solvent polarity, such as the dielectric constant, solubility parameter, surface tension, interfacial tension, and octanol–water partition coefficient. The aprotic co-solvents give a much higher degree of solubility than the amphiprotic co-solvents. This means that if a co-solvent can donate a hydrogen bond, it might be an important factor in determining whether it is a good co-solvent. Use of co-solvents with polar drugs can reduce the solubility.

On formulating parenteral dosage forms, the use of co-solvents to prevent precipitation can be hampered by the quantity of the allowed co-solvents in the formulation for toxicity and hemolysis considerations. Other considerations like dilution prior to administration, and the rate of administration (dilution factor) should also be simulated using in vitro techniques. Although co-solvents can increase the solubility of compounds, on certain occasions, they can have a detrimental effect on their stability. One point that is often overlooked when considering co-solvents is their influence on buffers or salts. As these are conjugate acid–base systems, it is not surprising that by introducing solvents into the solution, a shift in the pK_a of the buffer or salt can result. These effects are important in formulation terms, as many injectable formulations that contain co-solvents also contain a buffer to control the pH.

EMULSION FORMULATIONS

For drugs with poor water solubility, emulsion formulation, such as oil-in-water (O/W), where the drug has good partitioning in the oil phase chosen, often offers an excellent choice. The particle size of the emulsion and its stability (physical and chemical) then become significant factors, as larger globule sizes may lead to phlebitis. To achieve smaller particle size, the technique of microfluidization is often used among other available homogenization methods. The phospholipids added stabilize emulsions through surface charge changes and provide a good mechanical barrier.

The particle size of an emulsion is governed by the method used. Figure 7 shows the various particle sizes achieved using different methods.

The particle size is measured using photon correlation spectroscopy (PCS) (10), a technique for measuring particle size distributions. When fine particles are suspended in a fluid, they are constantly in random motion as a result of collisions with the molecules of the fluid. This is known as "Brownian Motion," and was first observed in the 1820s. When the suspension is irradiated by a beam

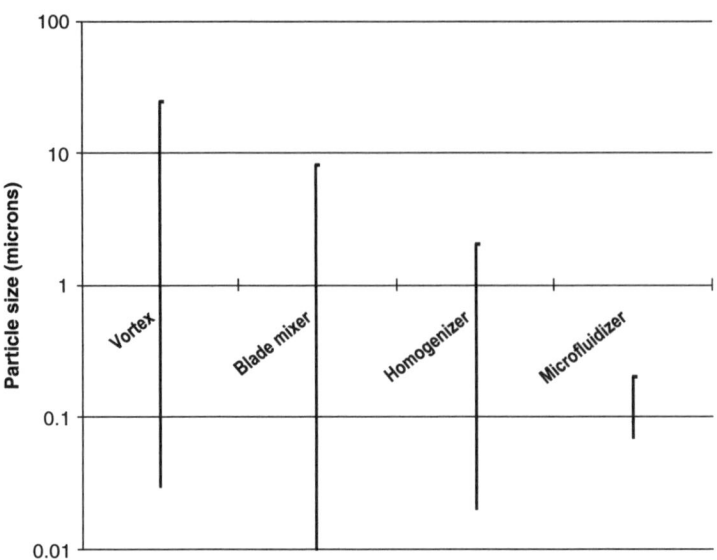

Figure 7 Emulsion particle size dependence on the method of manufacture.

of laser light, some of the light is scattered by the particles. Very fine particles exhibit wavelength smaller than that of light (typically 500–700 nm), and as they move relative to the light beam, the phase of the light scattered from each particle will vary. The intensity of the scattered light, measured at some fixed point, is the sum of the light scattered from all the individual particles, formed by constructive and destructive interference. This intensity will therefore vary as the particles move, and the rate of variation will depend on the speed of the movement of the particles, which is in turn related to their size. This is why PCS is also known as "dynamic light scattering." The advantages of the use of PCS include:

- direct measurement of particle size in the sub-micron range
- particles are observed in situ in the fluid matrix
- broad range—1 to 5000 nm (5 microns)
- reasonably fast

The disadvantages in the use of PCS include:

- requires very dilute dispersions and careful sample preparation;
- low resolution: provides only limited detail of the distributions;
- expensive equipment; and
- skill required to interpret results.

A large variety of instruments are available for PCS (11).

Also useful are the measurements of zeta potential. Particles in suspensions typically acquire a surface charge by the adsorption of ions on the surface, dissolution of the material, chemical reaction, or preferential adsorption of a specific additive or impurity ions from the solution. Surface charge of colloidal particles can be inferred through the measurement of the zeta potential (the particle charge at the shear plane). The zeta potential is an important particle property that affects interparticle forces, particle adhesion, suspension stability, and polymer/surfactant adsorption characteristics. A large zeta potential, either positive or negative, promotes a stable dispersion, while a small zeta potential signifies poor stability or an unstable dispersion. The zeta potential of the particles can be determined by measuring the mobility of the particles in an applied DC electric field (termed as electrophoretic mobility) or its sonic response to an alternating electric field (electrokinetic sonic amplitude effect). For large particles or flat surfaces, streaming potential measurements can be utilized to determine the zeta potential. In streaming potential measurements, the flow of electrolyte induces an electric field that can be measured. The reverse electro-osmosis phenomena can be used to determine the charge on large particles or surfaces that cannot be measured by electrophoretic methods. Other methods, such as atomic force microscope (AFM) measurements, can be used to measure the surface forces between particles and surfaces.

From this data, the surface charge on the particles can often be determined. The AFM is one of about two dozen types of scanned-proximity probe microscopes. All of these microscopes work by measuring a local property, such as height, optical absorption, or magnetism with a probe or "tip" placed very close to the sample. The small probe-sample separation (on the order of the instrument's resolution) makes it possible to take measurements over a small area. To acquire an image, the microscope raster scans the probe over the sample, at the same time as measuring the local property in question. The resulting image resembles an image on a television screen in that both consist of many rows or lines of information placed one above the other. Unlike traditional microscopes, scanned-probe systems do not use lenses; hence, the size of the probe, rather than diffraction effects, generally limits their resolution. The AFM operates by measuring attractive or repulsive forces between a tip and the sample. In its repulsive "contact" mode, the instrument lightly touches a tip at the end of a leaf spring or "cantilever" to the sample. As a raster-scan drags the tip over the sample, some sort of detection apparatus measures the vertical deflection of the cantilever, which indicates the local sample height. Thus, in contact mode, the AFM measures hard-sphere repulsion forces between the tip and the sample. In noncontact mode, the AFM derives topographic images from measurements of attractive forces; the tip does not touch the sample. The AFM can achieve a resolution of 10 Å, and unlike electron microscopes, can image samples in air and under liquids. In principle, AFM resembles the record player and the stylus profilometer. However, AFM incorporates a number of refinements that enable it to achieve atomic-scale resolution:

- sensitive detection
- flexible cantilevers
- sharp tips
- high-resolution tip-sample positioning
- force feedback

Several vendors provide additional details on the refined features of AFM (12–14).

The physical instability of emulsions involves creaming, flocculation, coalescence, or breaking, whereas the chemical instability can be a result of hydrolysis of the stabilizing moieties. In order to assess the stability of the emulsion, heating and freezing cycles and centrifugation and steam sterilization can be employed.

Generally, emulsions with a high negative zeta potential do not show any change in their particle-size distribution after autoclaving. Emulsions with a lower negative value, on the other hand, would generally separate into two phases during autoclaving. Because the stability of phospholipid-stabilized emulsions is dependent on the surface charge, these emulsions are normally autoclaved at pH 8–9.

Stability Considerations

Newer drug molecules appear to be more reactive and potent in their response. As a result, they are also more reactive in the dosage forms, and when the dosage forms are liquids, the stability becomes a serious challenge to overcome. This is one reason why solid dosage forms are preferred by formulation scientists. A basic matrix to evaluate stability potential can take scores of permutations and combinations of pH, ionic strength, dielectric constant, temperature, and the like. Given the small quantity of substance available at the preformualtion stage, it is unlikely to complete these studies with sufficient vigor.

The degradation kinetics of drugs in solution state has been a broad subject dealt with in major textbooks and a large number of detailed review articles. What needs to be understood at the preformulation level are gross observations as to whether a drug would sustain the solution or liquid environment for any significant length of time. Quick pH and accelerated testing is required, where the quantity of the substance available is sufficient for the purpose. At this stage, stability may be tested with most likely co-solvents, particularly those used in pediatric or geriatric dosage forms.

For parenteral formulations, sterility can be maintained either by sterile filtration, or by autoclaving. It is noteworthy that there is no history of any significant recalls of products that were autoclaved, and thus manufacturers prefer to use this method of sterilization where possible; obviously, it cannot be used where drugs are inherently unstable to heat, like the biological products. Autoclaving (usually 15–20 minutes at 121°C) at various pH values is a good test to study the impact on impurities, color, pH, and other degradation products. The autoclave cycle should ideally represent a real-time manufacturing process with its common fill, heat-up, peak-dwell, and cool-down steps.

Oxidation

Oxygen sensitivity is common for many molecules. Oxidation reactions are the most difficult reactions to understand, let alone prevent. As a result, oxidation-prone compounds are combined with antioxidants. In a solution form, the degradation is fast, particularly in the presence of trace metals.

The use of common antioxidants, such as water-soluble sodium bisulfate, sodium sulfite, sodium metabisulfite, sodium thiosulfate, sodium formaldehyde sulfoxylate, L- and D-ascorbic acid, acetylcysteine, cysteine, thioglycerol, thioglycollic acid, thiolactic acid, thiourea, dithithreitol or oil-soluble propyl gallate, butylated hydroxyanisole, butylated hydroxytoluene, ascorbyl palmitate, nordihydroguaiaretic acid, and α-tocopherol is widely made in pharmaceutical formulations. Oxygen-sensitive substances should be screened for their compatibility with a range of antioxidants. One of the most commonly used antioxidants is metabisulfite. It should be noted that bisulfite has also been known to catalyze hydrolysis reactions. The reaction of bisulfate with dissolved oxygen is given by

$$2HSO_3^- + O_2 \longrightarrow 2SO^{-4} + 2H^+ \tag{10}$$

Several newer techniques, such as cyclic voltammetry (CV) are now used to identify a proper choice of an antioxidant. CV is an electrolytic method that uses microelectrodes and an unstirred solution, so that the measured current is limited by analyte diffusion at the electrode surface. The electrode potential is ramped linearly to a more negative potential, and then ramped in reverse back to the starting voltage. The forward scan produces a current peak for any analyte that can be reduced through the range of the potential scan. The current will increase as the potential reaches the reduction potential of the analyte, but then falls off as the concentration of the analyte is depleted close to the electrode surface. As the applied potential is reversed, it will reach a potential that will reoxidize the product formed in the first reduction reaction, and produce a current of reverse polarity from the forward scan. This oxidation peak will usually have a similar shape to the reduction peak. The peak current, i_p, is described by the Randles–Sevcik equation:

$$i_p = (2.69 \times 10^5) n^{3/2} A C D^{1/2} v^{1/2} \tag{11}$$

where n is the number of moles of electrons transferred in the reaction, A is the area of the electrode, C is the analyte concentration (in mol/cm^3), D is the diffusion coefficient, and v is the scan rate of the applied potential.

The potential difference between the reduction and the oxidation peaks is theoretically 59 mV for a reversible reaction. In practice, the difference is typically 70–100 mV. Larger differences, or nonsymmetric reduction and oxidation peaks, are an indication of a nonreversible reaction. These parameters of cyclic voltammograms make CV most suitable for characterization and mechanistic studies of redox reactions at the electrodes.

The basic components of a modern electroanalytical system for voltammetry are a potentiostat, computer, and the electrochemical cell. In some cases, the potentiostat and the computer are bundled into one package, whereas in other systems the computer and the A/D and D/A converters and microcontroller are separate, and the potentiostat can operate independently.

Trace Metals

Trace metal ions can affect the stability, and can arise from the bulk drug, formulation excipients, or glass containers. Metal ions can also act as degradation catalysts by their involvement in the production of highly reactive free radicals, especially in the presence of oxygen. The formation of these radicals can be initiated by the action of light or heat, and propagate the reaction until they are destroyed by inhibitors or by side reactions that break the chain. Free-radical oxygen species can be generated by transition metals in solutions, such that reactions can be initiated. Because of the involvement of metal ions in degradation reactions, the inclusion of a chelating agent is often advocated.

Chelating agents are frequently used to protect substances from undergoing oxidation by removing metallic contaminants. A chelate is a chemical compound composed of a metal ion and a chelating agent. A chelating agent is a substance whose molecules can form several bonds to a single metal ion. In other words, a chelating agent is a multidentate ligand. An example of a simple chelating agent is ethylenediamine. A single molecule of ethylenediamine can form two bonds to a transition-metal ion, such as nickel (II), Ni^{2+}. The bonds form between the metal ion and the nitrogen atoms of ethylenediamine. The nickel (II) ion can form six such bonds; hence, a maximum of three ethylenediamine molecules can be attached to one Ni^{2+} ion. A chelating agent of particular economic significance is ethylenediaminetetraacetic acid (EDTA). Ethylenediaminetetraacetic acid is a versatile chelating agent. It can form four or six bonds with a metal ion, and it forms chelates with both transition-metal ions and main-group ions. Ethylenediaminetetraacetic acid is frequently used in pharmaceutical products. In addition to EDTA, β-hydroxyethylenediaminetriacetic acid (HEDTA), diethylenetriaminepentaacetic, for example, isoniazid solutions are also used.

Ethylenediaminetetraacetic acid has pK_a values of $pK_1 = 2.0$, $pK_2 = 2.7$, $pK_3 = 6.2$, and $pK_4 = 10.4$ at 20°C. Generally, the reaction of EDTA with metal ions can be described by

$$M^{n+} + Y^{4-} \longrightarrow MY^{(4-n)+} \tag{12}$$

In practice, however, the disodium salt is used because of its greater solubility, hence,

$$M^+ + H_2Y \longrightarrow MY^{(4-n)+} + 2H^+ \tag{13}$$

The dissociation (or equilibrium) is sensitive to the pH of the solution; therefore, this will have implications for the formulation. The stability of the complex formed by EDTA–metal ions is characterized by the stability or formation constant, K. This is derived from the reaction equation and is given by

$$K = \frac{[(MY)_{(n-4)+}]}{[M^+]} \tag{14}$$

Equation 14 assumes that the fully ionized form of $EDTA^{4-}$ is present in solution; however, at low pH, other species will be present, that is, $HEDTA^{3-}$, H_2EDTA^{2-}, and H_3EDTA^-, and the undissociated H_4EDTA. Thus, the stability constants become conditional upon pH.

The ratio can be calculated for the total uncombined EDTA (in all forms) to the form $EDTA^{4-}$. Thus, the apparent stability constant becomes K/α_L, such that

$$\alpha_L = \frac{[EDTA] \text{ all forms}}{[EDTA_{4-}]} \tag{15}$$

Thus,

$$K_H = \frac{K}{\alpha_L} \quad \text{or} \quad K_H = \log K - \alpha_L \tag{16}$$

where $\log K_H$ is known as the conditional stability constant. Fortunately, α_L can be calculated from the known dissociation constants of EDTA, and its value can be calculated from

$$\alpha_L = \left\{1 + \frac{[H+]}{K_4} + \frac{[H+]}{K_4 K_3} + \cdots \right\} = 1 + 10^{(pK_4 - pH)} + 10^{(pK_4 + pK_3 pH)} + \cdots \tag{17}$$

Thus at pH = 4, the conditional stability constants of some metal–EDTA complexes are calculated as follows:

$$\log K_H \text{ EDTA Ba}^{2+} = 0.6 \tag{18}$$

$$\log K_H \text{ EDTA Mg}^{2+} = 1.5 \tag{19}$$

$$\log K_H \text{ EDTA Ca}^{2+} = 3.4 \tag{20}$$

$$\log K_H \text{ EDTA Zn}^{2+} = 9.5 \tag{21}$$

$$\log K_H \text{ EDTA Fe}^{2+} = 17.9 \tag{22}$$

Thus, at pH = 4, the zinc and ferric complexes will exist; however, calcium, magnesium, and barium will be only weakly complexed, if at all.

The inclusion of EDTA is occasionally not advantageous, as there are a number of reports of EDTA catalyzing the decomposition of drugs. Citric acid, tartaric acid, glycerin, and sorbitol can also be considered as complexing agents; however, these are often ineffective. Interestingly, some Japanese formulators often resort to amino acids or tryptophan, because of a ban on EDTA in a particular country.

Photostability

For drug substances, photostability testing should consist of two parts: forced degradation testing and confirmatory testing. The purpose of forced degradation testing studies is to evaluate the overall photosensitivity of the material for method development purposes and/or degradation pathway elucidation. This testing may involve the drug substance alone and/or in simple solutions/suspensions to validate the analytical procedures. In these studies, the samples should be in chemically inert and transparent containers. In these forced degradation studies, a variety of exposure conditions may be used, depending on the photosensitivity of the drug substance involved and the intensity of the light sources used. For development and validation purposes, it is appropriate to limit the exposure and end the studies if extensive decomposition occurs. Under forced conditions, decomposition products may be observed that are unlikely to be formed under the conditions used for confirmatory studies. This information may be useful in developing and validating suitable analytical methods. If in practice it is demonstrated that they are not formed in the confirmatory studies, these degradation products need not be examined further. The forced degradation studies should be designed to provide suitable information to develop and validate test methods for the confirmatory studies. These test methods should be capable of resolving and detecting photolytic degradants that appear during the confirmatory studies. When evaluating the results of these studies, it is important to recognize that they form a part of the stress testing, and are not therefore designed to establish qualitative or quantitative limits for change.

Confirmatory studies should then be undertaken to provide the information necessary for handling, packaging, and labeling. Normally, only one batch of drug substance is tested during the development phase, and subsequently, the photostability characteristics should be confirmed on a single batch selected, as described in the parent guideline, if the drug is clearly photostable or photolabile. If the results of the confirmatory study are equivocal, testing of up to two additional batches should be conducted. Care should be taken to ensure that the physical characteristics of the samples under test are taken into account, and efforts should be made, such as cooling and/or placing the samples in sealed containers, to ensure that the effects of the changes in physical states, such as sublimation, evaporation, or melting are minimized. All such precautions should be chosen to provide minimal interference with the exposure of samples under test. Possible interactions between the samples and any material used for containers or for general protection of the sample should also be considered and eliminated, wherever not relevant to the test being carried out. The confirmatory studies should identify precautionary measures needed in manufacturing, or in the formulation of the drug product, and if light-resistant packaging is needed. When evaluating the results of confirmatory studies to determine whether the change caused by exposure to light is acceptable, it is important to consider the results from other formal stability studies in order to assure that the drug will be within the justified limits at the time of use.

As a direct challenge for samples of solid drug substances, an appropriate amount of sample should be taken and placed in a suitable glass or plastic dish, and protected with a suitable transparent cover, if considered necessary. Solid drug substances should be spread across the container to give a thickness of typically not more than 3 mm. In cases where solid drug substance samples are involved, sampling should ensure that a representative portion is used in individual tests. Similar sampling considerations, such as homogenization of the entire sample, apply to other materials that may not be homogeneous after exposure. The analysis of the exposed sample should be performed concomitantly with that of any protected samples used as dark control, if these are used in the test.

Drug substances that are liquids should be exposed in chemically inert and transparent containers. At the end of the exposure period, the samples should be examined for any changes in physical properties (e.g., appearance, clarity, or color of solution) and for assay. The degradants must be examined by a method suitably validated for products likely to arise from photochemical degradation processes.

Protection from light often offers excellent solutions to stabilize liquids; the use of amber-colored vials and ampuls is one example of precaution taken in reducing the effect of light waves.

Surface Activity

Many drugs show surface-active behavior, because they have the correct mixture of chemical groups that are characteristic of surfactants. The surface activity of drugs can be important if they show a tendency to, for example, adhere to surfaces, or if solutions, foam. Not all surface-active drugs form micelles, because of steric hindrances. The surface activity of compounds can be determined using a variety of techniques, such as surface tension measurements using a Du Nouy tensiometer (15), Wilhelmy plate, and conductance measurements. Sigma 703 is a simple, reliable instrument for the measurement of surface and interfacial tension by Du Nouy ring or Wilhelmy plate methods (16).

Osmolality

A 0.9% w/v NaCl solution is iso-osmotic with blood. The commonly used unit to express osmolality is the ion, and this is defined as the weight in grams per solute, existing in a solution as molecules, ions, macromolecules, and the like that is osmotically equivalent to the gram MW of an ideally behaving nonelectrolyte. This is an important consideration for parenteral and ophthalmic products. Extreme discomfort in the use of ophthalmic preparations is experienced when the osmolality is too high or too low. Osmolality is determined using a cryoscopic osmometer, which is calibrated with deionized water and solutions of sodium chloride of known concentrations (17). Using this technique, the sodium chloride equivalents and freezing point depressions for a large number of substances have been determined and reported.

FREEZE-DRIED FORMULATIONS

The stability of the solution forms intended for parenteral administration can be significantly improved by lyophilizing the solutions to dryness without the use of heat. The solution is frozen to a very low temperature, and vacuum is applied to remove water through sublimation. The cake left is easily dispersible, and thus offers a highly desirable dosage form that is reconstituted just prior to administration. Examples of lyophilized drugs include erythromycin, vancomycin, bacitracin, cyclophosphamide, cefazolin, infliximab, somatropin, trimetrexate glucuronate MVI, and doxorubicin. The Food and Drug Administration (FDA) classification for lyophilized products is as follows:

713	injection, powder, lyophilized, for liposomal suspension,
705	injection, powder, lyophilized, for solution,
706	injection, powder, lyophilized, for suspension,
712	injection, powder, lyophilized, for suspension, extended release.

Another advantage of lyophilization is that it allows the formulation without any additives, though some are often added to increase the bulk. As a result, each drug must be formulated with a highly specific temperature and vacuum cycle that is highly dependent on the nature of the drug, the quantity used, and the nature of the additives. The science of lyophilization is complex, and specialized training is needed to evaluate this potential of new molecules. If during freezing, the solutes crystallize, the first thermal event detected using DSC will be the endotherm that corresponds to the melting of the eutectic formed between ice and the solute. This is usually followed by an endothermic event, corresponding to the melting of ice. Normally, freeze-drying of these systems is carried out below the eutectic melting temperature. Another way of detecting whether a solute or formulation crystallizes on freezing is to conduct subambient XRD. If there is no crystallization upon cooling the solution of the drug, the supercooled liquid becomes more concentrated and viscous, leading to glass formation at temperature known as the glass transition temperature (T_g). Generally, the freeze drying should take place below this temperature to avoid the collapsing of the cake, wherein high residual water remains and requires prolonged reconstitution time. There is an increased degradation as a result of increased mobility of molecules above T_g.

Testing of lead compound would preferably involve freezing and studying using DSC. In some cases, an endotherm caused by stress relaxation may be superimposed on the glass transition. It is possible to resolve these events using the related technique, modulated DSC (MDSC) or dynamic DSC (DDSC). Differential scanning calorimetry is used to determine a wide range of physical properties of materials, including the glass transition temperature T_g, the melting temperature

T_m, and solid–solid transitions. In this technique, a sample and a reference material are subjected to a controlled temperature program. When a phase transition, such as melting, occurs in the sample, an input of energy is required to keep the sample and the reference at the same temperature. This difference in energy is recorded as a function of temperature to produce the DSC trace. The MDSC provides the same qualitative and quantitative information about the physical and chemical changes as conventional DSC, and in addition, provides unique thermochemical data that are unavailable from conventional DSC. The effects of baseline slope and curvature are reduced, increasing the sensitivity of the system. Overlapping events, such as molecular relaxation and glass transitions, can be separated. Heat capacity can be measured directly with MDSC in a minimum number of experiments. Both MDSC and DSC measure the difference in heat flow to a sample and to an inert reference. The sample and reference cells are identical. However, MDSC uses a different heating profile. The DSC measures heat flow as a function of a constant rate of change in temperature, whereas the MDSC superimpose a sinusoidal temperature modulation on this rate. The sinusoidal change in temperature permits the measurement of heat-capacity effects simultaneously with the kinetic effect. Typical experimental procedure for an initial MDSC experiment include a heating rate from isothermal to 5°C/min and a modulation amplitude from 0.01°C to 10°C. The modulation period can vary from 10 to 100 s, or is expressed as a frequency, from 10 MHz to 100 MHz.

The method of DDSC, such as using the Perkin-Elmer DSC 7 (18), along with the Pyris software platform, creates a modulated temperature profile applied to the sample, rather than a straight heating ramp, and the response of the sample analyzed by Fourier transformation. The DDSC is particularly useful for separating overlapping thermal events, such as melting and recrystallization. Subambient operation of the DSC 7 normally employs a Perkin-Elmer Intracooler II, which allows reliable data to be acquired down to approximately −40°C; for even lower temperatures, a liquid nitrogen bath can be employed, which allows the collection of data down to approxi-mately −150°C.

SUSPENSIONS

When the lead compound has limited solubility, and the efforts to enhance it fail, and when there is a tendency for fast crystallization from solutions, or even when chemical stability is a problem, often formulating suspension dosage forms obviates some of these drawbacks. However, suspensions, by nature, must have higher viscosity to prevent the settling of particles, and thus create problems in pourability, syringability, and so on. Appropriate selection of a vehicle that provides an ideal compromise among all characteristics thus becomes a critical factor, because the intent is to have as little solubility in the vehicle as possible to prevent crystallization from the solution that surrounds the suspended particles. As a result, weak acids and bases appear as poor choice for suspension formulation. In some instances, it may be possible to prepare a derivative with larger hydrophobic groups or salt formation that would have lower solubility, if preparing a suspension dosage form were particularly desired. Compounds that can form hydrates when in suspension state can create stability problems. A significant thermodynamic problem in suspension formulation comes from Ostwald ripening, crystal growth, not because of phase change, but as a result of the differences in the solubility as a function of crystal size:

$$\frac{RT}{M}\ln\left(\frac{S_2}{S_1}\right) = \frac{2\sigma}{\rho}\left(\frac{1}{r_1} - \frac{1}{r_2}\right) \tag{23}$$

where R is the gas constant, T is the absolute temperature, S_1 and S_2 are the solubilities of crystals of radii r_1 and r_2, respectively, σ is the specific surface energy, ρ is the density, and M is the MW of the solute molecules. Temperature fluctuation is obviously one factor that promotes Ostwald ripening. Although phase changes can be studied using standard techniques, such as DSC, hot-stage microscopy, or XRPD, Ostwald ripening is best studied using microscopic methods. The art of suspension formulation is complex, as a large number of factors including additives can have a significant influence on the crystal growth; for example, dye molecules often attach to high-energy points on crystals, affecting their growth. Similarly, it is reported that PVP, a common ingredient of many suspension formulations, inhibits crystal growth. Albumin is also known to have a similar impact. The choice of additives is also governed by the final form of suspension. If it has to be sterilized, the additives must be able to sustain autoclave temperatures; besides, autoclaving itself can affect both the physical and the chemical stabilities of the drug. Zeta potential measurements of suspensions often prove useful.

Shelf-life specifications of a suspension dosage form include redispersability upon storage. However, there is no official method to test this, and most manufacturers design their own methods, chiefly requiring some type of subjective shaking. The stability of the suspensions is partly dependent on the particle size in suspension. This can be measured using techniques, such as laser diffraction or Malvern Mastersize (19). As the stabilized suspension is mostly in a flocculated form due to the electrolytes added to it, it may be necessary to apply sonification in the study of particle sizes.

TOPICAL

Topical delivery of drugs using semisolid, controlled release patches, and many other dosage forms offers advantages, including reduced blood level fluctuation, obviating the first pass effect, and protection from gastrointestinal pH. In cases where localized action is desired, this dosage form offers remarkable opportunity for drug action. However, skin is a poor medium to deliver drugs, because by its very design it is supposed to prevent the entry of chemicals (though it fails miserably as we know from the chemical warfare agents). Generally, large polar molecules do not penetrate the stratum corneum well. The intrinsic physico-chemical properties of candidate drugs, important in expediting delivery across the skin include MW and volume, aqueous solubility, melting point, and log P. For weakly acidic or basic drugs, the skin pH will play a strong role in their transport. Drugs that form zwitter ions can be made more penetrable by using appropriate salt forms.

The formulation additives strongly impact on transdermal delivery as the variety of dosage forms, such as creams, ointments, lotions, gels, and patches offer a wide variety of formulation additives. The problems related to the crystallization of drugs, as discussed under suspension dosage forms, also apply here just as the considerations that optimize the physical and chemical stabilities. The entire textbook has been dedicated to the formulation of semisolid and topical delivery dosage forms that describe in detail how the choice of basic drug structure and additives affects the stability. Where salt forms are available, it is often difficult to predict the stability profile, including factors such as photostability, a test that must be conducted for all dosage forms. It is known that different salt forms can show differences in their photostability profile.

PULMONARY DELIVERY

The pressurized metered dose inhalers in the use of environmentally friendly propellants means the choice of hydrofluoroalkanes, wherein the dosage form can be a suspension of the solution form. The problems of formulating suspensions, as discussed earlier, apply here as well, but particularly with respect to interactions with the formulation components specific to pressurized inhaler systems.

Solution dosage forms require the selection of propellants, wherein the drug can dissolve without crystallizing, and may require the addition of surfactants and co-solvents. However, there are toxicological issues with the use of surfactants. The solubility of drugs in solvents is determined by filtering the suspension in a pressurized can into another can, and then evaporating the clear solution (bringing to room temperature) followed by determination of the amount of drug in it. High solubility in propellants can lead to crystal growth as propellants evaporate. Ostwald ripening, common to suspensions, applies to inhalation suspensions. The changes in the property of suspension can be studied by using microscopy, and observing the changes in the axial ratio of crystal.

Drugs for inhalation therapy in a powder form require a particular particle size, which is achieved by the process of micronization between 1 μm and 6 μm to allow deep penetration through the lung alveoli system. There are a number of devices that can deliver drugs to the lungs as dry powders, for example, Turbuhaler^{TM} or Diskhaler^{TM}. These dosage forms rely on a larger carrier particle, such as α-lactose monohydrate, to which the drug is attached. The lactose is usually fractionated, such that it lies in the size range 63–90 μm. Upon delivery, the drug detaches from the lactose, and because the drug is micronized, it is delivered to the lung, while the lactose is eventually swallowed. It should be realized that the polymorphic form of the lactose used could affect the aerosolization properties of the formulation. The β-forms were easily entrained, but held onto the drug particles most strongly when the flow properties were studied. The anhydrous α-form shows an opposite behavior, and the monohydrate α-form demonstrates intermediate behavior. Interactions with packaging materials can also alter the powder characteristics; for example, long contact times with polyvinyl chloride, polyethylene, or aluminum should be avoided, because the adhesion force between the drug and these surfaces is much higher than between it and the lactose carrier. Thus, detachment and loss of drug in the formulation could occur. As lactose is widely used as a carrier, its compatibility with the new drugs should be studied in detail, especially if there are any amino groups in the structure. The surface property of lactose is also important. With increasing specific surface area and roughness, the effective index of inhalation decreases as a result of the drug being held more tightly in the inhaled airstreams. Therefore, characterization of the carrier particles by, for example, surface area measurements, SEM, and other solid-state techniques are recommended preformulation activities.

The recent approval by the U.S. FDA of Exubera, an inhalation form of insulin, is a classical example where the dosage form is an integral part of drug action. Using the Nektar company's delivery system to create a fine powder mist, insulin in Exubera is absorbed as the mist of fine powder reaches into the deep portions of the lung structure, without getting impacted. Although reduction in particle size is pivotal to pulmonary delivery of drugs, micronization makes powders difficult to flow, and these changes should be studied using techniques, such as DVS, microcalorimetry, and IGC. The high energy at the surface of micronized powders can often be relieved by exposing it to air of higher humidity, which can crystallize the amorphous high-energy regions. As a result, the common preformulate stage evaluations include the measurements of the micromeretic, RH, and electrostatic properties of the powder. Different salt forms show variant flow properties; for example, stearate salts generally are better for aerosol formulation.

Nebulizer formulations are normally solutions, but suspensions (particle size of less than 2 µm) are also used. Important preformulation considerations include stability, solubility, viscosity, and surface tension of the solution of suspension.

GENERAL COMPATIBILITY

Some excipients are universal to specific dosage forms. General compatibility testing with components like lactose and other fillers, lubricants like magnesium stearate, and suspending agents like PVP and the like, can be done at the preformulation level if there is sufficient quantity of drug available. In cases where a specific dosage form is definitely desired, such as a pulmonary delivery aerosol, this testing becomes more important. Generally, chemical reactivity known between prominent function groups can be put to test in projecting the likely excipients. The testing involves making a binary mixture of the drug with the excipient in the ratio 1:1 or other similar ratios, followed by moistening and sealing into ampoules. Stored at suitable temperatures, such as 50°C, and analyzed at various time points—preferably after three weeks, using HPLC, DSC, and TGA, as appropriate—provides gross incompatibility profile. In cases where a larger number of excipients are to be tested, a factorial design can be used to minimize the testing samples. The purpose of testing at this stage is to determine if there were any phase changes, changes in the crystallinity, and so on.

REFERENCES

1. http://www.smsuk.co.uk/index.php.
2. http://www.vticorp.com/products/sga100.htm.
3. http://www.malvern.co.uk/home/index.htm.
4. http://www.iupac.org/reports/2001/colloid_2001/manual_of_s_and_t.pdf.
5. http://www.thermo.com/com/cda/product/detail/1,1055, 12967,00.html.
6. http://micromeritics.com.
7. http://www.ccdc.cam.ac.uk/.
8. http://www.skyscan.be/next/home.htm.
9. http://www.hunterlab.com/.
10. http://www.pcs-instruments.com/.
11. http://www.pcs-instruments.com/products.shtml.
12. http://www.veeco.com.

13. http://www.microphotonics.com.
14. http://www.asylumresearch.com/.
15. http://www.thomassci.com/catalog/product/25667.
16. http://www.ksvinc.com/tensiometers.htm.
17. http://www.rlinstruments.com/osmometer.htm.
18. http://www.princeton.edu/~polymer/dsc.html.
19. http://www.malvern.co.uk.
20. http://www.pharmquest.com/default.html.

BIBLIOGRAPHY

Abdul S, Poddar SS. A flexible technology for modified release of drugs: multi-layered tablets. J Control Release 2004, 7 Jul; 97(3):393–405.

Abrams J. Glyceryl trinitrate (nitroglycerin) and the organic nitrates. Choosing the method of administration. Drugs 1987 Sep; 34(3):391–403.

Aldridge MA, Ito MK. Colesevelam hydrochloride: a novel bile acid-binding resin. Ann Pharmacother 2001 Jul–Aug; 35(7–8):898–907.

Atkins PJ. Dry powder inhalers: an overview. Respir Care 2005 Oct; 50(10):1304–1312; discussion 1312.

Bach M, Lippold BC. Percutaneous penetration enhancement and its quantification. Eur J Pharm Biopharm 1998 Jul; 46(1):1–13.

Baker MT, Naguib M. Propofol: the challenges of formulation. Anesthesiology 2005 Oct; 103(4):860–876.

Baldi F. Lansoprazole oro-dispersible tablet: pharmacokinetics and therapeutic use in acid-related disorders. Drugs 2005; 65(10):1419–1426.

Baldi F, Malfertheiner P. Lansoprazole fast disintegrating tablet: a new formulation for an established proton pump inhibitor. Digestion 2003; 67(1–2):1–5.

Bang LM, Keating GM. Paroxetine controlled release. CNS Drugs 2004; 18(6):355–364; discussion 365–366.

Behar-Cohen F. [Drug delivery systems to target the anterior segment of the eye: fundamental bases and clinical applications]. J Fr Ophtalmol 2002 May; 25(5):537–544.

Behrend M, Braun F. Enteric-coated mycophenolate sodium: tolerability profile compared with mycophenolate mofetil. Drugs 2005; 65(8):1037–1050.

Bejan A, Turcu G. Liposomes: presentation and actual applicative trends in medicine. Rom J Intern Med 1995 Jul–Dec; 33(3–4):141–149.

Benninger MS. Amoxicillin/clavulanate potassium extended release tablets: a new antimicrobial for the treatment of acute bacterial sinusitis and community-acquired pneumonia. Expert Opin Pharmacother 2003 Oct; 4(10):1839–1846.

Bernkop-Schnurch A, Kast CE, Guggi D. Permeation enhancing polymers in oral delivery of hydrophilic macromolecules: thiomer/GSH systems. J Control Release 2003, 5 Dec; 93(2):95–103.

Bianchi Porro G, Parente F. Antacids for duodenal ulcer: current role. Scand J Gastroenterol (Suppl) 1990; 174:48–53.

Bittner B, Mountfield RJ. Intravenous administration of poorly soluble new drug entities in early drug discovery: the potential impact of formulation on pharmacokinetic parameters. Curr Opin Drug Discov Dev 2002 Jan; 5(1):59–71.

Bourquin J, Schmidli H, van Hoogevest P, Leuenberger H. Basic concepts of artificial neural networks (ANN) modeling in the application to pharmaceutical development. Pharm Dev Technol 1997 May; 2(2):95–109.

Brocklebank D, Ram F, Wright J, et al. Comparison of the effectiveness of inhaler devices in asthma and chronic obstructive airways disease: a systematic review of the literature. Health Technol Assess 2001; 5(26):1–149.

Budde K, Glander P, Diekmann F, et al. Review of the immunosuppressant enteric-coated mycophenolate sodium. Expert Opin Pharmacother 2004 Jun; 5(6): 1333–1345.

Bunker M, Davies M, Roberts C. Towards screening of inhalation formulations: measuring interactions with atomic force microscopy. Expert Opin Drug Deliv 2005 Jul; 2(4):613–624.

Butcher EC. Can cell systems biology rescue drug discovery? Nat Rev Drug Discov 2005 Jun; 4(6):461–467.

Cabrera J, Redondo P, Becerra A, et al. Ultrasound-guided injection of polidocanol microfoam in the management of venous leg ulcers. Arch Dermatol 2004 Jun; 140(6):667–673.

Chang TM. Artificial cells with emphasis on bioencapsulation in biotechnology. Biotechnol Annu Rev 1995; 1:267–295.

Chaubal MV. Application of drug delivery technologies in lead candidate selection and optimization. Drug Discov Today 2004 Jul 15; 9(14):603–609.

Chen D, Maa YF, Haynes JR. Needle-free epidermal powder immunization. Expert Rev Vaccines 2002 Oct; 1(3):265–276.

Chew NY, Chan HK. The role of particle properties in pharmaceutical powder inhalation formulations. J Aerosol Med 2002 Fall; 15(3):325–330.

Chourasia MK, Jain SK. Pharmaceutical approaches to colon targeted drug delivery systems. J Pharm Sci 2003 Jan–Apr; 6(1):33–66.

Cleland JL. Protein delivery from biodegradable microspheres. Pharm Biotechnol 1997; 10:1–43.

Cole P. Pharmacologic and clinical comparison of cefaclor in immediate-release capsule and extended-release tablet forms. Clin Ther 1997 Jul–Aug; 19(4):617–625; discussion 603.

Constantinides PP, Tustian A, Kessler DR. Tocol emulsions for drug solubilization and parenteral delivery. Adv Drug Deliv Rev 2004 May 7; 56(9):1243–1255.

Courrier HM, Butz N, Vandamme TF. Pulmonary drug delivery systems: recent developments and prospects. Crit Rev Ther Drug Carrier Syst 2002; 19(4–5): 425–498.

Dahlof B, Andersson OK. A felodipine-metoprolol extended-release tablet: its properties and clinical development. J Hum Hypertens 1995 Jul; 9(suppl 2):S43–S47.

Dando TM, Scott LJ. Abacavir plus lamivudine: a review of their combined use in the management of HIV infection. Drugs 2005; 65(2):285–302.

Darkes MJ, Perry CM. Clarithromycin extended-release tablet: a review of its use in the management of respiratory tract infections. Am J Respir Med 2003; 2(2):175–201.

Davis SS. Coming of age of lipid-based drug delivery systems. Adv Drug Deliv Rev 2004, May 7; 56(9):1241–1242.

Davis SS, Illum L. Absorption enhancers for nasal drug delivery. Clin Pharmacokinet 2003; 42(13):1107–1128.

De Moor R, Verbeeck R, Martens L. [Evaluation of long-term release of fluoride by type II glass ionomer cements with a conventional hardening reaction]. Rev Belge Med Dent 1996; 51(3):22–35.

Dempski RE, Scholtz EC, Oberholtzer ER, Yeh KC. Pharmaceutical design and development of a Sinemet controlled-release formulation. Neurology 1989 Nov; 39(11 suppl 2):20–24.

Devlin JW, Welage LS, Olsen KM. Proton pump inhibitor formulary considerations in the acutely ill. Part 1: Pharmacology, pharmacodynamics, and available formulations. Ann Pharmacother 2005 Oct; 39(10):1667–1677. Epub 2005, 23 Aug.

Dhaon NA. Amoxicillin tablets for oral suspension in the treatment of acute otitis media: a new formulation with improved convenience. Adv Ther 2004 Mar–Apr; 21(2): 87–95.

Digenis GA, Gold TB, Shah VP. Cross-linking of gelatin capsules and its relevance to their in vitro–in vivo performance. J Pharm Sci 1994 Jul; 83(7):915–921.

Doelker E. Recent advances in tableting science. Boll Chim Farm 1988 Feb; 127(2): 37–49.

Doelker E, Massuelle D. Benefits of die-wall instrumentation for research and development in tabletting. Eur J Pharm Biopharm 2004 Sep; 58(2):427–444.

Dowson AJ, Almqvist P. Part III: the convenience of, and patient preference for, zolmitriptan orally disintegrating tablet. Curr Med Res Opin 2005; 21(suppl 3): S13–S17.

Duchene D, Ponchel G. Principle and investigation of the bioadhesion mechanism of solid dosage forms. Biomaterials 1992; 13(10):709–714.

DuPont HL. Bismuth subsalicylate in the treatment and prevention of diarrheal disease. Drug Intell Clin Pharm 1987 Sep; 21(9):687–693.

DuPont HL. Nonfluid therapy and selected chemoprophylaxis of acute diarrhea. Am J Med 1985, 28 June; 78(6B):81–90.

Faassen F, Vromans H. Biowaivers for oral immediate-release products: implications of linear pharmacokinetics. Clin Pharmacokinet 2004; 43(15):1117–1126.

Fedorak RN, Bistritz L. Targeted delivery, safety, and efficacy of oral enteric-coated formulations of budesonide. Adv Drug Deliv Rev 2005, 6 Jan; 57(2):303–316.

Felt O, Buri P, Gurny R. Chitosan: a unique polysaccharide for drug delivery. Drug Dev Ind Pharm 1998 Nov; 24(11):979–993.

Femia RA, Goyette RE. The science of megestrol acetate delivery: potential to improve outcomes in cachexia. BioDrugs 2005; 19(3):179–187.

Figgitt DP, Plosker GL. Saquinavir soft-gel capsule: an updated review of its use in the management of HIV infection. Drugs 2000 Aug; 60(2):481–516.

Flores NA. Ezetimibe + simvastatin (Merck/Schering-Plough). Curr Opin Investig Drugs 2004 Sep; 5(9):984–992.

Frijlink HW, De Boer AH. Dry powder inhalers for pulmonary drug delivery. Expert Opin Drug Deliv 2004 Nov; 1(1):67–86.

Frishman WH, Sherman D, Feinfeld DA. Innovative drug delivery systems in cardiovascular medicine: nifedipine-GITS and clonidine-TTS. Cardiol Clin 1987 Nov; 5(4): 703–716.

Fu Y, Yang S, Jeong SH, Kimura S, Park K. Orally fast disintegrating tablets: developments, technologies, taste-masking and clinical studies. Crit Rev Ther Drug Carrier Syst 2004; 21(6):433–476.

Fuseau E, Petricoul O, Moore KH, Barrow A, Ibbotson T. Clinical pharmacokinetics of intranasal sumatriptan. Clin Pharmacokinet 2002; 41(11):801–811.

Gallen CC. Strategic challenges in neurotherapeutic pharmaceutical development. NeuroRx 2004 Jan; 1(1):165–180.

Garg S, Kandarapu R, Vermani K, et al. Development pharmaceutics of microbicide formulations. Part I: preformulation considerations and challenges. AIDS Patient Care STDS 2003 Jan; 17(1):17–32.

Garg S, Tambwekar KR, Vermani K, et al. Development pharmaceutics of microbicide formulations. Part II: formulation, evaluation, and challenges. AIDS Patient Care STDS 2003 Aug; 17(8):377–399.

Gehl J. Electroporation: theory and methods, perspectives for drug delivery, gene therapy and research. Acta Physiol Scand 2003 Apr; 177(4):437–447.

Gill J, Feinberg J. Saquinavir soft gelatin capsule: a comparative safety. Drug Saf 2001; 24(3):223–232.

Gillis JC, Benfield P, Goa KL. Transnasal butorphanol. A review of its pharmacodynamic and pharmacokinetic properties, and therapeutic potential in acute pain management. Drugs 1995 Jul; 50(1):157–175.

Goldenberg MM. An extended-release formulation of oxybutynin chloride for the treatment of overactive urinary bladder. Clin Ther 1999 Apr; 21(4):634–642.

Goldenheim PD, Conrad EA, Schein LK. Treatment of asthma by a controlled-release theophylline tablet formulation: a review of the North American experience with nocturnal dosing. Chronobiol Int 1987; 4(3):397–408.

Goldstein JL, Larson LR, Yamashita BD. Prevention of nonsteroidal anti-inflammatory drug-induced gastropathy: clinical and economic implications of a single- tablet formulation of diclofenac/misoprostol. Am J Manag Care 1998 May; 4(5):687–697.

Gooding OW. Process optimization using combinatorial design principles: parallel synthesis and design of experiment methods. Curr Opin Chem Biol 2004; 8(3): 297–304.

Goodnow RA Jr. Current practices in generation of small molecule new leads. J Cell Biochem (Suppl) 2001; suppl 37:13–21.

Gotfried MH. Clarithromycin (Biaxin) extended-release tablet: a therapeutic. Expert Rev Anti Infect Ther 2003 Jun; 1(1):9–20.

Gregoriadis G. Liposomes as a drug delivery system: optimization studies. Adv Exp Med Biol 1988; 238:151–159. No abstract available.

Grosdidier J, Boissel P, Bresler L, Vidrequin A. [Stenosing and perforated ulcers of the small intestine related to potassium chloride in enteric-coated tablets. Apropos of 11 cases]. Chirurgie 1989; 115(3):163–169. French.

Grzeszczak W. [Cardura XL—a unique drug formulation—doxazosine administered in a slow-release form (doxazosine GITS)]. Przegl Lek 2000; 57(11):643–654.

Gudgin Dickson EF, Goyan RL, Pottier RH. New directions in photodynamic therapy. Cell Mol Biol (Noisy-le-grand) 2002 Dec; 48(8):939–954.

Guyton JR. Extended-release niacin for modifying the lipoprotein profile. Expert Opin Pharmacother 2004 Jun; 5(6):1385–1398.

Hadgraft J. Passive enhancement strategies in topical and transdermal drug delivery. Int J Pharm 1999 Jul 5; 184(1):1–6.

Haefner S, Knietsch A, Scholten E, Braun J, Lohscheidt M, Zelder O. Biotechnological production and applications of phytases. Appl Microbiol Biotechnol 2005 Sep; 68(5):588–597. Epub 2005, 26 Oct.

Harashima H, Ishida T, Kamiya H, Kiwada H. Pharmacokinetics of targeting with liposomes. Crit Rev Ther Drug Carrier Syst 2002; 19(3):235–275.

Harashima H, Shinohara Y, Kiwada H. Intracellular control of gene trafficking using liposomes as drug carriers. Eur J Pharm Sci 2001 Apr; 13(1):85–89.

Hardy IJ, Fitzpatrick S, Booth SW. Rational design of powder formulations for tamp filling processes. J Pharm Pharmacol 2003 Dec; 55(12):1593–1599.

Harsch IA. Inhaled insulins: their potential in the treatment of diabetes mellitus. Treat Endocrinol 2005; 4(3):131–138.

Hatefi A, Amsden B. Camptothecin delivery methods. Pharm Res 2002 Oct; 19(10): 1389–1399.

Hausheer FH, Kochat H, Parker AR, et al. New approaches to drug discovery and development: a mechanism-based approach to pharmaceutical research and its application

to BNP7787, a novel chemoprotective agent. Cancer Chemother Pharmacol 2003 Jul; 52(suppl 1):S3–S15. Epub 2003, 18 June.
Heinig R. Clinical pharmacokinetics of nisoldipine coat-core. Clin Pharmacokinet 1998 Sep; 35(3):191–208. Erratum in Clin Pharmacokinet 1998 Nov; 35(5):390.
Hiestand EN. Mechanical properties of compacts and particles that control tableting success. J Pharm Sci 1997 Sep; 86(9):985–990.
Hirata K, Iwanami H. [The role of triptans and analgesics for primary headache treatment]. Nippon Rinsho 2005 Oct; 63(10):1797–1801. Japanese.
Hirschberger R. [Pharmacokinetics and dynamics of retard formation of isradipine. Summary of studies]. Fortschr Med 1993, 30 Oct; 111(30):481–484.
Huang Y, Leobandung W, Foss A, Peppas NA. Molecular aspects of muco- and bioadhesion: tethered structures and site-specific surfaces. J Control Release 2000, 1 Mar; 65(1–2):63–71.
Itkin YM, Trujillo TC. Intravenous immunoglobulin-associated acute renal failure: case series and literature. Pharmacotherapy 2005 Jun; 25(6):886–892.
Jones DH, Partidos CD, Steward MW, Farrar GH. Oral delivery of poly(lactide-co-glycolide) encapsulated vaccines. Behring Inst Mitt 1997 Feb; (98):220–228.
Jovanovic N, Bouchard A, Hofland GW, Witkamp GJ, Crommelin DJ, Jiskoot W. Stabilization of proteins in dry powder formulations using supercritical fluid technology. Pharm Res 2004 Nov; 21(11):1955–1969.
Kalia YN, Guy RH. Modeling transdermal drug release. Adv Drug Deliv Rev 2001, 11 June; 48(2–3):159–172.
Kanjickal DG, Lopina ST. Modeling of drug release from polymeric delivery systems—a Crit Rev Ther Drug Carrier Syst 2004; 21(5):345–386.
Karlsson L, Torstensson A, Taylor LT. The use of supercritical fluid extraction for sample preparation of pharmaceutical formulations. J Pharm Biomed Anal 1997 Feb; 15(5):601–611.
Kaur IP, Kanwar M. Ocular preparations: the formulation approach. Drug Dev Ind Pharm 2002 May; 28(5):473–493.
Kleinebudde P. Roll compaction/dry granulation: pharmaceutical applications. Eur J Pharm Biopharm 2004 Sep; 58(2):317–326.
Knop K. [Active ingredient release from solid drug forms—test methods, evaluation, influencing parameters]. Pharm Unserer Zeit 1999 Nov; 28(6):301–308. German.
Knox ED, Stimmel GL. Clinical review of a long-acting, injectable formulation of risperidone. Clin Ther 2004 Dec; 26(12):1994–2002.
Kostarelos K. Rational design and engineering of delivery systems for therapeutics: biomedical exercises in colloid and surface science. Adv Colloid Interface Sci 2003, 1 Dec; 106:147–168.
Kozubek A, Gubernator J, Przeworska E, Stasiuk M. Liposomal drug delivery, a novel approach: PLARosomes. Acta Biochim Pol 2000; 47(3):639–649.
Kuhlmann J. Alternative strategies in drug development: clinical pharmacological aspects. Int J Clin Pharmacol Ther 1999 Dec; 37(12):575–583.
Kulkarni SB, Betageri GV, Singh M. Factors affecting microencapsulation of drugs in liposomes. J Microencapsul 1995 May–Jun; 12(3):229–246.
Labiris NR, Dolovich MB. Pulmonary drug delivery. Part II: the role of inhalant delivery devices and drug formulations in therapeutic effectiveness of aerosolized medications. Br J Clin Pharmacol 2003 Dec; 56(6):600–612.
Laube BL. The expanding role of aerosols in systemic drug delivery, gene therapy, and vaccination. Respir Care 2005 Sep; 50(9):1161–176.

Leuenberger H. New trends in the production of pharmaceutical granules: batch versus continuous processing. Eur J Pharm Biopharm 2001 Nov; 52(3):289–296.
Levy RA. Therapeutic inequivalence of pharmaceutical alternates. Am Pharm 1985 Apr; NS25(4):28–39.
Lightman S. Somatuline Autogel: an extended release lanreotide formulation. Hosp Med 2002 Mar; 63(3):162–165.
Maa YF, Prestrelski SJ. Biopharmaceutical powders: particle formation and formulation considerations. Curr Pharm Biotechnol 2000 Nov; 1(3):283–302.
Machida Y. [Development of topical drug delivery systems utilizing polymeric materials]. Yakugaku Zasshi 1993 May; 113(5):356–368. Japanese.
Mainardes RM, Silva LP. Drug delivery systems: past, present, and future. Curr Drug Targets 2004 Jul; 5(5):449–455.
Man M, Rugo H. Paclitaxel poliglumex. Cell Therapeutics/Chugai Pharmaceutical. IDrugs 2005 Sep; 8(9):739–754.
Maroni A, Zema L, Cerea M, Sangalli ME. Oral pulsatile drug delivery systems. Expert Opin Drug Deliv 2005 Sep; 2(5):855–871.
McKay B, Hoogenraad M, Damen EW, Smith AA. Advances in multivariate analysis in pharmaceutical process development. Curr Opin Drug Discov Devel 2003 Nov; 6(6):966–977.
Mehnert W, Mader K. Solid lipid nanoparticles: production, characterization and applications. Adv Drug Deliv Rev 2001, 25 Apr; 47(2–3):165–196.
Melia CD, Davis SS. Review article: mechanisms of drug release from tablets and capsules. 2. Dissolution. Aliment Pharmacol Ther 1989 Dec; 3(6):513–525.
Michele TM, Knorr B, Vadas EB, Reiss TF. Safety of chewable tablets for children. J Asthma 2002 Aug; 39(5):391–403.
Mooradian AD. Towards single-tablet therapy for type 2 diabetes mellitus. Rationale and recent developments. Treat Endocrinol 2004; 3(5):279–287.
Moribe K, Maruyama K. Pharmaceutical design of the liposomal antimicrobial agents for infectious disease. Curr Pharm Des 2002; 8(6):441–454.
Morissette SL, Almarsson O, Peterson ML, et al. High-throughput crystallization: polymorphs, salts, co-crystals and solvates of pharmaceutical solids. Adv Drug Deliv Rev 2004, 23 Feb; 56(3):275–300.
Nagai T. [New drug development by innovative drug administration—"change" in pharmaceutical field]. Yakugaku Zasshi 1997 Nov; 117(10–11):963–971.
Nail SL, Jiang S, Chongprasert S, Knopp SA. Fundamentals of freeze-drying. Pharm Biotechnol 2002; 14:281–360.
Najib J. Fenofibrate in the treatment of dyslipidemia: a review of the data as they relate to the new suprabioavailable tablet formulation. Clin Ther 2002 Dec; 24(12):2022–2050.
Nishida K. [Development of drug delivery system based on a new administration route for targeting to the specific region in the liver]. Yakugaku Zasshi 2003 Aug; 123(8):681–689. Japanese.
Norman TR, Olver JS. New formulations of existing antidepressants: advantages in the management of depression. CNS Drugs 2004; 18(8):505–520.
Nuijen B, Bouma M, Schellens JH, Beijnen JH. Progress in the development of alternative pharmaceutical formulations of taxanes. Invest New Drugs 2001 May; 19(2):143–153.
Nunn T, Williams J. Formulation of medicines for children. Br J Clin Pharmacol 2005 Jun; 59(6):674–676.
Odegard PS, Capoccia KL. Inhaled insulin: Exubera. Ann Pharmacother 2005 May; 39(5):843–853. Epub 2005, 12 Apr.

O'Hagan DT, Singh M. Microparticles as vaccine adjuvants and delivery systems. Expert Rev Vaccines 2003 Apr; 2(2):269–283.
Okada N. [Design and creation of cytomedicine for application to cell therapy]. Yakugaku Zasshi 2005 Aug; 125(8):601–615. Japanese.
Ormrod D, Goa KL. Intranasal metoclopramide. Drugs 1999 Aug; 58(2):315–322; discussion 323–324.
Owens DR, Zinman B, Bolli G. Alternative routes of insulin delivery. Diabet Med 2003 Nov; 20(11):886–898.
Panchagnula R, Agrawal S, Ashokraj Y, et al. Fixed dose combinations for tuberculosis: lessons learned from clinical, formulation and regulatory perspective. Methods Find Exp Clin Pharmacol 2004 Nov; 26(9):703–721.
Pathak P, Meziani MJ, Sun YP. Supercritical fluid technology for enhanced drug delivery. Expert Opin Drug Deliv 2005 Jul; 2(4):747–761.
Patro SY, Freund E, Chang BS. Protein formulation and fill-finish operations. Biotechnol Annu Rev 2002; 8:55–84.
Pawar R, Ben-Ari A, Domb AJ. Protein and peptide parenteral controlled delivery. Expert Opin Biol Ther 2004 Aug; 4(8):1203–1212.
Perkins AC, Frier M. Nuclear medicine techniques in the evaluation of pharmaceutical formulations. Pharm World Sci 1996 Jun; 18(3):97–104.
Perkins AC, Frier M. Radionuclide imaging in drug development. Curr Pharm Des 2004; 10(24):2907–2921.
Pettipher R, Cardon LR. The application of genetics to the discovery of better medicines. Pharmacogenomics 2002 Mar; 3(2):257–263.
Pfister WR, Hsieh DS. Permeation enhancers compatible with transdermal drug delivery systems. Part I: selection and formulation considerations. Med Device Technol 1990 Sep–Oct; 1(5):48–55. Erratum in Med Device Technol 1990 Nov–Dec; 1(6):28–33.
Rabiskova M, Vostalova L, Medvecka G, Horackova D. [Hydrophilic gel matrix tablets for oral administration of drugs]. Ceska Slov Farm 2003 Sep; 52(5):211–217. Czech.
Ram CV, Featherston WE. Calcium antagonists in the treatment of hypertension. An overview. Chest 1988 Jun; 93(6):1251–1253.
Ramsay EC, Dos Santos N, Dragowska WH, Laskin JJ, Bally MB. The formulation of lipid-based nanotechnologies for the delivery of fixed dose anticancer drug combinations. Curr Drug Deliv 2005 Oct; 2(4):341–351.
Ranade VV. Drug delivery systems 5B. Oral drug delivery. J Clin Pharmacol 1991 Feb; 31(2):98–115.
Reddy KR. Controlled-release, pegylation, liposomal formulations: new mechanisms in the delivery of injectable drugs. Ann Pharmacother 2000 Jul–Aug; 34(7–8): 915–923.
Reeves RR, Wallace KD, Rogers-Jones C. Orally disintegrating olanzapine: a possible alternative to injection of antipsychotic drugs. J Psychosoc Nurs Ment Health Serv 2004 May; 42(5):44–48.
Richard A, Margaritis A. Poly(glutamic acid) for biomedical applications. Crit Rev Biotechnol 2001; 21(4):219–232.
Ritschel WA. Microemulsion technology in the reformulation of cyclosporine: the reason behind the pharmacokinetic properties of Neoral. Clin Transplant 1996 Aug; 10(4):364–373.
Sable D, Murakawa GJ. Quinolones in dermatology. Dis Mon 2004 Jul; 50(7):381–394.
Sams-Dodd F. Target-based drug discovery: is something wrong? Drug Discov Today 2005, 15 Jan; 10(2):139–147.

Schmidt PC, Christin I. [Effervescent tablets—a nearly forgotten drug form]. Pharmazie 1990 Feb; 45(2):89–101.

Schneider G, Fechner U. Computer-based de novo design of drug-like molecules. Nat Rev Drug Discov 2005 Aug; 4(8):649–663.

Schumacher HR Jr. Ketoprofen extended-release capsules: a new formulation for the treatment of osteoarthritis and rheumatoid arthritis. Clin Ther 1994 Mar–Apr; 16(2):145–159.

Seager H. Drug-delivery products and the Zydis fast-dissolving dosage form. J Pharm Pharmacol 1998 Apr; 50(4):375–382.

Sezaki H. [Drug delivery systems]. Gan To Kagaku Ryoho 1985 Nov; 12(11):2077–2082.

Sharma VK. Comparison of 24-hour intragastric pH using four liquid formulations of lansoprazole and omeprazole. Am J Health Syst Pharm 1999 Dec 1; 56(23 suppl 4): S18–S21. Erratum in Am J Health Syst Pharm 2000, 1 Apr; 57(7):699.

Sharpe M, Ormrod D, Jarvis B. Micronized fenofibrate in dyslipidemia: a focus on plasma high-density lipoprotein cholesterol (HDL-C) levels. Am J Cardiovasc Drugs 2002; 2(2):125–132; discussion 133–134.

Shigeyama M. [Preparation of a gel-forming ointment base applicable to the recovery stage of bedsore and clinical evaluation of a treatment method with different ointment bases suitable to each stage of bedsore]. Yakugaku Zasshi 2004 Feb; 124(2):55–67. Japanese.

Siepmann J, Gopferich A. Mathematical modeling of bioerodible, polymeric drug delivery systems. Adv Drug Deliv Rev 2001, 11 Jun; 48(2–3):229–247.

Singh B, Dahiya M, Saharan V, Ahuja N. Optimizing drug delivery systems using systematic "design of experiments." Part II: retrospect and prospects. Crit Rev Ther Drug Carrier Syst 2005; 22(3):215–294.

Singh B, Kumar R, Ahuja N. Optimizing drug delivery systems using systematic "design of experiments." Part I: fundamental aspects. Crit Rev Ther Drug Carrier Syst 2005; 22(1):27–105.

Singla AK, Chawla M. Chitosan: some pharmaceutical and biological aspects—an update. J Pharm Pharmacol 2001 Aug; 53(8):1047–1067.

Singla AK, Garg A, Aggarwal D. Paclitaxel and its formulations. Int J Pharm 2002, 20 Mar; 235(1–2):179–192.

Sinha VR, Kumria R. Microbially triggered drug delivery to the colon. Eur J Pharm Sci 2003 Jan; 18(1):3–18.

Smart JD. Buccal drug delivery. Expert Opin Drug Deliv 2005 May; 2(3):507–517.

Song H, Guo T, Zhang R, et al. Preparation of the traditional Chinese medicine compound recipe heart-protecting musk pH-dependent gradient-release pellets. Drug Dev Ind Pharm 2002 Nov; 28(10):1261–1273.

Speiser PP. Nanoparticles and liposomes: a state of the art. Methods Find Exp Clin Pharmacol 1991 Jun; 13(5):337–342.

Strickley RG. Solubilizing excipients in oral and injectable formulations. Pharm Res 2004 Feb; 21(2):201–230.

Sun Y, Peng Y, Chen Y, Shukla AJ. Application of artificial neural networks in the design of controlled release drug delivery systems. Adv Drug Deliv Rev 2003, 12 Sep; 55(9):1201–1215.

Swainston Harrison T, Keating GM. Extended-release carbamazepine capsules: in bipolar I disorder. CNS Drugs 2005; 19(8):709–716.

Takada S, Ogawa Y. [Design and development of controlled release of drugs from injectable microcapsules]. Nippon Rinsho 1998 Mar; 56(3):675–679.

Takakura Y, Nishikawa M, Yamashita F, Hashida M. Influence of physicochemical properties on pharmacokinetics of non-viral vectors for gene delivery. J Drug Target 2002 Mar; 10(2):99–104.
Takayama K, Fujikawa M, Nagai T. Artificial neural network as a novel method to optimize pharmaceutical formulations. Pharm Res 1999 Jan; 16(1):1–6.
Takayama K, Fujikawa M, Obata Y, Morishita M. Neural network based optimization of drug formulations. Adv Drug Deliv Rev 2003, 12 Sep; 55(9):1217–1231.
Takeuchi H, Thongborisute J, Matsui Y, Sugihara H, Yamamoto H, Kawashima Y. Novel mucoadhesion tests for polymers and polymer-coated particles to design optimal mucoadhesive drug delivery systems. Adv Drug Deliv Rev 2005, 3 Nov; 57(11):1583–1594. Epub 2005, 16 Sep.
Tobyn M, Staniforth JN, Morton D, Harmer Q, Newton ME. Active and intelligent inhaler device development. Int J Pharm 2004, 11 June; 277(1–2):31–37.
Todd PA, Faulds D. Felodipine: a review of the pharmacology and therapeutic use of the extended release formulation in cardiovascular disorders. Drugs 1992 Aug; 44(2):251–277.
Toguchi H, Ogawa Y, Okada H, Yamamoto M. [Once-a-month injectable microcapsules of leuprorelin acetate]. Yakugaku Zasshi 1991 Aug; 111(8):397–409.
Turker S, Onur E, Ozer Y. Nasal route and drug delivery systems. Pharm World Sci 2004 Jun; 26(3):137–142.
Tye H. Application of statistical "design of experiments" methods in drug discovery. Drug Discov Today 2004, 1 June; 9(11):485–491.
Valenta C, Auner BG. The use of polymers for dermal and transdermal delivery. Eur J Pharm Biopharm 2004 Sep; 58(2):279–289.
Wagstaff AJ, Figgitt DP. Extended-release metformin hydrochloride. Single-composition osmotic tablet formulation. Treat Endocrinol 2004; 3(5):327–332.
Wagstaff AJ, Goa KL. Once-weekly fluoxetine. Drugs 2001; 61(15):2221–2228; discussion 2229–2230.
Wassef NM, Alving CR, Richards RL. Liposomes as carriers for vaccines. Immunomethods 1994 Jun; 4(3):217–222.
Wellington K. Rosiglitazone/Metformin. Drugs 2005; 65(11):1581–1192; discussion 1593–1594.
Wernsdorfer WH. Coartemether (artemether and lumefantrine): an oral antimalarial drug. Expert Rev Anti Infect Ther 2004 Apr; 2(2):181–196.
White NS, Errington RJ. Fluorescence techniques for drug delivery research: theory and practice. Adv Drug Deliv Rev 2005, 2 Jan; 57(1):17–42.
Willems L, van der Geest R, de Beule K. Itraconazole oral solution and intravenous formulations: a review of pharmacokinetics and pharmacodynamics. J Clin Pharm Ther 2001 Jun; 26(3):159–169.
Wissing SA, Kayser O, Muller RH. Solid lipid nanoparticles for parenteral drug delivery. Adv Drug Deliv Rev 2004, 7 May; 56(9):1257–1272.
Wouters J, Ooms F. Small molecule crystallography in drug design. Curr Pharm Des 2001 May; 7(7):529–545.
Yalkowsky SH, Krzyzaniak JF, Ward GH. Formulation-related problems associated with intravenous drug delivery. J Pharm Sci 1998 Jul; 87(7):787–796.
Yeo Y, Park K. Control of encapsulation efficiency and initial burst in polymeric microparticle systems. Arch Pharm Res 2004 Jan; 27(1):1–12.
Yilmaz E, Borchert HH. Effect of lipid-containing, positively charged nanoemulsions on skin hydration, elasticity and erythema—an in vivo study. Int J Pharm 2006, 13 Jan; 307(2):232–238. Epub 2005, 11 Nov.

Young SS, Lam RL, Welch WJ. Initial compound selection for sequential screening. Curr Opin Drug Discov Dev 2002 May; 5(3):422–427.

Zargar A, Basit A, Mahtab H. Sulphonylureas in the management of type 2 diabetes during the fasting month of Ramadan. J Indian Med Assoc 2005 Aug; 103(8):444–446.

Zhang GG, Law D, Schmitt EA, Qiu Y. Phase transformation considerations during process development and manufacture of solid oral dosage forms. Adv Drug Deliv Rev 2004, 23 Feb; 56(3):371–390.

Zimmermann U, Mimietz S, Zimmermann H, et al. Hydrogel-based non-autologous cell and tissue therapy. Biotechniques 2000 Sep; 29(3):564–572, 574, and 576.

8 Chemical Drug Substance Characterization

INTRODUCTION

Lead drug substances might be derived from three sources: chemical, biological, and botanical (including minerals). These compounds (or a mixture of compounds) may be delivered to the preformulation team at a myriad of characterization stages. Whereas the drug discovery group has, by this time, established the pharmacological activity, the synthesis or extraction group has, by this time, not necessarily fully characterized the lead compound. There is also a proposal on the table regarding the prospective drug delivery systems and their routes of administration. The task of the preformulation group therefore starts with the development of a detailed plan for the complete characterization of the lead compound. The depth and breadth of the characterization would depend on the type and source of the compound as well as its destination—the dosage form. Whereas the formulation part is yet to come, the preformulation group must provide lead suggestions on the choice of the excipients through preliminary interaction trials. These studies must be conducted using analytical methods that are established though not necessarily fully validated at this stage.

The regulatory impact of preformulation studies is very significant; as in one format or another, the key component of all regulatory filings involves the complete characterization of the drug substance. These details are provided in the guidance provided by the United States Food and Drug Administration (U.S. FDA) on the preparation of the Chemistry, Manufacturing, and Control package or in the Common Technical Document package as required by the European Agency for Evaluation of Medicinal Products. Whereas the full scope of these documents covers the drug substance and the drug product, the studies conducted at the drug substance level are pivotal to further development. The aim of pharmaceutical development is to design a quality product and a manufacturing process to deliver the product in a reproducible manner. The information and knowledge gained from pharmaceutical development studies provide scientific understanding to support the establishment of specifications and manufacturing controls. Of greatest importance in the development of pharmaceutical studies are the sections devoted to characterization of drug substances.

The physico-chemical and biological properties of the drug substance that can influence the performance of the drug product and its manufacturability, or those that are specifically designed into the drug substance (e.g., crystal engineering, botanical extraction conditions, protein yields at folding stages, etc.), should be identified and discussed. However, the type of properties of the drug substance to be studied is highly dependent on the source of the drug. While a majority of the new drugs are still derived by chemical synthesis, more drugs are now beginning to be sourced from biological sources, particularly through recombinant DNA manufacturing and the recent acknowledgment by the regulatory authorities that botanical (herbal in Europe) drugs should be controlled has resulted in an organized sourcing through botanical means as well. This chapter describes the characterization of drug substances derived by chemical synthesis, wherein the molecules are well defined. The general characteristics described here may also apply to biological and botanical drugs as well; the specific differences will be discussed in a later chapter devoted to those drugs.

SCHEME OF CHARACTERIZATION

Systematic development cycles are more likely to be efficient and should result in a definite specification for the lead compound; Scheme 1 shows a typical flow chart for the characterization that leads to the development of specifications for the lead compound.

Examples of properties that are routinely examined include solubility, water content, particle size, crystal properties, biological structure, chirality, and so on. The compatibility of the drug substance with excipients should be discussed. For products that contain more than one drug substance, the compatibility of the drug substances with each other should also be evaluated. Whereas the dosage form considerations are still to evolve, based on a prospective dosage form, the specifications should include those parameters that may be relevant. For example, if the final dosage form intended is an injectable product, solubility and thermal stability (to autoclaving) are important considerations. Table 1 lists some common study protocols for different dosage forms.

The experience and data accumulated during the preformulation stage proves pivotal to the development of the dosage form based on the specifications developed. A specification is defined as a list of tests, references to analytical procedures, and appropriate acceptance criteria that are numerical limits, ranges, or other criteria for the tests described. It establishes the set of criteria to which a new drug substance or new drug product should conform to be considered acceptable for its intended use. Conformance to specifications means that the drug substance and/or drug product, when tested according to the listed analytical procedures, will meet the listed acceptance criteria. Specifications are critical quality standards

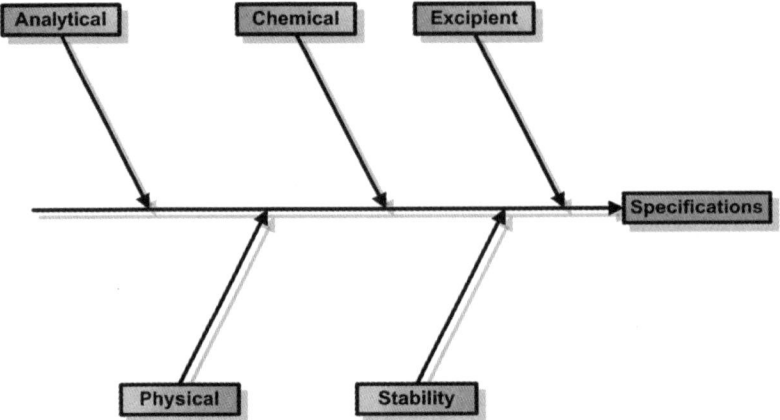

Scheme 1 Steps that lead to the development of specifications for new lead compounds.

Table 1 Study Factors for Various Prospective Dosage Forms

Prospective dosage form	Study factors
Parenteral	Solubility, micellization, thermal stability, chemical stability, packaging component interaction (glass, stoppers), photostability, physical stress (particularly for protein drugs), buffer interactions, and viscosity.
Oral solids	Solubility, dissolution, polymorphism, chirality, particle size, powder flow, chemical stability, photostability, compressibility, hygroscopicity, and excipient interactions.
Oral liquids	Solubility, polymorphic conversions, chirality, excipient interactions, chemical stability, photostability, pH effects, and container interactions (e.g., type III glass).
Semisolids	Solubility, dissolution, particle size, polymorphism, chirality, chemical stability, photostability, viscosity, and excipient interactions.

that are proposed and justified by the manufacturer and approved by regulatory authorities as conditions of approval. It is possible that, in addition to release tests, a specification may list in-process tests, periodic or skip tests, and other tests that are not always conducted on a batch-by-batch basis. When a specification is first proposed, justification should be presented for each procedure and each acceptance criterion included. The justification should refer to relevant development data, pharmacopoeia standards, test data for drug substances and drug products used in toxicology and clinical studies, and results from accelerated and long-term stability studies, as appropriate. Additionally, a reasonable range of expected analytical and manufacturing variability should be considered. Test results from stability and scale-up/validation batches, with emphasis on the primary stability batches, should be considered in setting and justifying specifications.

The U.S. FDA recommends the initiative, process analytical technology (PAT), which applies to both drug substances and drug products (1). PAT is a system for designing, analyzing, and controlling the manufacture through timely measurements (i.e., during processing) of critical quality and performance attributes of raw and in-process materials and processes with the goal of ensuring final product quality. The goal of PAT is to understand and control the manufacturing process, which is consistent with our current drug quality system: *quality cannot be tested into products; it should be built-in or should be by design.* It is important to note that the term *analytical* in PAT is viewed broadly to include chemical, physical, microbiological, mathematical, and risk analysis conducted in an integrated manner. There are many current and new tools available that enable scientific, risk-managed pharmaceutical development, manufacture, and quality assurance. These tools, when used within a system can provide effective and efficient means for acquiring information to facilitate process understanding, develop risk-mitigation strategies, achieve continuous improvement, and share information and knowledge. In the PAT framework, these tools can be categorized as:

- Multivariate data acquisition and analysis tools
- modern process analyzers or process analytical chemistry tools
- process and endpoint monitoring and control tools
- continuous improvement and knowledge management tools

An appropriate combination of some, or all, of these tools may be applicable to a single-unit operation, or to an entire manufacturing process and its quality assurance. A variety of sophisticated softwares, such as RAPID-Pharma (2) are now available to consolidate many functions required to manage the initiatives related to PAT.

Specifications
The following tests and acceptance criteria are considered generally applicable to all new drug substances.

Description
Description is a qualitative statement about the state (e.g., solids and liquids) and color of the new drug substance. If any of these characteristics change during storage, this change should be investigated and appropriate action needs to be taken.

Identification
Identification testing should optimally be able to discriminate between compounds of closely related structures that are likely to be present. Identification tests should be specific for the new drug substance, for example, IR spectroscopy. Identification solely by a single chromatographic retention time, for example, is not regarded as being specific. However, the use of two chromatographic procedures, where the separation is based on different principles or a combination of tests into a single procedure, such as high-pressure liquid chromatography (HPLC)/UV diode array, HPLC/mass spectroscopy (MS), or gas chromatography (GC)/MS is generally acceptable. If the new drug substance is a salt, identification testing should be specific for the individual ions. An identification test that is specific for the salt itself should suffice.

Chirality
New drug substances that are optically active may also need specific identification testing or performance of a chiral assay. For chiral drug substances that are developed as a single enantiomer, control of the other enantiomer should be considered in the same manner as for other impurities. However, technical limitations may preclude the same limits of quantification or qualification from being applied. Assurance of control also could be given by appropriate testing of a starting material or intermediate, with suitable justification. An enantioselective determination of the drug substance should be part of the specification. It is considered acceptable in order to achieve this either through use of a chiral assay procedure or by the combination of an achiral assay together with appropriate methods for controlling the enantiomeric impurity. For a drug substance developed as a single enantiomer, the identity test(s) should be capable of distinguishing both enantiomers and the racemic mixture. For a racemic drug substance, there are generally two situations where a stereospecific identity test is appropriate for the release/acceptance testing; one, where there is a significant possibility that the enantiomer might be substituted for the racemate, or second, when there is evidence that preferential crystallization may lead to unintentional production of a nonracemic mixture. Scheme 2 shows a decision-making tree on the studies that are needed when a chiral substance is suspected.

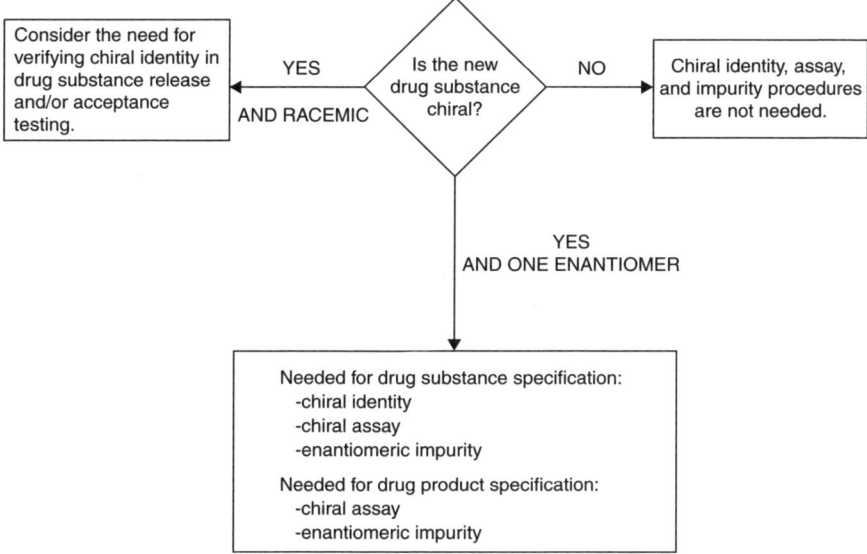

Scheme 2 Decision tree for chiral compound resolution. *Source*: Courtesy of Pharmquest Corporation, Mountain View, California, U.S.A.

Assay
A specific, stability-indicating procedure should be developed to determine the content of the new drug substance. In many cases, it is possible to employ the same procedure (e.g., HPLC) for both assay of the new drug substance and quantification of the impurities. In cases where use of a nonspecific assay is justified, other supporting analytical procedures should be used to achieve overall specificity. For example, where titration is adopted to assay the drug substance, the combination of the assay and a suitable test for impurities should be used. Assay methods clearly define the degradation products and their limits. Scheme 3 shows a decision tree for handling the degradation products in the drug substance.

Impurities
Organic and inorganic impurities and residual solvents are included in this category. Scheme 4 shows a decision tree that can be used to decide the disposition of impurities in the drug substance.

Physicochemical Properties
These are properties, such as the pH of an aqueous solution, melting point/range, and refractive index. The procedures used for the measurement of these properties are usually unique and do not need much elaboration, for example, capillary melting point and Abbe refractometry. The tests performed in this category should be determined by the physical nature of the new drug substance and its intended use.

Particle Size
For some new drug substances intended for use in solid or suspension drug products, particle size can have a significant effect on dissolution rates, bioavailability,

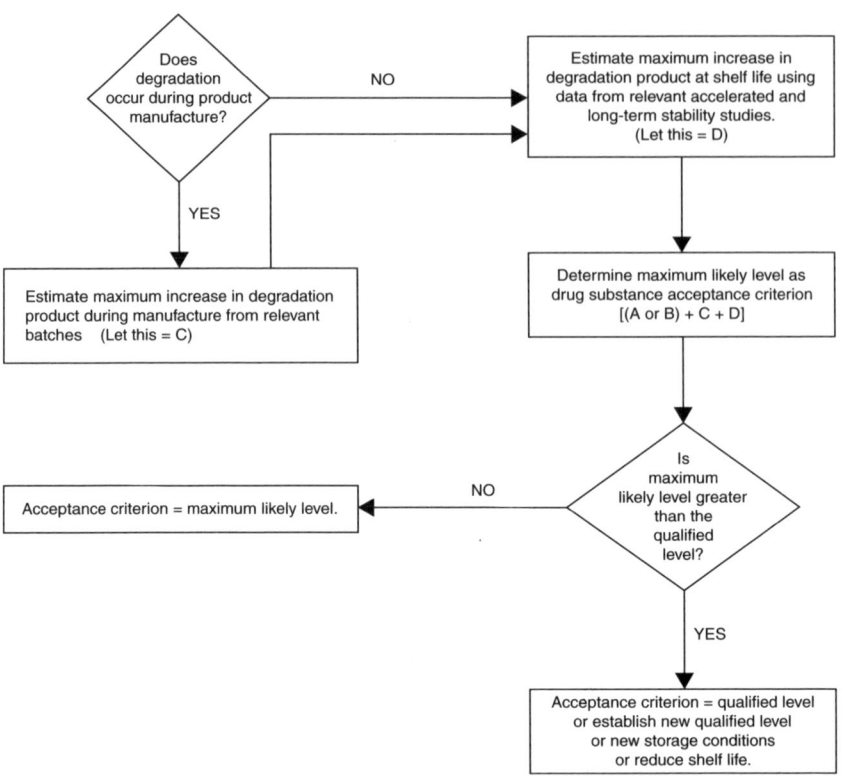

Scheme 3 Decision tree for disposition of degradation products. *Source*: Courtesy of Pharmquest Corporation, Mountain View, California, U.S.A.

and/or stability. In such instances, testing for particle size distribution should be carried out using an appropriate procedure, and acceptance criteria should be provided. Scheme 5 is a decision tree for the disposition of particle size variations in the drug substances.

Polymorphic Forms
Some new drug substances exist in different crystalline forms that differ in their physical properties. Polymorphism may also include solvation or hydration products (also known as pseudopolymorphs) and amorphous forms. Differences in these forms could, in some cases, affect the quality or performance of the new drug products. In such cases, the bioavailability stability can be altered requiring choice of specific stable solid dosage forms.

Microbiology
Microbiological attributes are required where preparation and storage can significantly compromise microbiological quality. Scheme 7 shows a decision tree for the disposition of microbiologically related attributes.

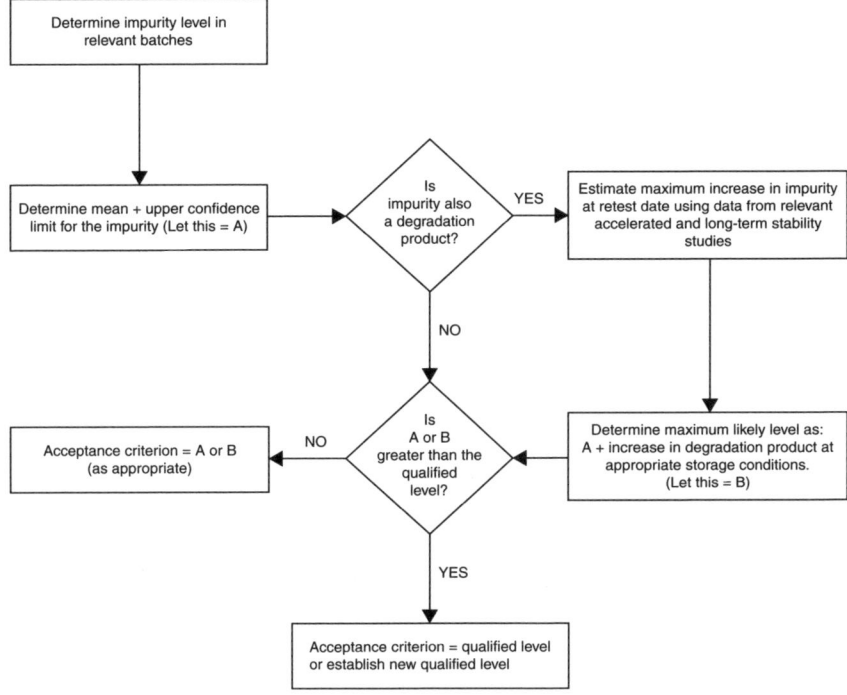

Scheme 4 Decision tree for disposition of impurities in the drug substances. *Source*: Courtesy of Pharmquest Corporation, Mountain View, California, U.S.A.

Excipients
The excipients chosen, their concentration, and the characteristics that can influence the drug product performance (e.g., stability and bioavailability) or manufacturability should be discussed relative to the respective function of each excipient. Compatibility of excipients with other excipients, where relevant (e.g., combination of preservatives in a dual preservative system), should be established. The ability of excipients (e.g., antioxidants, penetration enhancers, disintegrants, and release-controlling agents) to provide their intended functionality, and the intended drug product shelf life that is needed for performance throughout, should also be demonstrated. The information on excipient performance can be used, as appropriate, to justify the choice and quality attributes of the excipient, and to support the justification of the drug product specification. Information to support the safety of excipients, when appropriate, should be cross-referenced.

Stability Evaluation
The International Conference on Harmonization (ICH) guidelines regarding the quality of drug substances are specific and require frequent referral to keep the preformulation activities current with the regulatory requirements. The pertinent guidelines are listed in the following.

- Q1A(R2) Stability Testing of New Drug Substances and Products (Issued 11/2003, Posted 11/20/2003);

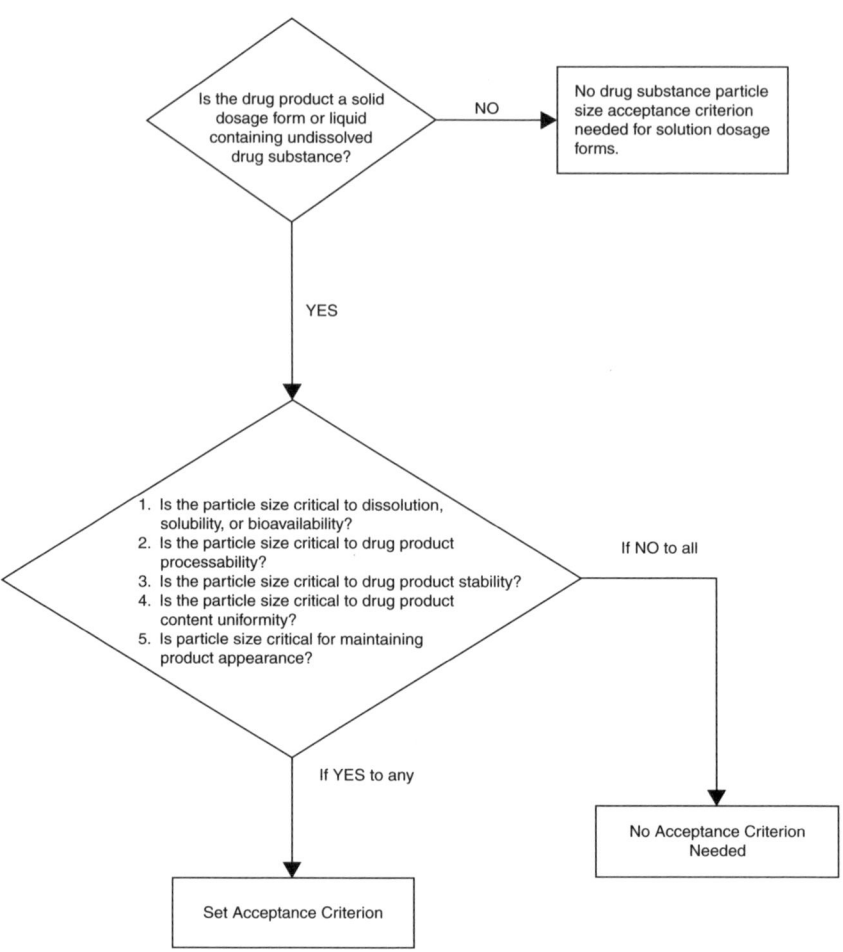

Scheme 5 Decision tree for the disposition of particle size variation. *Source*: Courtesy of Pharmquest Corporation, Mountain View, California, U.S.A.

- Q1B Photostability Testing of New Drug Substances and Products (Issued 11/1996, Reposted 7/7/1998);
- Q1D Bracketing and Matrixing Designs for Stability Testing of New Drug Substances and Products (Issued 1/2003, Posted 1/15/2003);
- Q1E Evaluation of Stability Data (Issued 6/2004, Posted 6/7/2004);
- Q3A Impurities in New Drug Substances (Issued 2/10/2003, Posted 2/10/2003);
- Q3B(R) Impurities in New Drug Products (Issued 11/2003, Posted 11/13/2003);
- Q5C Quality of Biotechnological Products: Stability Testing of Biotechnological/ Biological Products;
- Q6A ICH; Guidance on Q6A Specifications: Test Procedures and Acceptance Criteria for New Drug Substances and New Drug Products: Chemical Substances (12/29/2000);

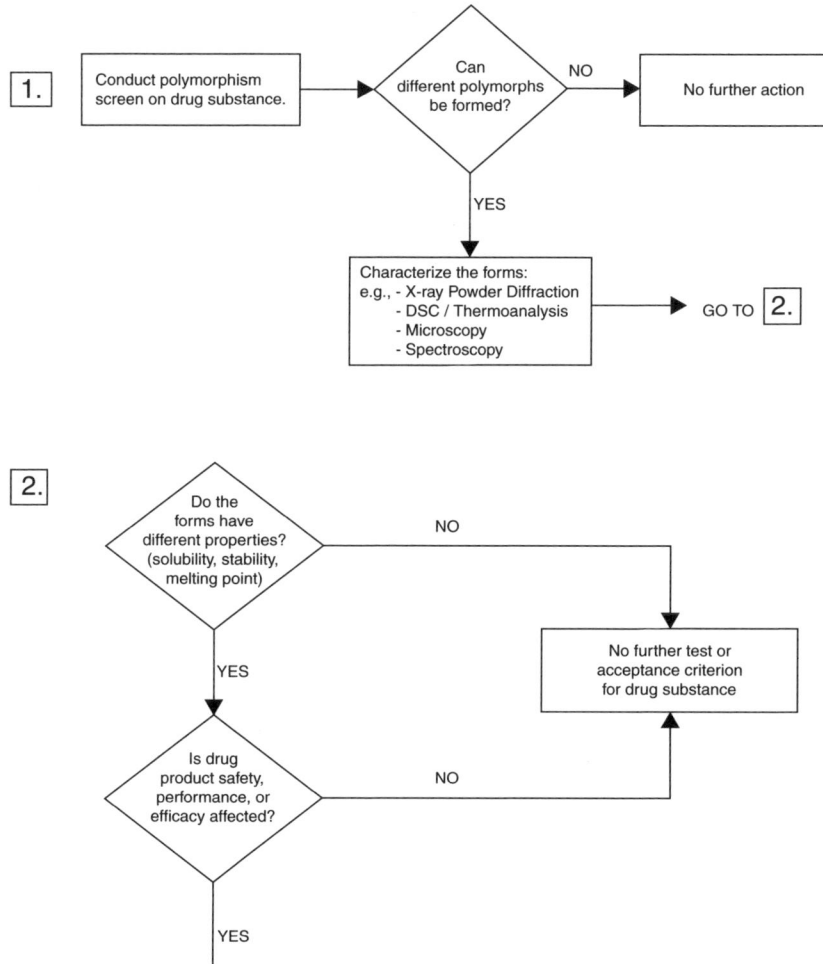

Scheme 6 Decision tree for the disposition of polymorphism. *Source*: Courtesy of Pharmquest Corporation, Mountain View, California, U.S.A.

- Q6B Specifications: Test Procedures and Acceptance Criteria for Biotechnological/Biological Products (Issued 8/1999, Posted 12/14/2001);
- Q7A Good Manufacturing Practice Guidance for Active Pharmaceutical Ingredients (Issued 8/2001, Posted 9/24/2001); ICH, "Q3A Impurities in New Drug Substances," 1996.

Knowledge about the chemical and physical stabilities of a candidate drug in the solid and liquid states is extremely important in drug development for a number of reasons. In the longer term, the stability of the formulation will dictate the shelf life

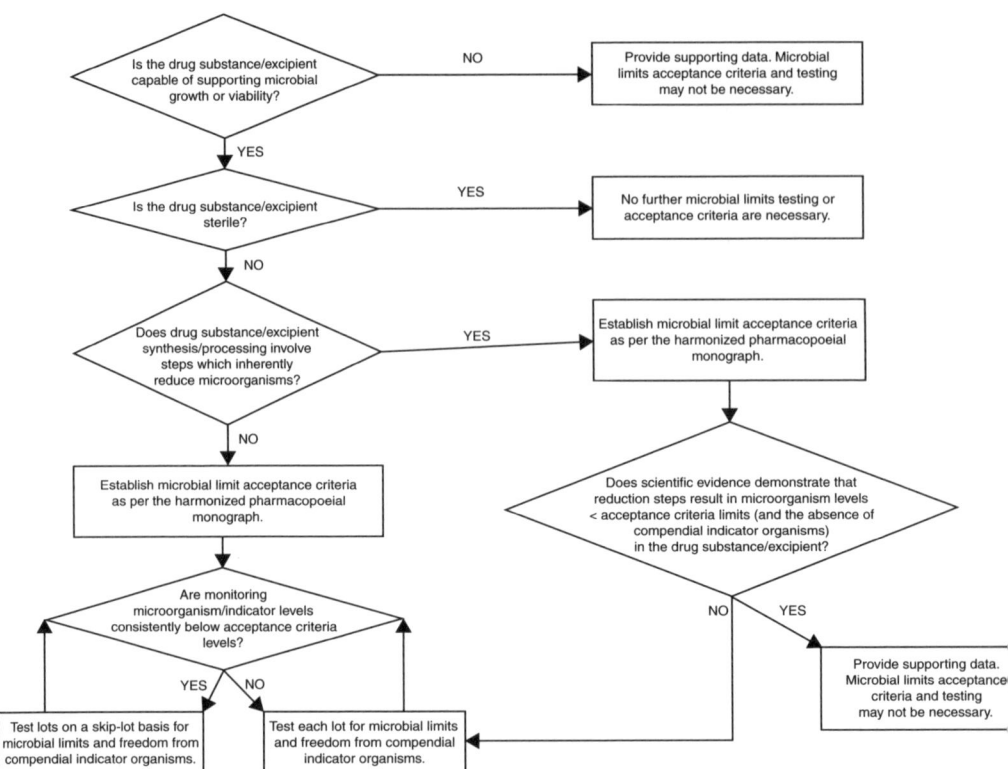

Scheme 7 Decision tree for the disposition of microbiological attributes. *Source*: Courtesy of Pharmquest Corporation, Mountain View, California, U.S.A.

of the marketed product; however, to achieve this stable formulation, careful preformulation work would have characterized the drug substance such that a rational choice of conditions and excipients is available to the formulation team.

Candidate drugs being evaluated for development are often one of a series of related compounds that may have similar chemical properties, that is, similar paths of degradation may be deduced. However, this rarely tells us the rate at which they decompose, which is of more importance to pharmaceutical development terms. To elucidate their stability with respect to temperature, pH, light, and oxygen, a number of experiments need to be performed. The major objectives of the preformulation team are, therefore, to identify conditions to which the compound is sensitive, and to identify degradation profiles under these conditions.

The major routes of drug degradation in solution are via hydrolysis, oxidation, or photochemical degradation.

Hydrolysis

Hydrolysis is a two-stage process, where a nucleophile, such as water or the hydroxyl ion adds to, for example, an acyl carbon, to form an intermediate from which the leaving group breaks away in the second stage. The structure of the compound affects the hydrolysis rate; the stronger the conjugate acid that leaves, the faster the reaction.

Many drugs contain the functional groups, which are extremely prone to hydrolysis; the replacement of the X group with the hydroxyl moiety from water takes place as shown in the following to yield the two types of hydrolysis reactions:

$$\underset{X}{\overset{R}{>}}C=O \quad \text{where } X = OR \text{ (ester)}, NR^1R^2 \text{(amide)} \\ \text{or related systems (Type I)}$$

$$\underset{X}{\overset{R}{>}}C=O \quad \text{where } X = \text{halogen or good leaving group (Type II)}$$

The hydrolysis of Type I compounds can be catalyzed both by acids and bases; hence, pH control of formulations has a strong influence on the rates of decomposition. An example of acid-catalyzed hydrolysis is expressed as follows.

(1)

The base-catalyzed hydrolysis takes place as follows.

(2)

Common examples of drugs with hydrolysis-prone function groups are shown in Table 2.

The common examples are ester (aspirin), thiol ester (spirolactone), amide (chloramphenicol), sulfonamide (sulfapyrazine), imide (phenobarbital), lactam (methicillin), lactone (spiranolactone), and halogenated aliphatic (chlorambucil).

Table 2 Common Functional Groups of Drugs

Functional group	Name	Example
—C— (single bonds)	Alkane	$CH_3CH_2CH_3$ (propane)
C=C	Alkene	$CH_3CH=CH_2$ (propene)
C≡CH	Alkyne	CH_3CCH (propyne)
F, Cl, Br, or I	Alkyl halide	CH_3Br (methyl bromide)
—OH	Alcohol	CH_3CH_2OH (ethanol)
—O—	Ether	CH_3OCH_3 (dimethyl ether)
—NH_2	Amine	CH_3NH_2 (methyl amine)
—C(=O)—H	Aldehyde	CH_3CHO (acetaldehyde)
—C(=O)—	Ketone	CH_3COCH_3 (acetone)
—C(=O)—Cl	Acyl chloride	CH_3COCl (acetyl chloride)
—C(=O)—OH	Carboxylic acid	CH_3CO_2H (acetic acid)
—C(=O)—O—	Ester	$CH_3CO_2CH_3$ (methyl acetate)
—C(=O)—NH_2	Amide	CH_3NH_2 (acetamide)

Examples of hydrolysis include drugs, such as atropine, procaine, and so on. Atropine undergoes hydrolysis at pH values higher than four as shown in the following.

$$\text{atropine} \xrightleftharpoons[-H_2O]{+H_2O} \text{tropine} + \text{tropic acid} \qquad (3)$$

This decomposition can be accounted for the base-catalyzed hydrolysis. At pH values below four, the acid-catalyzed decomposition that causes acceleration in hydrolysis is observed due to dehydration of the alcoholic portion of the molecule followed by hydrolysis.

$$\text{(4)}$$

As a consequence, it is recommended to preserve aqueous atropine solutions buffered at around pH 3.5–4. If stored in lime glass, the stability is reduced as a result of alkalinity of the glass.

Another good example of hydrolysis reaction is the anesthetic procaine, which is prone to specific acid-catalyzed decomposition (of the protonated form) at a pH value below 2.5, a specific base-catalyzed hydrolysis between pH 5.5 and 8.5, titration around pH 8.5–11.0 and finally, the base-catalyzed hydrolysis of the free base at a pH value above 12.0. Maximum stability occurs at pH 3.5.

$$\text{(5)}$$

Commercial preparations of procaine are limited to injectable forms. As water is the solvent of choice for injectable drugs, procaine can be protected from hydrolysis by increasing the solubility of procaine in a nonaqueous environment and protecting (shielding) the drug from hydrolysis by employing surfactants to micellize the drug.

Degradation by hydrolysis is affected by a number of factors including solution pH, buffer salts, ionic strength, and so on. In addition, the presence of co-solvents, complexing agent's surfactant can also affect this type of degradation. The most important factor is the solution pH as the hydroxyl ions are stronger nucleophiles than water; thus, degradation reactions are usually faster in alkaline solution than in water. At low pH, protons can also catalyze hydrolysis reactions as a result of specific acid–base catalysis. Besides these two ions, buffer ions such as acetate or citric acid can also catalyze degradation, where the effect is known as general acid–base degradation. Therefore, pH adjustment should make the right choice of buffers and where possible, in smallest quantities (strength).

Oxidation
The second most common way a compound can decompose in solution is via oxidation. Reduction/oxidation (redox) reactions involve either the transfer of oxygen or hydrogen atoms or the transfer of electrons. Oxidation is promoted by the presence of oxygen, and the reaction can be initiated by the action of heat, light, or trace metal ions that produce organic free radicals. These radicals propagate the oxidation reaction, which proceeds until inhibitors destroy the radicals or until side reactions eventually break the chain. A simple test to see if a drug substance is prone to oxidation, is to bubble air or oxygen or treat with hydrogen peroxide and estimate the level of degradation.

The sensitivity of each new drug entity to atmospheric oxygen must be evaluated to establish if the final product should be packaged under inert atmospheric conditions and if it should contain an antioxidant. Sensitivity to oxidation of a solid drug can be ascertained by investigating its stability in an atmosphere of high oxygen tension. Usually a 40% oxygen atmosphere allows for a rapid evaluation. Results should be compared against those obtained under inert or ambient atmospheres.

To test whether a compound is sensitive to oxygen, simply bubble air through the solution, or add hydrogen peroxide, and assess the amount of degradation that takes place.

Focusing on the functional groups in a molecule allows us to recognize patterns in the behavior of related compounds. Consider what we know about the reaction between sodium metal and water, as an example.

$$2Na(s) + 2H_2O(l) \rightarrow H_2(g) + 2Na^+(aq) + 2OH^-(aq) \tag{6}$$

We can divide this reaction into two half reactions. One involves the oxidation of sodium metal to form sodium ions.
Oxidation:

$$Na \rightarrow Na^+ + e^- \tag{7}$$

The other involves the reduction of an H^+ ion in water to form a neutral hydrogen atom that combines with another hydrogen atom to form a hydrogen molecule.
Reduction:

$$2\{H-O-H\} \rightarrow H^+ + {}^-O-H^- \tag{8}$$
$$2H^+ + 2e^- \rightarrow 2H \rightarrow H_2$$

Once we recognize that water contains an —OH functional group, we can predict what might happen when sodium metal reacts with an alcohol that contains the same functional group. Sodium metal reacts with methanol (CH_3OH), for example, to give hydrogen gas and a solution of the Na^+ and CH_3O^- ions dissolved in this alcohol.

$$2Na(s) + 2CH_3OH(l) \rightarrow H_2(g) + 2Na^+(alc) + 2CH_3O^-(alc) \tag{9}$$

Because of the involvement of transfer of electrons, the reaction between sodium metal and either water or an alcohol are examples of oxidation–reduction reactions.

But what about the following reaction, in which hydrogen gas reacts with an alkene in the presence of a transition metal catalyst to form an alkane?

$$\begin{array}{c} H_2C=CH_2 + H_2 \xrightarrow{Ni} H_3C-CH_3 \end{array} \quad (10)$$

There is no change in the number of valence electrons in any of the atoms in this reaction. Both before and after the reaction, each carbon atom shares a total of eight valence electrons and each hydrogen atom shares two electrons. Instead of electrons, this reaction involves the transfer of atoms—in this case, the hydrogen atoms. There are so many atom-transfer reactions that chemists developed the concept of *oxidation number* to extend the idea of oxidation and reduction to reactions in which electrons are not necessarily gained or lost. Oxidation involves an increase in the oxidation number of an atom whereas reduction occurs when the oxidation number of an atom decreases.

During the transformation of ethene into ethane, there is a *decrease* in the oxidation number of the carbon atom. This reaction therefore involves the *reduction* of ethene to ethane.

$$\begin{array}{c} \overset{-2}{H_2C}=CH_2 + H_2 \xrightarrow{Ni} H_3\overset{-3}{C}-CH_3 \end{array} \quad (11)$$

Reactions in which none of the atoms undergo a change in oxidation number are called *metathesis reactions*. Consider the reaction between a carboxylic acid and an amine, for example.

$$CH_3CO_2H + CH_3NH_2 \longrightarrow CH_3CO_2^- + CH_3NH_3^+ \quad (12)$$

Or the reaction between an alcohol and hydrogen bromide.

$$CH_3CH_2OH + HBr \longrightarrow CH_3CH_2Br + H_2O \quad (13)$$

These are metathesis reactions because there is no change in the oxidation number of any atom in either reaction.

The oxidation numbers of the carbon atoms in a variety of compounds are given in Table 3.

The oxidation numbers given in Table 3 can be used to classify organic reactions as either oxidation–reduction reactions or metathesis reactions. Because electrons are neither created nor destroyed, oxidation cannot occur in the absence of reduction, or vice versa. It is often useful, however, to focus attention on one component of the reaction and ask: Is that substance oxidized or reduced? Assigning oxidation numbers to the individual carbon atoms in a complex molecule can be difficult. Fortunately, there is another way to recognize oxidation–reduction reactions in organic chemistry.

Oxidation occurs when hydrogen atoms are removed from a carbon atom or when an oxygen atom is added to a carbon atom. Reduction occurs when hydrogen

Table 3 Typical Oxidation Numbers of Carbon

Functional group	Example	Oxidation number of carbon in the example
Alkane	CH_4	−4
Alkyl lithium	CH_3Li	−4
Alkene	$H_2C=CH_2$	−2
Alcohol	CH_3OH	−2
Ether	CH_3OCH_3	−2
Alkyl halide	CH_3Cl	−2
Amine	CH_3NH_2	−2
Alkyne	HC CH	−1
Aldehyde	H_2CO	0
Carboxylic acid	HCO_2H	2
Carbon dioxide	CO_2	4

atoms are added to a carbon atom or when an oxygen atom is removed from a carbon atom or when an oxygen atom is removed from a carbon atom. An alkene is reduced, for example, when it reacts with hydrogen to form the corresponding alkane.

$$CH_2=CHCH_3 + H_2 \xrightarrow{Ni} CH_3CH_2CH_3 \quad (14)$$

The scheme that follows Eq. 15 provides a useful guide to the oxidation–reduction reactions of organic compounds. Each of the arrows in this figure involves a two-electron oxidation of a carbon atom along the path toward carbon dioxide. A line is drawn through the first arrow because it is impossible to achieve this transformation in a single step.

$$\underset{-4}{CH_4} \not\rightarrow \underset{-2}{CH_3OH} \rightarrow \underset{0}{\overset{O}{\underset{\|}{HCH}}} \rightarrow \underset{+2}{\overset{O}{\underset{\|}{HCOH}}} \rightarrow \underset{+4}{CO_2} \quad (15)$$

Kinetics of Degradation
In real-time stability tests, a product is stored at recommended storage conditions and monitored for a period of time (t_{test}). The product will degrade below its specification, at some time, denoted t_s, and it must also be ensured that t_s is less than or equal to t_{test}. The estimated value of t_s can be obtained by modeling the degradation pattern. Good experimental design and practices are needed to minimize the risk of biases and reduce the amount of random error during data collection. Testing should be performed at time intervals that encompass the target shelf life and must be continued for a period after the product degrades below specification. It is also required that at least three lots of material be used in stability testing to capture lot-to-lot variation, which is an important source of product variability.

The true degradation pattern of a certain product, assuming that it degrades via a first-order reaction, can be described as follows.

$$D = \alpha \exp(-\delta t) \quad (16)$$

The observed result (Y) of each has a random component φ associated with it, and a random experimental error, ε.

$$Y = D + \phi + \varepsilon = \alpha \exp(-\delta t) + \phi + \varepsilon \quad (17)$$

Both α and δ represent the fixed parameters of the model that need to be estimated from the data, while φ and ε are assumed to be normally distributed with mean = 0, and standard deviations of $\sigma\varphi$ and $\sigma\varepsilon$, respectively. Equation 17 is a nonlinear mixed model.

Let C represent a critical level where the essential performance characteristics of the product are within the specification. A product is considered to be stable when $Y \geq C$. The product is not stable when $Y < C$, while $Y < C$ occurs at ts. The manufacturer determines the value of C. The estimated time during which the product is stable is calculated as:

$$t_s = \{\ln C - \ln a\}/-d \tag{18}$$

where a and d are the estimated values of the intercept and the degradation rate. The standard error of the estimated time can be obtained from the Taylor series approximation method and is used to calculate confidence limits. The labeled shelf life of the product is the lower confidence limit of the estimated time.

In accelerated stability testing, a product is stored at elevated stress conditions. Degradation at recommended storage conditions could be predicted based on the degradation at each stress condition and known relationships between the acceleration factor and the degradation rate. A product may be released based on accelerated stability data, but the real-time testing must be done in parallel to confirm the shelf life prediction. Sometimes, the amount of error of the predicted stability is so large that the prediction itself is not useful. The experiments have to be designed carefully to reduce this error. It is recommended that several products should be stored at various acceleration levels to reduce prediction error. Increasing the number of levels is a good strategy for reducing error.

Temperature is probably the most common acceleration factor used for chemicals, pharmaceuticals, and biological products as its relationship with the degradation rate is well characterized by the Arrhenius equation. This equation describes a relationship between temperature and the degradation rate as in Eq. 19.

$$\delta = A \exp\left(\frac{-E_a}{RT}\right) \tag{19}$$

where A is the Arrhenius factor, E_a is the activation energy, R is the gas constant, and T is the temperature.

This relationship can be used in accelerated stability studies when the following conditions are met:

- A zero- or first-order kinetic reaction takes place at each elevated temperature and at the recommended storage temperature.
- The same model is used to fit the degradation patterns at each temperature.

These requirements do not fully guarantee that the Arrhenius equation can be used to predict the degradation rate at storage temperature, but they are a good start. The analytical accuracy should not be compromised during the course of the study to distinguish between the degradation rates at each temperature.

Select temperature levels based on the nature of the product and the recommended storage temperature. The selected temperatures should stimulate relatively fast degradation and quick testing, but not destroy the fundamental characteristics of the product. It is not reasonable to test at very high temperatures for a very short period of time, as the mechanisms of degradation at high temperatures may be very different than those at the recommended storage temperature. Choose the adjacent levels appropriately so that degradation trends are larger than experimental variability.

Choosing levels depends on the nature of the product and analytical accuracy, but other practical implications also need to be considered. Testing should be performed at time intervals that encompass the target stability at each elevated temperature. Acquire some data below C so that the degradation trend can be determined. Humidity and pH can be used along with temperature to accelerate degradation, but modeling of multifactor degradation is very complex.

Assuming that the degradation pattern follows a first-order reaction as described in Eq. 17, the Arrhenius equation (Eq. 19) can be used to predict the degradation rate at the recommended storage temperature. First, an acceleration factor, λ, is calculated as the ratio of the degradation rate at elevated temperature to the degradation rate at storage temperature. This ratio, which can be worked out easily from Eq. 17, can be expressed as:

$$\lambda = \exp\left[\frac{E_a}{0.00199}\left(\frac{1}{T_s} - \frac{1}{T_e}\right)\right] \quad (20)$$

where T_e is the elevated temperature and T_s is the storage temperature.

The true degradation pattern at storage temperature can be expressed as:

$$D = \alpha \exp(-\delta\lambda t) \quad (21)$$

Where, λ is obtained from accelarated stability tests. The testing result (Y) will include random components representing a lot-to-lot variability and experimental error. Once the estimates of α and δ are obtained, stability time is calculated in a similar fashion as in real-time stability testing. Shelf life is the lower confidence limit of the estimated time.

Activation energy is usually estimated from the accelerated stability data. However, when the activation energy is known, the degradation rate at storage temperature may be predicted from data collected at only one elevated temperature. This practice is sometimes preferred in the industry as it reduces the size and time of accelerated stability tests. Experience indicates that some pharmaceutical analytes have activation energies in the range of 10–20 kcal/mol, but it is unlikely you will have precise information or be able to make assumptions about the activation energy of a certain product.

The bracket method is a straightforward application of the Arrhenius equation that can be used if the value of the activation energy is known. Assuming that stability of a product at 50°C is 32 days, and it will be stored at 25°C, then, $t_e = 32$ days, $T_e = 273 + 50°C = 323$ K, and $T_s = 273 + 25°C = 298$ K. We know that activation energy is $E_a = 10$ kcal/mol. Stability at recommended storage temperature is calculated as:

$$t_s = \lambda t_e = \exp\left[\frac{E_a}{0.00199}\left(\frac{1}{T_s} - \frac{1}{T_e}\right)\right]$$

$$t_e = \exp\left[\frac{10}{0.00199}\left(\frac{1}{298} - \frac{1}{323}\right)\right](32) = 118 \text{ days} \quad (22)$$

Calculated stability is highly dependent on the value of the activation energy. A stability of 435 days results when $E_a = 20$ kcal/mol.

The bracket method should not be confused with bracketing, which is an experimental design that allows one to test a minimum number of samples at extremes of certain factors, such as strength, container size, and container fill. Bracketing assumes that the stability of any intermediate level is represented by the stability of the extremes and testing at those extremes is performed at all time points.

The Q-Rule states that the degradation rate decreases by a constant factor when temperature is lowered by certain degrees. The value of Q is typically set at two, three, or four. This factor is proportional to the temperature change Q_n, where n equals the temperature change in °C divided by 10°C. As 10°C is the baseline temperature, the Q-Rule is sometimes referred to as Q_{10}.

To illustrate the application of the Q-Rule, let us assume that the stability of a product at 50°C is 32 days. The recommended storage temperature is 25°C and $n = (50 - 25)/10 = 2.5$. Let us set an intermediate value of $Q = 3$. Thus, $Q_n = (3)2.5 = 15.6$. The predicted shelf life is 32 days $\times$ 15.6 = 500 days. This approach is more conservative when lower values of Q are used. Both Q-Rule and the bracket methods are rough approximations of stability. They can be effectively used to plan elevated temperature levels and the duration of testing in the accelerated stability testing protocol.

Theoretically, the Arrhenius equation does not apply when more than one kind of molecule is involved in the reactions. However, if the degradation rate and temperature are linearly related, the prediction of shelf life can be approximated by the Arrhenius equation. In a polynomial model to fit the degradation,

$$D = \beta_0 + (\beta_1 + \beta_2 t)t \tag{23}$$

where β_0, β_1, and β_2 are the parameters of the second-degree polynomial and t is the time. The degradation rate is a function of time, which is not constant in this case.

$$\delta = \beta_1 + \beta_2 t \tag{24}$$

Degradation at storage temperature can be predicted from the degradation at elevated temperatures as

$$D = \beta_0 + (\beta_1 + \beta_2 t)t\lambda \tag{25}$$

The acceleration factor, λ, is based on the Arrhenius equation. Statistical tests indicated that the use of this equation was appropriate in this case. Shelf life predictions were also verified by real-time stability testing results.

Similar Products

When most of the assumptions required to use the Arrhenius equation are not satisfied, comparisons with a product of a known stability are performed to assess shelf life. This approach requires having a similar product with a known shelf life to be used as a control. The new or test product is expected to demonstrate similar behavior to the control as they belong to the same family and have the same kinetics of degradation. Side-by-side testing of the control and test products at different elevated temperatures is then performed. It is necessary to assume that the same model can represent the degradation pattern at each elevated and storage temperature.

If the degradation patterns of the test and control samples at the same elevated temperatures are not statistically different, it can be assumed that they will degrade similarly at the storage temperatures. The closer the elevated temperatures are to the storage temperatures, the more confident we can be in making this statement. The experimental protocols used are similar to the protocols used with the Arrhenius equation. Degradation patterns of a family of products at certain elevated temperatures can be modeled and used to check the behavior of a new product that belongs to the family.

The complications in calculations arise mainly because of degradation models that are usually nonlinear mixed models, where a lot-to-lot variability is the random component. Estimation of the parameters of the models is important for the accuracy of shelf life predictions. It is recommended to use the maximum likelihood (ML) approach to estimate these parameters. As no closed-form solutions for ML estimates exist, an iterative procedure is performed, starting with some initial values for the parameters and updating them until differences between consecutive iterations are minimal and the estimates converge to their final value. Initial values are usually chosen by experience. The closer these values are to the final values, the faster the model will converge. A suitable program is Nonlinear Mixed Model procedure of statistical analysis system for data analysis (3). Values of the real-time stability model converge relatively quickly, while several initial values for the parameters of the accelerated model are tried before they converge. Statistical theory and the applicability of ML estimation are common in the literature, and many computer routines are available to facilitate data analysis. However, experience with the modeling and estimation processes is necessary, as any unexpected results must be appropriately interpreted. It is quite easy to get useless numbers from a computer run.

Due to insufficient drug quantity, a complete degradation profile will probably not be possible to construct during prenomination studies, but it should be possible to assess the stability of the candidate drug at a few pH values (acid, alkaline, and neutral) to establish the approximate stability of the compound with respect to hydrolysis. To accelerate the reaction, temperature elevation will probably be necessary to generate the data. Although it is difficult to assign a definite temperature for these studies, 50–90°C in the first instance is a reasonable compromise. This should be followed by extrapolation via the Arrhenius equation to 25°C. Hydrolytic stability of >100 days at 25°C should be taken as a goal of these studies. In terms of candidate drug selection, if all other factors are equal, the compound that is most stable should be the one taken forward into development.

Normally, the stability of solutions is assessed using HPLC to determine the amount of decomposition of a compound with time. However, microcalorimetry can determine decomposition reactions with an annual degradation rate of 0.03%, that is, a half life of 2200 years! However, the nature and form of the reaction and the calorimetric output need to be assessed extremely carefully, and this can require careful work-up of the technique to bring it into routine use.

Solid-State Stability
The solid-state degradation of candidate drugs, particularly in the candidate selection phase, is an important consideration, as degradation rates as slow

as 0.5% per year at 25°C may affect the development of the compound. Solid-state degradation reactions can be complex and can involve both oxidation and hydrolysis together. As with solutions, solids can also exhibit instability due to the effects of light. This is further complicated by the fact that in solids, these reactions usually only occur on the surface. Three phases have been identified in solid-state degradation: the lag, acceleration, and deceleration phases. Depending on the conditions of temperature and the humidity to which the solid is exposed, the acceleration phase may follow zero-, first- or higher-orders. A general equation has been proposed to describe the process.

$$\frac{d[D]}{dt} = k\alpha 1 - x(1-\alpha)1 - y \tag{26}$$

where α is the fraction of the reaction that has occurred at time t, such that $\alpha = 0$ when $t = 0$, and $a = 1$ at $t = \infty$; k is the rate constant; and x and y are constants characteristic of the reaction rate law, that is, when $x = y = 1$, the reaction is zero order. If, however, $x = 1$ and $y = 0$, the reaction is first order. If x and y have fractional values, the reaction will be autocatalytic.

To accelerate the degradation so that the amount degraded becomes quantifiable in a typically calculation using the Arrhenius equation, the assumption made during these studies is that the degradation mechanism at higher temperature is the same as that at 25°C. However, this need not be the case, and a nonlinear Arrhenius plot may be an indication of change of mechanism as the temperature is increased. Furthermore, many compounds are hydrates that dehydrate at higher temperature, which can change the degradation mechanism in the solid state.

The effect of moisture uptake in solid stability state depends on the extent of available water. Where the quantity is limited, water is used up during the degradation reaction, and there is not enough present to degrade the compound completely. Where there is adequate water, the degradation can be significant and where there is excess water, that is, greater than needed to dissolve the drug, solution degradation kinetics is observed.

With regard to crystallinity, it should be noted that amorphous materials are generally less stable than the corresponding crystalline forms. Often, amorphous phases crystallize on exposure to moisture. In amorphous solids, the net effect of water sorption is to lower the glass transition temperature, T_g, and hence plasticize the material. In turn, this increases the molecular mobility and therefore, the chemical reactivity.

In prenomination studies, a useful protocol to assess the effects of these factors is as follows. The compound is accurately weighed into each of six open glass vials. These are then placed (in duplicate if possible) under the following conditions: light stress (5000 lux, 25°C), 40°C/75% relative humidity (RH), and 30°C/60% RH. Typically, the sample would be sampled as necessary up to three months to determine its stability. After each time point, all samples are assessed visually and with a suitable HPLC (or liquid chromatography tandem mass spectroscopy method that can detect degradation products). In addition, differential scanning calorimetry and X-ray powder diffraction can be used to detect phase changes.

Photostability
A wide range of drug types undergo photochemical degradation. Theoretically, candidate drugs with absorption maxima greater than 280 nm may decompose in sunlight. However, instability due to light will probably only be a problem if the drug significantly absorbs light with a wavelength greater than 330 nm and, even then, only if the reaction proceeds at a significant rate. Light instability is a problem in both the solid and solution states, and formulations therefore need to be designed to protect the compound from its deleterious effects. There are a number of chemical groups that might be expected to give rise to decomposition. These include the carbonyl group, the nitroaromatic group, the N-oxide group, the C=C bond, the aryl chloride group, groups with a weak C—H bond, sulfides, alkenes, polyenes, and phenols. The first evidence that compounds are light-sensitive is usually discovered during lead optimization studies. Thus, candidate drugs should be assessed in the prenomination phase with respect to light stability to alert the formulation team whether special measures are needed to protect the drug from light. Indeed, this could be used as a selection criterion in many cases to reject unsuitable candidate drugs.

The ICH guideline on photostability testing is a good document to follow wherein it is suggested that photostability testing should consist of forced degradation and confirmatory testing. The forced degradation experiments can involve the candidate drug alone, in solution or in suspension using exposure conditions that reflect the nature of the compound and the intensity of the light sources used. The samples are then analyzed at various time points using appropriate techniques (e.g., HPLC). In addition, changes in physical properties, such as appearance and clarity or color should be noted. Confirmatory studies involve exposing the compound to light whose total output is not less than 1.2 million lux hours, and has a near UV energy of not less than 200 watt hr/m^2. Light sources for testing photostability include artificial daylight tubes, xenon lamps, tungsten-mercury lamps, laboratory light, and natural light. Instruments from Thermometric (4) are useful in measuring solid-state stability. Up to four independent calorimeter units can be used simultaneously with thermal activity monitor (TAM) III. A unit can be a 4-ml nanocalorimeter for measurements, which require very high sensitivity, a 20-ml standard microcalorimeter, a multicalorimeter consisting of six minicalorimeters, or the semi-adiabatic solution calorimeter. The multicalorimeter increases sample throughput considerably and is used for applications where microwatt rather than nanowatt sensitivity is sufficient. TAM III also exists in a 48-channel version for screening applications, or in general, when high sample throughput is required. TAM III can be operated in isothermal, step-isothermal, or temperature scanning mode. The isothermal mode is the classical mode for microcalorimetric experiments. In the isothermal mode, the liquid thermostat is maintained at a constant temperature (± 50 μK). Any heat generated or absorbed by the sample as a consequence of any chemical or physical process is measured continuously as a function of time. Step-isothermal mode is used to perform isothermal experiments at a number of temperatures in one single experiment. This is of particular interest for extracting temperature-dependent kinetic behavior for different kinds of processes. In the scanning mode, the temperature is scanned linearly over a certain interval. As the scanning rate is very slow, the sample can be considered to be in virtual thermal, chemical, and physical equilibria during measurement. For sample handling, a variety of ampoules and microreaction systems can be inserted in the calorimetric units, the type of which is determined by the experimental application.

The terahertz pulsed imaging (TPI)^{TM} spectra 1000 system exploits the spectroscopic information within each TPI^{TM} waveform to determine the chemical composition and the structural features of a sample (5) using terahertz technology. Terahertz data are complementary to Raman spectroscopy. It also provides information on both high-frequency (just below IR) and low-frequency vibrational modes; the latter are difficult to assess in Raman due to the proximity to the visible excitation line. Terahertz spectral interpretation and instrumentation are similar to basic IR and therefore easy to understand. The sample preparation techniques are the same as those used in IR and Raman. The unique spectral imaging characteristics of combining TPI and terahertz pulsed spectroscopy (TPS) can be used to investigate applications of proteomics in the pharmaceutical industry. The TPI^{TM} spectra 1000 can assist pharmaceutical companies in the rapid characterization of the stability and polymorphic forms of drugs. Many drug molecules after purification can crystallize in many different forms. These are known as polymorphs. Terahertz technology provides a rapid technique to identify different polymorphs.

In terms of the kinetics, light degradation in dilute solution is first order; however, in more concentrated solutions, decomposition approaches pseudo-zero order. The reason for this is that as the solution becomes more concentrated, degradation becomes limited due to the limited number of incident quanta and quenching reactions between the molecules. It should be noted that ionizable compounds, for example, ciprofloxacin, can show large differences in photostability between the ionized and unionized forms.

Regulatory Consideration in Stability Testing

The purpose of stability testing is to provide evidence on how the quality of a drug substance or drug product varies with time under the influence of a variety of environmental factors, such as temperature, humidity, and light, and to establish a retest period for the drug substance or a shelf life for the drug product and recommended storage conditions.

The choice of test conditions defined in this guidance is based on an analysis of the effects of climatic conditions in the three regions of the European Union, Japan, and the United States. The mean kinetic temperature in any part of the world can be derived from climatic data, and the world can be divided into four climatic zones, I–IV. This guidance addresses climatic zones I and II. The principle has been established that stability information generated in any one of the three regions of the European Union, Japan, and the United States would be mutually acceptable to the other two regions, provided the information is consistent with this guidance and the labeling is in accord with national/regional requirements. Information on the stability of the drug substance is an integral part of the systematic approach to stability evaluation.

Stress Testing

Stress testing of the drug substance can help identify the likely degradation products, which can in turn help establish the degradation pathways and the intrinsic stability of the molecule, and validate the stability indicating the power of the analytical procedures used. The nature of the stress testing will depend on the individual drug substance and the type of the drug product involved.

Stress testing is likely to be carried out on a single batch of the drug substance. The testing should include the effect of temperatures [in 10°C increments (e.g., 50°C and 60°C) above that for accelerated testing], humidity (e.g., 75% RH or greater), oxidation, and photolysis where appropriate on the drug substance. The testing should also evaluate the susceptibility of the drug substance to hydrolysis across a wide range of pH values when in solution or suspension. Photostability testing should be an integral part of stress testing. The standard conditions for photostability testing are described in ICH *Q1B Photostability Testing of New Drug Substances and Products*.

Examination of degradation products under stress conditions is useful in establishing degradation pathways and developing and validating suitable analytical procedures. However, such examination may not be necessary for certain degradation products if it has been demonstrated that they are not formed under accelerated or long-term storage conditions.

Results from these studies will form an integral part of the information provided to regulatory authorities.

Selection of Batches

Data from formal stability studies should be provided on at least three primary batches of the drug substance. The batches should be manufactured to a minimum of pilot scale by the same synthetic route as production batches and using a method of manufacture and procedure that simulates the final process to be used for production batches. The overall quality of the batches of drug substance placed on formal stability studies should be representative of the quality of the material to be made on a production scale. Other supporting data can be provided.

Container Closure System

The stability studies should be conducted on the drug substance packaged in a container closure system that is the same as or simulates the packaging proposed for storage and distribution.

Specifications

Specification, which is a list of tests, references to analytical procedures, and proposed acceptance criteria, is addressed in ICH *Q6A Specifications: Test Procedures and Acceptance Criteria for New Drug Substances and New Drug Products: Chemical Substances* and *Q6B Specifications: Test Procedures and Acceptance Criteria for New Drug Substances and New Drug Products: Biotechnological/Biological Products*. In addition, specification for degradation products in a drug substance is discussed in ICH *Q3A Impurities in New Drug Substances*.

Stability studies should include testing of those attributes of the drug substance that are susceptible to change during storage and are likely to influence quality, safety, and/or efficacy. The testing should cover, as appropriate, the physical, chemical, biological, and microbiological attributes. Validated stability-indicating analytical procedures should be applied. Whether and to what extent replication should be performed should depend on the results from validation studies.

Testing Frequency

For long-term studies, frequency of testing should be sufficient to establish the stability profile of the drug substance. For drug substances with a proposed retest period of at least 12 months, the frequency of testing at the long-term storage condition should normally be every three months over the first year, every six months over the second year, and annually thereafter throughout the proposed retest period.

At the accelerated storage condition, a minimum of three time points, including the initial and final time points (e.g., zero, three, and six months), from a six-month study is recommended. Where an expectation (based on development experience) exists such that the results from accelerated studies are likely to approach significant change criteria, increased testing should be conducted either by adding samples at the final time point or including a fourth time point in the study design.

When testing at the intermediate storage condition is called for as a result of significant change at the accelerated storage condition, a minimum of four time points, including the initial and final time points (e.g., zero, six, nine, and 12 months), from a 12-month study is recommended.

Storage Conditions

In general, a drug substance should be evaluated under storage conditions (with appropriate tolerances) that test its thermal stability and, if applicable, its sensitivity to moisture. The storage conditions and the duration of studies chosen should be sufficient to cover storage, shipment, and subsequent use.

The long-term testing should cover a minimum of 12 months' duration on at least three primary batches at the time of submission and should be continued for a period of time sufficient to cover the proposed retest period. Additional data accumulated during the assessment period of the registration application should be submitted to the authorities if requested. Data from the accelerated storage condition and, if appropriate, from the intermediate storage condition can be used to evaluate the effect of short-term excursions outside the label storage conditions (such as might occur during shipping).

Long-term, accelerated, and, where appropriate, intermediate storage conditions for drug substances are detailed in Table 4. The general case should apply if the drug substance is not specifically covered by a subsequent section. Alternative storage conditions can be used if justified (Table 4).

If long-term studies are conducted at 25°C ± 2°C/60% RH ± 5% RH and *significant change* occurs at any time during the six months' testing at the accelerated storage condition, additional testing at the intermediate storage condition should be conducted and evaluated against significant change criteria. Testing at the

Table 4 General Case of Study, Storage Condition, and Time Covered

Study	Storage condition	Minimum time period covered by data at submission
Long-term[a]	25°C ± 2°C/60% RH ± 5% RH or 30°C ± 2°C/65% RH ± 5% RH	12 months
Intermediate[b]	30°C ± 2°C/65% RH ± 5% RH	Six months
Accelerated	40°C ± 2°C/75% RH ± 5% RH	Six months

[a] It is up to the applicant to decide whether long-term stability studies are performed at 25°C ± 2°C/60% RH ± 5% RH or 30°C ± 2°C/65% RH ± 5% RH.
[b] If 30°C ± 2°C/65% RH ± 5% RH is the long-term condition, there is no intermediate condition.

Table 5 Storage and Minimum Time Covered for Drug Substances Intended for Storage in a Refrigerator

Study	Storage condition	Minimum time period covered by data at submission
Long-term	5°C ± 3°C	12 months
Accelerated	25°C ± 2°C/60% RH ± 5% RH	Six months

intermediate storage condition should include all tests, unless otherwise justified. The initial application should include a minimum of six months' data from a 12-month study at the intermediate storage condition. *Significant change* for a drug substance is defined as failure to meet its specification (Table 5).

Data from refrigerated storage should be assessed according to the evaluation section of this guidance, except where explicitly noted in Table 6.

If significant change occurs between three and six months' testing at the accelerated storage condition, the proposed retest period should be based on the real-time data available at the long-term storage condition.

If significant change occurs within the first three months' testing at the accelerated storage condition, a discussion should be provided to address the effect of short-term excursions outside the label storage condition (e.g., during shipping or handling). This discussion can be supported, if appropriate, by further testing on a single batch of the drug substance for a period shorter than three months, but with more frequent testing than usual. It is considered unnecessary to continue to test a drug substance through six months when a significant change has occurred within the first three months.

For drug substances intended for storage in a freezer, the retest period should be based on the real-time data obtained at the long-term storage condition. In the absence of an accelerated storage condition for drug substances intended to be stored in a freezer, testing on a single batch at an elevated temperature (e.g., 5°C ± 3°C or 25°C ± 2°C) for an appropriate time period should be conducted to address the effect of short-term excursions outside the proposed label storage condition (e.g., during shipping or handling). Drug substances intended for storage below −20°C should be treated on a case-by-case basis.

Stability Commitment

When available long-term stability data on primary batches do not cover the proposed retest period granted at the time of approval, a commitment should be made to continue the stability studies postapproval to firmly establish the retest period.

Where the submission includes long-term stability data on three production batches covering the proposed retest period, a postapproval commitment is

Table 6 Storage Conditions and Minimum Time for Drug Substances Intended for Storage in a Freezer

Study	Storage condition	Minimum time period covered by data at submission
Long-term	−20°C ± 5°C	12 months

considered unnecessary. Otherwise, one of the following commitments should be made:

- If the submission includes data from stability studies on at least three production batches, a commitment should be made to continue these studies through the proposed retest period.
- If the submission includes data from stability studies on fewer than three production batches, a commitment should be made to continue these studies through the proposed retest period and to place additional production batches, to a total of at least three, on long-term stability studies through the proposed retest period.
- If the submission does not include stability data on production batches, a commitment should be made to place the first three production batches on long-term stability studies through the proposed retest period.

The stability protocol used for long-term studies for the stability commitment should be the same as that for the primary batches, unless otherwise scientifically justified.

Evaluation
The purpose of the stability study is to establish, based on testing a minimum of three batches of the drug substance and evaluating the stability information (including, as appropriate, results of the physical, chemical, biological, and microbiological tests), a retest period applicable to all future batches of the drug substance manufactured under similar circumstances. The degree of variability of individual batches affects the confidence that a future production batch will remain within specification throughout the assigned retest period.

The data may show very little degradation and so little variability that it is apparent from looking at the data that the requested retest period will be granted. Under these circumstances, it is normally unnecessary to go through the formal statistical analysis; providing a justification for the omission should be sufficient.

An approach for analyzing the data on a quantitative attribute that is expected to change with time is to determine the time at which the 95%, one-sided confidence limit for the mean curve intersects the acceptance criterion. If analysis shows that the batch-to-batch variability is small, it is advantageous to combine the data into one overall estimate. This can be done by first applying appropriate statistical tests (e.g., P values for the level of significance of rejection of more than 0.25) to the slopes of the regression lines and zero time intercepts for the individual batches. If it is inappropriate to combine data from several batches, the overall retest period should be based on the minimum time a batch can be expected to remain within the acceptance criteria.

The nature of any degradation relationship will determine whether the data should be transformed for linear regression analysis. Usually the relationship can be represented by a linear, quadratic, or cubic function on an arithmetic or logarithmic scale. Statistical methods should be employed to test the goodness of fit of the data on all batches and combined batches (where appropriate) to the assumed degradation line or curve.

Limited extrapolation of the real-time data from the long-term storage condition beyond the observed range to extend the retest period can be undertaken at approval time if justified. This justification should be based, for example, on what is known about the mechanism of degradation, the results of testing under accelerated conditions, the goodness of fit of any mathematical model, batch size, and/or existence of supporting stability data. However, this extrapolation assumes that the same degradation relationship will continue to apply beyond the observed data.

Any evaluation should cover not only the assay, but also the levels of degradation products and other appropriate attributes.

Statements/Labeling

A storage statement should be established for the labeling in accordance with relevant national/regional requirements. The statement should be based on the stability evaluation of the drug substance. Where applicable, specific instructions should be provided, particularly for drug substances that cannot tolerate freezing. Terms such as *ambient conditions* or *room temperature* should be avoided.

A retest period should be derived from the stability information, and a retest date should be displayed on the container label if appropriate (Table 7).

If long-term studies are conducted at 25°C ± 2°C/60% RH ± 5% RH and *significant change* occurs at any time during six months' testing at the accelerated storage condition, additional testing at the intermediate storage condition should be conducted and evaluated against significant change criteria. The initial application should include a minimum of six months' data from a 12-month study at the intermediate storage condition.

In general, *significant change* for a drug product is defined as one or more of the following (as appropriate for the dosage form):

- A 5% change in assay from its initial value, or failure to meet the acceptance criteria for potency when using biological or immunological procedures.
- Any degradation product exceeding its acceptance criterion.
- Failure to meet the acceptance criteria for appearance, physical attributes, and functionality test (e.g., color, phase separation, resuspendibility, caking, hardness, and dose delivery per actuation). However, some changes in physical attributes (e.g., softening of suppositories, melting of creams) may be expected under accelerated conditions.
- Failure to meet the acceptance criterion for pH.
- Failure to meet the acceptance criteria for dissolution for 12 dosage units.

Table 7 Retest Study, Storage Condition, and Minimum Time Period

Study	Storage condition	Minimum time period covered by data at submission
Long-term[a]	25°C ± 2°C/60% RH ± 5% RH or	12 months
	30°C ± 2°C/65% RH ± 5% RH	
Intermediate[b]	30°C ± 2°C/65% RH ± 5% RH	Six months
Accelerated	40°C ± 2°C/75% RH ± 5% RH	Six months

[a]It is up to the applicant to decide whether long-term stability studies are performed at 25°C ± 2°C/60% RH ± 5% RH or 30°C ± 2°C/65% RH ± 5% RH.
[b]If 30°C ± 2°C/65% RH ± 5% RH is the long-term condition, there is no intermediate condition.

Photostability Testing

The *ICH Harmonized Tripartite Guideline on Stability Testing of New Drug Substances* notes that light testing should be an integral part of stress testing. The intrinsic photostability characteristics of new drug substances should be evaluated to demonstrate that, as appropriate, light exposure does not result in unacceptable change. Normally, photostability testing is carried out on a single batch of material. Under some circumstances, these studies should be repeated if certain variations and changes are made to the product (e.g., formulation, packaging). Whether these studies should be repeated depends on the photostability characteristics determined at the time of initial filing and the type of variation and/or change made.

A systematic approach to photostability testing is recommended covering, as appropriate, studies such as light sources and the procedures used. The light sources described next may be used for photostability testing while maintaining the temperature to avoid these effects to confound light effects. Any light source that is designed to produce an output similar to the D65/ID65 emission standard, such as an artificial daylight fluorescent lamp combining visible (VIS) and UV outputs, xenon, or metal halide lamp. D65 is the internationally recognized standard for outdoor daylight as defined in ISO 10977 (1993). ID65 is the equivalent indoor indirect daylight standard. For a light source emitting significant radiation below 320 nm, an appropriate filter(s) may be fitted to eliminate such radiation. An alternate source of light is a cool white fluorescent lamp designed to produce an output similar to that specified in ISO 10977 (1993); and a near UV fluorescent lamp having a spectral distribution from 320 nm to 400 nm with a maximum energy emission between 350 nm and 370 nm; a significant proportion of UV should be in both bands of 320–360 nm and 360–400 nm.

For confirmatory studies, samples should be exposed to light providing an overall illumination of not less than 1.2 million lux hours and an integrated near UV energy of not less than 200 watt h/m^2 to allow direct comparisons to be made between the drug substance and the drug product. Samples may be exposed side-by-side with a validated chemical actinometric system to ensure that the specified light exposure is obtained, or for the appropriate duration of time when conditions have been monitored using calibrated radiometers/lux meters. If protected samples (e.g., wrapped in aluminum foil) are used as dark controls to evaluate the contribution of thermally induced change to the total observed change, these should be placed alongside the authentic sample.

The photostability testing should consist of two parts: forced degradation testing and confirmatory testing. The purpose of forced degradation testing studies is to evaluate the overall photosensitivity of the material for method development purposes and/or degradation pathway elucidation. This testing may involve the drug substance alone and/or in simple solutions/suspensions to validate the analytical procedures. In these studies, the samples should be in chemically inert and transparent containers. In these forced degradation studies, a variety of exposure conditions may be used, depending on the photosensitivity of the drug substance involved and the intensity of the light sources used. For development and validation purposes, it is appropriate to limit exposure and end the studies if extensive decomposition occurs. For photostable materials, studies may be terminated after an appropriate exposure level has been used. The design of these experiments is left to the applicant's discretion although the exposure levels used should be justified.

Under forced conditions, decomposition products may be observed that are unlikely to be formed under the conditions used for confirmatory studies. This information may be useful in developing and validating suitable analytical methods. If in practice it has been demonstrated that they are not formed in the confirmatory studies, these degradation products need not be examined further. Confirmatory studies should then be undertaken to provide the information necessary for handling, packaging, and labeling. Normally, only one batch of drug substance is tested during the development phase, and then the photostability characteristics should be confirmed on a single batch selected as described in the parent guideline if the drug is clearly photostable or photolabile. If the results of the confirmatory study are equivocal, testing of up to two additional batches should be conducted. Samples should be selected as described in the parent guideline.

Care should be taken to ensure that the physical characteristics of the samples under test are taken into account and efforts should be made, such as cooling and/or placing the samples in sealed containers, to ensure that the effects of the changes in physical states, such as sublimation, evaporation, or melting are minimized. All such precautions should be chosen to provide minimal interference with the exposure of samples under test. Possible interactions between the samples and any material used for containers or for general protection of the sample should also be considered and eliminated wherever not relevant to the test being carried out.

As a direct challenge for samples of solid drug substances, an appropriate amount of sample should be taken and placed in a suitable glass or plastic dish and protected with a suitable transparent cover if considered necessary. Solid drug substances should be spread across the container to give a thickness of typically not more than 3 mL. Drug substances that are liquids should be exposed in chemically inert and transparent containers.

At the end of the exposure period, the samples should be examined for any changes in physical properties (e.g., appearance, clarity, or color of the solution) and for assay and degradants by a method suitably validated for products likely to arise from photochemical degradation processes. Where solid drug substance samples are involved, sampling should ensure that a representative portion is used in individual tests. Similar sampling considerations, such as homogenization of the entire sample, apply to other materials that may not be homogeneous after exposure. The analysis of the exposed sample should be performed concomitantly with that of any protected samples used as dark control if these are used in the test.

The forced degradation studies should be designed to provide suitable information to develop and validate test methods for the confirmatory studies. These test methods should be capable of resolving and detecting photolytic degradants that appear during the confirmatory studies. When evaluating the results of these studies, it is important to recognize that they form part of the stress testing and are not therefore designed to establish qualitative or quantitative limits for change.

The confirmatory studies should identify precautionary measures needed in manufacturing or in formulation of the drug product, and whether light-resistant packaging is needed. When evaluating the results of confirmatory studies to determine whether change due to exposure to light is acceptable, it is important to consider the results from other formal stability studies in order to assure that the drug will be within justified limits at the time of use.

Bracketing

For the study of drug substances, matrixing is of limited utility and bracketing is generally not applicable.

IMPURITIES

Impurities can be classified into the following categories:

- Organic impurities (process- and drug-related);
- inorganic impurities; and
- residual solvents.

Organic impurities can arise during the manufacturing process and/or storage of the new drug substance. They can be identified or unidentified, volatile or nonvolatile, and include:

- Starting materials,
- by-products,
- intermediates,
- degradation products, and
- reagents, ligands, and catalysts.

There is a need to summarize the actual and potential impurities most likely to arise during the synthesis, purification, and storage of a new drug substance. This summary should be based on sound scientific appraisal of the chemical reactions involved in the synthesis, impurities associated with raw materials that could contribute to the impurity profile of the new drug substance, and the possible degradation products. This discussion can be limited to those impurities that might reasonably be expected based on knowledge of the chemical reactions and the conditions involved.

In addition, the applicant should summarize the laboratory studies conducted to detect impurities in the new drug substance. This summary should include test results of batches manufactured during the development process and batches from the proposed commercial process, as well as the results of stress testing [see ICH Q1A(R) on stability] used to identify potential impurities that arise during storage. The impurity profile of the drug substance batches intended for marketing should be compared with those used in development, and any differences discussed.

The studies conducted to characterize the structure of actual impurities present in a new drug substance at a level greater than the identification threshold calculated using the response factor of the drug substance is described seperately. Note that any impurity at a level greater than the identification threshold in any batch manufactured by the proposed commercial process should be identified. In addition, any degradation product observed in stability studies at recommended storage conditions at a level greater than the identification threshold should be identified. When identification of an impurity is not feasible, a summary of the laboratory studies demonstrating the unsuccessful effort should be included in the application. Where attempts have been made to identify impurities present at levels of not more than the identification thresholds, it is useful also to report the results of these studies.

Identification of impurities present at an apparent level of not more than the identification threshold is generally not considered necessary. However, analytical procedures should be developed for those potential impurities that are expected to be unusually potent, producing toxic or pharmacological effects at a level not more than the identification threshold. All impurities should be qualified:

- each specified, unidentified impurity;
- any unspecified impurity with an acceptance criterion of not more than the identification threshold;
- total impurities.

Inorganic impurities can result from the manufacturing process. They are normally known and identified and include:

- reagents, ligands and catalysts,
- heavy metals or other residual metals,
- inorganic salts, and
- other materials (e.g., filter aids, charcoal).

Inorganic impurities are normally detected and quantified using pharmacopoeial or other appropriate procedures. Carryover of catalysts to a new drug substance should be evaluated during development. The need for inclusion or exclusion of inorganic impurities in a new drug substance specification should be discussed. Acceptance criteria should be based on pharmacopoeial standards or known safety data.

Solvents are inorganic or organic liquids that are used as vehicles for the preparation of solutions or suspensions in the synthesis of a new drug substance. As these are generally of known toxicity, the selection of appropriate controls is easily accomplished (see ICH Q3C on Residual Solvents). The control of residues of the solvents used in the manufacturing process for a new drug substance should be discussed and presented according to ICH *Q3C Impurities: Residual Solvents*.

A registration application should include documented evidence that the analytical procedures are validated and suitable for the detection and quantification of impurities (see ICH Q2A and Q2B on Analytical Validation). Technical factors (e.g., manufacturing capability and control methodology) can be considered as part of the justification for selection of alternative thresholds based on manufacturing experience with the proposed commercial process. The use of two decimal places for thresholds does not necessarily reflect the precision of the analytical procedure used for routine quality control purposes. Thus, the use of lower precision techniques (e.g., thin-layer chromatography) can be appropriate where justified and appropriately validated. Differences in the analytical procedures used during development and those proposed for the commercial product should be discussed in the registration application. The quantification limit for the analytical procedure should be not more than the reporting threshold.

Organic impurity levels can be measured by a variety of techniques, including those that compare an analytical response for an impurity with that of an appropriate reference standard or with the response of the new drug substance itself. Reference standards used in the analytical procedures for control of impurities should be evaluated and characterized according to their intended uses. The drug substance can be used as a standard to estimate the levels of impurities. In cases where the response factors of a drug substance and the relevant impurity are not close, this practice can still be appropriate, provided a correction factor is applied or the impurities are, in fact, being overestimated. Acceptance criteria and analytical procedures used to estimate identified or unidentified impurities can be based on analytical assumptions (e.g., equivalent detector response). These assumptions should be discussed in registration applications.

The specification for a new drug substance should include a list of impurities. Stability studies, chemical development studies, and routine batch analyses can be used to predict those impurities that are likely to occur in the commercial product. The selection of impurities in a new drug substance specification should be based on the impurities found in batches manufactured by the proposed commercial process. Those individual impurities with specific acceptance criteria included in the specification for a new drug substance are referred to as *specified impurities* in this guidance. Specified impurities can be identified or unidentified.

A rationale for the inclusion or exclusion of impurities in a specification should be presented. The rationale should include a discussion of the impurity profiles observed in the safety and clinical development batches, together with a consideration of the impurity profile of batches manufactured by the proposed commercial process. Specified, identified impurities should be included along with specified, unidentified impurities estimated to be present at a level greater than the identification threshold given. For impurities known to be unusually potent or that produce toxic or unexpected pharmacological effects, the quantification/detection limit of the analytical procedures should be commensurate with the level at which the impurities should be controlled. For unidentified impurities, the procedure used and assumptions made in establishing the level of the impurity should be clearly stated. Specified, unidentified impurities should be referred to by an appropriate qualitative analytical descriptive label (e.g., "unidentified A," "unidentified with relative retention of 0.9"). A general acceptance criterion of not more than the identification threshold for any unspecified impurity and an acceptance criterion for total impurities should be included.

Acceptance criteria should be set no higher than the level that can be justified by safety data and should be consistent with the level achievable by the manufacturing process and the analytical capability. Where there is no safety concern, impurity acceptance criteria should be based on data generated on batches of a new drug substance manufactured by the proposed commercial process, allowing sufficient latitude to deal with normal manufacturing and analytical variations and the stability characteristics of the new drug substance. Although normal manufacturing variations are expected, significant variation in batch-to-batch impurity levels can indicate that the manufacturing process of the new drug substance is not adequately controlled and validated (see ICH Q6A guidance on specifications, decision tree #1, for establishing an acceptance criterion for a specified impurity in a new drug substance). The use of two decimal places for thresholds does not necessarily indicate the precision of the acceptance criteria for specified impurities and total impurities.

Qualification is the process of acquiring and evaluating the data that establishes the biological safety of an individual impurity or a given impurity profile at the level(s) specified. The applicant should provide a rationale for establishing impurity acceptance criteria that includes safety considerations. The level of any impurity present in a new drug substance that has been adequately tested in safety and/or clinical studies would be considered qualified. Impurities that are also significant metabolites present in animal and/or human studies are generally considered qualified. A level of a qualified impurity higher than that present in a new drug substance can also be justified based on an analysis of the actual amount of impurity administered in previous relevant safety studies.

If data are unavailable to qualify the proposed acceptance criterion of an impurity, studies to obtain such data can be appropriate when the usual qualification thresholds given are exceeded.

Higher or lower thresholds for qualification of impurities can be appropriate for some individual drugs based on the scientific rationale and level of concern, including drug class effects and clinical experience. For example, qualification can be especially important when there is evidence that such impurities in certain drugs or therapeutic classes have previously been associated with adverse reactions in patients. In these instances, a lower qualification threshold can be appropriate. Conversely, a higher qualification threshold can be appropriate for individual drugs when the level of concern for safety is less than the usual based on similar considerations (e.g., patient population, drug class effects, and clinical considerations). Proposals for alternative thresholds would be considered on a case-by-case basis.

The "decision tree for identification and qualification" (Scheme 8) describes considerations for the qualification of impurities when thresholds are exceeded. In some cases, decreasing the level of impurity to not more than the threshold can be simpler than providing safety data. Alternatively, adequate data could be available in the scientific literature to qualify an impurity. If neither is the case, additional safety testing should be considered. The studies considered appropriate to qualify an impurity will depend on a number of factors, including the patient population, daily dose, and route and duration of drug administration. Such studies can be conducted on the new drug substance containing the impurities to be controlled, although studies using isolated impurities can sometimes be appropriate.

Although this guidance is not intended to apply during the clinical research stage of development, in the later stages of development, the thresholds in this guidance can be useful in evaluating new impurities observed in drug substance batches prepared by the proposed commercial process. Any new impurity observed in later stages of development should be identified if its level is greater than the identification threshold. Similarly, the qualification of the impurity should be considered if its level is greater than the qualification threshold (Table 8). Safety assessment studies to qualify an impurity should compare the new drug substance containing a representative amount of the new impurity with previously qualified material. Safety assessment studies using a sample of the isolated impurity can also be considered.

Chemical Drug Substance Characterization

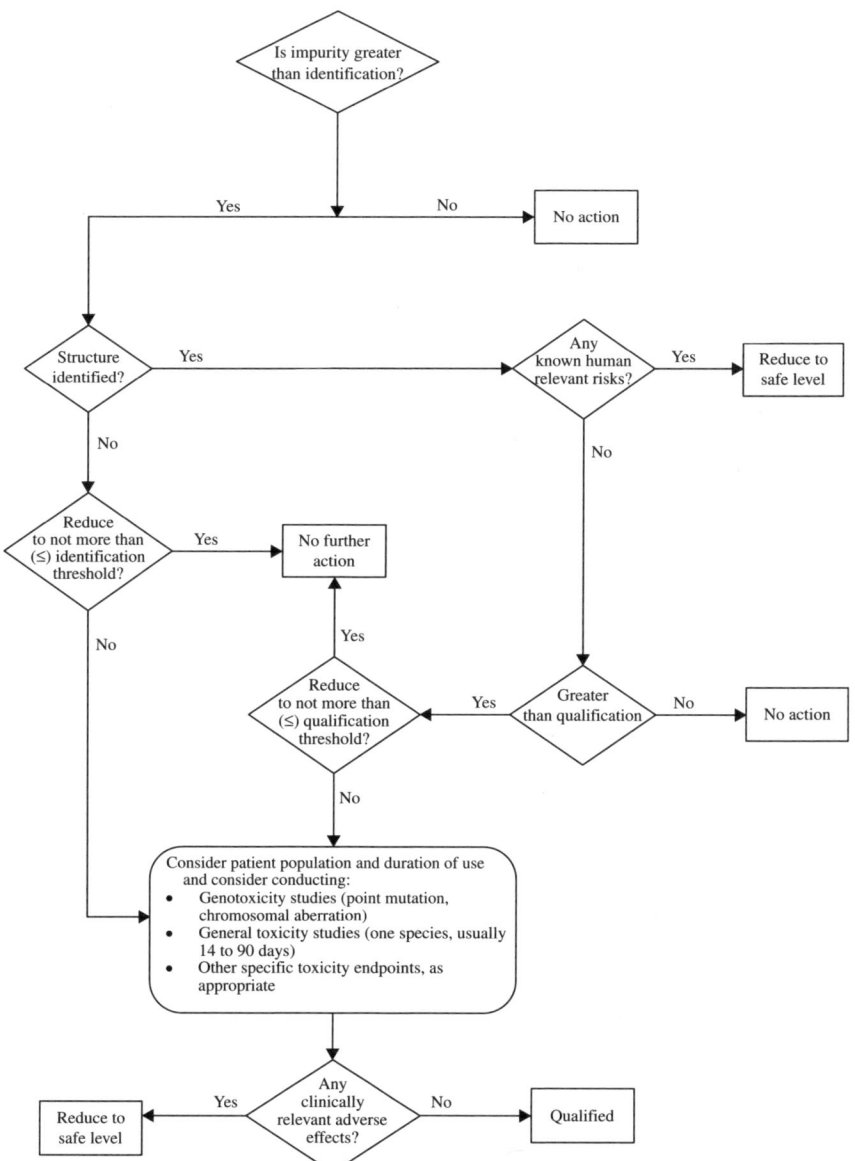

Scheme 8 Decision tree for impurities in drug substances. *Source*: Courtesy of Pharmquest Corporation, Mountain View, California, U.S.A.

If considered desirable, a minimum screen (e.g., genotoxic potential) should be conducted. A study to detect point mutations and one to detect chromosomal aberrations, both in vitro, are considered an appropriate minimum screen. If general toxicity studies are desirable, one or more studies should be designed to allow comparison of unqualified and qualified materials. The study duration should be based on available relevant information and performed in the species

Table 8 Thresholds of Impurities

Maximum daily dose[a]	Reporting threshold[b,c]	Identification threshold[c]	Qualification threshold[c]
≤2 g/day	0.05%	0.10% or 1.0 mg/day intake (whichever is lower)	0.15% or 1.0 mg/day intake (whichever is lower)
>2 g/day	0.03%	0.05%	0.05%

[a]The amount of drug substance administered per day.
[b]Higher reporting thresholds should be scientifically justified.
[c]Lower thresholds can be appropriate if the impurity is unusually toxic.

most likely to maximize the potential to detect the toxicity of an impurity. On a case-by-case basis, single-dose studies can be appropriate, especially for single-dose drugs. In general, a minimum duration of 14 days and a maximum duration of 90 days would be considered appropriate. Lower thresholds can be appropriate if the impurity is unusually toxic. For example, does known safety data for this impurity or its structural class preclude human exposure at the concentration present?

GOOD MANUFACTURING PRACTICE

The term manufacturing is defined as one that includes all operations of receipt of materials, production, packaging, repackaging, labeling, relabeling, quality control, release, storage, and distribution of active pharmaceutical ingredients (APIs) and the related controls. An API starting material is a raw material, an intermediate, or an API that is used in the production of an API and that is incorporated as a significant structural fragment into the structure of the API. API starting materials normally have defined chemical properties and structure.

The company should designate and document the rationale for the point at which production of the API begins. For synthetic processes, this is known as the point at which API starting materials are entered into the process. For other processes (e.g., fermentation, extraction, and purification), this rationale should be established on a case-by-case basis. Table 9 gives guidance on the point at which the API starting material is normally introduced into the process.

From this point on, appropriate good manufacturing practice (GMP) as defined in this guidance should be applied to these intermediate and/or API manufacturing steps. This would include the validation of critical process steps determined to impact the quality of the API. However, it should be noted that the fact that a company chooses to validate a process step does not necessarily define that step as critical.

The guidance in this document would normally be applied to the steps shown in gray in Table 9. However, all steps shown may not need to be completed. The stringency of GMP in API manufacturing should increase as the process proceeds from early API steps to the final steps, purification, and packaging. Physical processing of APIs, such as granulation, coating, or physical manipulation of particle size (e.g., milling, micronizing) should be conducted according to this guidance. This GMP guidance does not apply to steps prior to the introduction of the defined API starting material.

Table 9 Application of Good Manufacturing Practice to Active Pharmaceutical Ingredient Manufacturing

Type of manufacturing	Application zone in gray					
Chemical manufacturing	Production of the API starting material	Introduction of the API starting material into the process	Production of intermediate(s)	Isolation and purification	Physical processing and packaging	
API derived from animal sources	Collection of organ, fluid, or tissue	Cutting, mixing, and/or initial processing	Introduction of the API starting material into the process	Isolation and purification	Physical processing and packaging	
API extracted from plant sources	Collection of plant	Cutting and initial extraction(s)	Introduction of the API starting material into the process	Isolation and purification	Physical processing and packaging	
Herbal extracts used as API	Collection of plants	Cutting and initial extraction		Further extraction	Physical processing and packaging	
API consisting of comminuted or powdered herbs	Collection of plants and/or cultivation and harvesting	Cutting/comminuting			Physical processing and packaging	
Biotechnology: fermentation/cell culture	Establishment of master cell bank and working cell bank	Maintenance of working cell bank	Cell culture and/or fermentation	Isolation and purification	Physical processing and packaging	
"Classical" fermentation to produce an API	Establishment of cell bank	Maintenance of the cell bank	Introduction of the cells into fermentation	Isolation and purification	Physical processing and packaging	

Abbreviation: API, active pharmaceutical ingredient.

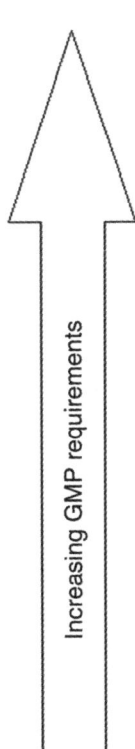

Increasing GMP requirements

Quality Management
The system for managing quality should encompass the organizational structure, procedures, processes and resources, as well as activities to ensure confidence that the API will meet its intended specifications for quality and purity. All quality-related activities should be defined and documented. There should be a quality unit(s) that is independent of production and that fulfills both quality assurance (QA) and quality control (QC) responsibilities. The quality unit can be in the form of separate QA and QC units or a single individual or group, depending upon the size and structure of the organization. The persons authorized to release intermediates and APIs should be specified. All quality-related activities should be recorded at the time they are performed. Any deviation from established procedures should be documented and explained. Critical deviations should be investigated, and the investigation and its conclusions should be documented.

No materials should be released or used before the satisfactory completion of evaluation by the quality unit(s) unless there are appropriate systems in place to allow for such use or the use of raw materials or intermediates pending completion of evaluation.

Procedures should exist for notifying responsible management in a timely manner of regulatory inspections, serious GMP deficiencies, product defects, and related actions (e.g., quality-related complaints, recalls, and regulatory actions).

Equipments used in the manufacture of intermediates and APIs should be of appropriate design and adequate size, and suitably located for its intended use, cleaning, sanitation (where appropriate), and maintenance. Equipment should be constructed so that surfaces that contact the raw materials, intermediates, or APIs do not alter the quality of the intermediates and APIs beyond the official or other established specifications. Production equipment should only be used within its qualified operating range. Major equipments (e.g., reactors and storage containers) and permanently installed processing lines used during the production of an intermediate or API should be appropriately identified.

Any substance associated with the operation of the equipment, such as lubricants, heating fluids, or coolants, should not contact the intermediates or APIs so as to alter the quality of APIs or intermediates beyond the official or other established specifications. Any deviations from this practice should be evaluated to ensure that there are no detrimental effects on the material's fitness for use. Wherever possible, food-grade lubricants and oils should be used.

Closed or contained equipments should be used whenever appropriate. Where open equipment is used, or equipment is opened, appropriate precautions should be taken to minimize the risk of contamination. A set of current drawings should be maintained for equipment and critical installations (e.g., instrumentation and utility systems).

For each batch of intermediate and API, appropriate laboratory tests should be conducted to determine conformance to specifications. An impurity profile describing the identified and unidentified impurities present in a typical batch produced by a specific controlled production process should normally be established for each API. The impurity profile should include the identity or some qualitative analytical designation (e.g., retention time), the range of each impurity observed, and classification of each identified impurity (e.g., inorganic, organic, and solvent). The impurity profile is normally dependent upon the production process and the origin of the API. Impurity profiles are normally not necessary for APIs from herbal or animal tissue origin. Biotechnology considerations are covered in ICH guidance Q6B. The impurity profile should be compared at appropriate intervals against the impurity profile in the regulatory submission or compared against historical data to detect changes to the API resulting from modifications in raw materials, equipment operating parameters, or the production process. Appropriate microbiological tests should be conducted on each batch of the intermediate and API where microbial quality is specified.

A documented, on-going testing program should be established to monitor the stability characteristics of APIs, and the results should be used to confirm appropriate storage conditions and retest or expiry dates. The test procedures used in stability testing should be validated and should be indicative of stability.

Stability samples should be stored in containers that simulate the market container. For example, if the API is marketed in bags within fiber drums, stability samples can be packaged in bags of the same material and in small-scale drums of similar or identical material composition to the market drums.

Normally, the first three commercial production batches should be placed on the stability monitoring program to confirm the retest or expiry date. However, where data from previous studies show that the API is expected to remain stable for at least two years, fewer than three batches can be used. Thereafter, at least one batch per year of API manufactured (unless none is produced that year) should be added to the stability monitoring program and tested at least annually to confirm the stability.

For APIs with short shelf lives, testing should be done more frequently. For example, for those biotechnological/biologic and other APIs with shelf lives of one year or less, stability samples should be obtained and should be tested monthly for the first three months, and at three-month intervals later. When data exist that confirm that the stability of the API is not compromised, elimination of specific test intervals (e.g., nine-month testing) can be considered. Where appropriate, the stability storage conditions should be consistent with the ICH guidances on stability.

When an intermediate is intended to be transferred outside the control of the manufacturer's material management system and an expiry or retest date is assigned, supporting stability information should be available (e.g., published data, test results).

An API expiry or retest date should be based on an evaluation of data derived from stability studies. Common practice is to use a retest date, not an expiration date.

Preliminary API expiry or retest dates can be based on pilot-scale batches if: (*i*) the pilot batches employ a method of manufacture and procedure that simulates the final process to be used on a commercial manufacturing scale and (*ii*) the quality of the API represents the material to be made on a commercial scale.

A representative sample should be taken for the purpose of performing a retest. The company's overall policy, intentions, and approach to validation, including the validation of production processes, cleaning procedures, analytical methods, in-process control test procedures, computerized systems, and persons responsible for design, review, approval, and documentation of each validation phase, should be documented. The critical parameters/attributes should normally be identified during the development stage or from historical data, and the necessary ranges for the reproducible operation should be defined. This should include:

- defining the API in terms of its critical product attributes;
- identifying process parameters that could affect the critical quality attributes of the API; and

- determining the range for each critical process parameter expected to be used during routine manufacturing and process control.

Validation should extend to those operations determined to be critical to the quality and purity of the API.

Before initiating process validation activities, appropriate qualification of critical equipments and ancillary systems should be completed. Qualification is usually carried out by conducting the following activities, individually or combined:

- *Design qualification (DQ)*: documented verification that the proposed design of the facilities, equipments, or systems is suitable for the intended purpose.
- *Installation qualification (IQ)*: documented verification that the equipments or systems, as installed or modified, comply with the approved design, the manufacturer's recommendations and/or user requirements.
- *Operational qualification (OQ)*: documented verification that the equipments or systems, as installed or modified, perform as intended throughout the anticipated operating ranges.
- *Performance qualification (PQ)*: documented verification that the equipment and ancillary systems, as connected together, can perform effectively and reproducibly based on the approved process method and specifications.

REFERENCES

1. http://www.fda.gov/cder/OPS/PAT.htm.
2. http://www.automsoft.com/products/solutionsforpat.asp.
3. http://www.sas.com/.
4. http://www.thermometric.se.
5. http://www.brukeroptics.com/terahertz/.

BIBLIOGRAPHY

Anderson G, Scott M. Determination of product shelf life and activation energy for five drugs of abuse. Clin Chem 1991; 37:398–429.

Buchacher A, Schulz P, Choromanski J, Schwinn H, Josic D. High-performance capillary electrophoresis for in-process control in the production of antithrombin III and human clotting factor IX. J Chromatogr A 1998, 10 Apr; 802(2):355–366.

Cimander C, Bachinger T, Mandenius CF. Integration of distributed multi-analyzer monitoring and control in bioprocessing based on a real-time expert system. J Biotechnol 2003, 15 Aug; 103(3):237–248.

Clarke FC, Hammon, SV, Mattisson, C. The development of NIR microscopy for process control in pharmaceutical manufacturing. J Process Anal Chem 2003; 7(3), http://www.infoscience.com/JPAC/JPACUserForm.html.

Creasy KE. Process analytical chemistry: smarter, faster, stronger. J Process Anal Chem 2003; 7(1), http://www.infoscience.com/JPAC/JPACUserForm.html.

Farquharson S, Smith W, Rose J, Shaw M. Correlations between molecular (Raman) and macroscopic (Rheology) data for process monitoring of thermoset composites. J Process Anal Chem 2003; 7(2), http://www.infoscience.com/JPAC/JPACUserForm.html.

FDA. Guidelines for submitting documentations for the stability of human drugs and biologics. Rockville (MD), 1987.

Frake P, Greenhalgh D, Grierson SM, Hempenstall JM, Rudd DR. Process control and endpoint determination of a fluid bed granulation by application of near infra-red spectroscopy. Intl J Pharm 1997; 51:75–80.

Gold TB, Buice RG Jr, Lodder RA, Digenis GA. Determination of extent of formaldehyde-induced crosslinking in hard gelatin capsules by near-infrared spectrophotometry. Pharm Res 1997 Aug; 14(8):1046–1050.

Groth T, Moden H. Related articles. A knowledge-based system for real-time quality control and fault diagnosis of multitest analyzers. Comput Methods Programs Biomed 1991 Feb–Mar; 34(2–3):175–190.

Gunnell JJ, van Vuuren P. Process analytical systems: a vision for the future. J Process Anal Chem 2002; 6(1), http://www.infoscience.com/JPAC/JPACUserForm.html.

Hamilton SJ, Lowell AE, Lodder RA. Hyperspectral techniques in analysis of oral dosage forms. J Biomed Opt 2002 Oct; 7(4):561–570.

Harthun S, Matischak K, Friedl P. Determination of recombinant protein in animal cell culture supernatant by near-infrared spectroscopy. Anal Biochem 1997, 15 Aug; 251(1):73–78.

Higgins JP, Arrivo SM, Thurau G, et al. Spectroscopic approach for on-line monitoring of particle size during the processing of pharmaceutical nanoparticles. Anal Chem 2003, 15 Apr; 75(8):1777–1785.

ICH. Bracketing and matrixing designs for stability testing of new drug substances and products. Q1D. Geneva, 2003a.

ICH. Stability data evaluation. Q1E. Geneva, 2003b.

ICH. Stability testing for new drug substances and products. Q1A(R). Geneva, 1994.

ICH. Stability testing for new drug substances and products. Q1A(R2). Geneva, 2003c.

King SY, Kung MS, Fung HL. Statistical prediction of drug stability based on nonlinear parameter estimation. J Pharm Sci 1984; 73:2332–2344.

Kirsch JD, Drennen JK. Nondestructive tablet hardness testing by near-infrared spectroscopy: a new and robust spectral best-fit algorithm. J Pharm Biomed Anal 1999 Mar; 19(3–4):351–362.

Leuenberger H. Scale-up in the 4th dimension in the field of granulation and drying or how to avoid classical scale-up. Powder Technol 2003; 130:225–230.

Lin TP, Hsu CC. Determination of residual moisture in lyophilized protein pharmaceuticals using a rapid and non-invasive method: near infrared spectroscopy. PDA J Pharm Sci Technol 2002 Jul–Aug; 56(4):196–205.

Liu Y, Wang F, Lee W. On-line monitoring and controlling system for fermentation processes. Biochem Eng J 2001 Jan; 7(1):17–25.

Magari R. Estimating degradation in real time and accelerated stability tests with random lot-to-lot variation: a simulation study. J Pharm Sci 2002; 91:893–899.

Magari RT, Murphy KP, Fernandez T. Accelerated stability model for predicting shelf-life. J Clin Lab Anal 2002; 16:221–226.

McGill CA, Alison Nordon A, Littlejohn D. Potential applications of process analytical analysis by NMR spectrometry. J Process Anal Chem 2002; 6(1), http://www.infoscience.com/JPAC/JPACUserForm.html.

Morisseau KM, Rhodes CT. Near-infrared spectroscopy as a nondestructive alternative to conventional tablet hardness testing. Pharm Res 1997 Jan; 14(1):108–111.

O'Neil AJ, Jee RD, Moffat AC. Measurement of the cumulative particle size distribution of microcrystalline cellulose using near infrared reflectance spectroscopy. Analyst 1999 Jan; 124(1):33–36.

O'Neil AJ, Jee RD, Moffat AC. The application of multiple linear regression to the measurement of the median particle size of drugs and pharmaceutical excipients by near-infrared spectroscopy. Analyst 1998 Nov; 123(11):2297–2302.

Otsuka M, Kato F, Matsuda Y. Comparative evaluation of the degree of indomethacin crystallinity by chemoinfometrical Fourier-transformed [corrected] near-infrared spectroscopy and conventional powder X-ray diffractometry. AAPS PharmSci 2000; 2(1):E9.

Patel AD, Luner PE, Kemper MS. Low-level determination of polymorph composition in physical mixtures by near-infrared reflectance spectroscopy. J Pharm Sci 2001 Mar; 90(3):360–370.

Rantanen J, Rasanen E, Tenhunen J, Kansakoski M, Mannermaa J, Yliruusi J. In-line moisture measurement during granulation with a four-wavelength near infrared sensor: an evaluation of particle size and binder effects. Eur J Pharm Biopharm 2000 Sep; 50(2):271–276.

Reif OW, Freitag R. Control of the cultivation process of antithrombin III and its characterization by capillary electrophoresis. J Chromatogr A 1994, 7 Oct; 680(2):383–394.

SAS Institute. SAS/stat user's guide, vers. 8. Cary (NC), 2000.

Some IT, Bogaerts P, Hanus R, Hanocq M, Dubois J. Stability parameter estimation at ambient temperature from studies at elevated temperatures. J Pharm Sci 2001; 90:1759–1766.

Strege MA, Lagu AL. Capillary electrophoresis of biotechnology-derived proteins. Electrophoresis 1997 Nov; 18(12–13):2343–2345.
Tsong Y. Recent issues in stability testing. J Biopharm Stat 2003; 13:vii–ix.
Tsong Y, Chen WJ, Lin TY, Chen CW. Shelf life determination based on equivalence assessment. J Biopharm Stat 2003; 13:431–449.
Tydeman MS, Kirkwood TBL. Design and analysis of accelerated degradation tests for the stability of biological standards. 1. Properties of maximum likelihood estimators. J Biol Stand 1984; 12:195–206.
Watano S, Numa T, Koizumi I, Osako Y. Feedback control in high shear granulation of pharmaceutical powders. Eur J Pharm Biopharm 2001 Nov; 52(3):337–345.
Watano S, Numa T, Miyanami K, Osako Y. A fuzzy control system of high shear granulation using image processing. Powder Technol 2001; 115:124–130.
Webster GK, Farrand DA, Litchman MA. A chemical sensor based scheme to evaluate packaging containment effectiveness—Part 1: development and optimization. J Process Anal Chem 8(1): xx.
Westerhuis JA, Coenegracht PMJ, Lerk CF. Multivariate modelling of the tablet manufacturing process with wet granulation for tablet optimization and in-process control. Int J Pharm 1997; 156:109–117.
Workman J Jr, Creasy KE, Doherty S, et al. Process analytical chemistry. Anal Chem 2001; 73(12):2705–2718.
Workman J Jr, Veltkamp DJ, Doherty S, et al. Process analytical chemistry. Anal Chem (Rev) 1999; 71(12):121–180.
Yeung KS, Hoare M, Thornhill NF, Williams T, Vaghjiani JD. Near-infrared spectroscopy for bioprocess monitoring and control. Biotechnol Bioeng 1999, 20 Jun; 63(6):684–693.
Yu LX, Lionberger RA, Rawa AS, D'Costa R, Wu H, Hussain AS. Applications of process analytical technology to crystallization processes. Center for Drug Evaluation and Research, Food and Drug Administration, 8/2003.

9 Characterization of Biopharmaceutical Drugs

INTRODUCTION

An early characterization of biopharmaceuticals (proteins) is needed to evaluate the comparability of materials, which is more complicated due to the inherently heterogeneous nature of many biologicals. This includes factors such as microheterogeneity of glycosylation, differential proteolytic processing during cellular production, or variations in post-translational modifications, factors which are not common to small molecule characterization and evaluation for interaction. This requires availability of highly specific discriminating methodologies, such as spectrophotometric, chromatographic, electrophoretic methods, and mass spectroscopy (MS), often combined with liquid chromatography (LC).

Unlike small molecule drugs, there are three-dimensional (3D) and four-dimensional (4D) considerations (aggregates) with almost endless variation of polypeptides and proteins that make them a challenge to develop into products. Whereas in small molecule drugs there can be classes of drugs with common elements, this is not the case with protein drugs as each one of them offers a unique structure requiring techniques of production and purification specific to the protein. The same holds true for the stability profile of these compounds. Specification of biopharmaceutical drugs also includes elements not found in small molecules like virus clearance, aggregate formation, and so on. As a result, the regulatory authorities worldwide treat biopharmaceutical drugs under separate administration wherein a high level of expertise is inducted to evaluate these products.

Marketing authorization approvals for biological products are subject to a similar process as adopted for chemical drugs; however, the nature of these products mandates special evaluation and monitoring techniques. As a result, historically, the United States Food and Drug Administration (U.S. FDA) has established separate sections for these products. Title 21 of the Code of Federal Regulations (CFR) concerns food and drugs. Table 1 lists those parts that particularly pertain to products under the purview of the Center for Biologics Evaluation and Research (CBER) of the U.S. FDA. The CFR (1996–2005) is available for browsing and/or searching (1).

On June 30, 2003, FDA transferred some of the therapeutic biological products that had been reviewed and regulated by the CBER to the Center for Drug Evaluation and Research (CDER). CDER now has regulatory responsibility, including premarket review and continuing oversight, over the transferred products. In regulating the products assigned to them, CBER and CDER will consult with each other regularly and whenever necessary (Table 2).

The categories of therapeutic biological products transferred to CDER (2) are as follows.

Table 1 Parts of Title 21 of the Code of Federal Regulations Relevant to Biological Drugs

Parts	
1–99	Includes general enforcement regulations, product jurisdiction, enforcement policy, hearings, protection of human subjects, financial disclosure by clinical investigators, institutional review boards, and good laboratory practice for nonclinical laboratory studies.
200–299	Includes labeling, advertising, registration, medication guides, GMP, product official, and established names.
300–320	Includes combination drugs, new drugs, IND application, application for approval to market, orphan drugs, bioavailability, and bioequivalence.
600–680	Includes information for biological products: general, licensing, GMP for blood and blood components, establishment registration for manufacturers of blood and blood products, product standards, requirements for human blood and blood products.
800	Includes information for medical devices: general, labeling, reporting, in vitro diagnostic products, investigational device exemptions, premarket approval, postmarket surveillance, and classification procedures.
1270	Human tissue intended for transplantation.
1271	Human cells, tissues and cellular and tissue-based products.

Abbreviations: GMP, good manufacturing practices; IND, investigational new drug.

- Monoclonal antibodies (MAbs) for in vivo use.
- Proteins intended for therapeutic use, including cytokines (e.g., interferons), enzymes (e.g., thrombolytics), and other novel proteins, except for those that are specifically assigned to CBER (e.g., vaccines and blood products). This category includes therapeutic proteins derived from plants, animals, or microorganisms, and recombinant versions of these products.
- Immunomodulators (nonvaccine and nonallergenic products intended to treat disease by inhibiting or modifying a pre-existing immune response).
- Growth factors, cytokines, and MAbs intended to mobilize, stimulate, decrease, or otherwise alter the production of hematopoietic cells in vivo.

The categories of therapeutic biological products that remain in the CBER are as follows.

- Cellular products, including products composed of human, bacterial or animal cells (such as pancreatic islet cells for transplantation), or from physical parts of those cells (such as whole cells, cell fragments, or other components intended for use as preventative or therapeutic vaccines).
- Gene therapy products. Human gene therapy/gene transfer is the administration of nucleic acids, viruses, or genetically engineered microorganisms that mediate their effect by transcription and/or translation of the transferred genetic material, and/or by integration into the host genome. Cells may be modified in these ways ex vivo for subsequent administration to the recipient, or altered in vivo by gene therapy products administered directly to the recipient.
- Vaccines (products intended to induce or increase an antigen-specific immune response for prophylactic or therapeutic immunization, regardless of the composition or method of manufacture).

Table 2 Approved Products Transferred from Center for Biologics Evaluation and Research to Center for Drug Evaluation and Research

Product name	Proprietary name	Applicant name
Abciximab	ReoPro®	Centocor B.V.
Adalimumab	Humira®	Abbott Laboratories
Agalsidase beta	Fabrazyme®	Genzyme Corp.
Aldesleukin	Proleukin®	Chiron Corp.
Alefacept	Amevive®	Biogen, Inc.
Alemtuzumab	Campath®	ILEX Pharmaceuticals LP
Alteplase	Activase®	Genentech, Inc.
Anakinra	Kineret®	Amgen, Inc.
Anistreplase	Eminase®	Wulfing Pharma GmbH
Arcitumomab	CEA-Scan®	Immunomedics, Inc.
Asparaginase	Elspar®	Merck & Co., Inc.
Basiliximab	Simulect®	Novartis Pharmaceuticals Corp.
Becaplermin	Regranex®	OMJ Pharmaceuticals, Inc.
Becaplermin Concentrate		Chiron Corp.
Botulinum Toxin Type A	Botox®	Allergan, Inc.
Botulinum Toxin Type B	Myobloc®	Elan Pharmaceuticals
Capromab Pendetide	ProstaScint®	Cytogen Corp.
Collagenase	Santyl®	Advance Biofactures Corp.
Daclizumab	Zenapax®	Hoffmann-La Roche, Inc.
Darbepoetin alfa	Aranesp®	Amgen, Inc.
Denileukin diftitox	Ontak®	Seragen, Inc.
Dornase alfa	Pulmozyme®	Genentech, Inc.
Drotrecogin alfa	Xigris®	Eli Lilly & Co.
Epoetin alfa	Epogen®	Amgen, Inc.
Epoetin alfa	Eprex®	Ortho Biologics LLC
Etanercept	Enbrel®	Immunex Corp.
Filgrastim	Neupogen®	Amgen, Inc.
Ibritumomab tiuxetan	Zevalin®	IDEC Pharmaceuticals Corp.
Indium In-111 Chloride	Indium Chloride®	In-111 Mallinckrodt Medical, Inc.
Indium In-111 Chloride	Indiclor®	Medi-Physics, Inc.
Infliximab	Remicade®	Centocor, Inc.
Interferon alpha-2a	Roferon A®	Hoffmann-La Roche, Inc.
Interferon alpha-2b	Intron A®	Schering Corp.
Interferon alfacon-1	Infergen®	InterMune, Inc.
Interferon alpha-n3 (human leukocyte derived)	Alferon N Injection®	Interferon Sciences, Inc.
Interferon beta-1a	Avonex®	Biogen, Inc.
Interferon beta-1a	Rebif®	Serono, Inc.
Interferon beta-1b	Betaseron®	Chiron Corp.
Interferon gamma-1b	Actimmune®	InterMune, Inc.
Laronidase	Aldurazyme®	Biomarin Pharmaceutical, Inc.
Muromonab-CD3	Orthoclone OKT3®	Ortho Biotech Products, LP
Nofetumomab	Verluma®	Boehringer Ingelheim Pharma
Omalizumab	Xolair®	Genentech, Inc.
Oprelvekin	Neumega®	Genetics Institute, Inc.
Palivizumab	Synagis®	MedImmune, Inc.
Pegaspargase	Oncaspar®	Enzon, Inc.
Pegfilgrastim	Neulasta®	Amgen, Inc.
Peginterferon alpha-2a	Pegasys®	Hoffman-La Roche, Inc.

(*Continued*)

Table 2 Approved Products Transferred from Center for Biologics Evaluation and Research to Center for Drug Evaluation and Research (*Continued*)

Product name	Proprietary name	Applicant name
Peginterferon alpha-2b	Polyethylene glycol (PEG)-Intron®	Schering Corp.
Rasburicase®	Elitek®	Sanofi-Synthelabo, Inc.
Reteplase	Retavase®	Centocor, Inc.
Rituximab	Rituxan®	Genentech, Inc.
Rituximab Formulated Bulk		IDEC Pharmaceuticals Corp.
Sargramostim	Leukine®	Berlex Laboratories, Inc.
Satumomab Concentrate		LONZA Biologics PLC
Satumomab Pendetide	OncoScint CR/OV®	Cytogen Corp.
Streptokinase	Streptase®	Aventis Behring GmbH
Tenecteplase	TNKase®	Genentech, Inc.
Tositumomab and Iodine I 131 Tositumomab	Bexxar®	Corixa Corp.
Trastuzumab	Herceptin®	Genentech, Inc.
Urokinase	Abbokinase®	Abbott Laboratories

- Allergenic extracts used for the diagnosis and treatment of allergic diseases and allergen patch tests.
- Antitoxins, antivenins, and venoms.
- Blood, blood components, plasma-derived products (e.g., albumin, immunoglobulins, clotting factors, fibrin sealants, proteinase inhibitors), including recombinant and transgenic versions of plasma derivatives (e.g., clotting factors), blood substitutes, plasma volume expanders, human, or animal polyclonal antibody preparations including radiolabeled or conjugated forms, and certain fibrinolytics, such as plasma-derived plasmin and red cell reagents.

Significant developments over the past couple of decades have resulted in the development of a new class of biopharmaceutical products based on recombinant DNA technology that has placed a greater burden on the developer to provide characterization protocols that take into account modifications induced by the recombinant techniques. Generally, the analytical precision for these molecules has been not as sophisticated as available for small molecules; a few thousand Daltons seemed to be the limit of analytical accuracy; the characterization method requiring essentially biological methods that are by nature more variable. However, recent developments in both in vivo and in vitro studies needed to ensure comparable safety and efficacy as well as the powerful techniques like the high-resolution tandem mass spectrometry, circular dichroism and chromatographic media make it possible to provide complete covalent structure for proteins over 100,000 Da with less than one Dalton change in proteins routinely detectable. The newer sensitive methods now detect differences in the higher order structure as well and for all practical purposes, it is reasonable to conclude that the historic differences between characterization specifications have been removed. This is an important consideration as the regulatory authorities worldwide begin to develop standards for biogeneric or biosimilar products. The recent migration of biopharmaceutical products from the CBER to CDER at the FDA reflects this consideration and points to a level of comfort at the regulatory authorities that in vivo studies would soon become redundant in characterizing biopharmaceutical drugs. The new guidelines issued by the European Agency for the Evaluation of Medicines on the development of biosimilar products point out that several in situ or in vitro tests can be used to establish activity and safety of these molecules.

PREFORMULATION STUDIES

The unique nature of characteristics of biological products requires early integration of scientific activity bringing onboard protein biochemists, purification scientists, quality control and regulatory affairs personnel, clinical investigators, manufacturing technicians, marketing specialists, and managers.

The diversity of preformulation studies required to characterize biological drugs requires application of a comprehensive array of multiple sensitive and selective analytical methods to several batches. This requires use of orthogonal methods, to study virtually every observable property of a protein including covalent structure, conformation, pI, aggregation, charge, mass, fragmentation, surface structure, hydrophobicity, spectrophotometric, magnetic resonance, fluorescence, light scattering, sedimentation, electrophoretic properties, charge, immunological properties, enzyme activity, biological potency, substituent patterns, and so on. The methods currently available adequately characterize covalent structure using peptide mapping with high-resolution (LC/MS) (Q-TOF) to locate every atom in peptide backbone and other sensitive methods as circular dichroism (CD), high-performance liquid chromatography (HPLC), Fourier transform-infrared (FT-IR), fluorescence, nuclear magnetic resonance (NMR), size exclusion chromatography (SEC), light scattering, AUC, enzyme-linked immunosorbent assay (ELISA), and SPR. The impurities are easily detected using HPLC, SEC, isoelectric chromatography, sodium dodecyl sulfate-polyacrylamide gel electrophoresis (SDS-PAGE), and isoelectric focusing (IEF). Multidimensional analytical methods (e.g., LC/MS, 2D NMR) provide greater information than the 1D methods, particularly when combined with multivariate mathematical methods for analyzing complex mixture data. The preformulation studies further encompass the scope of studies that are required to prove equivalence of products in the new generation of biogeneric (or biosimilar or follow-on) products. The unique set of activities related to overcoming the inherent instability of the drug is referred to as formulation development and would not normally fall under preformulation exercises, but recently the boundary between preformulation and formulation studies of biological drugs is disappearing.

A biopharmaceutical drug often undergoes development before its mechanism of action or other properties are established. The new protein may be identified through genomics or proteomics activities or through more traditional medical research. It may initially be associated with a particular disease process or a certain metabolic event. In any case, its mechanism of action—as well as many of its structural characteristics and biochemical properties—may be unknown. One of the more challenging aspects of developing protein pharmaceuticals is dealing with and overcoming the inherent physical and chemical instabilities of proteins. This inherent instability has the potential to alter the state of the protein from the desired (native) form to an undesirable form (upon storage), compromising patient safety and drug efficacy. While there are numerous ways for a protein to lose its stability, the three most commonly encountered modes of denaturation and degradation are aggregation, oxidation, and deamidation. The commonly accepted strategy for rational formulation development relies on identifying mechanisms of denaturation and degradation in order to develop effective countermeasures. Once the specifics of any particular degradation pathway are understood, a more informed choice regarding excipients and formulation can be made, accelerating product development.

Preformulation is an exploratory activity that begins early in biopharmaceutical development. Preformulation studies are designed to determine the compatibility of initial excipients with the active substance for a biopharmaceutical, physicochemical, and analytical investigation in support of promising experimental formulations. Data from preformulation studies provide the necessary groundwork for formulation attempts. Successful formulations take into account a drug's interactions with the physico-chemical properties of other ingredients (and their interactions with each other) to produce a safe, stable, beneficial, and marketable product. One factor that narrows the scope of studies is the route of administration as most biological drugs are likely to be administered parenterally; this obviates the need and investment in characterization vis-à-vis other dosage form presentations. However, some other considerations become involved such as the dosage required; for example, while many drugs are administered at the microgram level, many MAbs are given in much larger quantities (hundreds of milligrams per dose) and are normally delivered intravenously. The drive to reduce healthcare costs has created a need to administer MAb therapeutics more conveniently, at home, subcutaneously. Thus, MAbs must be available at high concentrations (~200 mg/mL) in the vial. At these high concentrations, MAb-containing solutions are viscous, making them difficult to administer conveniently. Hence, a preformulation activity that needs to be considered is a concentration study investigating solubility behavior, effect of concentration on viscosity, and increased potential for aggregation. These studies have the potential to strongly influence the target product profile and the design of the clinical trial. Similarly, such questions as to whether a drug is going to be administered in a freeze-dried form or liquid form focus the studies accordingly. These considerations are mostly determined by stability considerations as freeze-drying imparts greater stability; however, some recent studies suggest that the reconstitution process can alter the 3D structure of proteins making them more immunogenic.

Preformulation begins with thorough characterization of proteins including their pharmacokinetics and physico-chemical characterization. Ideally, this information should be in hand at the beginning of the product development program, but this is unrealistic as product characterization is an evolving process that involves contributions at different stages of the product life cycle.

Stability

The most difficult aspect of biopharmaceutical stabilization is the ease with which these products begin to show aggregation, something which is often difficult to predict. As a result, evaluation of biopharmaceutical products focuses on both physical and chemical stability studies that investigate temperature dependencies to convert in vivo to native forms (if it has denatured). Bioassays study protein activity, identity, and critical pathways. Chemical degradation changes the primary structure of a protein. Bond cleavage will create an entirely new molecule. Such chemical degradation is usually preceded by a causal physical process, typically unfolding, which makes available residues that are usually inaccessible for chemical reactions with their environment. Physical degradation changes only the higher-order structure (secondary, tertiary, quaternary) of the polypeptide, not necessarily creating a brand new molecule. Such degradation includes aggregation, adsorption, unfolding, and precipitation.

Because proteins and peptides are such large molecules and exist to interact with their environment, they are somewhat fragile. They must be protected from denaturation and degradation until they can be delivered to their site of action in a patient's body.

The biopharmaceutical development process does not allow enough time to confirm stability requirements for a final formulation (which could take two years) before the company is otherwise ready to apply to market the product. Initial indications should be developed during clinical studies, so formulators must begin with three-, six-, and nine-month tests of their molecule's structure and innate stability using various analytical methods. "Accelerated" stability tests subject products to various stresses: a range of pH values, heat, light, freezing and thawing conditions, additives, and surface materials and interfaces. The test for agitation-induced denaturation is performed by swirling (creating a vortex inside the vials), rotating vials at elevated temperatures, and testing the surface tension of the liquid formulation.

The stress tests form the basis of many directions in which the preformulation studies can assist in the final dosage formulation. The common tests include the shake test (agitation), surfactant test, freeze-thaw test, and heating experiments (limited because of denaturation). Each formulation configuration is shaken in a vial to determine whether it forms aggregates. Then a surfactant (usually a polysorbate detergent, such as Tween 80®) may be selected to prevent formation of precipitants by making it harder for proteins to aggregate. Human albumin is a frequent additive, but its use is discouraged because of supply constraints and viral clearance requirements. Formulations are checked through multiple freeze-thaw cycles (which can take about a week) to check for the effects of temperature and freezing-process stresses. Most proteins are stable around 2–8°C, but few are stable at room temperature. Heating experiments help scientists examine degradation at temperature extremes by heating them to 30°C (about 86°F), and maybe even 45°C (about 113°F). At high temperatures, different mechanisms of protein denaturation may arise.

Most prominent stability reactions include oxidation, hydrolysis, and disulfide exchange. Stability of protein products can be significantly enhanced if the oxygen is removed from the headspace of the unit pack as oxygen induces specific degradation reactions, which are complex and difficult to study. In most instances, this will speed up the development time as fewer matrices would have to be evaluated for factors affecting stability. Certain amino acids (tryptophan, methionine, cysteine, histidine, and tyrosine) are susceptible to oxidation. Metal ions like copper and iron can accelerate the process of oxidation as does the higher pH, fluorescent light and hydrogen peroxide. If the amino acids along a polypeptide chain are deformed by oxidation, the molecule can be irreversibly altered, and the new molecule will likely be inactive. Antioxidants help protect against oxidation by scavenging oxygen for themselves. Ascorbic acid is used, but citric acid is preferred, and it can be used as a pH adjuster as well.

Hydrolysis of a side-chain amide on a polypeptide's glutamine or asparagine residues can yield a carboxylic acid. The process, called deamidation, is facilitated by elevated temperature and pH, resulting in loss of activity. The peptide bonds that hold amino acids together in the chain can also be severed by hydrolysis—particularly where aspartic acid residues are located. This effect is usually due to heat or to low pH.

Cysteine residues form disulfide bonds, which are important to protein structural integrity. Shuffling of these bonds, where two sulfur atoms from two different amino acid molecules link up, often changes the 3D structure, causing a loss of activity.

Excipients

Salts and nonelectrolytes (such as ammonium sulfate and glycerol) help stabilize proteins in high temperatures and low pH when freezing is not an option, but they still require low-temperature storage. Sometimes, they must be removed before the drug is used, which can be inconvenient, time-consuming, and expensive. Also, the active ingredient must be diluted, allowing further waste and variability in the final product just as in reconstituting freeze-dried products.

The biological drug is not likely to be stable without the addition of additives like buffers, albumin, or surfactants if kept in a liquid form; otherwise, an early conversion of the drug to a lyophilized form is completed. Most formulations would not include any preservative. Table 3 lists the recombinant products approved by the U.S. FDA and their composition. This table should serve as a good source to scout the most common additives found in these formulations.

An early preformulation decision is generally made regarding the final formulation, whether it will be liquid or lyophilized. It is this a choice that determines which excipients will be needed (Table 4), as liquid and lyophilized products require different excipients. Several products are available in both forms. As clinical supplies will often require a placebo, unless the product is developed as a generic equivalent, these formuations will be made without the active ingredient.

There are common ingredients to specific type of products, for example, vaccines would be stabilized and formulated differently requiring adsorption ingredients or the use of zinc in insulin to prolong its action, and so on. It is important to know that despite the large number of products available in the market, the formulation of these injectable products includes a rather limited number of components as anticipated; therefore, any preformulation study may be extended to these interaction studies. As stabilized commercial products have been in use, it becomes easier to select an excipient that would be compatible with packaging commodities, particularly syringes, rubber stoppers, and so on.

Prospective formulations will depend to a great degree on the pharmacokinetics, dosing levels, and indications for use. Unlike other drugs where the pharmacology of the product may be a new discovery, many biological products, especially the recombinant DNA products, focus on providing the endogenous compounds or antigens to promote antibody production. As a result, the indications may well be known at the preformulation stages and interaction studies to obtain a delivery system can be made by taking into account the pharmacokinetics of the drug. It is also important if the product will be administered daily, weekly, or on longer bases, or whether it would be a single-dose packaging or a multiple-dose packaging. Multiple-dose packages require use of preservatives and there is much disagreement among the regulatory authorities on how the preservative efficacy test should be conducted and also what type of preservatives to use; for example, phenol can not be used in products destined for Japan.

(*Text continues on page 348.*)

Characterization of Biopharmaceutical Drugs

Table 3 Composition of Approved Biological Products

Product	Composition
Abciximab is chimeric human-murine MAb at 47,615 Da.	Each single-use vial contains 2 mg/mL of Abciximab in a buffered solution (pH 7.2) of 0.01 M sodium phosphate, 0.15 M sodium chloride, and 0.001% polysorbate 80 in water for injection. No preservatives are added.
Adalimumab is a recombinant human IgG1 MAb that consists of 1330 amino acids and has a MW of approximately 148 kDa.	Each syringe delivers 0.8 mL (40 mg) of DP. Each vial contains approximately 0.9 mL of solution to deliver 0.8 mL (40 mg) of DP. Each 0.8 mL contains 40 mg of adalimumab, 4.93 mg of sodium chloride, 0.69 mg of monobasic sodium phosphate dihydrate, 1.22 mg of dibasic sodium phosphate dihydrate, 0.24 mg of sodium citrate, 1.04 mg of citric acid monohydrate, 9.6 mg of mannitol, 0.8 mg of polysorbate 80 and water for injection, USP. Sodium hydroxide added as necessary to adjust pH.
Aldesleukin, an interleukin-2 product with MW of approximately 15,300 Da. The chemical name is des-alanyl-1, serine-125 human interleukin-2.	Each milliliter contains 18 million International Unit (IU) (1.1 mg) of aldesleukin, 50 mg mannitol, and 0.18 mg sodium dodecyl sulfate, buffered with approximately 0.17 mg monobasic and 0.89 mg dibasic sodium phosphate to a pH of 7.5 (range 7.2–7.8).
Alemtuzumab is a recombinant DNA-(rDNA) derived humanized MAb (Campath-1H) that has an approximate MW of 150 kDa.	Each single use ampule of Campath contains 30 mg Alemtuzumab, 24.0 mg sodium chloride, 3.5 mg dibasic sodium phosphate, 0.6 mg potassium chloride, 0.6 mg monobasic potassium phosphate, 0.3 mg polysorbate 80, and 0.056 mg disodium edetate. No preservatives are added.
Alteplase is a glycoprotein of 527 amino acids.	Each 100-mg vial contains alteplase (100 mg), L-arginine (3.5 g), phosphoric acid (1 g), polysorbate 80 (>11 mg); the 50-mg vial is packed under vacuum.
Anakinra is a nonglycosylated form of the human interleukin-1 and consists of 153 amino acids and has a MW of 17.3 kDa.	Each 1-mL prefilled glass syringe contains 0.67 mL (100 mg) of anakinra in a solution (pH 6.5) containing sodium citrate (1.29 mg), sodium chloride (5.48 mg), disodium EDTA (0.12 mg), and polysorbate 80 (0.70 mg) in water for injection, USP.
AHF (recombinant), a glycoprotein consisting of multiple peptides including an 80 kDa and various extensions of the 90 kDa.	It is formulated with sucrose (0.9–1.3%), glycine (21–25 mg/mL), and histidine (18–23 mM) as stabilizers in the final container in place of albumin. The final product also contains calcium chloride (2–3 mM), sodium (27–36 mEq/L), chloride (32–40 mEq/L), polysorbate 80 [not more than (NMT) 35 μg/mL], imidazole (NMT 20 μg/1000 IU), tri-n-butyl phosphate (NMT 5 μg/1000 IU), and copper (NMT 0.6 μg/1000 IU). The product contains no preservatives. The albumin formulation contains 12.5 mg/mL albumin (human), 1.5 mg/mL polyethylene glycol (PEG) (3350), 180 mEq/L sodium, 55 mM histidine, 1.5 pg/AHF IU polysorbate-80, and 0.20 mg/mL calcium.
Antihemophilic Factor (recombinant), Plasma/Albumin-Free Method (rAHF-PFM) is a purified glycoprotein consisting of 2332 amino acids.	When reconstituted, the product contains the following stabilizers in maximal amounts: 38 mg/mL mannitol, 10 mg/mL trehalose, 108 mEq/L sodium, 12 mM histidine, 12 mM Tris, 1.9 mM calcium, 0.17 mg/mL polysorbate-80, and 0.10 mg/mL glutathione. In another formulation, each vial contains nominally 250, 500, or 1000 IU per vial and upon reconstitution contains sodium chloride, sucrose, L-histidine, calcium chloride, and polysorbate 80.

(Continued)

Table 3 Composition of Approved Biological Products (Continued)

Product	Composition
Arcitumomab, formulated to be labeled with Technetium Tc 99 m, is a Fab' fragment generated from IMMU-4, a murine IgG1 MAb produced in murine ascitic fluid and is of 50,000 Da.	Each vial contains the nonradioactive materials necessary to prepare one patient dose, which is a sterile, lyophilized formulation, containing 1.25 mg of Arcitumomab and 0.29 mg of stannous chloride per vial, with potassium sodium tartrate tetrahydrate, sodium acetate trihydrate, sodium chloride, acetic acid, glacial, hydrochloric acid, and sucrose. Technetium Tc 99 m Arcitumomab, is formed by reconstitution of the contents of this vial with 30 mCi of Tc 99 m sodium pertechnetate in 1 mL of sodium chloride for injection, USP.
Basiliximab is a chimeric (murine/human) MAb (also known as CD25 antigen) on the surface of activated T-lymphocytes; the calculated MW of the glycoprotein is 144 kDa.	Each vial contains 20 mg basiliximab, 7.21 mg monobasic potassium phosphate, 0.99 mg disodium hydrogen phosphate (anhydrous), 1.61 mg sodium chloride, 20 mg sucrose, 80 mg mannitol, and 40 mg glycine, to be reconstituted in 5 mL of SWFI, USP. No preservatives are added.
Becaplermin has a MW of approximately 25 kDa and is a homodimer composed of two identical polypeptide chains that are bound together by disulfide bonds.	Each gram of gel contains 100 μg of becaplermin, sodium carboxymethylcellulose, sodium chloride, sodium acetate trihydrate, glacial acetic acid, water for injection, and methylparaben, propylparaben, and *m*-cresol as preservatives, and L-lysine hydrochloride as a stabilizer.
Capromab pendetide is the murine MAb, 7E11-C5.3, conjugated to the linker-chelator, glycyl-tyrosyl-(*N*-diethylenetriaminepentaacetic acid)-lysine hydrochloride (GYK-DTPA-HCl). Given with Indium In 111.	Each vial contains 0.5 mg of capromab pendetide in 1 mL of sodium phosphate-buffered saline solution adjusted to pH 6; the sodium acetate solution must be added to the sterile, nonpyrogenic high-purity Indium In 111 chloride solution to buffer it prior to radiolabeling.
Choriogonadotropin alpha is a water-soluble glycoprotein consisting of two noncovalently linked subunits—designated (alpha) and (beta)—consisting of 92 and 145 amino acid residues, respectively, with carbohydrate moieties linked to ASN-52 and ASN-78 (on alpha subunit) and ASN-13, ASN-30, SER-121, SER-127, SER-132, and SER-138 (on beta subunit).	Each vial contains 285 mcg of choriogonadotropin alpha, 30 mg sucrose, 0.98 mg phosphoric acid, and sodium hydroxide (for pH adjustment) which, when reconstituted with the diluent, will deliver 250 mcg of recombinant human chorionic gonadotropin. The pH of the reconstituted solution is 6.5–7.5.
Coagulation Factor IX (recombinant) is a glycoprotein with an approximate molecular mass of 55,000 Da consisting of 415 amino acids in a single chain.	Each vial contains 250, 500, or 1000 IU of Coagulation Factor IX (recombinant). After reconstitution of the lyophilized DP, the concentrations of excipients in the 500- and 1000-IU dosage strengths are 10 mM L-histidine, 1% sucrose, 260 mM glycine, and 0.005% polysorbate 80. The concentrations after reconstitution in the 250-IU dosage strength are half those of the other two dosage strengths. The 500- and 1000-IU dosage strengths are isotonic after reconstitution, and the 250-IU dosage strength has half the tonicity of the other two dosage strengths after reconstitution. All dosage strengths yield a clear, colorless solution upon reconstitution.

Characterization of Biopharmaceutical Drugs

Coagulation Factor VIIa (rFVIIa) is a vitamin K-dependent glycoprotein consisting of 406 amino acid residues (MW is 50 kDa).	Each vial contains 1.2 mg rFVIIa, 5.85 mg sodium chloride, calcium chloride (2.94 mg), glycylglycine (2.64 mg), Tween 80 (0.14 mg), and manitol (60 mg); the sodium content is 0.44 mEq and calcium content is 0.06 mEq.
Daclizumab is humanized IgG1 MAb with MW of approximately 144 kDa.	Each milliliter contains 5 mg of Daclizumab and 3.6 mg of sodium phosphate monobasic monohydrate, 11 mg of sodium phosphate dibasic heptahydrate, 4.6 mg of sodium chloride, 0.2 mg of polysorbate 80 and may contain hydrochloric acid or sodium hydroxide to adjust the pH to 6.9. No preservatives are added.
Darbepoetin alpha is a 165-amino acid protein that differs from recombinant human erythropoietin in containing five N-linked oligosaccharide chains, whereas recombinant human erythropoietin contains three. The two additional N-glycosylation sites result from amino acid substitutions in the erythropoietin peptide backbone. The additional carbohydrate chains increase the approximate MW of the glycoprotein from 30,000 to 37,000 Da.	Two formulations contain excipients as follows: polysorbate solution contains 0.05 mg of polysorbate 80, 2.12 mg of sodium phosphate monobasic monohydrate, 0.66 mg of sodium phosphate dibasic anhydrous, and 8.18 mg of sodium chloride in water for injection, USP (per 1 mL) at pH 6.2 ± 0.2. Albumin solution contains 2.5 mg of albumin (human), 2.23 mg of sodium phosphate monobasic monohydrate, 0.53 mg of sodium phosphate dibasic anhydrous, and 8.18 mg of sodium chloride in water for injection, USP (per 1 mL) at pH 6.0 ± 0.3.
Denileukin diftitox, a rDNA-derived cytotoxic protein composed of the amino acid sequences for diphtheria toxin fragments A and B (Met 1-Thr 387)-His followed by the sequences for interleukin-2 (IL-2; Ala 1-Thr 133) and has a MW of 58 kDa.	Each 2-mL vial contains 300 mcg of recombinant denileukin diftitox in a sterile solution of citric acid (20 mM), EDTA (0.05 mM), and polysorbate 20 (<1%) in water for injection, USP. The solution has a pH of 6.9–7.2.
Dornase alpha, rhDNase, a glycoprotein that contains 260 amino acids with an approximate MW of 37,000 Da.	The aqueous solution contains 1.0 mg/mL of dornase alpha, 0.15 mg/mL of calcium chloride dihydrate, and 8.77 mg/mL of sodium chloride. The solution contains no preservative. The nominal pH of the solution is 6.3 and is used as nebulizer.
Drotrecogin alpha (activated) is a glycoprotein of approximately 55 kDa MW, consisting of a heavy chain and a light chain linked by a disulfide bond.	Each vial contains 5 and 20 mg drotrecogin alpha and 40.3 and 158.1 mg of sodium chloride, 10.9 and 42.9 mg of sodium citrate, and 31.8 and 124.9 mg of sucrose, respectively.
Efalizumab is humanized IgG1 kappa isotype, monoclonal, and has a MW of approximately 150 kDa.	Each single-use vial contains 150 mg of efalizumab, 123.2 mg of sucrose, 6.8 mg of L-histidine hydrochloride monohydrate, 4.3 mg of L-histidine and, 3 mg of polysorbate 20 and is designed to deliver 125 mg of efalizumab in 1.25 mL.

(Continued)

Table 3 Composition of Approved Biological Products (Continued)

Product	Composition
Erythropoietin is a glycoprotein (Epoetin alpha), a 165 amino acid glycoprotein with MW of 30,400 Da.	Each 1 mL of single dose solution contains 2000, 3000, 4000 or 10,000 units of Epoetin alpha, 2.5 mg of albumin (human), 5.8 mg of sodium citrate, 5.8 mg of sodium chloride, and 0.06 mg of citric acid in water for injection, USP (pH 6.9 ± 0.3). This formulation contains no preservative. The 2-mL (20,000 units, 10,000 units/mL) multidose vial, each 1 mL of solution contains 10,000 units of Epoetin alpha, 2.5 mg of albumin (human), 1.3 mg of sodium citrate, 8.2 mg of sodium chloride, 0.11 mg of citric acid, and 1% benzyl alcohol as preservative in water for injection, USP (pH 6.1 ± 0.3).
Etanercept is a dimeric fusion protein consisting of the extracellular ligand-binding portion of the human 75 kDa (p75) TNFR linked to the Fc portion of human IgG1. The Fc component of etanercept contains the CH_2 domain, the CH_3 domain, and hinge region, but not the CH_1 domain of IgG1. It consists of 934 amino acids and has an apparent MW of approximately 150 kDa.	After reconstitution with 1 mL of the supplied sterile BWFI, USP (containing 0.9% benzyl alcohol), solution is clear and colorless, with a pH of 7.4 ± 0.3. Each single-use vial of etanercept contains 25 mg of etanercept, 40 mg of mannitol, 10 mg of sucrose, and 1.2 mg of tromethamine.
Filgrastim is a human G-CSF, and is a 175 amino acid protein that has a MW of 18,800 Da.	The single-use 300-mcg prefilled syringe (0.5 mL) contains acetate (0.295 mg), sorbital (25 mg), Tween 80 (0.004%), and sodium (0.0175 mg); higher concentration of 480 mcg contains proportional amounts of additives.
Follicle-stimulating hormone (FSH) consists of two noncovalently linked, nonidentical glycoproteins designated as the (alpha)- and (beta)-subunits. The (alpha)- and (beta)-subunits have 92 and 111 amino acids, respectively.	For subcutaneous injection after reconstitution with either SWFI, USP for single-dose ampules or BWFI (0.9% benzyl alcohol), USP for multiple-dose vials. Each ampule contains either 37.5, 75, or 150 IU rFSH, 30 mg of sucrose, 1.11 mg of dibasic sodium phosphate, and 0.45 mg of monobasic sodium phosphate monohydrate. O-phosphoric acid and/or sodium hydroxide may be used prior to lyophilization for pH adjustment. It may contain up to 15% of oxidized FSH.
hFSH, a glycoprotein hormone, which is manufactured by rDNA technology, a dimeric structure containing two glycoprotein subunits (alpha and beta). Both the 92 amino acid alpha-chain and the 111 amino acid beta-chain have complex heterogeneous structures arising from two N-linked oligosaccharide chains.	Each vial contains 75 IU of FSH activity plus 25.0 mg of sucrose, NF, 7.35 mg of sodium citrate dihydrate, USP, 0.10 mg of polysorbate 20, NF, and hydrochloric acid, NF and/or sodium hydroxide, NF to adjust the pH to a sterile, lyophilized form. The pH of the reconstituted preparation is approximately 7.0.

Galactosidase alpha is an enzyme, which is a homodimeric glycoprotein with a MW of approximately 100 kDa.	Each vial contains 37 mg of alpha-galactosidase beta as well as 222 mg of mannitol, 20.4 mg of sodium phosphate monobasic monohydrate, and 59.2 mg of sodium phosphate dibasic heptahydrate. Following reconstitution as directed, 35 mg of agalactosidase beta (7 mL) may be extracted from each vial.
Gemtuzumab ozogamicin is a humanized IgG4, kappa antibody conjugated with a cytotoxic antitumor or antibiotic, calicheamicin via a bifunctional linker. It has a MW of 151–153 kDa.	Each vial contains 5 mg of drug conjugate (protein equivalent) in a 20-mL amber vial. The inactive ingredients are: dextran 40, sucrose, sodium chloride, monobasic and dibasic sodium phosphate. Light-sensitive.
Glucagon (rDNA origin) has the empirical formula of $C_{153}H_{225}N_{43}O_{49}S$, and a MW of 3483. It is a single-chain polypeptide containing 29 amino acid residues and a MW of 3483.	Each vial contains Glucagon as hydrochloride, 1 mg (corresponding to 1 IU). Other ingredient is lactose monohydrate (107 mg).
Hemophilus b Conjugate (Meningococcal Protein Conjugate) and Hepatitis B (Recombinant) Vaccine is a sterile bivalent vaccine.	Each 0.5 mL contains 7.5-mcg dose conjugated to approximately 125 mcg of outer membrane protein complex, 5 mcg of HBsAg, approximately 225 mcg of aluminum as amorphous aluminum hydroxyphosphate sulfate, and 35 mcg of sodium borate (decahydrate) as a pH stabilizer, in 0.9% sodium chloride. The vaccine contains not more than 0.0004% (w/v) residual formaldehyde.
Hepatitis B (recombinant) Vaccine is a noninfectious rDNA hepatitis B vaccine containing purified surface antigen of the virus obtained by culturing genetically engineered *Saccharomyces cerevisiae* cells, which carry the surface antigen gene of the hepatitis B virus.	Each 0.5 mL of the vaccine consists of 10 mcg of hepatitis B surface antigen adsorbed on 0.25 mg of aluminum as aluminum hydroxide with a trace amount of thimerosal (<0.5 mcg of mercury) from the manufacturing process, sodium chloride (9 mg/mL), and phosphate buffers (disodium phosphate dihydrate, 0.98 mg/mL; sodium dihydrogen phosphate dihydrate, 0.71 mg/mL). Each 1-mL adult dose consists of 20 mcg of hepatitis B surface antigen adsorbed on 0.5 mg of aluminum as aluminum hydroxide. The adult vaccine is formulated without preservatives. The adult formulation contains a trace amount of thimerosal (<1.0 mcg of mercury) from the manufacturing process, sodium chloride (9 mg/mL), and phosphate buffers (disodium phosphate dihydrate, 0.98 mg/mL; sodium dihydrogen phosphate dihydrate, 0.71 mg/mL).
Iinsulin glargine (rDNA origin) injection is human insulin analog that differs from human insulin in that the amino acid asparagine at position A21 is replaced by glycine and two arginines are added to the C-terminus of the B-chain. Chemically, it is 21 A-Gly-30 B a-L-Arg-30 B b-L-Arg-human insulin and has the empirical formula $C_{267}H_{404}N_{72}O_{78}S_6$ and a MW of 6063.	Each milliliter contains 100 IU (3.6378 mg) of insulin glargine, 30 mcg of zinc, 2.7 mg of *m*-cresol, 20 mg of glycerol 85%, and water for injection. The pH is adjusted by addition of aqueous solutions of hydrochloric acid and sodium hydroxide and has a pH of approximately 4.

(Continued)

Table 3 Composition of Approved Biological Products (*Continued*)

Product	Composition
Imiglucerase is an analog of the human enzyme, (beta)-glucocerebrosidase, a monomeric glycoprotein of 497 amino acids, containing four N-linked glycosylation sites (MW = 60,430).	Each vial contains imiglucerase (212 units), mannitol (170 mg), sodium citrates (70 mg), trisodium citrate (52 mg), disodium hydrogen citrate (18 mg), polysorbate 80 (0.53 mg). Citric acid and/or sodium hydroxide may have been added at the time of manufacture to adjust pH. Haemaccel® (cross-linked gelatin polypeptides) is used as a stabilizing agent.
Infliximab is a chimeric IgG1 k MAb with an approximate MW of 149,100 Da. It is composed of human constant and murine variable regions.	Each single-use vial contains 100 mg of infliximab, 500 mg of sucrose, 0.5 mg of polysorbate 80, 2.2 mg of monobasic sodium phosphate, monohydrate, and 6.1 mg of dibasic sodium phosphate, dihydrate. No preservatives are present.
Insulin aspart (rDNA origin) homologous with regular human insulin with the exception of a single substitution of the amino acid proline by aspartic acid in position B28 with the empirical formula $C_{256}H_{381}N_{65}O_{79}S_6$ and a MW of 5825.8.	Each milliliter contains insulin aspart (B28 asp regular human insulin analog), 100 units/mL, glycerin (16 mg/mL), phenol (1.50 mg/mL), metacresol (1.72 mg/mL), zinc (19.6 μg/mL), disodium hydrogen phosphate dihydrate (1.25 mg/mL), and sodium chloride (0.58 mg/mL). It has a pH of 7.2–7.6. Hydrochloric acid 10% and/or sodium hydroxide 10% may be added to adjust pH.
Insulin glulisine (rDNA origin) differs from human insulin in that the amino acid asparagine at position B3 is replaced by lysine and the lysine in position B29 is replaced by glutamic acid. Chemically, it is 3B-lysine-29B-glutamic acid-human insulin, and has the empirical formula $C_{258}H_{384}N_{64}O_{78}S_6$ and a MW of 5823.	Each milliliter of APIDRA (insulin glulisine injection) contains 100 IU (3.49 mg) of insulin glulisine, 3.15 mg of *m*-cresol, 6 mg of trometamine, 5 mg of sodium chloride, 0.01 mg of polysorbate 20, and water for injection.
Insulin lispro (rDNA origin) is Lys(B28), Pro(B29) human insulin analog, created when the amino acids at positions 28 and 29 on the insulin B-chain are reversed. It has the empirical formula $C_{257}H_{383}N_{65}O_{77}S_6$ and a MW of 5808, both identical to that of human insulin.	Each milliliter contains insulin lispro 100 units, 16 mg of glycerin, 1.88 mg of dibasic sodium phosphate, 3.15 mg of *m*-cresol, zinc oxide content adjusted to provide 0.0197 mg of zinc ion, trace amounts of phenol, and water for injection. Insulin lispro has a pH of 7.0–7.8. Hydrochloric acid 10% and/or sodium hydroxide 10% may be added to adjust pH.
Interferon alpha-2a contains 165 amino acids, and it has an approximate MW of 19,000 Da.	Each milliliter contains 3 MIU of interferon alpha-2a, recombinant, 7.21 mg of sodium chloride, 0.2 mg of polysorbate 80, 10 mg of benzyl alcohol as a preservative, and 0.77 mg of ammonium acetate.

Characterization of Biopharmaceutical Drugs

Interferon alpha-2b has a MW of 19,271 Da.	Each milliliter contains 3, 5, 18, 25, or 50 MIU and also contains 20 mg of glycine, 2.3 mg of sodium phosphate dibasic, 0.55 mg of sodium phosphate monobasic, and 1.0 mg of human albumin after reconstitution. The solution formulation contains besides the active drug, 7.5 mg of sodium chloride, 1.8 mg of sodium phosphate dibasic, 1.3 mg of sodium phosphate monobasic, 0.1 mg of edetate disodium, 0.1 mg of polysorbate 80, and 1.5 mg of m-cresol as a preservative. The multidose preparations also contain 1.5 mg of cresol as preservative.
Interferon alphacon-1 has a 166-amino acid sequence and differs from interferon alpha-2b at 20/166 amino acids (88% homology), and comparison with interferon-beta shows identity over 30% of the amino acid positions and has a MW of 19,434 Da.	Each vial and prefilled syringe contains 0.03 mg/mL of interferon alphacon-1, 5.9 mg/mL of sodium chloride, and 3.8 mg/mL of sodium phosphate in water for injection, USP.
Interferon beta-1a is a 166-amino acid glycoprotein with a MW of approximately 22,500 Da.	Each 0.5 ml contains either 44 mcg or 22 mcg of interferon beta-1a, 4 or 2 mg of albumin (human) USP, 27.3 mg of mannitol USP, 0.4 mg of sodium acetate, and water for injection, USP.
Interferon beta-1a is a 166-amino acid glycoprotein with a predicted MW of approximately 22,500 Da.	Each 1.0 mL (1.0 cc) of reconstituted solution contains 30 mcg of interferon beta-1a, 15 mg of albumin (human), USP, 5.8 mg of sodium chloride, USP, 5.7 mg of dibasic sodium phosphate, USP, and 1.2 mg of monobasic sodium phosphate, USP, at a pH of approximately 7.3.
Interferon beta-1b is a protein that has 165 amino acids and an approximate MW of 18,500 Da. It does not include the carbohydrate side-chains found in the natural material.	Dextrose and albumin (human), USP (15 mg each/vial) are added as stabilizers. Lyophilized Betaseron is a sterile, white to off-white powder intended for subcutaneous injection after reconstitution with the diluent supplied (sodium chloride, 0.54% solution).
Interferon gamma-1b is a single-chain polypeptide containing 140 amino acids consisting of noncovalent dimers of two identical 16,465-Da monomers.	Each 0.5 mL contains: 100 mcg (two million IU) of interferon gamma-1b formulated in 20 mg of mannitol, 0.36 mg of sodium succinate, 0.05 mg of polysorbate 20, and SWFI.
Interleukin eleven (IL-11) is a thrombopoietic growth factor and has a molecular mass of approximately 19,000 Da, and is nonglycosylated. The polypeptide is 177 amino acids in length.	Each vial contains 5 mg of IL-11 with 23 mg of glycine, USP, 1.6 mg of dibasic sodium phosphate heptahydrate, USP, and 0.55 mg of monobasic sodium phosphate monohydrate, USP. When reconstituted with 1 mL of SWFI, USP, the resulting solution has a pH of 7.0 and a concentration of 5 mg/mL.

(Continued)

Table 3 Composition of Approved Biological Products (Continued)

Product	Composition
Laronidase is a glycoprotein with a MW of approximately 83 kDa. The recombinant protein is comprised of 628 amino acids after cleavage of the N-terminus and contains six N-linked oligosaccharide modification sites. Two oligosaccharide chains terminate in mannose-6-phosphate sugars.	Must be diluted prior to administration in 0.9% sodium chloride injection, USP containing 0.1% of albumin (human). The solution in each vial contains a nominal laronidase concentration of 0.58 mg/mL and a pH of approximately 5.5. The extractable volume of 5.0 mL from each vial provides 2.9 mg of laronidase, 43.9 mg of sodium chloride, 63.5 mg of sodium phosphate monobasic monohydrate, 10.7 mg of sodium phosphate dibasic heptahydrate, and 0.05 mg of polysorbate 80.
Lepirudin (rDNA) is: [Leu1, Thr2]-63-desulfohirudin, a polypeptide composed of 65 amino acids and has a MW of 6979.5 Da.	Each vial contains 50 mg of lepirudin, 40 mg of mannitol and sodium hydroxide for adjustment of pH to approximately 7.
Muromonab-CD3 is a murine MAb to the CD3 antigen of human T cells. The antibody is a biochemically purified IgG2a immunoglobulin with a heavy chain of approximately 50,000 Da and a light chain of approximately 25,000 Da.	Each ampule contains a buffered solution (pH 7.0 + 0.5) of monobasic sodium phosphate (2.25 mg), dibasic sodium phosphate (9.0 mg), sodium chloride (43 mg), and polysorbate 80 (1.0 mg) in water for injection.
Nesiritide is a hBNP with a MW of 3464 g/mol and an empirical formula of C143H244N50O42S4.	Each 1.5-mg vial contains nesiritide (1.58 mg), mannitol (20.0 mg), citric acid monohydrate (2.1 mg), and sodium citrate dihydrate (2.94 mg).
Omalizumab is a MAb with a MW of approximately 149 kDa.	Each vial contains 202.5 mg of omalizumab, 145.5 mg sucrose, 2.8 mg L-histidine hydrochloride monohydrate, 1.8 mg L-histidine, and 0.5 mg polysorbate 20, and is designed to deliver 150 mg of omalizumab in 1.2 mL after reconstitution with 1.4 mL SWFI, USP.
Palivizumab is a humanized MAb composed of two heavy chains and two light chains and has a MW of approximately 148,000 Da.	Upon reconstitution, it contains the following excipients: 47 mM of histidine, 3.0 mM of glycine and 5.6% mannitol, and the active ingredient, palivizumab, at a concentration of 100 mg/mL solution.
Parathyroid hormone (1–34) has a MW of 4117.8 Da.	Each prefilled delivery device is filled with 3.3 mL to deliver 3 mL. Each milliliter contains 250 mcg of teriparatide (corrected for acetate, chloride, and water content), 0.41 mg of glacial acetic acid, 0.10 mg of sodium acetate (anhydrous), 45.4 mg of mannitol, 3.0 mg of m-cresol, and water for injection. In addition, hydrochloric acid solution 10% and/or sodium hydroxide solution 10% may have been added to adjust the product to pH 4. Each cartridge preassembled into a pen device delivers 20 mcg of teriparatide per dose each day for up to 28 days.

Drug	Formulation
Pegfilgrastim is a covalent conjugate of recombinant methionyl human G-CSF (filgrastim) and monomethoxypoly-ethylene glycol. Filgrastim is a water-soluble 175 amino acid protein with a MW of approximately 19 kDa. To produce pegfilgrastim, a 20-kDa monomethoxypolyethylene glycol (PEG) molecule is covalently bound to the N-terminal methionyl residue of filgrastim. The average MW of pegfilgrastim is approximately 39 kDa.	Each syringe contains 6 mg of pegfilgrastim (based on protein weight), in a sterile, clear, colorless, preservative-free solution (pH 4.0) containing acetate (0.35 mg), sorbitol (30.0 mg), polysorbate 20 (0.02 mg), and sodium (0.02 mg) in water for injection.
Peginterferon alpha-2a is a covalent conjugate of recombinant alpha-2a interferon (approximate MW is 20,000 Da) with a single branched *bis*-monomethoxy PEG chain (approximate MW is 40,000 Da). Peginterferon alpha-2a has an approximate MW of 60,000 Da.	Each vial contains approximately 1.2 mL of solution to deliver 1.0 mL of DP. Subcutaneous administration of 1.0 mL delivers 180 mcg of DP (expressed as the amount of interferon alpha-2a), 8.0 mg of sodium chloride, 0.05 mg of polysorbate 80, 10.0 mg of benzyl alcohol, 2.62 mg of sodium acetate trihydrate, and 0.05 mg of acetic acid. The solution is colorless to light yellow and the pH is 6.0 ± 0.01.
Peginterferon alpha-2b is a covalent conjugate of recombinant alpha interferon with monomethoxy PEG. The MW of the PEG portion of the molecule is 12,000 Da. The average MW of the PEG-intron molecule is approximately 31,000 Da.	Each vial contains either 74, 118.4, 177.6, or 222 μg of PEG-intron, and 1.11 mg of dibasic sodium phosphate anhydrous, 1.11 mg of monobasic sodium phosphate dihydrate, 59.2 mg of sucrose, and 0.074 mg of polysorbate 80. Following reconstitution with 0.7 mL of the supplied diluent (SWFI, USP), each vial contains peginterferon at strengths of either 100, 160, 240, or 300 μg/mL.
Pegvisomant is a protein containing 191 amino acid residues to which several PEG polymers are covalently bound (predominantly 4 to 6 PEG/protein molecule). The MW of the protein of pegvisomant is 21,998 Da. The MW of the PEG portion of pegvisomant is approximately 5000 Da. The predominant MWs of pegvisomant are thus approximately 42,000, 47,000, and 52,000 Da.	Each vial also contains 1.36 mg of glycine, 36.0 mg of mannitol, 1.04 mg of sodium phosphate dibasic anhydrous, and 0.36 mg of sodium phosphate monobasic monohydrate.

(Continued)

Table 3 Composition of Approved Biological Products (*Continued*)

Product	Composition
Rasburicase is a recombinant urate-oxidase enzyme produced by a genetically modified *Saccharomyces cerevisiae* strain. It is a tetrameric protein with identical subunits of a molecular mass of about 34 kDa. The molecular formula of the monomer is $C_{1523}H_{2383}N_{417}O_{462}S_7$. The monomer, made up of a single 301 amino acid polypeptide chain, has no intra- or interdisulfide bridges and is N-terminal-acetylated.	Each 3-mL vial containing 1.5 mg of rasburicase, 10.6 mg of mannitol, 15.9 mg of L-alanine, and between 12.6 and 14.3 mg of dibasic sodium phosphate. The diluent solution for reconstitution, supplied in a 2-mL clear, glass ampule, is composed of 1.0 mL of SWFI, USP, and 1.0 mg of Poloxamer 188.
Reteplase is a nonglycosylated deletion mutein of tPA, containing the kringle 2 and the protease domains of human tPA. It contains 355 of the 527 amino acids of native tPA (amino acids 1–3 and 176–527). The MW is 39,571 Da.	Each vial contains 10.4 units (18.1 mg), tranexamic acid (8.32 mg), dipotassium hydrogen phosphate (136.24 mg), phosphoric acid (51.27 mg), sucrose (364 mg), and polysorbate 88 (5.2 mg).
Rituximab is a chimeric murine/human MAb composed of two heavy chains of 451 amino acids and two light chains of 213 amino acids.	The product is formulated for intravenous administration in 9.0 mg/mL of sodium chloride, 7.35 mg/mL of sodium citrate dihydrate, 0.7 mg/mL of polysorbate 80, and SWFI. The pH is adjusted to 6.5.
Sargramostim is a rhu GM-CSF, a glycoprotein of 127 amino acids characterized by three primary molecular species having molecular masses of 19,500, 16,800, and 15,500 Da.	Liquid formulation contains 500 mcg (2.8×10^6 IU/mL) of sargramostim and 1.1% benzyl alcohol in 1 mL solution. Lyophilized vial contains 25 mcg (1.4×10^6 IU/vial) of sargramostim. Both contain 40 mg/mL of mannitol, 10 mg/mL of sucrose, and 1.2 mg/mL of tromethamine.
Sermorelin acetate is the acetate salt of an amidated synthetic 29-amino acid peptide (GRF 1–29 NH2) consisting of 44 amino acid residues. The free base of sermorelin has the empirical formula $C_{149}H_{246}N_{44}O_{42}S$ and a MW of 3358 Da.	Each vial contains 0.5 mg of sermorelin (as the acetate) and 5 mg of mannitol. The pH is adjusted with dibasic sodium phosphate and monobasic sodium phosphate buffer. Each 1.0-mL vial contains 1.0 mg of sermorelin (as the acetate) and 5 mg of mannitol. The pH is adjusted with dibasic sodium phosphate and monobasic sodium phosphate buffer.

Characterization of Biopharmaceutical Drugs

Somatropin (rDNA origin) is a polypeptide hormone of rDNA origin. It has 191 amino acid residues and a MW of 22,124 Da.	A dose of 1.5 mg is dispensed in a two-chamber cartridge. The front chamber contains recombinant somatropin (1.5 mg) (approximately 4.5 IU), glycine (27.6 mg), sodium dihydrogen phosphate anhydrous (0.3 mg), and disodium phosphate anhydrous (0.3 mg); the rear chamber contains 1.13-mL water for injection. The 5.8-mg dose system has in the rear chamber 0.3% m-cresol (as a preservative) and mannitol, 45 mg in 1.14-mL water for injection. Long acting contains 13.5 mg of somatropin, 1.2 mg of zinc acetate, 0.8 mg of zinc carbonate, and 68.9 mg of poly-L-glutamate. In another formulation, each vial contains 8.8 mg of somatropin (approximately 26.4 IU), 60.2 mg of sucrose and 2.05 mg of O-phosphoric acid. The pH is adjusted with sodium hydroxide or O-phosphoric acid. The diluent is BWFI, USP containing 0.9% benzyl alcohol added as an antimicrobial preservative.
Tenecteplase is a 527-amino acid glycoprotein.	Each vial contains 52.5 mg of Tenecteplase, 0.55 g of L-arginine, 0.17 g of phosphoric acid, and 4.3 mg of polysorbate 20, which includes a 5% overfill.
Thyrotropin alpha is a heterodimeric glycoprotein comprised of two noncovalently linked subunits, an alpha subunit of 92 amino acid residues containing two N-linked glycosylation sites and a beta subunit of 118 residues containing one N-linked glycosylation site.	Each vial contains 1.1 mg thyrotropin alpha ($\geq$4 IU), 36 mg of mannitol, 5.1 mg of sodium phosphate, and 2.4 mg of sodium chloride. After reconstitution with 1.2 mL of SWFI, USP, the thyrotropin alpha concentration is 0.9 mg/mL. The pH of the reconstituted solution is approximately 7.0.
Tositumomab and iodine I_{131} Tositumomab. Tositumomab is composed of two murine gamma 2a heavy chains of 451 amino acids each and two lambda light chains of 220 amino acids each. The approximate MW of Tositumomab is 150 kDa.	It is supplied at a nominal concentration of 14 mg/mL Tositumomab in 35 mg and 225 mg single-use vials. The formulation contains 10% (w/v) maltose (145 mM). The formulation for the dosimetric and the therapeutic dosage forms contains 5.0–6.0% (w/v) of povidone, 1–2 mg/mL of maltose (dosimetric dose) or 9–15 mg/mL of maltose (therapeutic dose), 0.85–0.95 mg/mL of sodium chloride, and 0.9–1.3 mg/mL of ascorbic acid. The pH is approximately 7.0.
Trastuzumab is a rDNA-derived humanized MAb, an IgG1 kappa that contains human framework regions with the complementarity-determining regions of a murine antibody (4D5) that binds to HER2.	Each vial contains 440 mg of Trastuzumab, 9.9 mg of L-histidine HCl, 6.4 mg of L-histidine, 400 mg of (alpha)-trehalose dihydrate, and 1.8 mg of polysorbate 20, USP. Reconstitution with only 20 mL of the supplied BWFI, USP, containing 1.1% of benzyl alcohol as a preservative yields a multidose solution containing 21 mg/mL of Trastuzumab, at a pH of approximately 6.

Abbreviations: AHF, antihemophilic factor; BWFI, bacteriostatic water for injection; DP, drug product; EDTA, ethylenediaminetetraacetic acid; FSH, follicle-stimulating hormone; G-CSF, granulocyte colony-stimulating factor; hBNP, human B-type natriuretic peptide; IgG1, immunoglobulin G1; IU, International Unit; MAb, monoclonal antibody; MW, molecular weight; rDNA, recombinant DNA; rhDNase, recombinant human deoxyribonuclease I; rhu GM-CSF, recombinant human granulocyte-macrophage colony-stimulating factor; SWFI, sterile water for injection; TNFR, tumor necrosis factor receptor; tPA, tissue plasminogen activator.

Table 4 Common Excipients in Biological Products

Albumin, human	Maltose
Aluminum hydroxide	Mannitol
Amorphous aluminum hydroxyphosphate	Parabens
Benzyl alcohol	Polysorbate 80
Cresol	Sodiumdodecyl sulphate
Ethylenediamineteraacetic acid	Sucrose
Glutathione	Trehalose dehydrate
Glycerol	Tri-N-butyl phosphate
Glycine	Tromethamine
Imidazole	Zinc
L-histidine	

PACKAGING AND MATERIALS

Packaging material is an important part of formulation because of possible interactions. Protein may stick to the walls of a vial or other containers, particularly in low-concentration formulations. Glass is also reactive; it can be delaminated, for example, through contact with a formulation. Certain kinds of plastic are less reactive than glass. The importance of packaging materials was highlighted in the case of erythropoietin formulation sold outside of United States under the brand name of Eprex®. Since December 2002, there has been a significant and sustained worldwide decrease in newly reported cases of erythropoietin antibody-positive pure red cell aplasia (PRCA) in chronic renal failure patients who receive the recombinant erythropoietin medication, Eprex (epoetin alpha). The most probable product-specific cause of the increased incidence of PRCA in chronic renal failure patients being treated with Eprex has been identified as the so-called leachates. By 2005, the incidence of PRCA was reduced to its pre-2002 incidence when steps were taken to remove the causes of PRCA. These included the use of more immune response-causing subcutaneous route in place of intravenous use and the replacement of albumin in the formulation with Tween 80. There was also identified an interaction between uncoated rubber stoppers previously used in prefilled syringes of Eprex and the stabilizer polysorbate 80. This interaction resulted in the presence of organic compounds—called leachates—in prefilled syringes with uncoated rubber stoppers. Coated stoppers, which prevent the interaction with the stabilizer, are now used in all Eprex prefilled syringes. These observations emphasize the need to investigate possible interactions between the formulation and the stopper or elastomer since such contact could even destroy the protein. To test the reactivity of stoppers, vials are inverted—to give their rubber stoppers total contact with the formulation—and then stored horizontally to create more surface area—and more chances for the protein to degrade. More oxidation of the protein can occur as the surface area grows larger.

The replacement of albumin in Eprex example was made in response to the concern that the human serum albumin used in the manufacture of the drug could transmit a variant of Creutzfeldt-Jakob disease. This example shows remarkably how subtle change in the formulation can produce significant changes in the toxicity of the drug; generally, this would not be the case with small molecule drugs. Table 3 lists eight products wherein human albumin is used as a stabilizer compared with 31 products wherein polysorbate surfactants are used; it is noteworthy that it was not the surfactant but its interaction with the packaging commodity that resulted, at least in part, in the adverse reaction.

Dosage Form and Storage

Storage conditions and handling of product make much difference in how the product is marketed: refrigerated, frozen, freeze-dried, or kept in reduced-light yellow or brown vials, for example. Less stringent storage conditions are preferred. The shelf life of the product is preferably one to two years, but it is not possible to wait that long at the prescribed temperature in early development stage; generally, after three months of testing the data will be sent to CBER and it may be necessary to change the formulations as more stability data becomes available; while formulation changes prove expensive, if these result in extension of shelf life it is almost always a good investment.

Proteins are most stable in solid-state form and show less impact of storage temperature (though it still requires keeping at a cool temperature). The liquid formulations are more advanced formulations that are generally stabilized and must be stored at controlled temperature throughout the use chain. As the bulk manufacturing of proteins results in a concentrated solution state, this must be processed to concentrate and eventually solidify it through such processes as, cryopreservation, cryogranulation, spray-drying, undercooling and lyophilization.

Cryopreservation

Freezing can extend the shelf life of unstable products and improve containment if the freeze-thaw process is consistent. Frozen products can be transported safely for final formulation elsewhere and stockpiled to optimize the fill and finish process, which can reduce processing costs. Generally, freeze-thaw cycles result in significant loss of activity and many believe that proteins cannot be frozen and thawed without damage because no practical methods exist for doing so on a large scale. The usual method, which is both slow and nonreproducible, involves small volumes in bags, bottles, or vials freezing at $-0°C$ or below. A slow freeze alters the physical properties (pH, diffusion, and reaction rates) of the aqueous solvent medium or mixture, which can denature proteins, particularly over extended time. The solution is then thawed at room temperature, with some components thawing before others. Ice recrystallization in the thawing process creates mechanical stress.

Cryogranulation

It is a frequently used technique for small-molecule drugs, but cryogranulation of proteins using liquid nitrogen has not been very successful.

Spray Drying

It is a dehydration process that uses heat from a hot-air stream to evaporate dispersed droplets created by atomization of a continuous liquid feed. Products dry within a few seconds into fine particles (powders). It is similar to freeze-drying. Important data for formulators working with any type of freezing process will be the glass transition temperature (T_g) of each component and the solution. At this temperature, ice crystal formation decreases to undetectable levels, and the freeze is an amorphous glass from which water will sublime. Spray-drying offers some advantages over freeze-drying: shorter process times, lower capital investment in equipment, and lower energy input to the solution, which can lessen the chances for protein denaturation. Disadvantages include the brief but high air-temperature exposure during the rapid-drying step (around 100°C), and shear stresses caused by spraying a formulation through a nozzle. The product temperature reaches about 80°C but only for a fraction of a second. Particle characteristics and size distribution can be closely controlled in spray-drying by changing process variables, such as solution composition and feed rate, atomizing gas pressure, air flow rate, and the inlet and outlet gas temperatures. Physical characteristics of the final product (size, morphology, surface area, and density of particles) may be manipulated, as well as biochemical characteristics (purity, potency, solubility, and stability of the formulation), and process yields. The goal is a dry, free-flowing powder with well-defined particle characteristics, consistent purity and presentation, and an active, stable, and acceptable dissolution profile, all obtained by as simple a process as possible.

Undercooling
It is another process wherein under certain conditions, a liquid can be cooled to temperatures below its freezing point (supercooling or undercooling). If a solution is separated into droplets, as in a mist or a water-in-oil emulsion, then only those drops with particles in them will freeze if cooled, rather than catalyzing a chain reaction of freezing throughout the solution. For protein formulations, the product in aqueous solution can be dispersed as microparticles through an oil-phase carrier (liquid at the mixing temperature, solid if stored at $-20°C$), so that the drops are locked in as liquid. Reconstitution is easier than with a lyophilized or spray-dried product as the product is warmed up at the point of use until the phases separate. This undercooling process may be used for even high-concentration protein formulations, requiring no additives like glycerol, which must be filtered out before the product can be used. Undercooling may be a particularly good choice for products that are susceptible to freeze damage.

Lyophilization
This technique produces stress conditions that may cause certain stresses on proteins that can result in unfolding (though generally completely refolded when reconstituted); so, specific conditions must be determined and stabilizing additives must be appropriate for use during both the freezing and drying stages to allow at least two years of shelf life with stable cake morphology, dispersability, and dissolution. The presence of solutes lowers the freezing temperature of an aqueous solution making the solution more viscous at even below water freezing temperature, when the solute becomes supersaturated and releases latent heat; this point is known as the eutectic point, T_e; when the physical state changes from elastic liquid to brittle but amorphous solid glass, this is called glass transition temperature or T_g at which point ice formation ceases. For pure water, T_g is $-134°C$; the T_g for solutes is generally higher and for solutions it is somewhere in between. The T_g is an important temperature as below this temperature, the mobility of a solution is greatly reduced and all degradation reactions slow down, but not reactions like oxidation. Presence of some moisture reduces T_g and presence of components like sugar raises it substantially.

Freezing forms an amorphous solid of the protein and excipients with associated water in crystalline form. Annealing, an optional step, increases ice-crystal size and allows crystallization of bulking agents (such as glycine or mannitol), removing them from the amorphous portion and increasing T_g. Primary drying sublimes water ice at temperatures lower than T_g (to avoid collapse of the cake). The higher the T_g, the higher can be the mixture's temperature and sublimation rate. Increasing the protein to the excipient ratio will increase T_g. Secondary drying removes water from the amorphous phase by an increase in sample temperature, which still may not exceed T_g. Luckily, as water leaves the amorphous phase, T_g increases. Sample temperatures play a larger role than the duration of secondary drying for determining final water content of the lyophilized cake. Low moisture increases T_g and thus increases the temperature at which the product can be stored.

A drug solution is first frozen at atmospheric pressure, and then water is removed by a reduction of pressure in the lyophilizer chamber, collecting the water as ice on a condenser. Samples are placed in glass vials and frozen, either before being put in the lyophilizer or on the lyophilizer shelves. The samples contain ice crystals, unfrozen water, amorphous solids (including the therapeutic protein), and crystalline additives. Pressure is reduced, and the ice crystals sublime. This constitutes the primary drying.

It is harder to remove the unfrozen water trapped in an amorphous solid. So after primary drying, a secondary drying stage removes that water by increasing the temperature. The final temperature of this secondary process is the key factor in determining residual moisture in the dried cake. The pressure is kept the same for secondary and primary drying to avoid protein collapse. The ideal result is a porous cake with little residual moisture. Porosity is important in later reconstituting the product.

Sometimes an annealing step (in which the product is kept at a set temperature) is added before the primary drying or near the end of secondary drying to crystallize excipients. This assures that crystallization and moisture release do not happen in an uncontrolled fashion later on during shipping and storage. Phase changes (e.g., crystallization of formulation sugars) during shipping or storage can be disastrous. Moisture can even transfer from rubber vial-stoppers during storage unless savvy formulators plan for and prevent it.

The criteria for dried-protein stability include minimum lyophilization-induced unfolding, with proteins native in the dried solid; a powder with T_g higher than the desired storage temperature; low residual moisture (<1%) in the cake; and formulation conditions (such as pH) that inhibit chemical degradation reactions unaffected by glass transition (such as oxidation). The goal is to design the fastest and most robust (acceptable quality even with variations in operating parameters) processing cycle: one that consumes the least amount of energy, does not compromise product quality, and produces a mechanically strong, rapidly insoluble cake. The cycle must be controlled for reproducibility and rapid correction of any problems that develop.

The use of FT-IR spectroscopy shows that almost all proteins except G-CSF (filgrastim) unfold during lyophilization and while they do refold upon reconstitution, it is often necessary to add ingredients and stabilizers to keep them from unfolidng in the solid state. Some of the ingredients used are listed in Table 4. However, the use of additives, when combined, may show unexpected interactions, synergism, enhanced instability, or even altered immunogenicity. For example, sugars (or saccharides) raise T_g and act as stabilizers. Dextran, lactose, maltose, sucrose, and trehalose are used, but the latter two are preferred. Acidic amino acids (glutamic acid, glycine, histidine, and threonine) and alkaline amino acids (arginine, lysine) are used to adjust pH besides the use of buffering systems. A nonionic surfactant (usually a polysorbate) is often added to inhibit protein aggregation in the early stage of processing and given its innocuous nature, it is left in the formulation while other contaminants like the peroxides are removed. Also included in the formulations are other surfactants and bulking agents, such as mannitol and certain biodegradable polymers mainly in low-concentration formulations.

A patented variation of the classical lyophilization technology is applied as VitriLife (3) wherein the product is lyophilized using a mixture of sugars; this formulation can be prepared in much shorter time and is different from classic lyophilization in that no sublimation is involved and the product undergoes fast drying; the resultant product can be kept at room temperature, obviating the need for the cold chain for these products. The method has been applied to cholera vaccine and other vaccines and is still under development.

Stabilization Through PEGylation

Where stabilization of native proteins is of great importance while in the formulation, the use of PEGylation is made to prolong the disposition half lives of proteins in the body. It is a relatively new technique that requires fusing PEG molecule with the drug (Fig. 1) and features:

- improved bioavailability, including longer circulation time and slower clearance
- optimized pharmacokinetics resulting in sustained duration
- improved safety profile with lower toxicity, immunogenicity, and antigenicity
- increased efficacy
- decreased dosing frequency
- improved drug solubility and stability
- reduced proteolysis
- controlled drug release

There are different types of PEGs. Linear PEGs are straight-chained PEGs that are monofunctional, homobifunctional, or heterobifunctional. Linear monofunctional PEGs (mPEG-X) have one reactive moiety at one end of the PEG with the other end considered nonreactive (typically end-capped with a methoxy group). Linear homobifunctional PEGs (X-PEG-X) contain the same reactive moiety at each end of the PEG. Linear heterobifunctional PEGs (X-PEG-Y) contain a different reactive moiety at each end of the PEG. Branched PEGs (PEG2-X), also referred to as "Y-shaped" branched PEGs, contain two PEGs attached to a central core, from which extends a tethered reactive moiety. Forked PEGs (PEG-X2) contain a PEG with one end having two or more tethered reactive moieties extending from a central core. Multiarm PEGs (two-, three-, four-, and eight-arm PEG-Xs) are based upon ethyoxylation of either glycerine (three-arm), pentaerythritol (four-arm), or hexaglycerine (eight-arm). The two-arm PEG was previously noted

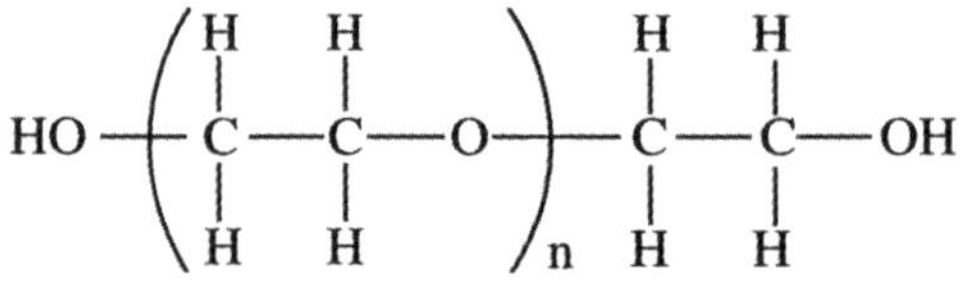

Figure 1 Structure of polyethylene glycol.

under the linear homobifunctional and heterobifunctional PEGs. Each arm has a tethered reactive group on the end. These multifunctional PEGs offer the potential to increase potency of the resulting conjugate by attaching multiple drug molecules to each arm of the PEG. Multifunctional PEGs have several applications, including linking macromolecules to surfaces (for immunoassays, biosensors, or various probe applications), hydrogel formation, and drug targeting, as well as targeting liposomes and viruses.

All PEGs have some variance in the number of ethylene oxide units; this is a result of anionic polymerization. Polydispersity (PD) is a ratio that represents the broadness of a MW distribution. PD is the ratio of the number average MW (M_n) to the weight average molecular weight (M_w) (PD = M_w/M_n). If the PD is equal to one, then M_n equals M_w and the polymer is said to be monodispersed. Typically, polymers are not truly monodispersed, although PEGs made anionically do have a low PD (1.01–1.08). As M_n changes with M_w, the PD changes (PD will always be greater than one for polymers).

The process of PEGylation is complex to control the location of the PEGylation bond requiring detailed studies on the effect of ionic strength, pH; for isotonicity purposes, glycine or sodium chloride are also added; when using cryoprotection, the use of sucrose and mannitol is recommended and for general purpose, polysorbate 80 works well in most situations. As mPEG fuses to lysine reside on polypeptide chains, the 3D structure of protein must be well understood in native conformation. One of the negative aspects of PEGylation is the heterogeneity of the product that requires more expensive separation procedures and the final product is more expensive. When the formulation is well-balanced, the in vivo stability is improved and effect of agitation on the formulation is minimized; how PEGylation protects against agitation is not known. The effect of agitation was noted when the formulations of erythropoietin were changed from protein stabilization of surfactant stabilization.

Nektar Advanced PEGylation (4) is a major supplier of activated large PEGs that provide several advantages over early stage low MW PEGylation technology. The advanced PEGylation yields single PEGylation, more stable product, higher bioactivity, and high product purity (Fig. 2).

In an aqueous medium, the long, chain-like PEG molecule is heavily hydrated and in rapid motion. This rapid motion causes the PEG to sweep out a large volume and prevents the approach and interference of other molecules. As a result, when attached to a drug, PEG polymer chains can protect the drug molecules from

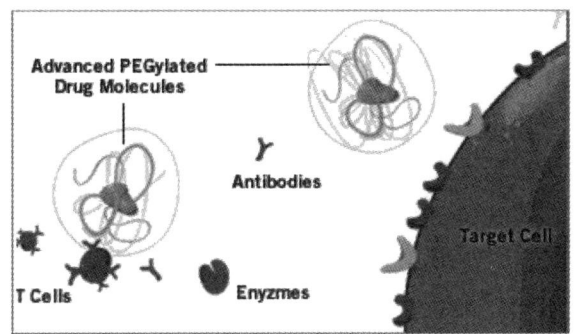

Figure 2 Improved drug performance using advanced PEGylation. *Source*: Courtesy of Ref. 4.

immune response and other clearance mechanisms, sustaining drug bioavailability. Covalent drug modification with PEG enables improved drug performance by optimizing pharmacokinetics, increasing bioavailability, decreasing immunogenicity, and decreasing dosing frequency using stable PEG linkages. These activated high-MW PEGs can be linked site-specifically to drug molecules. The following is a list of commercial products that were developed in a PEGylated form:

- Neulasta® (pegfilgrastim) by Amgen;
- Somavert® (pegvisomant) by Pfizer;
- PEGASYS® (peginterferon alpha-2a) by Roche;
- PEG-INTRON® (peginterferon alpha-2b) by Schering-Plough;
- Definity® (perflutren lipid microsphere) by Bristol-Myers Squibb;
- Macugen® (pegabtanib) by Eyetech & Pfizer;
- DuraSeal™ (PEG hydrogel) by Confluent Surgical.

Numerous studies have demonstrated that PEGylation of biologically active agents is an effective way to prolong the half-life of the drug in the circulation, alter the pattern of drug distribution, and camoflage the drug, thereby reducing immunogenicity and protecting it from biological degradation. The MW of PEG and PEG-conjugates, when injected intravenously, has a great effect on the time course of serum circulation. In general, the serum half-life of PEG extends from 18 minutes to 20 hours as the PEG MW increases from 5 to 190 kDa with a leveling-off of the serum half-life period at 20–24 hours for PEG and PEG-conjugates having a MW more than 30 kDa or a molecular size more than 8 nm. Renal clearance rate of PEGs is controlled by the glomerular filtration rate in a normal kidney. The vascular wall of the renal glomeruli functions as a filter for ionic and nonionic substances that may accumulate in the kidney through blood circulation. The excretion of these molecules can be a function of molecular size (3–5 nm) and electric charge. The glomerular filtration for the kidneys is less than 70 kDa for proteins (due to charge and molecular size) and less than 30 kDa for PEGs (due to molecular size only for nonionic, randomly coiled molecules). Short linear strands of PEG have a high clearance rate, but large linear PEGs, multiarm PEGs, and PEGylated proteins have a slower clearance rate. This difference in renal clearance rate can be attributed to an increase in structure size, hydrodynamic volume, and a change in the total charge of the molecule. In the case of the liver, PEGs with MW less than 50 kDa have decreased hepatic clearance with increasing MW (similar to renal clearance), but the liver clearance increases when the MW is more than 50 kDa.

Amine PEGylation is one of the most common techniques used and N-hydroxysuccinimide (NHS) ester of PEG carboxylic acid remains the most popular derivative for coupling PEG to proteins (Fig. 3), liposomes, soluble and insoluble polymers, and other biological molecules.

The branched activated PEG ester (mPEG2NHS) has several advantages over linear-based PEGs. First, a branched PEG "acts" as if it were larger than a corresponding linear PEG of the same MW. Second, the compound is purely monofunctional, because the intermediate acid is chromatographically purified. Third, proteins modified with branched PEG possess a greater stability from enzymatic degradation and pH degradation, thereby reducing its antigenicity and likelihood of destruction. Moreover, proteins modified with the branched PEG reagent may retain more activity when compared with modification with a

Characterization of Biopharmaceutical Drugs

```
    O    O
    ‖    ‖
mPEG-C-O-N⟨  ⟩   +   H₂N-Protein
         ‖
         O
   1 - 10 parts      1 part
                     0.5 - 5mg/mL

          │  15 - 120 min
          │  pH7 - 9
          │  4 - 25°C
          ▼

         O
         ‖
    mPEG-C-NH-Protein
```

Figure 3 m-PEGylation using N-hydroxysuccinimide ester.

linear PEG counterpart. A branched PEG reagent was used in the creation of PEGASYS (PEGylated-interferon alpha) by Roche and on aptamers. Substitution values are 95% or greater.

Characterization Methods
Spectroscopy
Spectrophotometric analyses are the most common method to characterize proteins. The use of ultraviolet-visible (UV-VIS) spectroscopy is typically used for the determination of protein concentration by using either a dye-binding assay (e.g., the Bradford or Lowry method) or by determining the absorption of a solution of protein at one or more wavelengths in the near UV region (260–280 nm). Another spectroscopic method used in the early-phase characterization of biopharmaceuticals is CD.

The Bradford method, which is more sensitive and less affected by most common detergents or other common biochemicals than the Lowry method, is the most widely used dye-binding method. There are two common Bradford methods: the standard assay Bradford method, with a range of 10–100 mg, and the microassay method, which is linear between 1 mg and 10 mg. A standard curve is constructed with a common protein that is readily available in pure form, such as bovine serum albumin or bovine gamma globulin. The standards and the sample are then reacted with a solution of Coomassie Brilliant Blue G250 in an acidic solution, and the absorbance is measured at 595 nm. The protein concentration of the sample is then calculated against the constructed curve. This value is an approximation of the protein concentration because different proteins react differently with the Bradford reagent. Further on in development, the calibration curve should be determined using the protein of interest.

Direct determination of the absorbance of a protein solution requires no other reagents or standards. Two solutions are prepared, one of the samples and one blank solution of water or containing all the buffer components. After zeroing the spectrophotometer at the wavelengths to be measured using the blank solution, the analyst measures the absorbance of the protein solution. For relatively pure solutions, measuring the absorbance at 280 nm (A280) is usually sufficient. However, for protein solutions containing significant amounts of nucleic acid (as little as a few percent), it is best also to determine the absorbance at 260 nm (A260), to correct for the presence of nucleic acids. The protein concentration is then determined using the following equation:

$$\text{Protein (mg/mL)} = 1.55(A280) - 0.76(A260) \qquad (1)$$

If the extinction coefficient has not yet been determined, a more absolute concentration is determined using the following equation:

$$A280(\text{mg/mL}) = (5690*\text{Trp} + 1280*\text{Tyr} + 120*\text{Cys})/\text{protein MW} \qquad (2)$$

where Trp is the number of tryptophan residues in the protein and similarly, Tyr is the number of tyrosines, Cys is the number of cysteine residues, and MW is the molecular weight of the protein.

Spectrophotometric methodologies are used less commonly in late-stage development of proteins, but are very helpful in early development of biopharmaceuticals. For example, CD can be used to study the tertiary structure of proteins. Use of CD does not require the highly pure concentrated protein solutions needed to prepare protein crystals for X-ray crystallography. A protein's specific CD spectrum in the near UV region (250–340 nm) is determined by its regular 3D structure in solution. By comparing the CD spectra of a protein in both a denaturing and nondenaturing solvent, some estimate can be made regarding the conformational stability of the protein. Because the protein concentration needed to perform CD studies is relatively low, these studies can be undertaken early in development with small amounts of manually purified protein. Because interpretation of the spectra is often difficult, in many cases CD spectroscopy analyses are sent to laboratories experienced in utilizing these techniques.

FT-IR can also be used to determine the tertiary structure of a protein. It does not require the protein to be in solution, and it can often be used to support early formulation development for either liquid or lyophilized proteins.

Electrophoresis

Electrophoresis is the separation of charged molecules in an electric field. In PAGE, the electric field is formed within the pores of a polyacrylamide gel that are filled with a running buffer. Addition of SDS to the sample preparation buffer as well as to the running buffer is often used to pretreat the protein prior to electrophoresis, hence the term SDS-PAGE. In SDS-PAGE, the SDS molecules interact with the protein, unfolding it, and adding multiple charges to the molecule from the associated sulfate groups. Complete unfolding of a protein may require the addition of a reducing agent as well as the SDS. Proteins migrate through the polyacrylamide gel and are separated according to their MW in SDS-PAGE.

Another common technique is to run native or nondenaturing PAGE. In native gel electrophoresis, the migration of the protein through the gel is affected by both the charge and the shape of the protein, as well as the size. While SDS-PAGE is commonly used to determine MW of proteins, it would be incorrect to use native PAGE for weight determination. Both methods are used to assess purity of a protein.

Protein is invisible in the gels and must be stained for detection. The most commonly used visualization techniques are silver and Coomassie blue stains. While silver is more sensitive, the intensity of silver stains is affected by the proteins and is not linear with the concentration of protein, as is Coomassie blue staining. If the intention is to quantify the relative amounts of each protein band, Coomassie blue staining should be used.

In addition to the determination of MW, SDS-PAGE is used to examine the presence of aggregates. Samples can be prepared with and without the reducing agent, either mercaptoethanol or dithiothreitol. Comparison of reduced and nonreduced gel patterns allows the analyst to determine whether the higher MW aggregates seen are due to intermolecular disulfide bridges. Additionally, SDS-PAGE provides information about the purity of the protein. After scanning Coomassie blue-stained gels and calculating the area or relative intensity of each band seen in a sample, the percentage of the total protein can be determined. Most laboratories have scanning software capable of performing both image analysis and quantification. Many software programs can also determine the MW using results from the standards run on the same gels.

IEF is another electrophoretic separation method. In this method, the polyacrylamide gel or another support layer also contains a pH gradient. This is a powerful method for investigating the charge differences among proteins. In IEF, each protein migrates through the support layer until it is "trapped" at the point where the pI of the protein is the same as the pH gradient formed in the support media. At this point, the charge on the protein is zero and it no longer migrates but focuses. Separated proteins need to be stained to be visualized. The pI of a protein can be determined in an IEF separation either by comparison with standards run simultaneously, or by measuring the pH of the band with a special pH electrode. In proteins with multiple glycosylated forms, it is often difficult to determine the pI because the multiple forms may run as a smear across the gel. In such cases, the carbohydrates could be enzymatically removed, yielding a single protein form.

An electrophoretic method that is used increasingly in early-stage characterization of proteins is 2D electrophoresis. This method separates the proteins in one dimension based solely on charge (IEF), and in the second dimension by size (SDS-PAGE). This powerful method can determine whether a protein that is a single band on SDS-PAGE co-migrates with another protein. A new use of 2D electrophoresis is for determination of host-cell proteins. This method can often identify host-cell proteins, which co-migrate with the protein of interest in SDS-PAGE gels. The protein is separated in a thin IEF gel, the lane is then placed across the top of an SDS-PAGE gel, and a second electrophoresis is run. After staining, the gel contains one or more spots. The gel can be scanned on a densitometer, and the relative intensity of each spot can be used to determine the percentage of protein that is not the product.

Chromatography

High-performance liquid chromatography is a core technique in characterization of proteins. These separations are coupled with detectors that are sensitive to the proteins eluted during chromatographic separation. The most common detector used in HPLC measures the UV absorption of the elute at one or more wavelengths or, in the case of a diode array detector, it can scan all the wavelengths simultaneously and provide a clear quantification of each separated protein. Other detection methods sometimes used with HPLC separations are evaporative light scattering and refractive index. The three most common types of HPLC are SEC, which separates based on the size or MW of the protein; ion-exchange (IEX) chromatography, which separates based on the charge of the protein; and reverse-phase (RP) chromatography, which separates based on the hydrophobicity of the protein. RP is such a common HPLC method that when people do not specify a particular HPLC method, they usually refer to RP-HPLC.

In RP-HPLC, separation of proteins is accomplished by differential interaction with the column matrix and the column buffer. Two buffers, called the aqueous buffer and the organic buffer (thus identifying the most important attribute of each) are used, and the separation is done with a gradient of these buffers. The most common organic buffers are based on acetonitrile, though other organic solvents, such as methanol or tetrahydrofuran may be used. The column used for the RP-HPLC separation is most commonly a silica base, coated with hydrocarbon chains of varying sizes, such as C4, C8, and C18. RP columns built on polymer backbones are becoming more readily available. To minimize any nonspecific interaction between the protein and the column matrix, an ion-pairing component, frequently trifluoroacetic acid, is added to both the aqueous and organic buffers. After the column is equilibrated with either the aqueous buffer or a defined mixture of the aqueous and organic buffers, the sample is loaded onto the column as an aqueous solution. Separation of the varying proteins is done by running a gradient of increasing organic buffer; proteins are resolubilized when the hydrophobic nature of the particular protein partitions into the buffer. The use of highly hydrophobic buffers for RP-HPLC usually precludes the presence of large amounts of salt, which destabilizes some proteins. Additionally, proteins are denatured in RP-HPLC, and so tertiary and quaternary structures are lost. The subunits of multi-subunit proteins will usually elute separately. Multiple forms of a protein can usually be separated in RP-HPLC by their small differences of hydrophobicity and sometimes MW. RP-HPLC is often considered to be a good method to separate related isoforms of a protein.

IEX-HPLC separates molecules based on charge. The protein interacts with the charged moiety on the column and is then eluted with either salt or pH gradients. Elution from the column is from the weakest to the strongest bound. The protein solution is loaded onto a column that has been charged with the counter ion, and is then equilibrated with the starting buffer. Proteins are eluted from the column by a gradient of either salt or pH. If the column with a second buffer contains salt, this disrupts the protein interaction with the column and replaces the protein with the counter ion. If the second buffer changes the pH, this alters the charge on the protein and decreases the interaction of the protein with the column. IEX columns can be either anionic or cationic. The most common anion exchangers are quaternary ammonium, diaminoethyl, and quaternary aminoethyl, and the most common cation exchangers are sulfopropyl, methyl sulfonate, and carboxymethyl. By using buffers above or below the protein's pI, the same protein can be analyzed on both anionic and cationic columns. Because of the ionic nature of the interaction between the protein and the column, the size of the protein does not affect binding. Additionally, because IEX is run in an aqueous environment, the protein is not denatured and maintains its structure, which renders the method more sensitive to differences, such as oxidation in surface amino acids. While this method can be used to assess purity, it is generally less sensitive to purity than RP-HPLC because proteins can remain associated during the separation.

SEC-HPLC is different from RP-HPLC and IEX in two major ways. The first difference is that separation is based on size only, with a small impact on the shape of the molecule. The second difference is that during SEC, the protein does not adsorb or bind to the separation media. In some cases, a protein will nonspecifically bind to the column. In these instances, it is important to use a different column matrix or to change the composition of the running buffer. The MW of a protein is determined by the comparison of its elution time with the elution time of the standard proteins of known MW. Because there is no binding of the protein with the column matrix, the protein separation is sensitive to the sample's volume. After running a set of known proteins through an SEC column, the protein of interest is loaded onto the column in a small volume and eluted under the same conditions. A standard curve is constructed, based on the MW of the standard proteins and elution time. This curve is used to determine the MW of the eluted sample. Size exclusion chromatograph columns are available that can separate proteins with MWs as high as 1,000,000, allowing SEC to be used to identify and quantify the size and amount of aggregates in a protein preparation. Unlike SDS-PAGE, the protein is not denatured before separation, so that non-cross-linked aggregates are not disrupted and can be identified. Purity (or percent aggregation) determined by these two methods (SDS-PAGE and SEC) often differs considerably due to the detection of the additional aggregate forms in SEC.

Mass Spectroscopy
MS is a method used increasingly to characterize proteins, in the early stages as well as through commercial manufacture. The popularity of this very sensitive technique has increased as it has become more available in analytical laboratories, and the methods to use it have become more robust. This separates proteins based on their mass-to-charge ratio. To separate by MS, a protein is ionized in one of the several ways; then it is accelerated by electric or magnetic fields. In some cases, the charged protein will break apart to produce ions. The pattern of ions produced is dependent on the structure of the protein so that they may be used to determine the primary structure of the protein. Most MS instruments in use today ionize proteins in ways that minimize protein fragmentation to allow a true mass determination.

The information lost by reducing fragmentation in standard MS can be determined using MS/MS. In MS/MS, specific ions are subjected to an additional energy by collision, and the resulting daughter ions allow even more structural information to be determined, even to the level of the amino acid sequence. This technique is especially useful for determining post-translational changes to the protein. MS/MS can also be used to sequence the structure of carbohydrate side-chains on glycosylated proteins and to identify the micro-heterogeneity that they introduce.

With large proteins, the determination of the primary sequence and post-translational modifications is most efficiently done after digestion with trypsin or another protease to generate smaller peptides. In this case, the peptides are first separated by HPLC, most commonly RP-HPLC, and the column eluant is directed into the MS. In this hyphenated method, known as LC-MS or liquid chromatography-tandem mass spectroscopy (LC-MS/MS), the individual peptides are analyzed, allowing the identification of post-translational modification sites. In some cases, there are potentially multiple sites in a single peptide that may be modified.

Absolute identification of the modified amino acids may require more than one enzyme digest to produce different peptides. Some kinds of modifications that are easily identified by MS include: phosphorylation of threonine or serine; sulfation or phosphorylation of tyrosine; deamidation of asparagine or glutamine; O- or N-linked glycosylation; oxidation of methionine or cysteine; and N-terminal modification by formylation or prenylation. Combining enzymatic maps (tryptic mapping) with MS/MS may identify single amino acid variants of the protein that cannot otherwise be seen.

MS is often used as part of hyphenated methods, such as LC-MS, where the proteins are separated by a chromatographic method, and the column eluant is then directed to the mass spectrometer for additional characterization. One of the common confusions experienced when evaluating the results of MS analyses of proteins involves equating the observed size of the ion current peak (for a particular ion species) with the amount of the species present. The size of the peak is sensitive to several things and cannot be used for quantification. For this reason, the use of LC separation and quantification "front-end" to the MS allows the relative amounts to be determined.

Validation
Synthetic drugs are readily characterized by established analytical methods. Biologics, on the other hand, are complex, high-molecular-weight products, and analytical methods have limited abilities to completely characterize them and their impurity profiles. Regulation of biologics includes not only final product characterization but also characterization and controls on raw materials and the manufacturing process. The FDA has defined process validation as "establishing documented evidence which provides a high degree of assurance that a specific process will consistently produce a product meeting its predetermined specifications and quality attributes." This involves supporting product and manufacturing process claims with documented scientific studies. Protocols, results with statistical analysis, authorizations, and approvals must be available to regulatory inspectors. Process validation is part of current good manufacturing practices (cGMP) and is required in the United States and Europe for a manufacturing license.

Various types of validation generally required in biopharmaceutical manufacturing include process validation, facility and equipment validation, analytical method validation, software validation, cleaning validation and expression system characterization. Combined with other elements of cGMP, including lot release testing, raw material testing, vendor quality certifications, and vendor audits, the quality of product can be consistently assured.

Process Validation
Process validation involves the identification, monitoring, and control of sources of variation that can contribute to changes in the product. It starts with process characterization studies using scale-down models for optimization, operating range specification, extractables and leachables characterization, and clearance studies. Such work depends on validated assays and representative scale-down models.

Process development normally involves identifying critical variables, defining setpoints for each unit operation, and establishing operating ranges (deviations from the setpoint). Maximum operating range (MOR) limits are typically set during

Phase II or III. If they are exceeded, an investigation is necessary to determine if product quality remains acceptable.

Normal operating range (NOR) limits are determined by run-to-run reproducibility with scale-down models and trending with control charts at production scale. NOR limits lie within MOR limits, which must allow for normal variability while maintaining acceptable operation.

Facility and Equipment Validation

Facility and equipment validation is normally divided into design qualification (DQ), installation qualification (IQ), operational qualification (OQ), and performance qualification (PQ). Equipment validation begins with pilot production of clinical materials for Phase II.

DQ provides documented evidence that the proposed design of the facilities, equipment, and systems are suitable for the intended purpose. DQ must compare the design to a set of well-defined user requirements relating to product safety, identity, strength, purity, and quality.

IQ provides documented evidence that the system is assembled, installed, plumbed, and wired according to the user's design specifications, vendor recommendations, and appropriate codes and standards. Vendors typically provide much of the hardware documentation.

OQ provides documented evidence that the system performs as expected throughout its intended operating ranges, including all the system's different functions and all its components (hardware, monitoring instruments, controls, alarms, and recorders). Elements of OQ testing and documentation may be part of the factory acceptance test at the vendor's site. Integration with plant utilities and component installation must be verified at the factory. Hardware cleanliness must also be assessed after cleaning.

PQ is documented by processing actual feedstock by trained operators using buffers and utilities at the factory. Full-scale process validation includes testing the consistency of batch production.

Analytical Methods

Several methods are used to measure product characteristics important for therapeutic safety and efficacy during preclinical and early Phase I studies. Additional tests are developed for final product release and in-process sampling of the final manufacturing process. These measure characteristics such as molecular identity, purity, potency, and safety. The number of tests should be sufficient to show manufacturing consistency and the impact of manufacturing changes. Once a test is made a formal part of the manufacturing process, it is almost impossible to remove. Test methods are evaluated for different attributes such as accuracy, precision, range, selectivity, recovery, calibration (detection and quantitation limits), assay sampling, robustness, and stability.

Test method validation is needed to conduct clinical trials. Specifications should start off wide for Phase I and narrow to tighter values in the license application. Relaxing established specifications is very difficult.

Software Validation

Software validation operates under the principle that quality should not be diminished if a manual process is replaced with an automated process. Software must be developed and tested under a quality system with defined user requirements, change-control procedures, provisions for authorization of operators for data entry and data checking, data archiving, software backup, provisions for system crashing, and procedures for monitoring and correcting software problems. 21 CFR 11 defines requirements for maintaining the integrity of data and software and handling electronic signatures for traceability.

Cleaning Validation

Cleaning validation demonstrates the ability of cleaning procedures to permit reuse of processing components and equipment without a concomitant deterioration of product quality. Batch-to-batch carryover is of particular concern in multi-use plants making more than one product.

Consistency of product quality is demonstrated by showing operating consistency and product quality from batch-to-batch, processing with only buffer (blank runs) with assays for contaminants, examination of cleaned surfaces and materials, and extended scale-down clearance studies on reused materials. Disposable processing components that eliminate the need for cleaning validation are increasingly used at small scale.

Expression System Characterization

This is performed before Phase I studies in humans to insure safety. Concerns include the presence of contaminating organisms, tumorigenic cells, proteins, nucleic acids, retroviruses, or other pathogens. Taking tissue culture as an example, characterization includes the source, raw materials used, selection methods, number of generations, transfection or fusion methods used, procedures for establishing working cell banks, facilities, identity, homogeneity, absence of contaminating pathogens, tumorigenicity, and stability.

Stability Considerations

Commercial viability of recombinant production processes depends on the final product yield. This is a particularly more significant issue as biogeneric manufacturers bring out their line of products, which will be sold at a lower price than the innovator's products; the issue of yield becomes more important now. A primary cause of poor yield is neither the quality of the gene construct nor the nature of the molecule, but the degradation of the product during the manufacturing process. Protein degradation therefore becomes a key factor that must be thoroughly understood and steps must be taken to minimize this degradation step wherever possible. This chapter deals with this significant issue and makes suggestions on how to avoid the degradation of proteins in the downstream processing.

Proteolysis

In contrast to the cellular environment, where enzymatic degradation of proteins is highly controlled, extracellular proteases are the cause of uncontrolled protein degradation. The result of this proteolytic attack may vary from complete hydrolysis, single breaks within the peptide chain, or loss of a few N- or C-terminal amino acid residues. Besides losing the product, presence of truncated forms may seriously challenge the purification design.

Proteolytic enzymes are released to the medium because of cell death, mechanical stress, or induced cell lysis. Their presence is expected during fermentation and initial downstream unit operations. Most enzymes of the vacuoles and lysozomes are minimally active at slightly alkaline pH (7–9), a pH interval strongly recommended for extraction of proteins expressed in bacteria.

Proteins are probably more resistant toward proteolytic attacks in their native state and stabilizing factors (e.g., co-factor, correct parameter interval, co-solvent) are always considered optimized. Use of protein inhibitors is not recommended for safety reasons. The primary mechanism of proteolysis is the enzymatic hydrolysis of the peptide bond. The indicators of this reaction taking place in the system include loss of product or poor yield, lack of expected activity, changes in specific activity, change in MW, high background staining in 1D SDS electrophoresis, smeared bands, and many lower MW bands of poor resolution, disappearance of bands, and discrepancies in MW. The preventive actions taken to prevent proteolysis are listed in Table 5.

The use of enzyme inhibitors is not recommended as they are harmful to human beings. It should be ascertained that the degradation observed is not a function of the analytical assay. Enzyme inhibitors can be used for prevention of enzymatic activity in analytical assays. Mild denaturation may accelerate enzymatic digestion. Selective removal (e.g., affinity chromatography) of specific enzymes should be considered, too.

Deamidation

Two amino acid residues are involved in the deamidation reaction: asparagyl and glutamyl. The conversion to the corresponding carboxylic acid residues results in

Table 5 Preventive Actions Against Proteolysis

Factor	Comment
pH	There is no specific pH range in which all enzymes are considered inactive; at slightly alkaline pH, the nonspecific enzymes of the vacuoles and lysozomes are minimally active. Use strong buffers for extraction to prevent unintended shift in pH as a result of cell disruption. Some yeast enzymes are least active in the pH range 4–5, but active in the pH range 7–9. Phosphate may exhibit a stabilizing effect on proteins.
Temperature	Low temperature decreases the proteolytic activity. It is recommended to store harvest at 4–8°C or frozen.
Time	Enzymatic protein degradation is a function of time. Lengthy procedures and long storage times should be avoided during harvest, capture, and initial purification steps.
Conductivity	Noncritical.
Redox potential	Reducing and oxidizing conditions may alter the disulfide bridge arrangement and state of free cysteine residues, thus influencing the secondary and tertiary structures of protein.
Co-solvents	Presence of other proteins in excess (e.g., albumin) will reduce the proteolytic damage. Co-solvents, such as glycerol or dimethylsulfoxide may have a stabilizing effect, but will probably be too expensive for large-scale operations.
Low MW compounds	Substrates, substrate analogs, and co-factors can help in stabilizing the protein. Potential proteinase activators (e.g., divalent metal ions) are excluded from the extraction buffer.
Techniques	Careful cell disruption and specific extraction procedures may lower the enzymatic cleavage.
Denaturation	The proteolytic enzymes lose their biological activity upon denaturation. However, some enzymes are stable under mild denaturing conditions that lead to increased activity if the target protein is partly denatured under the same conditions.

Abbreviation: MW, molecular weight.
Source: From Ref. 5.

a shift in net charge of the protein at pH above the pK_a. As the deamidation may influence the biological activity and the stability of the molecule, the maximal content of des-amido forms in bulk materials and in biopharmaceutical preparations is constantly being debated. The list of proteins that undergo deamidation is comprehensive and includes well-known proteins, such as insulin, human growth hormone, and cytochrome C.

Asparagyl residues tend to be more susceptible to deamidation than glutamyl residues. Further, the deamidation reaction is strongly sequence-specific in model peptides with the half life of the -Asn-Pro- sequence being 100-fold greater than that of -Asn-Gly-. To some extent these observations can also be used on proteins that take the structural steric factors and nearby amino acid residues into consideration.

At pH above five, deamidation of asparagyl or glutamyl occurs via a relatively slow intermediate succinimide formation. The succinimidyl derivative is rapidly hydrolyzed at either the α- or β-carbonyl group to generate a mixture of normal- and iso-residues. Under strongly acidic conditions, asparagyl or glutamyl residues are hydrolyzed to the corresponding carboxyl residue. The indicators of deamidation include extra bands in electrophoresis and extra peaks in chromatographic recordings. Table 6 lists preventive actions against deamidation.

Oxidation

The amino acid residues histidyl, methionyl, cysteinyl, tryptophanyl, and tyrosinyl are potential oxidation sites at neutral or at slightly alkaline conditions. Oxidation of the said residues often results in loss of immunological and/or biological activity. The list of proteins that have been oxidized is comprehensive and includes biopharmaceutical products, such as albumin, growth hormone, glucagons, interleukin-1β and -2. In many cases, the immunological and/or biological activity was only partially lost. In general, oxidation of methionyl to methionyl sulfoxide does not affect protein antigenicity, probably because the conformational structure of the oxidized protein is close to the native structure. On the other hand, oxidation of a single amino acid residue often causes changes in the biological activity and all efforts should be taken to minimize oxidation reactions.

The mechanism of oxidation involves methionyl residues, which are converted to methionyl sulfoxide residues under mild oxidizing conditions. The most

Table 6 Preventive Actions Against Deamidation

Factor	Comment
pH	Deamidation is expected above pH 5. The optimal working range in which to avoid deamidation is probably between 3.0 and 5.0.
Temperature	The deamidation rate increases with increasing temperature.
Time	The deamidation rate is a function of time. Presence of des-amido forms is a marker for drug product stability and shelf life.
Conductivity	The ionic strength of the solution is kept low. At high ionic strength, the deamidation reaction can be fast even at neutral pH.
Redox potential	Nonessential parameter.
Co-solvents	In general, the buffer species and the buffer strength will influence the rate of deamidation. High solvent dielectrics favor deamidation.
	In model peptides, the protein stability was higher in Tris buffer than in phosphate buffer.

Source: From Ref. 5.

reactive residues are those exposed to the solvent, while those residues buried within the hydrophobic regions are fairly inert to oxidation (e.g., methionine residues in myoglobin and trypsin). Methionyl residues are susceptible to autooxidation, chemical oxidation, and photo oxidation.

The cysteinyl residues are easily oxidized and the reaction is usually accelerated at alkaline pH, where the thiol group is deprotonated. Under mild oxidizing conditions, the reactions are: oxidation of cysteinyl residues to sulfenic/sulfonic acid (alkaline conditions), cysteinyl residues to dehydroalanyl residues (alkaline conditions), and cysteinyl to cystine residues (neutral to alkaline conditions). In the absence of a thiol reagent or a nearby thiol, the cysteine may instead oxidize to sulfenic acid.

The oxidation reaction is strongly catalyzed by divalent metal ions (e.g., copper). The indicators of oxidation include extra bands in gel electrophoresis and extra peaks in chromatographic recordings. Preventive actions against oxidation are listed in Table 7.

The degradation rate is often governed by trace amounts of peroxides, divalent metal ions, light, base, and free radicals. There are three classes of antioxidants:

- *Phenolic compounds*: butylated hydroxytoluene (BHT), butylated hydroxyanisole (BHA), propyl gallate, and vitamin E
- *Reducing agents*: cysteine, dithiothreitol (DTT), methionine, ascorbic acid, sodium sulfite, thioglycolic acid, and thioglycerol
- *Chelating agents*: EDTA, citric acid, and thioglycolic acid

Carbamylation

Cyanate is able to react with amino, sulfhydryl, carboxyl, phenolic hydroxyl, imidazole, and phosphate groups in proteins according to the general scheme, $RXH + HNCO = RXCONH_2$. Cyanate is easily soluble in water. Most reactions have a pH optimum around seven. Acidic pH should be avoided as acidic conditions are ideal for modifications of carboxyl groups. For the same reason, reactions with cyanate should not be terminated with acid. At high concentrations, cyanate may react with itself to form cyanuric acid and cyamelide, and it is recommended to work at concentrations at about 0.2 M. Cyanate reacts rapidly with amino groups. At neutral pH and below, the α-amino group can be expected to react about 100 times faster than the ε-amino group. The resulting carbamoylamino

Table 7 Preventive Actions Against Oxidation

Factor	Comment
pH	The oxidation rate is assumed low at slightly acidic pH.
Temperature	Working at low temperatures decreases the rate of oxidation.
Time	The oxidation reaction is a function of time.
Conductivity	No data available.
Redox potential	Disulfide bond formation will take place at a redox potential above 0 mV. A high redox potential indicates presence of oxidizing agents.
Co-solvents	Avoid oxidizing agents and protect against light. Addition of chelating agents (EDTA, citric acid, thioglycolic acid), antioxidants (BHT, BHA, propyl gallate, vitamin E), and/or reducing agents (cysteine, DTT, methionine, ascorbic acid, sodium sulfite, thioglycolic acid, thioglycerol) may reduce oxidation.

Abbreviations: BHA, butylated hydroxyanisole; BHT, butylated hydroxytoluene; DTT, dithiothreitol; EDTA, ethylenediaminetetraacetic acid.
Source: From Ref. 5.

groups are stable even in dilute NaOH. Typical reaction conditions are 3 mg/mL protein, 0.1 M cyanate at pH 8, 25°C for 1 hour. Cyanate also reacts even more rapidly with sulfhydryl groups than amino groups resulting in the formation of S-carbamylcysteine residues. As cyanate reacts rapidly with sulfhydryl groups, labile disulfide bonds may be ruptured. The resulting carbamylmercaptans decompose readily to free mercaptan and cyanate at alkaline pH. Consequently, cyanate can be used as reversible blocking agent for −SH groups. At acidic pH, cyanate reacts with carboxylic groups resulting in the formation of a mixed anhydride, which can react with many nucleophiles (e.g., formation of amides). The reaction can be avoided entirely at pH 7–8. Aliphatic hydroxyls are resistant to carbamylation even at high cyanate concentrations at low pH. However, the reactive hydroxyl groups of chymotrypsin and other proteases react with cyanate to give urethans. Phenolic hydroxyl groups react more readily than aliphatic groups in a reversible reaction that is quite analogous to the one that occurs with −SH groups.

Cyanate present in aqueous urea solutions reacts with the free amino and sulfhydryl groups of proteins. Urea is often tacitly assumed to be a reagent, which alters the structure of the protein and may be used to keep target proteins on their monomeric form during purification. However, at pH 6 and above urea hydrolyses under formation of cyanate leading to carbamylation reactive groups in proteins.

The equilibrium $(NH_2)_2CO = NH_4CNO$ between an undissociated urea and dissociated cyanate in aqueous urea solutions is the main course of unintended carbamylation of primary amino groups in proteins. As the protein concentration normally is from 0.1 to 30 mg/mL corresponding to the micromolar range, a considerable part of the protein mass is expected to undergo carbamylation under these conditions. Thus, the exposure of ribonuclease to cyanate in aqueous solution leads to a considerable loss of enzymatic activity. Formation of cyanate is prevented by storage of neutral urea solutions at 4°C or by buffering the solution at pH 4.7. Thus, acifidication of urea solutions just before use will decompose any cyanate present. Cyanate can be removed from urea solutions by mixed IEX. The method of Salinas describes a sensitive and specific method for quantitative estimation of carbamylation in proteins (see Bibliography).

β-Elimination

The β-elimination reaction is caused by the abstraction of a β-hydrogen from cysteinyl, seryl, and threonyl residues under alkaline conditions. The cystinyl residue decomposes as a result of β-elimination under formation of HS⁻ and free sulfur, thus affecting the redox potential of the solution. Several studies indicate that the rate of the reaction is proportional to the hydroxide ion concentration and pH should consequently be kept low (use dilute NaOH solutions to adjust pH preferably below 0.1 M NaOH). In alkaline solutions, the abstraction of β-hydrogen from cystinyl, seryl, and threonyl residues results in formation of a carban ion. Depending on the nature of the side-chain, the carban ion can rearrange to form an unsaturated derivative (dehydroalanine or β-methyl-dehydroalanine) or add a proton to give the L- and D-amino acid residues (racemization). The derivatives formed are reactive with a number of nucleophilic protein groups. The reaction is independent of the primary structure of the protein. The indicators of β-elimination include degradation of the protein, cleavage of disulfide bridges, and smell of sulfur. Preventive actions against β-elimination are listed in Table 8.

Table 8 Preventive Actions Against β-Elimination

Factor	Comment
pH	pH is kept below 10. Do not use NaOH solutions above 0.1 M to adjust pH.
Temperature	High temperature even at pH 4–8 results in β-elimination.
Time	The β-elimination reaction is a function of time.
Conductivity	Increased ionic strength increases the rate of β-elimination.
Redox potential	Cystinyl-rich proteins may decompose under formation of HS$^-$, which will lower the redox potential of the solution. Reduction of disulfide bonds may result.
Co-solvents	Removal of divalent metal ions with EDTA.

Abbreviation: EDTA, ethylenediaminetetraacetic acid.
Source: From Ref. 5.

Racemization

All amino acid residues except glycine are subject to racemization at alkaline pH resulting in formation of the D-enantiomers of the residue. Racemization is inevitably associated with conformational changes and thereby loss of function. The racemization of proteins has been described in several reports. The initial step of the reaction is abstraction of the β-hydrogen by hydroxide ions. By uptake of a proton, this will result in either the L- or D-amino acid residue. The carban ion formed may also undergo β-elimination. At pH 5–12, Asn, Asp, Gln, and Glu may modify via a succinimidyl intermediate resulting in both the D- and L-derivatives. The indicators of racemization include change of protein structure and loss of biological activity. The change in optical rotation correlates with the rate of racemization. The amino acid residues undergo racemization at different rates. Preventive actions against racemization are listed in Table 9.

Cysteinyl Residues

The reactive site of the cysteinyl residue is the thiol group, which is deprotonated at alkaline pH (pK_a around 8.5). The residue under oxidizing conditions (and neutral to alkaline pH) is able to react with a similar residue under formation of a disulfide bond. Many proteins are stabilized by intramolecular disulfide bonds (e.g., insulin, growth hormone, IGF-1), but intermolecular bonds may also result from the reaction under formation of aggregates. In order to avoid unintended disulfide bond formation/cleavage, the redox potential of the solution must be monitored and controlled. In practice, aqueous buffers contain micromolar amounts of dissolved oxygen assuring a redox potential of 200–600 mV, which is sufficient to maintain the intramolecular disulfide bonds. Proteins with free cysteines may

Table 9 Preventive Actions Against Racemization

Factor	Comment
pH	High pH will favor abstraction of the β-hydrogen under formation of a carban ion. pH is kept below 10 and use of NaOH in concentration above 0.1 M is avoided when adjusting pH.
Temperature	The temperature is kept low.
Time	The reaction is a function of time.
Conductivity	No data are available.
Redox potential	No data are available.
Co-solvents	No data are available.

Source: From Ref. 5.

Table 10 Preventive Actions Against Cysteinyl Residue Loss

Factor	Comment
pH	Minimum reactivity is expected in the pH range 3–7. The reactivity of the –SH group is at maximum above the pK_a (8.5), where the group is deprotonated. In strongly acidic media the reaction is expected to take place via a sulfenium cation by an electrophile displacement.
Protein concentration	The intramolecular disulfide bond formation is a first-order reaction and thus independent of protein concentration. Intermolecular reactions via the cysteinyl residue may be affected by the protein concentration (aggregation). The aggregation rate is favored by high protein concentration.
Temperature	The temperature is kept low (4–20°C) especially at pH above 9.5.
Time	The reaction is a function of time.
Conductivity	No data available.
Redox potential	Reducing conditions favor free cysteinyl residues. Oxidizing conditions favor disulfide bonds. The redox potential is a function of pH (60 mV/pH unit).
Co-solvents	Cysteine (nonanimal origin) is recommended as a reducing agent for large-scale operations. Divalent metal ions are removed by EDTA.

Abbreviation: EDTA, ethylenediaminetetraacetic acid.
Source: From Ref. 5.

prefer slightly reducing conditions, which can be obtained by addition of micromolar amounts of reducing agent (e.g., cysteine or DTT). The number of proteins containing both –SH groups and disulfide bonds are relatively small (e.g., albumin, β-lactoglobulin). In many cases, the disulfide bond stabilization is essential for maintaining the biological activity. Ribonuclease, for example, loses almost all activity when the four disulfide bonds are reduced. The mechanism of reaction involves cysteinyl and cystinyl residues in the disulfide bond formation by oxidation or reduction, conversion of a cystinyl residue to a cysteinyl residue and a sulfenic/sulfonic acid residue at alkaline pH and decomposition to a dehydroalanine residue at alkaline pH (β-elimination reaction). Disulfide bond formation is often catalyzed by the presence of a mercapto reagent (e.g., DTT, cysteine) in millimolar concentrations (typically 1–10 mM). Controlled disulfide bond formation has gained much attention in the biopharmaceutical industry in connection with in vitro folding of proteins expressed in *Escherichia coli*. Presence of divalent metal ions (typically Cu^{2+}) may result in oxidation of cysteinyl residues by an ill-defined reaction mechanism. Cleavage of the disulfide bond is initiated by an attack on a sulfur atom by a nucleophile reagent (HS^-, RS^-, CN^-, SO_3^-, OH^-). The reaction, which takes place at neutral to alkaline pH, consists of two steps with a formation of a mixed disulfide as the intermediary step. The indicators of cysteinyl residues include intermolecular disulfide bond formation resulting in aggregation; under reducing conditions the disulfide bonds destabilize resulting in conversion of cystinyl to cysteinyl residues (in vitro refolding may be the only solution to re-establish the correct disulfide bonds), presence of scrambled and structural altered forms, and smell of sulfur. Be careful when adjusting pH with high concentrations of NaOH. Locally high pH may facilitate β-elimination. Preventive actions against cysteinyl residue loss are described in Table 10.

Hydrolysis

The peptide bond does not undergo significant hydrolysis in the pH interval (3–9.5) and is usually used in industrial downstream processing. However, in dilute acid, where the carboxyl group of aspartyl residues is not dissociated, the peptide bond is cleaved 100 times faster than the other peptide bonds and especially the -Asp-Pro- sequence is prone to degradation. The guanidinium group of arginine is hydrolyzed by OH$^-$ to give ornithine and possibly some citrulline, depending on the nature of the protein. The mechanism of hydrolysis includes hydrolysis of the peptide bond, hydrolysis of the amide group of Asn and Gln, and hydrolysis of the guanidine group from Arg residues resulting in the formation of ornithine residues (hydroxide ion catalyzed). The indicators of hydrolysis include formation of split products of identical MW (the peptide fragments are linked via disulfide bonds). Preventive actions against hydrolysis are described in Table 11.

Denaturation

The native protein molecule loses its tertiary structure upon denaturation, resulting in a population of partially unfolded molecules. In practice, the denaturation process will lead to a mixture of more or less unfolded molecules comprising of residual secondary structure elements (helix, β-sheet, β-turn, *cis–trans* isomers around the prolinyl residue). A population of random coil molecules is not expected even under strong denaturing and reducing conditions. Upon denaturation, the inner hydrophobic core of the protein molecule is exposed to the hydrophilic environment (solvent water) often resulting in (irreversible) aggregation of the target protein. The cooperativity of the denaturation process results in an abrupt transition from the native to the unfolded state within a narrow range of pH, temperature, ionic strength, and denaturant concentration meaning that protein denaturation may come fast and unexpected. As globular proteins are only marginally stable in aqueous solutions, parameter interactions should be well understood and described using, for example, factorial design experiments. Proteins with disulfide bonds may undergo unfolding under reducing conditions, where the covalent bond is cleaved.

A denatured protein may be brought back to its native form by in vitro folding. The folding process is often slow and yields can be poor. As each protein is unique, the in vitro folding conditions must be determined case by case often using specific co-solvents as additives. An example is the group of proteins where disulfide bonds must be re-established as part of the renaturation process.

Hydrogen bonds and intramolecular interactions (electrostatic, van der Waals) stabilize the native structure of the protein in a co-operative manner. Upon denaturation, the co-operative effect is lost, resulting in unfolding of the molecule and exposure of the inner hydrophobic core to the hydrophilic aqueous environment. For small globular proteins, denaturation is an almost all-or-none

Table 11 Preventive Actions Against Hydrolysis

Factor	Comment
pH	The Asp-peptide bonds are prone to degradation at acidic pH. Deamidation of Asn and Gln occurs at pH above 5. Arg is converted to ornithine by OH$^-$ in a concentration-dependent manner.
Temperature	The deamidation rate increases with increasing temperature.
Time	The degradation reactions are a function of time.
Conductivity	The ionic strength of the solution is low in order to prevent deamidation. At high ionic strength, the deamidation reaction can be fast even at neutral pH.
Redox potential	No data available.
Co-solvents	No data available.

Source: From Ref. 5.

Table 12 Preventive Actions Against Denaturation

Factor	Comment
pH	Loss of tertiary structure is expected at pH above 9.5. Proteins tend to be most stable near the isoelectric point.
Temperature	The unfolding process is a function of temperature. Many proteins have optimal stability in the temperature range 10–30°C. Loss of structure is expected both at low temperatures (cold denaturation) and at elevated temperatures. The reason that many protein biopharmaceuticals are stored at low temperatures is to minimize chemical degradation (e.g., deamidation).
Time	The denaturation reaction can be very fast.
Conductivity	No data available.
Redox potential	Cleavage of disulfide bonds is expected under reducing conditions. A redox potential below 100 mV is considered unstable for some proteins (e.g., insulin). Not all proteins undergo conformational changes upon reduction of the disulfide bond(s).
Co-solvents	Sucrose, mannose, glucose, glycine, alanine, glutamine, and ammonium sulfate are examples of compounds acting as protein stabilizers (weak or not binding to the protein surface).
	Magnesium sulfate, guanidinium sulfate, sodium chloride, and other weakly interacting salts exhibit an effect depending on protein charge and concentration.
	PEG and MPD act as stabilizers due to steric exclusion and repulsion from charged groups. Both PEG and MPD may destabilize the protein under certain circumstances, where binding is favored over exclusion.
	Co-solvents, such as urea or guanidinium chloride, which bind strongly to the protein surface, are strong denaturants.

Abbreviations: MPD, 2-methyl-2,4 pentanediol; PEG, polyethylene glycol.
Source: From Ref. 5.

process approximated rather well by the two-state transition. Thermodynamically, the denaturation process can be observed by an increase of molar heat capacity and a rapid enthalpy increase with increasing temperature. The primary structure (amino acid sequence) is not affected by denaturation. The indicators of denaturation include loss of structure and loss of biological activity and aggregation. Preventive actions against denaturation are described in Table 12.

Aggregation
Protein aggregation is a major problem in the purification and formulation of protein biopharmaceuticals. Two types of intermolecular reactions dominate: aggregation resulting from hydrophobic interactions, and aggregation stemming from intermolecular disulfide bond formation between cysteinyl residues.

Proteins exposed to even mildly denaturing conditions may partially unfold, resulting in exposure of hydrophobic residues to the aqueous solvent favoring aggregation. The aggregation process is assumed to be controlled by the initial dimerization step in a second-order reaction. Consequently, high-protein concentrations will increase the aggregation rate.

Intermolecular disulfide bond formation between cysteinyl residues takes place at alkaline pH under oxidizing conditions. Proteins with reactive free thiol groups should be purified under reducing conditions (typically 1–10 mM reducing agent) in the presence of EDTA. Even proteins with disulfide bonds may participate in intermolecular disulfide bond reactions due to disulfide bond shuffling at neutral and alkaline pH.

Table 13 Preventive Actions Against Aggregation

Factor	Comment
pH	High pH is avoided in order to prevent protein unfolding. Some proteins do change conformation as a function of pH and certain pH intervals are avoided. pH < 7 protects the protein from intermolecular disulfide bond formation.
Temperature	The unfolding process is a function of temperature. Many proteins have optimal stability in the temperature range 10–30°C. Loss of structure is expected both at low temperatures (cold denaturation) and at elevated temperatures.
Protein concentration	Low protein concentrations are favored.
Time	The aggregation reaction can be very fast.
Conductivity	No data available.
Redox potential	Oxidizing conditions result in formation of disulfide bonds and intermolecular interactions are expected. Reducing conditions will prevent intermolecular disulfide bond formation.
Co-solvents	Denaturing or destabilizing agents (e.g., urea, certain alcohols, and organic solvents) should be used with care. Detergents may prevent aggregation, but they often bind strongly to the protein.

Source: From Ref. 5.

Expression of proteins in *E. coli* often results in the formation of insoluble aggregates called inclusion bodies, probably comprising fully or partially unfolded proteins. Inclusion bodies are brought to their monomeric form by extraction with a denaturant (e.g., 8 M urea) under reducing conditions (e.g., 0.1 M cysteine).

The mechanism of the reaction is primarily hydrophobic interaction or interaction via disulfide formation. Exposure of hydrophobic residues to the surface of the molecule leads to disorganization of the surrounding water molecules, thus increasing the entropy of the system. In order to avoid the change of the hydration shell structure, the protein molecules are forced to aggregate. The aggregation reaction can be very fast and will, in severe cases, lead to formation of insoluble polymers. The dominant mechanism is presumably specific interaction of certain conformations of intermediates rather than nonspecific co-aggregation. Proteins comprising of free thiol groups may form intermolecular disulfide bonds leading to aggregation of the protein. The reaction takes place at alkaline pH (presence of $-S^-$) under oxidizing conditions. Indicators of aggregation include loss of structure and loss of biological activity, turbid solution, presence of fibrils in the solution, precipitation, and formation of gels. Formation of inclusion bodies in *E. coli* is an example of in vivo protein aggregates. Hydrophobic protein aggregates will often dissolve at high pH (>11). Intermolecular disulfide aggregates may dissolve under reducing conditions (presence of DTT, cysteine, or the like). Preventive actions against aggregation are described in Table 13.

Precipitation

Protein precipitates are aggregates large enough to be visible. However, in practice aggregates not visible to the naked eye may result in severe problems as filters and chromatographic columns can be blocked.

Protein precipitation is typically observed at high ionic strength, in the presence of organic solvents or close to the isoelectric point, where solubility is low due to the zero net charge of the protein. Presence of precipitates is not always easily observed. Typical markers are presence of large and often white particles or flocculates, a turboid appearance, fibrils, or increased viscosity.

Unintended precipitation can be difficult to predict as the effect depends on a combination of the distribution of hydrophilic and hydrophobic residues on the protein surface, the pH, ionic strength, protein concentration, temperature, and composition of the aqueous phase. A perfectly clear solution may gradually become turbid during application to a chromatographic column that results in column blocking.

The mechanism of precipitation involves salting out, iso-precipitation, or the presence of polar solvents. In most cases, high salt concentrations will lead to precipitation of the protein. The process is largely dependent on the hydrophobicity of the protein, and the optimal salts are those favoring dehydration of the nonpolar regions without binding to the protein. At zero net charge of the protein, the electrostatic repulsion between the molecules is minimal. Therefore, proteins tend to precipitate near the pI of the molecule. Additions of nonpolar organic solvents reduce the water activity. The organic solvent reducing the hydrophobic attraction will displace the water molecules around the hydrophobic areas. The principal forces that lead to precipitation are, therefore, likely to be electrostatic forces and di-polar van der Waals' forces. The indicators of precipitation include cloudy solution and precipitation of solid material (the precipitates often appear white). The Hoffmeister series provides the impact of various cations and anions:

Cations:

$$NH_4^+ > K^+ > Na^+ > Li^+ > Mg^{2+} > Ca^{2+} > Gdn^+ \qquad (3)$$

Anions:

$$SO_4^- > HPO_4^{-2} > CH_3COO^- > Cl^- > NO_3^- > SCN^- \qquad (4)$$

The ions to the left in the series exert a stabilizing effect on proteins. The ions to the right may bind to the protein surface and thereby destabilize the protein. The effect of the ions is additive. Ammonium sulfate is a stabilizing salt often used for precipitation of proteins (2–3-M solution). Gdn-sulfate is a stabilizing salt, while Gdn-chloride is a strong denaturant.

Precipitation is commonly used as a purification tool in downstream processing as the biological activity is rarely affected by this procedure (organic solvents may result in denaturation). Iso-precipitation becomes more effective by adding alcohols or polyalcohols to the solvent. pH adjustment may result in unintended iso-precipitation of the protein. Passing the iso-electric point does not affect the biological activity or stability of the protein (in most cases) and the protein will normally enter into solution again, 2–3 pH units from the pI. Precipitates may form hours after the protein solution has been prepared or adjusted. The precipitates are not always visible. As the particles may result in blockage of filters and chromatographic columns, unintended precipitation in application samples constitutes a great problem in downstream processing. Preventive actions taken against precipitation are described in Table 14.

The guidance stated in the International Conference on Harmonization (ICH) harmonized tripartite guideline entitled "Stability Testing of New Drug Substances and Products" (issued by ICH on October 27, 1993) applies in general to biotechnological/biological products. However, biotechnological/biological products have distinguishing characteristics to which consideration should be given in any well-defined testing program designed to confirm their stability during the intended storage period. For such products in which the active components are

Table 14 Preventive Actions Against Precipitation

Factor	Comment
pH	The protein solubility is minimal near the iso-electric point. A change in pH may affect the redox potential of the solution, the protein solubility, and the protein stability.
Temperature	High temperature increases the conformational flexibility and, for example, organic solvents may more easily penetrate the internal structure of the protein.
Protein concentration	Low protein concentration protects the protein from precipitation; concentrations below 0.1 mg/mL may be necessary to avoid precipitation.
Time	Precipitation is a function of time. Always determine the holding time for a given sample to assure precipitates are not formed during storage.
Conductivity	High ionic strength normally results in protein precipitation. Keep the salt concentration low to moderate, but at the same time, be aware of the "salting in" effect. Ions such as NH_4^+ do stabilize the protein upon precipitation.
Co-solvents	Typical protein precipitation agents are salts (e.g., ammonium sulfate), PEG (e.g., PEG 20,000), polyelectrolytes (e.g., carboxymethyl cellulose), and organic solvents (e.g., acetone).

Abbreviation: PEG, polyethylene glycol.
Source: From Ref. 5.

typically proteins and/or polypeptides, maintenance of molecular conformation and, hence, of biological activity, is dependent on noncovalent and covalent forces. The products are particularly sensitive to environmental factors, such as temperature changes, oxidation, light, ionic content, and shear. To ensure maintenance of biological activity and to avoid degradation, stringent conditions for their storage are usually necessary.

The evaluation of stability may necessitate complex analytical methodologies. Assays for biological activity, where applicable, should be part of the pivotal stability studies. Appropriate physico-chemical, biochemical, and immunochemical methods for the analysis of the molecular entity and the quantitative detection of degradation products should also be part of the stability program whenever purity and molecular characteristics of the product permit use of these methodologies.

With these concerns in mind, the applicant should develop the proper supporting stability data for a biotechnological/biological product and consider many external conditions that can affect the product's potency, purity, and quality. Primary data to support a requested storage period for either drug substance (DS) or drug product (DP) should be based on long-term, real-time, real-condition stability studies. Thus, the development of a proper long-term stability program becomes critical to the successful development of a commercial product. The purpose of this document is to give guidance to applicants regarding the type of stability studies that need to be provided in support of marketing applications. It is understood that during the review and evaluation process, continuing updates of initial stability data may occur.

Where bulk material is to be stored after manufacture, but before formulation and final manufacturing, stability data are provided on at least three batches for which manufacture and storage are representative of the manufacturing scale of production. A minimum of six months stability data at the time of submission should be submitted in cases where storage periods greater than six months are requested. For DS with storage periods of less than six months, the minimum amount of stability data in the initial submission should be determined on a case-by-case basis. Data from pilot plant scale batches of DS produced at a reduced scale of fermentation and purification may be provided at the time the dossier is submitted to the regulatory agencies with a commitment to place the first three manufacturing scale batches into the long-term stability program after approval.

The quality of the batches of DS placed into the stability program are representative of the quality of the material used in preclinical and clinical studies and of the quality of the material to be made at the manufacturing scale. In addition, the DS (bulk material) made at pilot plant scale should be produced by a process and stored under conditions representative of that used for the manufacturing scale. The DS entered into the stability program should be stored in containers that properly represent the actual holding containers used during manufacture. Containers of reduced size may be acceptable for DS stability testing provided that they are constructed of the same material and use the same type of container/closure system that is intended to be used during manufacture.

During manufacture of biotechnological/biological products, the quality and control of certain intermediates may be critical to the production of the final product. In general, the manufacturer should identify intermediates and generate in-house data and process limits that assure their stability within the bounds of the developed process. Although the use of pilot plant scale data is permissible, the manufacturer should establish the suitability of such data using the manufacturing scale process.

On the whole, there is no single stability-indicating assay or parameter that profiles the stability characteristics of a biotechnological/biological product. Consequently, the manufacturer should propose a stability-indicating profile that provides assurance that changes in the identity, purity, and potency of the product will be detected.

At the time of submission, applicants should have validated the methods that comprise the stability-indicating profile, and the data should be available for review. The determination of which tests should be included will be product-specific. The items emphasized in the following subsections are not intended to be all-inclusive, but represent product characteristics that should typically be documented to demonstrate product stability adequately.

For the purpose of stability testing of the products described in this guideline, purity is a relative term. Because of the effect of glycosylation, deamidation, or other heterogeneities, the absolute purity of a biotechnological/biological product is extremely difficult to determine. Thus, the purity of a biotechnological/biological product should be typically assessed by more than one method and the purity value derived is method dependent. For the purpose of stability testing, tests for purity should focus on methods for determination of degradation products.

The degree of purity, as well as the individual and total amounts of degradation products of the biotechnological/biological product entered into the stability studies, should be reported and documented whenever possible. Limits of acceptable degradation should be derived from the analytical profiles of batches of the DS and DP used in the preclinical and clinical studies.

The use of relevant physico-chemical, biochemical, and immunochemical analytical methodologies should permit a comprehensive characterization of the DS and/or DP (e.g., molecular size, charge, and hydrophobicity) and the accurate detection of degradation changes that may result from deamidation, oxidation, sulfoxidation, aggregation, or fragmentation during storage. As examples, methods that may contribute to this include electrophoresis (SDS09Page, immunoelectrophoresis, Western blot, IEF), high-resolution chromatography (e.g., RP chromatography, gel filtration, IEX, affinity chromatography), and peptide mapping.

Wherever significant qualitative or quantitative changes indicative of degradation product formation are detected during long-term, accelerated, and/or stress stability studies, consideration should be given to potential hazards and the need for characterization and quantification of degradation products within the long-term stability program. Acceptable limits should be proposed and justified, taking into account the levels observed in material used in preclinical and clinical studies.

For substances that cannot be properly characterized or products for which an exact analysis of the purity cannot be determined through routine analytical methods, the applicant should propose and justify alternative testing procedures.

Forced Degradation Studies

Stress testing studies are conducted to challenge specificity of stability-indicating and impurity-monitoring methods as part of the validation protocol. The current regulatory guidances governing forced degradation studies of biological pharmaceuticals are extremely general. They itemize broad principles and approaches with few practical instructions. There is no single document that comprehensively addresses issues related to stress studies, such as objectives, timing, selection of stress conditions, and extent of degradation. We will attempt to fill this gap by summarizing regulatory guidance for stress studies of biological products and present some examples of their practical applications. Stress testing studies are conducted to challenge the specificity of stability-indicating and impurity-monitoring methods as part of the validation protocol. Another major goal is to investigate degradation products and pathways. The results of the forced degradation studies are required to be included in a Phase III investigational new drug (IND) filing. It is important to start the study as early as possible to be able to provide valuable information that can be used to improve formulations and the manufacturing process.

The choice of stress conditions should be consistent with product decomposition under normal manufacturing, storage, and use conditions. Recommended stress factors include high and low pH, elevated temperature, photolysis, and oxidation. The extent of the stress applied in forced degradation studies should ensure formation of the desired amount (usually 10–20%) of degradation.

The complexity of biological macromolecules when compared with small molecule therapeutics, differences in manufacturing, and the broad variety of potential degradation pathways lead to special requirements in quality assurance and analytical testing of pharmaceutical proteins. The product-related impurities are molecular variants formed during manufacture, storage, or use, and their properties are different from the desired product with respect to activity, efficacy, and safety.

Forced degradation studies are designed to generate product-related variants and develop analytical methods to determine the degradation products formed during accelerated pharmaceutical studies and long-term stability studies. Any significant degradation product should be evaluated for potential hazards and the need for characterization and quantification. Forced degradation or stress testing studies are part of the development strategy and are also an integral component of validation of analytical methods that indicate stability and detect impurities. This relates to the specificity section of the validation studies. It is important to recognize that forced degradation studies are not designed to establish qualitative or quantitative limits for change in DS or DP. Testing of stressed samples is required to demonstrate the following abilities of analytical techniques employed in stability studies:

- to evaluate stability of DS and DP in solution
- to determine structural transformations of the DS and DP
- to detect low concentrations of potential degradation products
- to detect unrelated impurities in the presence of the desired product and product-related degradants
- to separate product-related degradants from those derived from excipients and intact placebo
- to elucidate possible degradation pathways
- to identify degradation products that may be spontaneously generated during drug storage and use
- to facilitate improvements in the manufacturing process and formulations in parallel with accelerated pharmaceutical studies

Stress studies may be useful in determining whether accidental exposures to conditions other than normal ranges (e.g., during transportation) are deleterious to the product, and also for evaluating which specific test parameters may be the best indicators of product stability.

Stress Testing
This involves establishing a stability-indicating analytical procedure that will detect significant changes in the quality of DS at the Phase II stage of IND. Stress studies on DS and DP should be completed during Phase III and significant impurities should be identified, qualified, and quantified. Starting forced degradation experiments before Phase II is highly encouraged and should be conducted on DS with multiple aims: to provide timely recommendations for improvements in the manufacturing process; to ensure proper selection of stability-indicating analytical techniques; and to assure sufficient time for degradation product identification, degradation pathways elucidation, and optimization of stress conditions.

Every change in stability-indicating analytical methods, manufacturing processes, or formulation requires revalidation of analytical methods; therefore, full validation commences only after the manufacturing process is finalized, formulations established, and test procedures are developed and qualified. However, method validation must be completed before a formal long-term stability study begins. These limitations impose time constraints on all method-validation activities including stressed sample development and testing. Consequently, all preliminary work on optimization of stress conditions must be completed at the earlier stages, even though results of forced degradation studies are not required to be reported until the Phase III stage of IND application.

The question of how much degradation is sufficient to meet the objectives of stress studies is widely discussed, especially with respect to conventional therapeutics. A degradation level of 10–15% is considered adequate for validation of a chromatographic purity assay. Chromatographic methods for product-related impurities (including degradants) should be validated by spiking experiments within the range of 0–20% if the expected range of impurities is 0–10%. It is also suggested that DS spiked with a mixture of known degradation products can be used to challenge methods employed for monitoring stability of DP. The apparent consensus among pharmaceutical scientists is that samples degraded ~10% are optimal for use in analytical method validation. These considerations apply to small organic pharmaceuticals for which stability is dictated by the typical pharmaceutical limit of 90% of label claim.

No such limits for physico-chemical changes, losses of activity, or degradation during shelf life have been established for individual types or groups of biological products. In general, international and national regulations for biological products provide little guidance with respect to stability-related issues. These issues should be considered on a case-by-case basis.

As a group, biological products form a wide variety of product-related degradants under stress conditions. In cases with multiple degradation pathways, it appears to be beneficial to develop multiple product-related variants to challenge the specificity of analytical methods, even when some of the degradants may be present at concentrations that exceed 10%. This is done when accelerated stability studies do not provide clear indication of the degradation pathways. When a stress factor generates only one degradation product, for example, higher MW noncovalent aggregates, 10–15% level of aggregation may be sufficient to challenge the specificity of methods, such as SEC or light scattering.

The forced degradation experiments do not necessarily result in product decomposition. The study can be stopped if no degradation is observed after DS or DP has been exposed to a stress that exceeds conditions of accelerated stability protocol. Protocols for generation of product-related degradation may differ for DS and DP due to differences in matrices and concentrations. For example, sugar additives often present in DP are known to stabilize proteins vis-à-vis denaturing conditions.

Selection of Stress Conditions

Forced degradation is normally carried out under more severe conditions than those used for accelerated studies. The choice of stress conditions should be consistent with the product's decomposition under normal manufacturing, storage, and use conditions, which are specific in each case. The ICH guidance recognizes it is impossible to provide strict degradation guidelines and allows certain freedom in the selection of stress conditions for biologics. The choice of forced degradation conditions should be based on data from accelerated pharmaceutical studies and sound scientific understanding of the product's decomposition mechanism under typical use conditions. A minimal list of stress factors suggested for forced degradation studies must include acid and base hydrolyses, thermal degradation, photolysis, oxidation, and may include freeze-thaw cycles and shear.

Regulatory guidance does not specify pH, temperature ranges, specific oxidizing agents, or conditions to use, the number of freeze-thaw cycles, or specific wavelengths and light intensities. The design of photolysis studies is left to the applicant's discretion although Q1B recommends that the light source should produce combined VIS and UV (320–400 nm) outputs, and that exposure levels should be justified. Consult the appropriate regulatory authorities on a case-by-case basis to determine guidance for light-induced stress.

Degradation products that arise in significant amounts during manufacture and storage should be identified, tested for, and monitored against appropriately established acceptance criteria. Examination of some degradation products generated under stress conditions may not be necessary for certain degradants if it has been demonstrated that they are not formed under accelerated or long-term storage conditions.

The forced degradation studies should be part of impurity characterization. When identification of the impurity is not feasible, incorporate the description of unsuccessful experiments (including those conducted in stress-testing studies) in the text of the application. The most frequently encountered protein variants include truncated fragments, deamidated, oxidized, isomerized, aggregated forms, and mismatched disulfide links.

Degradation pathways for proteins can be separated into two distinct classes that involve chemical and physical instabilities. Chemical instability is any process that yields a new chemical entity including modification of the protein (via individual amino acid alteration), covalent bond formation, or cleavage. Physical instability refers to changes in the higher order structures (secondary and above). Noncovalent aggregation usually results from partial or full unfolding, which enhances hydrophobic interactions between protein molecules. It may also lead to denaturation, adsorption to surfaces, and precipitation. Aggregation presents a significant patient risk because protein aggregates are frequently immunogenic; therefore, analytical methods employed in stability testing should detect low concentrations of aggregates. This is generally tested by providing stress conditions in the downstream processing to study if any unusual decomposition products are formed; a placebo is used as control. Potential degradation pathways are extensively researched, and methods for their detection are well established. A number of comprehensive reviews on this topic are available in the literature.

The proposed stability-indicating methodologies should provide assurance that changes in the identity, purity, and potency of the product will be detected. The selection of tests is product-specific. Stability-indicating methods will characterize potency, purity, and biological activity. As examples, stability-indicating methods may include electrophoresis (SDS-PAGE, immunoelectrophoresis, Western blot, IEF), high-resolution chromatography (e.g., RP chromatography, SEC, gel filtration, IEX, and affinity chromatography), and peptide mapping.

The selected set of methods must be able to detect, separate, and quantify all observed degradation products, although it is recognized that identification and characterization of the appropriate variants may require use of additional analytical methodologies. New analytical technologies and modifications of existing technologies are continuously being developed and should be utilized when appropriate. The list of assays challenged by stressed samples should include analytical methods employed in the stability program and those monitoring impurities.

Specifications

Specifications are one part of a total control strategy designed to ensure product quality and consistency. Other parts of this strategy include thorough product characterization during development, upon which many of the specifications are based, adherence to GMPs, a validated manufacturing process, raw materials testing, in-process testing, stability testing, and so on. Specifications are chosen to confirm the quality of the DS and DP rather than establishing full characterization and should focus on those molecular and biological characteristics found to be useful in ensuring the safety and efficacy of the product.

Characterization of a biotechnological or biological product (which includes the determination of physico-chemical properties, biological activity, immuno-chemical properties, purity, and impurities) by appropriate techniques is necessary to allow relevant specifications to be established. Acceptance criteria should be established and justified based on data obtained from lots used in preclinical and/or clinical studies, data from lots used for demonstration of manufacturing consistency, data from stability studies, and relevant development data.

Extensive characterization is performed in the development phase and, where necessary, following significant process changes. At the time of submission, the product should have been compared with an appropriate reference standard, if available. When feasible and relevant, it should be compared with its natural counterpart. Also, at the time of submission, the manufacturer should have established appropriately characterized in-house reference materials which will serve for biological and physico-chemical testing of production lots. New analytical technology and modifications to existing technology are continually being developed and will be utilized when appropriate.

Physico-Chemical Properties

A physico-chemical characterization program will generally include a determination of the composition, physical properties, and primary structure of the desired product. In some cases, information regarding higher-order structure of the desired product (the fidelity of which is generally inferred by its biological activity) may be obtained by appropriate physico-chemical methodologies.

An inherent degree of structural heterogeneity occurs in proteins due to the biosynthetic processes used by living organisms to produce them; therefore, the desired product can be a mixture of anticipated post-translationally modified forms (e.g., glycoforms). These forms may be active and their presence may have no deleterious effect on the safety and efficacy of the product. The manufacturer should define the pattern of heterogeneity of the desired product and demonstrate consistency with that of the lots used in preclinical and clinical studies. If a consistent pattern of product heterogeneity is demonstrated, an evaluation of the activity, efficacy, and safety (including immunogenicity) of individual forms may not be necessary.

Heterogeneity can also be produced during manufacture and/or storage of the DS or DP. As the heterogeneity of these products defines their quality, the degree and profile of this heterogeneity should be characterized to ensure lot-to-lot consistency. When these variants of the desired product have properties comparable with those of the desired product with respect to activity, efficacy, and safety, they are considered product-related substances. When process changes and degradation products result in heterogeneity patterns that differ from those observed in the material used during preclinical and clinical development, the significance of these alterations should be evaluated.

Biological Activity

Assessment of the biological properties constitutes an equally essential step in establishing a complete characterization profile. An important property is the biological activity that describes the specific ability or capacity of a product to achieve a defined biological effect.

A valid biological assay to measure the biological activity should be provided by the manufacturer. Examples of procedures used to measure biological activity include:

- animal-based biological assays, which measure an organism's biological response to the product;
- cell culture-based biological assays, which measure biochemical or physiological response at the cellular level; and
- biochemical assays, which measure biological activities, such as enzymatic reaction rates or biological responses induced by immunological interactions.

Other procedures, such as ligand and receptor-binding assays, may be acceptable.

Potency (expressed in units) is the quantitative measure of biological activity based on the attribute of the product that is linked to the relevant biological properties, whereas quantity (expressed in mass) is a physico-chemical measure of protein content. Mimicking the biological activity in the clinical situation is not always necessary. A correlation between the expected clinical response and the activity in the biological assay should be established in pharmacodynamic or clinical studies.

The results of biological assays should be expressed in units of activity calibrated against an international or national reference standard, when available and appropriate for the assay utilized. Where no such reference standard exists, a characterized in-house reference material should be established and assay results of production lots should be reported as in-house units.

Often, for complex molecules, the physico-chemical information may be extensive but unable to confirm the higher-order structure which, however, can be inferred from the biological activity. In such cases, a biological assay, with wider confidence limits, may be acceptable when combined with a specific quantitative measure. Importantly, a biological assay to measure the biological activity of the product may be replaced by physico-chemical tests only in those instances where:

- sufficient physico-chemical information about the drug, including higher-order structure, can be thoroughly established by such physico-chemical methods, and relevant correlation to biologic activity demonstrated;
- there exists a well-established manufacturing history;
- where physico-chemical tests alone are used to quantify the biological activity (based on appropriate correlation), results should be expressed in mass.

For the purpose of lot release, the choice of relevant quantitative assay (biological and/or physicochemical) should be justified by the manufacturer.

Immunochemical Properties

When an antibody is the desired product, its immunological properties should be fully characterized. Binding assays of the antibody to purified antigens and defined regions of antigens should be performed, as feasible, to determine affinity, avidity, and immunoreactivity (including cross-reactivity). In addition, the target molecule bearing the relevant epitope should be biochemically defined and the epitope itself defined, when feasible.

For some DS or DP, the protein molecule may need to be examined using immunochemical procedures [e.g., enzyme-linked immunosorbent assay (ELISA), Western-blot] and utilizing antibodies that recognize different epitopes of the protein molecule. Immunochemical properties of a protein may serve to establish its identity, homogeneity, or purity, or serve to quantify it.

If immunochemical properties constitute lot release criteria, all relevant information pertaining to the antibody should be made available.

Purity, Impurities, and Contaminants
Purity
The determination of absolute, as well as relative, purity presents considerable analytical challenges, and the results are highly method-dependent. Historically, the relative purity of a biological product has been expressed in terms of specific activity (units of biological activity per milligram of product), which is also highly method-dependent. Consequently, the purity of the DS and DP is assessed by a zcombination of analytical procedures.

Due to the unique biosynthetic production process and molecular characteristics of biotechnological and biological products, the DS can include several molecular entities or variants. When these molecular entities are derived from anticipated post-translational modification, they are part of the desired product. When variants of the desired product are formed during the manufacturing process and/or storage and have properties comparable with the desired product, they are considered product-related substances and not impurities.

Individual and/or collective acceptance criteria for product-related substances should be set, as appropriate.

For the purpose of lot release, an appropriate subset of methods should be selected and justified for determination of purity.

Impurities
In addition to the evaluation of the purity of the DS and DP, which may be composed of the desired product and multiple product-related substances, the manufacturer should also assess the impurities, which may be present. Impurities may be either process- or product-related. They can be of known structure, may be partially characterized, or unidentified. When adequate quantities of impurities can be generated, these materials should be characterized to the extent possible and, where possible, their biological activities should be evaluated.

Process-related impurities encompass those that are derived from the manufacturing process, that is, cell substrates (e.g., host cell proteins, host cell DNA), cell culture (e.g., inducers, antibiotics, or media components), or downstream processing. Product-related impurities (e.g., precursors and certain degradation products) are molecular variants arising during manufacture and/or storage that do not have properties comparable with those of the desired product with respect to activity, efficacy, and safety.

Further, the acceptance criteria for impurities should be based on data obtained from lots used in preclinical and clinical studies and manufacturing consistency lots.

Individual and/or collective acceptance criteria for impurities (product-related and process-related) should be set, as appropriate. Under certain circumstances, acceptance criteria for some selected impurities may not be necessary.

Contaminants

Contaminants in a product include all adventitiously introduced materials not intended to be part of the manufacturing process, such as chemical and biochemical materials (e.g., microbial proteases) and/or microbial species. Contaminants should be strictly avoided and/or suitably controlled with appropriate in-process acceptance criteria or action limits for DS or DP specifications. For the specialcase of adventitious viral or mycoplasma contamination, the concept of action limits is not applicable, and the strategies proposed in ICH Guidances *Q5A Quality of Biotechnological/Biological Products: Viral Safety Evaluation of Biotechnology Products Derived from Cell Lines of Human or Animal Origin* and *Q5D Quality of Biotechnological/Biological Products: Derivation and Characterization of Cell Substrates Used for Production of Biotechnological/Biological Products* should be considered.

Quantity

Quantity, usually measured as protein content, is critical for a biotechnological and biological product and should be determined using an appropriate assay, usually physico-chemical in nature. In some cases, it may be demonstrated that the quantity values obtained may be directly related to those found using the biological assay. When this correlation exists, it may be appropriate to use measurement of quantity rather than the measurement of biological activity in manufacturing processes, such as filling.

Analytical Considerations

Reference Standards and Reference Materials

For drug applications of new molecular entities, it is unlikely that an international or national standard will be available. At the time of submission, the manufacturer should have established an appropriately characterized in-house primary reference material, prepared from lot(s) representative of production and clinical materials. In-house working reference material(s) used in the testing of production lots should be calibrated against this primary reference material. Where an international or national standard is available and appropriate, reference materials should be calibrated against it. While it is desirable to use the same reference material for both biological assays and physico-chemical testing, in some cases, a separate reference material may be necessary. Also, distinct reference materials for product-related substances, product-related impurities, and process-related impurities may need to be established. When appropriate, a description of the manufacture and/or purification of reference materials should be included in the application. Documentation of the characterization, storage conditions, and formulation supportive of reference material(s) stability should also be provided.

Validation of Analytical Procedures

At the time the application is submitted to the regulatory authorities, applicants should have validated the analytical procedures used in the specifications in accordance with the ICH Guidances *Q2A Validation of Analytical Procedures: Definitions and Terminology* and *Q2B Validation of Analytical Procedures: Methodology*, except where there are specific issues for unique tests used for analyzing biotechnological and biological products.

Process Controls
Adequate design of a process and knowledge of its capability are part of the strategy used to develop a manufacturing process that is controlled and reproducible, yielding a DS or DP that meets the specifications. In this respect, limits are justified based on critical information gained from the entire process spanning the period from early development through commercial-scale production.

For certain impurities, testing of either the DS or the DP may not be necessary and may not need to be included in the specifications if efficient control or removal of acceptable levels is demonstrated by suitable studies. This testing can include verification at commercial scale in accordance with regional regulations. It is recognized that only limited data may be available at the time of submission of an application. This concept may, therefore, sometimes be implemented after marketing authorization, in accordance with regional regulations.

In-process tests are performed at critical decision-making steps and at other steps where data serve to confirm consistency of the process during the production of either the DS or the DP. The results of in-process testing may be recorded as action limits or reported as acceptance criteria. Performing such testing may eliminate the need for testing of the DS or DP. In-process testing for adventitious agents at the end of cell culture is an example of testing for which acceptance criteria should be established.

The use of internal action limits by the manufacturer to assess the consistency of the process at less critical steps is also important. Data obtained during development and validation runs should provide the basis for provisional action limits to be set for the manufacturing process. These limits, which are the responsibility of the manufacturer, may be used to initiate investigation or further action. They should be further refined as additional manufacturing experience and data are obtained after product approval.

The quality of the raw materials used in the production of the DS (or DP) should meet standards appropriate for their intended use. Biological raw materials or reagents may require careful evaluation to establish the presence or absence of deleterious endogenous or adventitious agents. Procedures that make use of affinity chromatography (e.g., employing MAbs) should be accompanied by appropriate measures to ensure that such process-related impurities or potential contaminants arising from their production and use do not compromise the quality and safety of the DS or DP. Appropriate information pertaining to the antibody should be made available.

The quality of the excipients used in the DP formulation (and in some cases, in the DS), as well as the container/closure systems, should meet pharmacopoeial standards, where available and appropriate. Otherwise, suitable acceptance criteria should be established for the nonpharmacopoeial excipients.

Release Limits vs. Shelf-Life Limits
The concept of release limits versus shelf-life limits may be applied where justified. This concept pertains to the establishment of limits that are tighter for the release than for the shelf life of the DS or DP. Examples where this may be applicable

include potency and degradation products. In some regions, the concept of release limits may only be applicable to in-house limits and not to the regulatory shelf-life limits.

Appropriate statistical analysis should be applied, when necessary, to quantitative data reported. The methods of analysis, including justification and rationale, should be described fully. These descriptions should be sufficiently clear to permit independent calculation of the results presented.

Justification of Specifications

The setting of specifications for DS and DP is part of an overall control strategy, which includes control of raw materials and excipients, in-process testing, process evaluation or validation, adherence to GMPs, stability testing, and testing for consistency of lots. When combined in total, these elements provide assurance that the appropriate quality of the product will be maintained. As specifications are chosen to confirm the quality rather than to characterize the product, the manufacturer should provide the rationale and justification for including and/or excluding testing for specific quality attributes. The following points should be taken into consideration when establishing scientifically justifiable specifications.

- Specifications are linked to a manufacturing process.
- Specifications should be based on data obtained from lots used to demonstrate manufacturing consistency. Linking specifications to a manufacturing process is important, especially for product-related substances, product-related impurities, and process-related impurities. Process changes and degradation products produced during storage may result in heterogeneity patterns that differ from those observed in the material used during preclinical and clinical development. The significance of these alterations should be evaluated.
- Specifications should account for the stability of DS and DP.
- Degradation of DS and DP, which may occur during storage, should be considered when establishing specifications. Due to the inherent complexity of these products, there is no single stability-indicating assay or parameter that profiles the stability characteristics. Consequently, the manufacturer should propose a stability-indicating profile. The result of this stability-indicating profile will then provide assurance that changes in the quality of the product will be detected. The determination of which tests should be included will be product-specific. The manufacturer is referred to the ICH guidance *Q5C Stability Testing of Biotechnological/Biological Products*.
- Specifications are linked to preclinical and clinical studies.
- Specifications should be based on data obtained for lots used in preclinical and clinical studies. The quality of the material made at commercial scale should be representative of the lots used in preclinical and clinical studies.
- Specifications are linked to analytical procedures.

Critical quality attributes may include items, such as potency, the nature and quantity of product-related substances, product-related impurities, and process-related impurities. Such attributes can be assessed by multiple analytical procedures, each yielding different results. In the course of product development, it is not unusual for the analytical technology to evolve in parallel with the product. Therefore, it is important to confirm that data generated during development correlate with those generated at the time the marketing application is filed.

PHYSIO-CHEMICAL CHARACTERIZATION TESTS
Structural Characterization and Confirmation
Amino Acid Sequence
The amino acid sequence of the desired product should be determined to the extent possible using customary approaches and then compared with the sequence of the amino acids deduced from the gene sequence of the desired product.

Amino Acid Composition
The overall amino acid composition is determined using various hydrolytic and analytical procedures and compared with the amino acid composition deduced from the gene sequence for the desired product, or the natural counterpart, if considered necessary. In many cases, amino acid composition analysis provides some useful structural information for peptides and small proteins, but such data are generally less definitive for large proteins. Quantitative amino acid analysis data can also be used to determine protein content in many cases.

Terminal Amino Acid Sequence
Terminal amino acid analysis is performed to identify the nature and homogeneity of the amino- and carboxy-terminal amino acids. If the desired product is found to be heterogeneous with respect to the terminal amino acids, the relative amounts of the variant forms should be determined using an appropriate analytical procedure. The sequence of these terminal amino acids should be compared with the terminal amino acid sequence deduced from the gene sequence of the desired product.

Peptide Map
Selective fragmentation of the product into discrete peptides is performed using suitable enzymes or chemicals, and the resulting peptide fragments are analyzed by HPLC or other appropriate analytical procedures. The peptide fragments should be identified to the extent possible using techniques, such as amino acid compositional analysis, N-terminal sequencing, or MS. Peptide mapping of the DS or DP using an appropriately validated procedure is a method that is frequently used to confirm desired product structure for lot release purposes.

Sulfhydryl Group(s) and Disulfide Bridges
If, based on the gene sequence for the desired product, cysteine residues are expected, the number and positions of any free sulfhydryl groups and/or disulfide bridges should be determined, to the extent possible. Peptide mapping (under reducing and nonreducing conditions), MS, or other appropriate techniques may be useful for this evaluation.

Carbohydrate Structure
For glycoproteins, the carbohydrate content (neutral sugars, amino sugars, and sialic acids) is determined. In addition, the structure of the carbohydrate chains, the oligosaccharide pattern (antennary profile), and the glycosylation site(s) of the polypeptide chain are analyzed, to the extent possible.

Physico-Chemical Properties

Molecular Weight or Size
Molecular weight (or size) is determined using SEC, SDS-PAGE (under reducing and/or nonreducing conditions), MS, and other appropriate techniques.

Isoform Pattern
This is determined by IEF or other appropriate techniques.

Extinction Coefficient (or Molar Absorptivity)
In many cases, it will be desirable to determine the extinction coefficient (or molar absorptivity) for the desired product at a particular UV/VIS wavelength (e.g., 280 nm). The extinction coefficient is determined using UV/VIS spectrophotometry on a solution of the product having a known protein content as determined by techniques, such as amino acid compositional analysis or nitrogen determination. If UV absorption is used to measure protein content, the extinction coefficient for the particular product should be used.

Electrophoretic Patterns
Electrophoretic patterns and data on identity, homogeneity, and purity can be obtained by PAGE, IEF, SDS-PAGE, Western blot, capillary electrophoresis, or other suitable procedures.

Liquid Chromatographic Patterns
Chromatographic patterns and data on the identity, homogeneity, and purity can be obtained by size exclusion chromatography, reverse-phase liquid chromatography, ion-exchange liquid chromatography, affinity chromatography, or other suitable procedures.

Spectroscopic Profiles
The UV and VIS absorption spectra are determined as appropriate. The higher-order structure of the product is examined using procedures, such as CD, NMR, or other suitable techniques as appropriate.

Process-Related Impurities and Contaminants

These are derived from the manufacturing process and are classified into three major categories: cell substrate-derived, cell culture-derived, and downstream-derived.

1. Cell substrate-derived impurities include, but are not limited to, proteins derived from the host organism and nucleic acid (host cell genomic, vector, or total DNA). For host cell proteins, a sensitive assay, for example, immunoassay, capable of detecting a wide range of protein impurities is generally utilized. In the case of an immunoassay, a polyclonal antibody used in the test is generated by immunization with a preparation of a production cell minus the product-coding gene, fusion partners, or other appropriate cell lines. The level of DNA from the host cells can be detected by direct analysis on the product (such as hybridization techniques). Clearance studies, which could include spiking experiments at the laboratory scale, to demonstrate the

removal of cell substrate-derived impurities, such as nucleic acids and host cell proteins may sometimes be used to eliminate the need for establishing acceptance criteria for these impurities.
2. Cell culture-derived impurities include, but are not limited to, inducers, antibiotics, serum, and other media components.
3. Downstream-derived impurities include, but are not limited to, enzymes, chemical and biochemical processing reagents (e.g., cyanogen bromide, guanidine, oxidizing and reducing agents), inorganic salts (e.g., heavy metals, arsenic, nonmetallic ion), solvents, carriers, ligands (e.g., MAbs), and other leachables.

For intentionally introduced, endogenous, and adventitious viruses, the ability of the manufacturing process to remove and/or inactivate viruses should be demonstrated as described in ICH guidance *Q5A Viral Safety Evaluation of Biotechnology Products Derived from Cell Lines of Human or Animal Origin*.

Product-Related Impurities Including Degradation Products

The following represents the most frequently encountered molecular variants of the desired product and lists relevant technology for their assessment. Such variants may need considerable effort in isolation and characterization in order to identify the type of modification(s). Degradation products arising in significant amounts during manufacture and/or storage should be tested for and monitored against appropriately established acceptance criteria.

Truncated Forms
Hydrolytic enzymes or chemicals may catalyze the cleavage of peptide bonds. These may be detected by HPLC or SDS-PAGE. Peptide mapping may be useful, depending on the property of the variant.

Other Modified Forms
Deamidated, isomerized, mismatched S–S linked, oxidized, or altered conjugated forms (e.g., glycosylation, phosphorylation) may be detected and characterized by chromatographic, electrophoretic, and/or other relevant analytical methods (e.g., HPLC, capillary electrophoresis, MS, and CD).

Aggregates
The category of aggregates includes dimers and higher multiples of the desired product. These are generally resolved from the desired product and product-related substances, and quantified by appropriate analytical procedures (e.g., SEC and capillary electrophoresis).

DESIGN OF PREFORMULATION STUDIES

A typical preformulation study would include a statistical design method to understand the effects of buffer strength, sodium chloride concentration, and pH on conformation and stability of the protein and the related interactions. It was also important to elucidate interactions between these factors. A central composite design (CCD) using a two-level full-factorial study can be used. Secondary structure can be evaluated using CD. Stability toward unfolding is investigated using high-sensitivity differential scanning calorimetry. DePEGylation (where applicable), aggregation, and protein loss are evaluated using SEC-HPLC with online light scattering, at time zero and after a two-week stability study. Response surface plots are used to show optimal pH, sodium chloride, and buffer conditions. Interactions between pH and sodium chloride as well as between pH and buffer concentration occur. T_m is predictive of compound stability. Statistical analyses can be performed using standard statistical programs. Generally, a CCD can be set up using a two-level full-factorial design with axial and center points. Factorial design is a statistical tool that allows experimentation on several factors simultaneously. A two-level design involves the evaluation of two or more factors (e.g., sodium chloride concentration, buffer concentration, and pH) at two levels—high and low. A factorial design evaluating three factors at all combinations of high and low levels for each factor will result in a full-factorial design consisting of $2^3 = 8$ runs experiments. Addition of center points allows detection of nonlinearity in the responses. Axial points are also included in the CCD. These points are levels of the factors under investigation, located at a distance outside the original high- and low-factor range. The CCD, therefore, contains five levels of each factor: low-axial, low-factorial, center, high-factorial, and high-axial. With these many levels, enough information is generated to fit second-order polynomials. The total number of runs becomes 2^3+6 runs with all factors set at the center points + 6 additional runs (one factor set at high- or low-axial condition with all others set at center points) = 20 runs. Actual fitting of the model is computed using the statistical software.

REFERENCES
1. http://www.gpoaccess.gov/cfr/index.html.
2. http://www.fda.gov/cber/transfer/tranfer.htm.
3. http://www.avantimmune.com.
4. http://www.nektar.com.
5. Handbook of Biogeneric Therapeutic Proteins: Regulatory, Manufacturing, Testing and Intellectual Property Issues, Taylor and Francis, Boca Raton, FL 2005.

BIBLIOGRAPHY
Anderson MJ, Witcomb PJ. DOE Simplified: Practical Tools for Effective Experimentation. Portland, OR: Productivity Inc, 2000.
Arakawa T, Bhat R, Timasheff SN. Preferential interactions determine protein solubility in three component solutions: the $MgCl_2$ system. Biochem 1990; 29:1914–1923.
Arakawa T, Timasheff SN. Mechanism of protein salting in and salting out by divalent cation salts: balance between hydration and salt binding. Biochem 1984; 23: 5912–5923.
Bichsel VE, Curico V, Gassmann R, Otto H. Requirements for the quality control of chemically synthesized peptides and biotechnologically produced proteins. Pharmaceutica Acta Helvetiae 1996; 71:439–446.
Boccu E, Velo GP, Veronese FM. Pharmacokinetic properties of polyethylene glycol derivatized superoxide dismutase. Pharmacol Res Commun 1982; 14:113–120.
Box GE, Hunter WG, Hunter JS. Statistics for Experimenters: An Introduction to Design, Data Analysis and Model Building. New York, NY: John Wiley and Sons, 1978.
Brenner BM, Hostetter TH, Humes HD. Glomerular permselectivity: barrier function based on discrimination of molecular size and charge. Am J Renal Physiol 1978; 234:F445.
Carr GP, Wahlich JC. A practical approach to method validation in pharmaceutical analysis. J Pharm Biomed Anal 1990; 86:613–618.
CDER. Reviewer Guidance. Validation of Chromatographic Methods. 1994, Nov.
FDA. Guidance for Industry. Analytical Procedures and Methods Validation: Chemistry, Manufacturing, and Controls Documentation. Draft Guidance, 2000, Aug.

FDA. Guidance for Industry. INDs for Phase 2 and 3 Studies of Drugs, Including Specified Therapeutic Biotechnology-Derived Products. Draft Guidance, 1999, Feb.
Francis GE, Delgado C, Fisher D. PEG modified proteins. In: Ahern TJ, Manning MC, eds. Stability of Protein Pharmaceuticals. Part B. New York, NY: Plenum Press, 1992: 235–263.
Handbook of Biogeneric Therapeutic Proteins: Regulatory, Manufacturing, Testing and Intellectual Property Issues, Taylor and Francis, Boca Raton, FL 2005.
Harris JM, Martin NE, Madi M. Clin. Pegylation: A novel process for modifying pharmacokinetics. Pharmacokinet 2001; 40:539–551.
Harris JM, Veronese FM, eds. Peptide and Protein PEGylation II—clinical evaluation. Adv Drug Deliv Rev 2003; 55:1259–1350.
Harris JM, Zalipsky S, eds. Poly(ethylene glycol), Chemistry and Biological Applications, ACS Symposium Series 680 (1997). NOTE: This book is available from Nektar for $160.00.
Healy JM, Lewis SD, Kurz M, Boomer RM, Thompson KM, Wilson C, McCauley TG, Pharm Res 2004; 21:2234–2245.
Hu R, Zhai Q, He W, Mei L, Liu W. Bioactivities of ricin retained and its immunoreactivity to anti-ricin polyclonal antibodies alleviated through pegylation. Int J Biochem Cell Biol 2002; 34(4):396–402.
ICH. Final Guidance on Stability Testing of Biotechnological/Biological Products; Availability. Federal Register 61FR, 1996:36466–36469.
ICH. Guidance for Industry. Q1A Stability Testing of New Drug Substances and Products. ICH-Q1A, 2001, Aug.
ICH. Guidance for Industry, Q1B Photostability Testing of New Drug Substances and Product. ICH-Q1B, 1996, Nov.
ICH. Guidance for Industry, Q2B Validation of Analytical Procedures: Methodology. ICH-Q2B, 1996, Nov.
ICH. Guidance for Industry, Q3A Impurities in New Drug Substances. ICH-Q3A, 2003, 11 Feb.
ICH. Guidance for Industry, Q3B(R). Impurities in New Drug Products. ICH-Q3B. 2003, 14 Nov.
ICH. Guidance for Industry, Q5C. Quality of Biotechnological Products: Stability Testing of Biotechnological/Biological Products. ICH-Q5C, 1996, 10 July.
ICH. Guidance for Industry, Q6B. Specifications: Test Procedures and Acceptance Criteria for Bio-technological/Biological Products. ICH-Q6B, 1999 18 Aug.
Jenke DR. Chromatographic method validation: a review of current practices and procedures. II. Guidelines for primary validation parameters. J Liq Chromatogr 1996; 19: 737–757.
Johnson WC Jr. Protein secondary structure and circular dichroism: a practical guide. Proteins 1990; 7:205–214.
Klibanov AL, Maruyama K, Torchilin VP, Huang L. Amphipathic polyethyleneglycols effectively prolong the circulation time of liposomes. FEBS Lett 1990; 268:235–237.
Manning MC et al. Stability of protein pharmaceuticals. Pharmaceutical Res 1989; 6: 903–918.
Mehvar R. Modulation of the pharmacokinetics and pharmacodynamics of protein by polyethilene glycol conjugation. Pharm Pharm Sci 2000; 3:125–136.
Monfardini C, Schiavon O, Caliceti P, Morpurgo M, Harris JM, Veronese FM. Bioconjugate Chem 1995; 6:62–69.
Nakaoka R et al. J Cont Release 1997; 46:253–261.
Papahadjopoulos D, Allen T, Gabizon A, Mayhew E, Redemann KC, Martin FJ. Sterically stabilized liposomes: improvements in pharmacokinetics and antitumor therapeutic efficacy. Proc Natl Acad Sci USA 1991; 88:11460–11464.
Pasut G, Guiotto A, Veronese FM. Expert Opin Ther Patents 2004; 14:1–36.
Reubsaet JLE et al. Analytical techniques used to study the degradation of proteins and peptides: chemical instability. J Pharm Biomed Anal 1988; 17:955–978.
Reubsaet JLE et al. Analytical techniques used to study the degradation of proteins and peptides: physical instability. J Pharm Biomed Anal 1998; 17:979–984.
Reynolds DW, Facchine KL, Mullaney JF, Alsante KM, Hatajik TD, Motto MG. Available guidance and best practices for conducting forced degradation studies. Pharm Technol 2002, 1 Feb.

Rouan SK. Biotechbased pharmaceuticals. In: Baker GS, Rhodes CT, eds. Modern Pharmaceutics. New York, NY: Marcel Dekker, 1995:843–873.

Syto R, Murgolo NJ, Braswell EH, Mui P, Huang E, Windsor WT. Structural and biological stability of the human Interleukin 10 homodimer. Biochemistry 1998; 37:16943–16951.

Szepesi G et al. Selection of high-performance liquid chromatographic methods in pharmaceutical analysis. III Method validation. J Chromatogr 1989; 464:265–278.

Timasheff SN. Stabilization of protein structure by solvent additives. In: Ahern TJ, Manning MC, eds. Stability of Protein Pharmaceuticals. Part B. New York, NY: Plenum Press; 1992:265–285.

Tsai PK, Volkin DB, Dabora JM et al. Formulation design of acidic fibroblast growth factor. Pharm Res 1993; 10:649–659.

Venyaminov S Yu, Yang JT. Determination of protein secondary structure. In: Fasman GD, ed. Circular Dichroism and the Conformational Analysis of Biomolecules. New York, NY: Plenum Press, 1996:69–108.

Veronese F, Harris JM, eds. Peptide and protein PEGylation. Adv Drug Deliv Rev 2002; 54:453–609.

Veronese FM, Caliceti P. J Bioact Compat Plym 1997; 12:196–207.

Volkin DB et al. "Degradative covalent reactions important to protein stability. Mol Biotechnol 1997; 8:105–122.

Wang W. Instability, stabilization and formulation of liquid protein pharmaceuticals. Int J Pharm 1999; 185:129–188.

Yamaoka T, Tabata Y, Ikada Y. Distribution and tissue uptake of poly(ethylene glycol) with different molecular weights after intravenous administration to mice. J Pharm Sci 1994; 83:601–606.

10 Characterization of Phytomedicines

INTRODUCTION

Drugs derived from plant sources are generally labeled as phytomedicines, botanical products, natural products, and so on. *Botanical products* are finished, labeled products that contain vegetable matter as ingredients. A botanical product may be a food (including a dietary supplement), a drug (including a biological drug), a medical device (e.g., gutta-percha), or a cosmetic. The term *botanical* includes plant materials, algae, macroscopic fungi, and combinations thereof.

The World Health Organization (WHO) estimates that four billion people—80% of the world population—use herbal medicine for some aspect of primary healthcare. Herbal medicine is a major component in all indigenous peoples' traditional medicine and is a common element in Ayurvedic, homeopathic, naturopathic, traditional oriental, and Native American Indian medicine. Opinions about the safety, efficacy, and appropriateness of medicinal herbs vary widely among medical and health professionals in countries where herbal remedies are used. Some countries' professionals accept historical, empirical evidence as the only necessary criterion for herbal medicine's efficacy. Others would ban all herbal remedies as dangerous or of questionable value.

Early humans recognized their dependence on nature in both health and illness. Led by instinct, taste, and experience, primitive men and women treated illness by using plants, animal parts, and minerals that were not part of their usual diet. Physical evidence of use of herbal remedies goes back some 60,000 years to a burial site of a Neanderthal man uncovered in 1960. All cultures have long folk medicine histories that include the use of plants. The invention of writing was a focus around which herbal knowledge could accumulate and grow. The first written records detailing the use of herbs in the treatment of illness are the Mesopotamian clay tablet writings and the Egyptian papyrus, which contains 876 prescriptions made up of more than 500 different substances, including many herbs. The Middle Eastern era is followed by the Greco-Roman era that saw the writing of the De Materia Medica, which contains 950 curative substances, of which 600 are plant products and the rest are of animal or mineral origin. The Arab medicine was built on Greco-Roman and the text of Jami of Ibn Baiar (died 1248 AD), which lists more than 2000 substances, including many plant products. India's Ayurvedic book on internal medicine, the Characka Samhita, describes 582 herbs. In China, the Classic of the Materia Medica, compiled no earlier than the first century AD focuses on the description of individual herbs. It includes 252 botanical substances, 45 mineral substances, and 67 animal-derived substances. Traditional Chinese medicine was brought to Japan via Korea, and Chinese-influenced Korean medicine was adapted by the Japanese during the reign of Emperor Ingyo (411–453 AD).

In North America, early explorers traded knowledge with the Native American Indians. In 1716, French explorer Lafitau found a species of ginseng, *Panax quinquefolius* L., growing in Iroquois territory in the New World. This American ginseng soon became an important item in world herb commerce. The Jesuits dug up the plentiful American ginseng, sold it to the Chinese, and used the money to build schools and churches. Even today, American ginseng is a sizable crude U.S. export. Whereas herbal medicines played a significant role in the lives of Americans, they have lost touch with it with the onslaught of allopathic medicines. One of the most significant reports on the use of botanical drugs in North America is the Baseline Natural Health Products Survey among consumers, March 2005, conducted by Health Canada. This survey concluded that 71% have used botanical, 38% use it on a daily basis, 37% seasonally, 11% weekly, 57% use vitamins, 15% Echinacea, and 11% use other herbal remedies, algal and fungal products. Almost 80% of North Americans believe that botanical drugs are safer and their use is likely to increase in the future. However, despite a long history of use, botanical drugs are generally considered to be anecdotal and ineffective by the regulatory agencies and allowed for sale only as food supplements. Recently, this trend of rejecting botanical drugs was reversed, mainly due to pressures from consumers, that the United States Food and Drug Administration (U.S. FDA) issued its first botanical guideline and established a separate division within the agency to evaluate and approve botanical products as drugs, both prescription and over-the-counter (OTC).

REGULATORY STATUS

The legal process of regulation and legislation of herbal medicines changes from country to countr (Table 1). The reason for this involves mainly cultural aspects and also the fact that herbal medicines are rarely studied scientifically. Thus, few herbal preparations have been tested for safety and efficacy. The WHO has published guidelines in order to define basic criteria for evaluating the quality, safety, and efficacy of herbal medicines aimed at assisting national regulatory authorities, scientific organizations, and manufacturers in this particular area. Furthermore, the WHO has prepared Pharmacopeia monographs on herbal medicines and the basis of guidelines for the assessment of herbal drugs.

CHARACTERISTICS OF PHYTOMEDICINES

Phytotherapeutic agents or phytomedicines are standardized herbal preparations consisting of complex mixtures of one or more plants, which are used in most countries for the management of various diseases. According to the WHO definition, herbal drugs contain as active ingredients plant parts or plant materials in the crude or processed state plus certain excipients, that is, solvents, diluents, or preservatives. Usually, the active principles responsible for their pharmacological action are unknown. One basic characteristic of phytotherapeutic agents is the fact that they normally do not possess an immediate or strong pharmacological action. For this reason, phytotherapeutic agents are not used for emergency treatment. Other characteristics of herbal medicines are their wide therapeutic use and great acceptance by the population. In contrast to modern medicines, herbal medicines are frequently used to treat chronic diseases. Combinations with chemically defined active substances or isolated constituents are not considered to be herbal medicines.

Characterization of Phytomedicines

Table 1 Herbal Drug Approval Rules in Different Countries

Argentina	The Herboristerias are authorized for sale as plant drugs but not as mixtures. Mixtures of plant drugs are controlled (Law No. 16.463). In 1993, a Ministry of Health regulation determined the obligatory registration of medicinal herbs. The Argentinian National Pharmacopeia established control over the existence of crude extracts, extracts, or fractions of complex chemical composition, and pure active principles. About 889 monographs exist in Argentina. About 56 describe crude drugs alone and 33 describe extracts or fractions. However, there is lack of control of raw materials, lack of control over the wild plants, lack of scientific criteria for the collection of plants, and lack of control over methods of drying, conservation, or grinding.
Australia	The Australian Parliament established the Working Party on Natural and Nutritional Supplements to review the quality, safety, efficacy, and labeling of herbal and related products (Therapeutic Good Act, 1990). The act provides: "that traditional claims for herbal remedies be allowed, providing general advertising requirements are complied with and providing such claims are justified by literature references."
Brazil	In 1994, the Ministry of Health created a commission to evaluate the situation of phytotherapeutic agents in Brazil. The commission proposed a directive based mainly on German and French regulations and on WHO guidelines for herbal drugs. In 1995, "Directive Number 6" established the legal requirement for the registration of herbal drugs and defined the phytopharmaceutical product as "a processed drug containing as active ingredients exclusively plant material and/or plant drug preparations. They are intended to treat, cure, alleviate, prevent, and diagnose diseases." Poor enforcement of ill-defined laws.
Canada	In 1986, the Canadian HPB constituted a special committee (three pharmacists, two herbalists, one nutritionist, and one physician) and classified herbal drugs as "Folk Medicine." The regulation is based on traditional uses, as long as the claim is validated by scientific studies. In 1990, the HPB listed 64 herbs that were considered to be unsafe. In 1992, the HPB submitted a regulatory proposal to the Canadian Parliament and listed another 64 herbs that were considered to be adulterants. The Canadian regulatory system is consistent with the WHO guidelines for the assessment of herbal medicines.
Chile	In 1992, the Unidad de Medicina Tradicional was established with the objective of incorporating traditional medicine with proven efficacy into health programs (Law No. 19.253, October 1993). Directive No. 435/81 defined herbal drugs with therapeutic indication claims and/or dosage recommendations as being drugs, restricted for sale in pharmacies and drug stores. Registration for marketing authorization is needed for herbal products. Natural products are legally differentiated as follows: (*i*) drugs intended to cure, alleviate, or prevent disease; (*ii*) food products for medicinal use and with therapeutic properties; and (*iii*) food products for nutritional purposes.
China	In China, until 1984 there was virtually no regulation of pharmaceuticals or herbal preparations. In 1984, the People's Republic implemented the Drug Administration Law, which said that traditional herbal preparations were generally considered "old drugs" and, except for new uses, were exempt from testing for efficacy or side effects. The Chinese Ministry of Public Health would oversee the administration of new herbal products.

(*Continued*)

Table 1 Herbal Drug Approval Rules in Different Countries (*Continued*)

England	England generally follows the rule of prior use, which says that hundreds of years of use with apparent positive effects and no evidence of detrimental side effects are enough evidence—in lieu of other scientific data—that the product is safe. To promote the safe use of herbal remedies, the Ministry of Agriculture, Fisheries, and Food and the Department of Health jointly established a database of adverse effects of nonconventional medicines at the National Poisons Unit. These products are distinguished from approved pharmaceutical drugs by labels stating "Traditionally used for …" Consumers understand this to mean that indications are based on historical evidence and have not necessarily been confirmed by modern scientific experimentation.
Europe	Drug approval considerations in Europe are the same as those for new drugs in the United States, where drugs are documented for safety, effectiveness, and quality. But historically Europeans have been more understanding of the value of phytomedicines and as a result, it is cheaper to secure approval of phytomedicines specially if there is a long history of their anecdotal use. The EEC, recognizing the need to standardize approval of herbal medicines, developed a series of guidelines, The Quality of Herbal Remedies Directive (EEC Directive, 75/318/EEC adopted November 1988), which outlines standards for quality, quantity, and production of herbal remedies and provides labeling requirements that member countries must meet. The EEC guidelines are based on the principles of the WHO's Guidelines for the Assessment of Herbal Medicines (1991). According to these guidelines, a substance's historical use is a valid way to document safety and efficacy in the absence of scientific evidence to the contrary. The guidelines suggest the following as a basis for determining product safety: A guiding principle should be that if the product has been traditionally used without demonstrated harm, no specific restrictive regulatory action should be undertaken unless new evidence demands a revised risk-benefit assessment. Prolonged and apparently uneventful use of a substance usually offers testimony of its safety. With regard to efficacy, the guidelines state the following: For treatment of minor disorders and for nonspecific indications, some relaxation is justified in the requirements for proof of efficacy, taking into account the extent of traditional use; the same considerations may apply to prophylactic use (WHO, 1991). The WHO guidelines give further advice for basing approval on existing monographs: If a pharmacopoeia monograph exists, it should be sufficient to make reference to this monograph. If no such monograph is available, a monograph must be supplied and should be set out in the same way as in an official pharmacopoeia. To further the standardization effort and to increase European scientific support, the phytotherapy societies of Belgium, France, Germany, Switzerland, and the United Kingdom founded the ESCOP. ESCOP's approach to eliminating problems of differing quality and therapeutic use within EEC is to build on the German scientific monograph system to create "European" monographs. In Europe, herbal remedies fall into three categories. The most rigorously controlled are the prescription drugs, which include injectable forms of phytomedicines and those used to treat life-threatening diseases. The second category is OTC phytomedicines, similar to American OTC drugs. The third category is traditional herbal remedies, products that typically have not undergone extensive clinical testing but are judged safe on the basis of generations of use without serious incident.

France	Approximately 200 herbs are approved as OTC in France with varying claims. Licensing approval for phytomedicines is subject to regulations generally required for all drugs. There is only one type of license, but for some plant drugs and preparations, this license is granted on the basis of an adapted documentation and an abridged application. In 1990, 115 herbs plus 31 laxatives were involved in this approval procedure. Currently, about 205 herbal drugs are listed. France, where traditional medicines can be sold with labeling based on traditional use, requires licensing by the French Licensing Committee and approval by the French Pharmacopoeia Committee.
Germany	Germany's Commission E (phytotherapy and herbal substances) was established in 1978. It is an independent division of the German Federal Health Agency that collects information on herbal medicines and evaluates them for safety and efficacy. The following methods and criteria are followed by Commission E: (*i*) traditional use; (*ii*) chemical data; (*iii*) experimental, pharmacological, and toxicological studies; (*iv*) clinical studies; (*v*) field and epidemiological studies; (*vi*) patient case records submitted from physician's files; and (*vii*) additional studies, including unpublished proprietary data submitted by manufacturers. Two kinds of monographs are prepared: monopreparations and fixed combinations. The composition of Commission E is as follows: physicians, pharmacists, pharmacologists, toxicologists, industry representatives, and laypersons, for a total of 24 members. Three possibilities for marketing herbal drugs exist: (*i*) temporary marking authorization for old herbal drugs until they are evaluated for safety and efficacy; (*ii*) monographs of standardized marketing authorization; and (*iii*) individual marketing authorization. Evaluations are published in the form of monographs that approve or disapprove the herbal drugs for over-the-counter use. Herbal medicines are sold in pharmacies, drug stores, and health food stores. Some herbal medicines are controlled by a physician's prescription. Commission E has published about 300 monographs: 200 "positives" and 100 "negatives." About 600–700 plants are sold in Germany. Approximately 70% of physicians prescribe registered herbal drugs. Part of the annual sales is paid for by government health insurance. Germany considers whole herbal products as a single active ingredient; this makes it simpler to define and approve the product. The German Federal Health Office regulates such products as ginkgo and milk thistle extracts by using a monograph system that results in products whose potency and manufacturing processes are standardized. The monographs are compiled from scientific literature on a particular herb in a single report and are produced under the auspices of the Ministry of Health Committee for Herbal Remedies (Kommission E). Approval of such remedies requires more scientific documentation than traditional remedies, but less than new pharmaceutical drug approvals.
India	India has thousands of years of history of use of ayurvedic medicine; almost 70–80% of the rural population of India depends on this mode of traditional medicine; no significant control on quality of drug of botanical origin exists in India.
Japan	Traditional Japanese medicine, called kampo, is similar to and historically derived from Chinese medicine but includes traditional medicines from Japanese folklore. Kampo declined when Western medicine was introduced between 1868 and 1912, but by 1928 it had begun to revive. Today almost half of Japan's Western-trained medical practitioners prescribe kampo medicines, and Japanese national health insurance pays for these medicines. In 1988, the Japanese herbal medicine industry established regulations to manufacture and control the quality of extract products in kampo medicine. These regulations comply with the Japanese government's Regulations for Manufacturing Control and Quality Control of Drugs.

(*Continued*)

Table 1 Herbal Drug Approval Rules in Different Countries (*Continued*)

U.S.A.	Since 1994, herbal medicines have been regulated under the "Dietary Supplement Health and Education Act of 1994." On the basis of this law, herbal medicines are not evaluated by the FDA and, most important, these products are not intended to diagnose, treat, cure, or prevent diseases. The U.S. FDA issued its first botanical drugs approval guidelines in August 2000. The mission of the NCCAM is, in part, to conduct rigorous research on CAM practices to evaluate the benefits and risks of CAM agents so as to optimize their effect on human diseases or conditions. NCCAM groups CAM practices within five major domains: biologically based therapies, manipulative and body-based methods, mind-body interventions, energy therapies, and alternative medical systems. Biologically based CAM agents are regulated under the of DSHEA 1994. This regulation includes botanicals and their constituents, vitamins, minerals, and amino acids. The U.S. FDA characterizes botanicals and other dietary agents according to their use, not according to their composition. If the intended use is to "promote health," the agent is viewed as a dietary supplement; if the intended use is to treat or prevent a disease, the agent is considered to be a drug.

Abbreviations: CAM, complementary and alternative medicine; DSHEA, Dietary Supplement Health and Education Act; EEC, European Economic Community; ESCOP, European Societies' Cooperative of Phytotherapy; HPB, Health Protection Branch; NCCAM, National Center for Complementary and Alternative Medicine; OTC, over the counter; U.S. FDA, United States Food and Drug Administration; WHO, World Health Organization.

It is important to note that, although homeopathic preparations may frequently contain plants, they are also not considered to be herbal medicines.

Compared with well-defined synthetic drugs, herbal medicines exhibit some marked differences, namely:

- the active principles are frequently unknown;
- standardization, stability, and quality control are feasible but not easy;
- the availability and quality of raw materials are frequently problematic;
- well-controlled double-blind clinical and toxicological studies to prove their efficacy and safety are rare;
- empirical use in folk medicine is a very important characteristic;
- they have a wide range of therapeutic use and are suitable for chronic treatments;
- the occurrence of undesirable side effects seems to be less frequent with herbal medicines, but well-controlled randomized clinical trials have revealed that they also exist; and
- they usually cost less than synthetic drugs.

Specifications

The setting of specifications for a herbal preparation (herbal substance) and herbal medicinal product is part of an overall control strategy, which includes control of raw materials and excipients, in-process testing, process evaluation/validation, stability testing, and testing for consistency of batches. When combined in total, these elements provide assurance that the appropriate quality of the product will be maintained. As specifications are chosen to confirm the quality rather than to characterize the product, the manufacturer should provide the rationale and justification for including and/or excluding testing for specific quality attributes. The following points should be taken into consideration when establishing scientifically justifiable specifications.

Specifications for herbal substances are linked to:

- botanical characteristics of the plant (genus, species, variety, chemotype; usage of genetically modified organisms), parts of the plants;
- macroscopical and microscopical characterization, phytochemical characteristics of the plant part constituents with known therapeutic activity or marker substances, toxic constituents (identity, assay, limit tests);
- biological/geographical variation;
- cultivation/harvesting/drying conditions (microbial levels, aflatoxins, heavy metals, and so on);
- pre-/postharvest chemical treatments (pesticides, fumigants); and
- profile and stability of the constituents.

Specifications for herbal preparations are linked to:

- quality of the herbal substance (as discussed previously);
- definition of the herbal preparation [drug extract ratio, extraction solvent(s)];
- method of preparation from the herbal substance;
- constituents—constituents with known therapeutic activity or marker substances;
- other constituents (identification, assay, limit tests);
- drying conditions (e.g., microbial levels, residual solvents in extracts);

- profile and stability of the constituents;
- microbial purity on storage; and
- batches used in preclinical/clinical testing (safety and efficacy considerations).

Specifications for herbal medicinal products are linked to:

- quality of the herbal substance and/or herbal preparation;
- manufacturing process (temperature effects, residual solvents);
- profile and stability of the active constituents/formulation in packaging; and
- batches used in preclinical/clinical testing (safety and efficacy considerations).

Specifications should be based on data obtained from lots used to demonstrate manufacturing consistency. Linking specifications to a manufacturing process is important, especially with regard to product-related substances, product-related impurities, and process-related impurities. Historical batch data should be taken into account where available.

Changes in the manufacturing process and degradation products produced during storage may result in a product that differs from that used in preclinical and clinical development. The significance of these changes should be evaluated.

Because of the inherent complexity of herbal products, there may be no single stability-indicating assay or parameter that profiles the stability characteristics. Consequently, the applicant should propose a series of product-specific, stability-indicating tests, the results of which will provide assurance that changes in the quality of the product during its shelf life will be detected. The determination of which tests should be included will be product-specific. Applicants are referred to the "Note for guidance on stability testing of new drug substances and products" [committee for proprietary medicinal products (CPMP)/ICH/2736/99], the "Guideline on stability testing of new veterinary drug substances and medicinal products" [Committee for Veterinary Medicinal Products (CVMP)/International Cooperation on Harmonisation of Technical Requirements for Registration of Veterinary Medicinal Products (VICH)/899/99], and the "Note for guidance on stability testing of existing active substances and related finished products" (CPMP/QWP/122/02 rev. 1 and European Agency for the Evaluation of Medicinal Products (EMEA)/CVMP/846/99).

Standardization

Plants contain several hundred constituents and some of them are present at very low concentrations. In spite of the modern chemical analytical procedures available, only rarely do photochemical investigations succeed in isolating and characterizing all secondary metabolites present in the plant extract. Apart from this, plant constituents vary considerably depending on several factors that impair the quality control of phototherapeutic agents. Quality control and standardization of herbal medicines involve several steps. However, the source and quality of raw materials play a pivotal role in guaranteeing the quality and stability of herbal preparations. Other factors, such as the use of fresh plants, temperature, light exposure, water availability, nutrients, period and time of collection, method of collecting, drying, packing, storage, and transportation of raw material, age and part of the plant collected, and so on, can greatly affect the quality and consequently, the therapeutic value of herbal medicines. Some plant constituents are heat labile and the plants containing them need to be dried at low temperatures. Also, other active principles are destroyed by enzymatic processes that continue for long periods of time after plant collection. This explains why frequently the composition of herbal drugs is quite variable. Thus, proper standardization and quality control of raw material and the herbal preparations themselves should be permanently carried out. In the cases where the active principles are unknown, marker substance(s) should be established for analytical purposes. However, in most cases these markers have never been tested to see whether they really account for the therapeutic action reported for the herbal drugs. As pointed out before, apart from these variable factors, others such as the method of extraction and contamination with microorganisms, heavy metals, pesticides, and the like, can also interfere with the quality, safety, and efficacy of herbal drugs. For these reasons, pharmaceutical companies prefer using cultivated plants instead of wild-harvested plants because they show smaller variation in their constituents. Furthermore and certainly more relevant, when medicinal plants are produced by cultivation, the main secondary metabolites can be monitored and this permits definition of the best period for harvesting.

The recent advances that occurred in the processes of purification, isolation, and structure elucidation of naturally occurring substances have made it possible to establish appropriate strategies for the analysis of quality and the process of standardization of herbal preparations in order to maintain as much as possible the homogeneity of the plant extract. Among others, thin-layer chromatography (TLC), gas chromatography (GC), high-performance liquid chromatography (HPLC), mass spectrometry (MS), infrared spectrometry (IRS), ultraviolet/visible (UV/VIS) spectrometry, and the like, used alone or in combination, can be successfully used for standardization and to control the quality of both the raw materials and the finished herbal drugs.

EFFICACY AND SAFETY

Although clinical trials with herbal drugs are feasible, few well-controlled double-blind (placebo-controlled) trials have been carried out with herbal medicines. Several factors might contribute to the explanation of such discrepancies, for example:

- lack of standardization and quality control of the herbal drugs used in clinical trials
- use of different dosages of herbal medicines
- inadequate randomization in most studies, and patients not properly selected
- numbers of patients in most trials insufficient for the attainment of statistical significance
- difficulties in establishing appropriate placebos because of the tastes, aromas, and so on
- wide variations in the duration of treatments using herbal medicines

However, a large number of clinical trials have been performed with some herbal drugs, including:

- *Ginkgo biloba* (used for the treatment of central nervous system and cardiovascular disorders);
- *Hypericum perforatum* (St. John's wort, used as an antidepressant);
- *Panax ginseng* (ginseng) herbs used as a tonic;
- *Tanacetun parthenium* (feverfew) used to treat migraine headache;
- *Allium sativum* (garlic) used to lower low-density protein cholesterol and some cardiovascular disturbances;

- *Matricaria chamomilla* (chamomile) recommended as a carminative, anti-inflammatory and antispasmodic;
- *Silybium marianun* (milk thistle) used for repairing liver function including cirrhosis;
- *Valeriana officinalis* (valerian) used as a sedative and sleeping aid;
- *Piper methysticum* (Kava kava) used as an anxiolytic;
- *Aesculus hippocastanum* (horse chestnut) used for the treatment of chronic venous insufficiency;
- *Cassia acutfolia* (Senna) and *Rhamnus purshiana* (cascara sagrada), which are used as laxatives;
- *Echinacea purpura* (Echinacea) used as an anti-inflammatory and immunostimulant;
- *Arnica montana* (arnica) used to treat post-traumatic and postoperative conditions; and
- *Serenoa repens* (saw palmetto) used for the treatment of benign prostatic hyperplasia.

REGULATORY FILING PROCEDURE

In general Chemistry, Manufacturing, and Control (CMC) requirements for standard synthetic/semisynthetic drugs are:

- synthesis of the drug;
- manufacturing of the product that is administered to the patient; and
- control of these processes.

Thus, the drug and the product are made reproducibly to provide assurance that active ingredients are administered to patients and toxic contaminants are not.

More specific CMC requirements, for example, for a plant substance that is later made into a pure drug (e.g., digitalis for heart failure, artemisinins for malaria), are shown in column 3 of Table 2.

Plant Substance
- description of the plant
- procedure by which a part of the plant is extracted
- quantity of active ingredient in the extract
- how the active ingredient is identified
- stability of the active ingredient (at least over the time of the trial)

Product (Capsule, Tablet, IV Formulation)
- how the product is manufactured
- quantity of active ingredient in the product
- how the active ingredient is identified
- impurities in the product, including microbials, pesticides, heavy metals, and adulterants
- storage conditions and physical–chemical stability of the active ingredient during storage
- specifications of a reference batch and the controls during the manufacturing process such that each batch is similar to the reference batch
- bioavailability of the active ingredient (disintegration and dissolution, or breakdown, in physiologic solutions in vitro, absorption in vivo)
- whether the environment is contaminated, for example, with carcinogens, as the product is being made

Table 2 Chemistry–Manufacturing–Control Considerations for National Center for Complementary and Alternative Medicine Clinical Trials

Subject of study	Study parameters	Study details	Data required for proposed NCCAM trial	
			Phase I/II trial	Phase III trial
Plant substance	Starting material	Botanical description	X	Expanded
		Extraction procedure	X	Expanded
		Quantity of active moiety		X
		Identity: chemical/biologic assay		X
		Stability		X
Plant product	Manufacturing	Reagents/process		X
	Finished product	Quantity of active moiety	X	X
	Product assay	Methods/specifications		X
		Identity: chemical/biologic assay	X	X
		Purity		X
	Storage	Describe conditions	X	X
	Stability	Light/heat/time	X	X
	Excipients	List		X
	Impurities	List/analyze	X	X
	Reference standard	Standard batch		X
	In-process controls	Standard operating procedures		X
	Bioavailability	Disintegration/dissolution rate	X	X
	Microbiology	Contamination		X
	Environmental	Assessment		X

Note: X means applicable.

Overview of Chemistry, Manufacturing, and Control Evidence Needed to Support Clinical Trials for Botanical Drugs

Unlike standard drugs, botanicals have been in use before they were studied in a clinical trial. Prior human use gives some assurance that the product will be safe and effective. Some of the CMC information needed for a standard drug is also needed for botanical drugs.

Unlike synthetic drugs, botanical drugs are mixtures of uncharacterized constituents. It is postulated that a mixture provides a therapeutic advantage. For example, unknown constituents may combine in an additive or synergistic way with the known constituents to provide greater efficacy than would be provided by the known constituent alone. For botanical drugs, analysis of the active ingredient(s) may be best approached by analysis for one or more hypothesized active ingredients, such as

- a chemical constituent that constitutes a sizable percentage of the total ingredients or
- a chemical fingerprint of the total ingredients.

The latter two analyses are surrogates for analysis of the unknown constituents that contribute to efficacy.

Information on a Plant Product that Was the Subject of Prior Human Use

Plants are often extracted and processed nonreproducibly. It is important that the product that is produced for a clinical trial is similar in analysis to the original product that has been used in humans.

Consider the following example of a botanical drug with three components. Component 1 is potentially toxic and components 2 and 3 are potentially effective at low doses, but also potentially toxic at high doses. Table 3 shows the number of units of each component for two lots of the drug.

If Lot 1 is administered but the lots previously used were comparable with Lot 2, the participants will suffer toxicity from the higher doses of components 2 and 3. If Lot 2 is administered, but the lots previously used were comparable with Lot 1, the participants will suffer toxicity from the higher dose of component 1 and efficacy will diminish because of the lower doses of components 2 and 3.

This example illustrates the wide variation in composition that may be found in identically labeled botanical products. For this reason, analysis of the product that is proposed for study in a clinical trial must be performed and shown to be similar to the analysis of the botanical with prior human experience, if prior clinical data are used to justify the proposed trial.

Therefore, the following information is needed for products previously used in humans:

- Plant substance:
  - description of the plant;
  - genus;
  - species (cultivar if appropriate);
  - country(s) of origin; and
  - plant extraction procedure.
- Plant product:
  - analysis of commonly accepted or supposed active ingredient(s) via chemical or biological parameters;
  - analysis of a sizeable chemical constituent (analytical marker compound); and
  - analysis via chemical fingerprint (analytical markers).

Information on the Plant Product Proposed for Phase I/II Studies (Table 2, Column 4)

Plant Substance

Description of the plant including genus, species (cultivar if appropriate), country(s) of origin, time of harvest, and plant extraction procedure.

Table 3 Example of Component Combinations for a Botanical Drug

	Component 1 • potentially toxic	Component 2 • low dose: effective • high dose: toxic	Component 3 • low dose: effective • high dose: toxic
Lot 1	1 unit	122 units	48 units
Lot 2	12 units	11 units	10 units

Plant Product
- Analysis of commonly accepted or supposed active ingredient(s) via chemical or biological parameters;
- analysis of a sizeable chemical constituent (analytical marker compound);
- analysis via chemical fingerprint (analytical markers);
- analysis for lack of contamination by pesticides, heavy metals, and synthetic drug adulterants;
- the breakdown or dissolution of the analyzed components in physiological solutions;
- list of inert substances (excipients) added to the product; and
- storage conditions and stability over the length of the trial.

Information on the Plant Product Proposed for Phase III Studies (Table 2, Column 5)
Plant Substance
- Botanical description;
- statement that the plant is cultivated according to Good Agricultural Practices or harvested according to Good Wildcrafting Practices;
- extraction procedure;
- quantity and identity of active ingredient(s) and of sizeable chemical constituent; and
- statement that extraction and analytic procedures are performed under Good Manufacturing Practices (GMP) (e.g., that the manufacturing processes and their controls provide the appropriate levels of assurance for the important quality characteristics of the product).

Plant Product
- Manufacturing methods;
- analysis of commonly accepted or supposed active ingredient(s) via chemical or biological parameters;
- analysis of a sizeable chemical constituent (analytical marker compound);
- analysis via chemical fingerprint (analytical markers);
- analysis for lack of contamination by pesticides, heavy metals, synthetic drug adulterants;
- the breakdown or dissolution of the analyzed components in physiological solutions;
- in-process controls for manufacturing process;
- list of inert substances (excipients) added to the product;
- description of the reference batch;
- storage conditions and stability over the length of the trial;
- environmental impact statement; and
- statement that the plant product is manufactured and analyzed according to GMP (e.g., that the manufacturing processes and their controls provide appropriate levels of assurance for the important quality characteristics of the product).

STARTING MATERIAL
Consistent quality for products of herbal origin can only be assured if the starting materials are defined in a rigorous and detailed manner, particularly the specific botanical identification of the plant material used. It is also important to know the geographical source and the conditions under which the herbal substance is obtained to ensure material of consistent quality.

"Standardized extracts" are herbal preparations adjusted within an acceptable tolerance to a given content of constituents with known therapeutic activity; standardization is achieved by adjustment of the herbal preparations with inert material or by blending batches of herbal preparations. "Quantified extracts" are herbal preparations adjusted to a defined range of constituents; adjustments are made by blending batches of herbal preparations. "Other extracts" are herbal preparations essentially defined by their production process and their specifications.

In the case of a herbal substance or a herbal preparation consisting of comminuted or powdered herbal substances, the quantity of the herbal substance or the herbal preparation shall be given as a range corresponding to a defined quantity of constituents with known therapeutic activity, or the quantity of the herbal substance or the quantity of the native herbal preparation shall be stated if constituents with known therapeutic activity are unknown.

Examples of how the composition is listed include: *Sennae folium*, 415–500 mg, corresponding to 12.5 mg of hydroxyanthracene glycosides, calculated as Sennoside B or *Valerianae radix* 900 mg.

In the case of an herbal preparation produced by steps which exceed comminution, the nature and concentration of the solvent and the physical state of the extract have to be given. Furthermore, the following have to be indicated:

- *Standardized extracts*: If the constituents with known therapeutic activity are known, the equivalent quantity or the ratio of the herbal substance to the herbal preparation shall be stated and the quantity of the herbal preparation may be given as a range corresponding to a defined quantity of these constituents (see example); or
- *Quantified extracts*: In the case of quantified extracts, the equivalent quantity or the ratio of the herbal substance to the herbal preparation shall be stated. Furthermore, content of the quantified substance(s) shall be specified in a range; or
- *Other extracts*: The equivalent quantity or the ratio of the herbal substance to the herbal preparation shall be stated if constituents with known therapeutic activity are unknown.

The composition of any solvent or solvent mixture and the physical state of the extract must be indicated. If any other substance is added during the manufacture of the herbal preparation to adjust the preparation to a defined content of constituents with known therapeutic activity, or for any other purpose, the added substance must be mentioned as an "other substance" and the genuine extract as the "active substance." However, where different batches of the same extract are used to adjust constituents with known therapeutic activity to a defined content, or, for any other purpose, the final mixture shall be regarded as the genuine extract and listed as the "active substance" in the unit formula. Full details of production and control must however be provided in the dossier.

Examples of representation may include:

- *Sennae folium* 50–65 mg, corresponding to 12.5 mg of dry extract ethanolic 60% (V/V) hydroxyanthracene glycosides, calculated as [(a–b): 1] Sennoside B.
- *Ginkgo biloba* L. *folium* 60 mg, containing 13.2–16.2 mg of flavanoid dry extract acetonic 60% (V/V) expressed as flavone glycosides [(a–b): 1] 1.68–2.04 mg ginkgolides A, B, and C and 1.56–1.92 mg bilobalide.
- *Valerianae radix* 125 mg dry extract ethanolic 60% (V/V) or *Valerianae radix* 125 mg dry extract ethanolic 60% (V/V) equivalent to x–y mg *Valerianae radix*.

Control of Herbal Substances and of Herbal Preparations

As a general rule, herbal substances must be tested, unless otherwise justified, for microbiological quality and for residues of pesticides and fumigation agents, toxic metals, likely contaminants, and adulterants, and so on. The use of ethylene oxide is prohibited for the decontamination of herbal substances in the International Conference on Harmonization (ICH) guidelines. Radioactive contamination should be tested for if there are reasons for concerns. Specifications and descriptions of the analytical procedures must be submitted, together with the limits applied. Analytical procedures not given in a Pharmacopoeia should be validated in accordance with the ICH guideline "Validation of analytical procedures: methodology" (CPMP/ICH/281/95) or the corresponding VICH guideline (CVMP/VICH/591/98).

Reference samples of the herbal substances must be available for use in comparative tests, for example, macroscopic and microscopic examination, chromatography, and so on.

If the herbal medicinal product contains a preparation, rather than merely the herbal substance itself, the comprehensive specification for the herbal substance must be followed by a description and validation of the manufacturing process for the herbal preparation. The information may be supplied either as part of the marketing authorization application or by using the European Active Substance Master File procedure. If the latter route is chosen, the documentation should be submitted in accordance with the "Guideline on active substance master file procedure" (EMEA/CPMP/QWP/227/02 and EMEA/CVMP/134/02).

Where the preparation is the subject of a European Pharmacopoeia monograph, the European Directorate for the Quality of Medicines Certification procedure (for Certificates of Suitability, CEPs) can be used to demonstrate compliance with the relevant European Pharmacopoeia monograph.

For each herbal preparation, a comprehensive specification is required. This should be established on the basis of recent scientific data and should give particulars of the characteristics, identification tests, and purity tests. Appropriate chromatographic methods should be used. If deemed necessary by analysis of the starting material, tests on microbiological quality, residues of pesticides, fumigation agents, solvents, and toxic metals should be performed. Radioactivity should be tested if there are reasons for concern. A quantitative determination (assay) of markers or of substances with known therapeutic activity is also required. The content should be indicated with the lowest possible tolerance (the narrowest possible tolerance with both upper and lower limits stated). The test methods should be described in detail.

If preparations from herbal substances with constituents of known therapeutic activity are standardized (i.e., adjusted to a defined content of constituents with known therapeutic activity), it should be stated how such standardization is achieved. If another substance is used for these purposes, it is necessary to specify as a range the quantity that can be added.

Control of Vitamins and Minerals (if Applicable)

Vitamin(s) and mineral(s), which could be ancillary substances in traditional herbal medicinal products for human use, should fulfill the requirements of the "Guideline on summary of requirements for active substances in the quality part of the dossier" (CHMP/QWP/297/97 Rev. 1 corr).

Control of Excipients

Excipients, including those added during the manufacture of the herbal preparations, should be described according to the "Note for guidance on excipients in the dossier for application for marketing authorization of a medicinal product" (Eudralex 3AQ 9A) or the "Note for guidance on excipients in the dossier for application for marketing authorisation of veterinary medicinal products" (EMEA/CVMP/004/98), as appropriate. For novel excipients, the dossier requirements for active substances apply (refer to Directive 2001/83/EC as amended for human medicinal products and Directive 2001/82/EC as amended for veterinary medicinal products).

The control tests on the finished product should allow the qualitative and quantitative determination of the composition of the active substance(s). A specification should be provided and this may include the use of markers where constituents with known therapeutic activity are unknown. In the case of herbal substances or herbal preparations with constituents of known therapeutic activity, these constituents should be specified and quantitatively determined. For traditional herbal medicinal products for human use containing vitamins and/or minerals, the vitamins and/or minerals should also be specified and quantitatively determined.

If a herbal medicinal product contains a combination of several herbal substances or preparations of several herbal substances, and if it is not possible to perform a quantitative determination of each active substance, the determination may be carried out jointly for several active substances. The need for this procedure should be justified.

The criteria given by the European Pharmacopoeia to ensure the microbiological quality should be applied unless justified. The frequency of testing for microbial contamination should be justified.

STABILITY TESTING

As the herbal substance or herbal preparation in its entirety is regarded as the active substance, a mere determination of the stability of the constituents with known therapeutic activity will not suffice. The stability of other substances present in the herbal substance or in the herbal preparation, should, as far as possible, also be demonstrated, for example, by means of appropriate fingerprint chromatograms. It should also be demonstrated that their proportional content remains constant.

If a herbal medicinal product contains combinations of several herbal substances or herbal preparations, and if it is not possible to determine the stability of each active substance, the stability of the medicinal product should be determined by appropriate fingerprint chromatograms, appropriate overall methods of assay and physical and sensory tests or other appropriate tests. The appropriateness of the tests shall be justified by the applicant.

In the case of a herbal medicinal product containing a herbal substance or herbal preparation with constituents of known therapeutic activity, the variation in content during the proposed shelf life should not exceed $\pm 5\%$ of the initial assay value, unless justified. In the case of a herbal medicinal product containing a herbal substance or herbal preparation where constituents with known therapeutic activity are unknown, a variation in marker content during the proposed shelf life of $\pm 10\%$ of the initial assay value can be accepted if justified by the applicant.

In the case of traditional herbal medicinal products for human use containing vitamins and/or minerals, the stability of the vitamins and/or minerals should be demonstrated.

Testing Criteria
Implementation of the recommendations in the following section should take into account the ICH/VICH Guidelines "Validation of analytical methods: definitions and terminology" (CPMP/ICH/381/95 and CVMP/VICH/590/98) and "Validation of analytical procedures: methodology" (CPMP/ICH/281/95 and CVMP/VICH/591/98).

Herbal Substances
Herbal substances are a diverse range of botanical materials including leaves, herbs, roots, flowers, seeds, bark, and so on. A comprehensive specification must be developed for each herbal substance even if the starting material for the manufacture of the finished product is a herbal preparation. In the case of fatty or essential oils used as active substances of herbal medicinal products, a specification for the herbal substance is required unless justified. The specification should be established on the basis of recent scientific data and should be set out in the same way as the European Pharmacopoeia monographs. The general monograph "Herbal drugs" (herbal substances) of the European Pharmacopoeia should be consulted for interpretation of the following requirements.

The following tests and acceptance criteria are considered generally applicable to all herbal substances:

1. *Definition*: a qualitative statement of the botanical source, plant part used, and its state (e.g., whole, reduced, powdered, fresh, dry). It is also important to know the geographical source(s) and the conditions under which the herbal substance is obtained.
2. *Characteristics*: a qualitative statement about the organoleptic character(s) including macro- and microscopic botanic properties of the herbal substance are described.
3. *Identification*: identification testing optimally should be able to discriminate between related species and/or potential adulterants/substitutes, which are likely to be present. Identification tests should be specific for the herbal substance and are usually a combination of three or more of the following: macroscopic, microscopic, chromatographic, and chemical.
4. *Tests*:
    - Foreign matter
    - Total ash
    - Ash insoluble in hydrochloric acid
    - Water soluble extractive and
    - Extractable matter
    - *Particle size*: For some herbal substances intended for use in herbal teas or solid herbal medicinal products, particle size can have a significant effect on dissolution rates, bioavailability, and/or stability. In such instances, testing for particle size distribution should be carried out using an appropriate procedure, and acceptance criteria should be provided. Particle size can also affect the disintegration time of solid dosage forms.
    - *Water content*: This test is important when the herbal substances are known to be hygroscopic. For nonpharmacopoeial herbal substances, acceptance

criteria should be justified by data on the effects of moisture absorption. A loss on drying procedure may be adequate; however, in some cases (essential oil-containing plants), a detection procedure that is specific for water is required.
- *Inorganic impurities, toxic metals*: The need for inclusion of tests and acceptance criteria for inorganic impurities should be studied during development and based on knowledge of the plant species, its cultivation, and the manufacturing process. Acceptance criteria will ultimately depend on safety considerations. Where justified, procedures and acceptance criteria for sulfated ash/residue on ignition should follow pharmacopoeia precedents; other inorganic impurities may be determined by other appropriate procedures, for example, atomic absorption spectroscopy.
- *Microbial limits*: There may be a need to specify the total count of aerobic micro-organisms, the total count of yeasts and molds, and the absence of specific objectionable bacteria. The source of the herbal material should be taken into account when considering the inclusion of other possible pathogens (e.g., *Campylobacter* and *Listeria* species) in addition to those specified in the European Pharmacopoeia. Microbial counts should be determined using pharmacopoeial procedures or other validated procedures. The European Pharmacopoeia gives guidance on acceptance criteria.
- *Mycotoxins*: The potential for mycotoxins contamination should be fully considered. Wherever necessary, suitable validated methods should be used to control potential mycotoxins and the acceptance criteria should be justified.
- *Pesticides, fumigation agents, and the like*: The potential for residues of pesticides, fumigation agents, and the like should be fully considered. Wherever necessary, suitable validated methods should be used to control potential residues and the acceptance criteria should be justified. In the case of pesticide residues the method, acceptance criteria, and guidance on the methodology of the European Pharmacopoeia should be applied unless fully justified.
- Other appropriate tests (e.g., swelling index).
5. *Assay*: In the case of herbal substances with constituents of known therapeutic activity, assays of their content are required with details of the analytical procedure. Wherever possible, a specific, stability-indicating procedure should be included to determine the content of the herbal substance. In cases where use of a nonspecific assay is justified, other supporting analytical procedures should be used to achieve overall specificity. For example, where determination of essential oils is adopted to assay the herbal substance, the combination of the assay and a suitable test for identification (e.g., fingerprint chromatography) can be used. In the case of herbal substances where the constituents responsible for the therapeutic activity are unknown, assays of marker substances or other justified determinations are required. The appropriateness of the choice of marker substance should be justified. For example, reference to the assay of a marker substance in the relevant monograph of the European Pharmacopoeia is an appropriate justification.

Herbal Preparations

Herbal preparations are also diverse in character ranging from simple, comminuted plant material to extracts, tinctures, oils, and resins. A comprehensive specification must be developed for each herbal preparation based on recent scientific data. The general monograph "Herbal drug preparations" (herbal preparations) of the European Pharmacopoeia should be consulted for the interpretation of the following requirements.

The following tests and acceptance criteria are considered generally applicable to all herbal preparations.

1. *Definition*: a statement of the botanical source, and the type of preparation (e.g., dry or liquid extract). The ratio of the herbal substance to the herbal preparation must be stated.
2. *Characters*: a qualitative statement about the organoleptic characters of the herbal preparation where characteristic.
3. *Identification*: Identification tests should be specific for the herbal preparation, and optimally should be discriminatory with regard to substitutes/adulterants that are likely to occur. Identification solely by chromatographic retention time, for example, is not regarded as being specific; however, a combination of chromatographic tests (e.g., HPLC and TLC-densitometry) or a combination of tests into a single procedure, such as HPLC/UV-diode array, HPLC/MS, or GC/MS may be acceptable.
4. *Tests*:
    - *Residual solvents*: Refer to the European Pharmacopoeia general text on Residual Solvents (01/2005: 50400) for detailed information;
    - *Water content*: This test is important when the herbal preparations are known to be hygroscopic. The acceptance criteria may be justified with data on the effects of hydration or moisture absorption. A loss on drying procedure may be adequate; however, in some cases (essential oil-containing preparations), a detection procedure that is specific for water is required;
    - *Inorganic impurities, toxic metals*: The need for inclusion of tests and acceptance criteria for inorganic impurities should be studied during development and based on knowledge of the plant species, its cultivation, and the manufacturing process. The potential for manufacturing process to concentrate toxic residues should be fully addressed. If the manufacturing process will reduce the burden of toxic residues, the tests with the herbal substance may be sufficient. Acceptance criteria will ultimately depend on safety considerations. Where justified, procedures and acceptance criteria for sulfated ash/residue on ignition should follow pharmacopoeial precedents; other inorganic impurities may be determined by other appropriate procedures, for example, atomic absorption spectroscopy.
    - *Microbial limits*: There may be a need to specify the total count of aerobic micro-organisms, the total count of yeasts and molds, and the absence of specific objectionable bacteria. These limits should comply with those in the European Pharmacopoeia.
    - *Mycotoxins*: The potential for mycotoxins contamination should be fully considered. Wherever necessary, suitable validated methods should be used to control potential mycotoxins and the acceptance criteria should be justified.
    - *Pesticides, fumigation agents, and the like*: The potential for residues of pesticides, fumigation agents, and the like should be fully considered. Wherever necessary, suitable validated methods should be used to control potential residues and the acceptance criteria should be justified. In the case of pesticide residues the method, acceptance criteria, and guidance on the

methodology of the European Pharmacopoeia should be applied unless fully justified.

5. *Assay*: In the case of herbal preparations with constituents of known therapeutic activity, assays of their content are required with details of the analytical procedure. Wherever possible, a specific, stability-indicating procedure should be included to determine the content of the herbal substance in the herbal preparation. In cases where use of a nonspecific assay is justified, other supporting analytical procedures should be used to achieve overall specificity. For example, where a UV/VIS spectrophotometric assay is used for anthraquinone glycosides, a combination of the assay and a suitable test for identification (e.g., fingerprint chromatography) can be used. In the case of herbal preparations where the constituents responsible for the therapeutic activity are not known, assays of marker substances or other justified determinations are required. The appropriateness of the choice of marker substance should be justified.

BIBLIOGRAPHY

Abt AB, Oh JY, Huntington RA, Burkhart KK. Chinese herbal medicine induced acute renal failure. Arch Intern Med 1995; 155:211–212.
Ackerknecht EH. Therapeutics: from the Primitives to the Twentieth Century. New York: Hafner Press, 1973.
Adams D. Herbal tea made woman's face "stop light," court told. Age (Melbourne) 1995, 12 Dec; 7.
Akamasu E. Modern Oriental Drugs. Tokyo: Yishiyakusha.
Akerele O. Summary of WHO guidelines for the assessment of herbal medicines. HerbalGram 1993; 28:13–19.
Anonymous. Program profile: international liaison brings global vision to OAM. Complementary and Alternative Medicine at the NIH 1996; 3:3.
Anonymous. Royal jelly: warning label required. TGA News 1994; 16:4.
Armstrong NC, Ernst E. The treatment of eczema with Chinese herbs: a systematic review of randomized clinical trials. Br J Clin Pharmacol 1999; 48:262–264.
Artiges A. What are the legal requirements for the use of phytopharmaceutical drugs in France? J Ethnopharmacol 1991; 32:231–234.
Arvigo R, Balick M. Rainforest Remedies: 100 Healing Herbs of Belize. Twin Lakes, WI: Lotus Press, 1993.
Bagheri H, Broué P, Lacroix I, et al. Fulminant hepatic failure after herbal medicine ingestion in children. Térapie 1998; 53:77–83.
Baisi F. Report on Clinical Trial of Bilberry Anthocyanocides in the Treatment of Venous Insufficiency in Pregnancy and of Postpartum Hemorrhoids. Presidio Ospedaliero di Livorno, Italy, 1987.
Baker ME. Evolution of regulation of steroid-mediated intercellular communication in vertebrates: insights from flavonoids, signals that mediate plant-rhizobia symbiosis. J Steroid Biochem Mol Biol 1992; 41(3–8):301–308.
Bauer R, Tittel G. Quality assessment of herbal preparations as a precondition of pharmacological and clinical studies. Phytomedicine 1996; 2:193–198.
Bauer V, Jurcie K, Puhlmann J, Wagner H. Immunologische in-vivo und in-vitro untersuchungen mit echinacea-extrakten (Immunologic in vivo and in vitro studies on echinacea extracts). Arzneimittelforschung 1988; 38(2):276–281.
Bayly GR, Braithwaite RA, Sheehan TMT, et al. Lead poisoning from Asian traditional remedies in the West Midlands—report of a series of five cases. Hum Exp Toxicol 1995; 14:24–28.
Beckstrom-Sternberg SM, Duke JA. Potential for synergistic action of phytochemicals in spices. In: Charalambous G, ed. Spices, Herbs, and Edible Fungi. New York: Elsevier Sciences B.V., 1994.

Bedi KL, Zutshi U, Chopra CL, Amla V. Picrorhiza kurroa, an Ayurvedic herb, may potentiate photochemotherapy in vitiligo. J Ethnopharmacol 1989; 27:347–352.
Bell LP, Hectorne K, Reynolds H, Balm TK, Hunninghake DB. Cholesterol-lowering effects of psyllium hydrophilic mucilloid. Adjunct therapy to a prudent diet for patients with mild to moderate hypercholesterolemia. JAMA 1989; 261:3419–3423.
Bensky D, Gamble A. Chinese Herbal Medicine: Materia Medica (revised edition). Seattle: Eastland Press Inc., 1993.
Bensoussan A, Myers SP. Towards a safer choice. The Practice of Traditional Chinese Medicine in Australia. Sydney: Faculty of Health, University of Western Sydney (Macarthur), 1996:54.
Bensoussan A, Talley NJ, Hing M, Menzies R, Guo A, Ngu M. Treatment of irritable bowel syndrome with Chinese herbal medicine: a randomized controlled trial. JAMA 1998; 280:1585–1589.
Bisset NG, ed. Herbal Drugs and Phytopharmaceuticals. Stuttgart: Medpharm Scientific Publishers, 1994.
Blumenthal M. Herb industry sees mergers, acquisitions, and entry by pharmaceutical giants in 1998. HerbalGram 1999; 45:67–68.
Blumenthal M, Brusse WR, Goldberg A, et al. The Complete German Commission E Monographs. Therapeutic Guide to Herbal Medicines. Austin, TX, USA: The American Botanical Council, 1998.
Bode JC, Schmidt U, Durr HK. Silymarin for the treatment of acute viral hepatitis? Med Klin 1977; 72:513–518.
Borchers AT, Hackman RM, Keen CL, Stern JS, Gershwin ME. Complementary medicine: a review of immunomodulatory effects of Chinese herbal medicines. Am J Clin Nutr 1997; 66:1303–1312.
Braunig B, Dorn M, Limburg E Knick, Bausendorf. Echinacea purpureae radix for strengthening the immune response in flu-like infections. Zeitschrift fhr Phytotherapie 1992; 13:7–13.
Brautigam MRH, Blommaert FA, Verleye G, Castermans J, Steur ENHJ, Kleijnen J. Treatment of age-related memory complaints with Ginkgo biloba extract: a randomized double blind placebo-controlled study. Phytomedicine 1998; 5:425–434.
Braz J. Efficacy, safety, quality control, marketing, and regulatory guidelines for herbal medicines (phytotherapeutic agents) Med Biol Res 2000 Feb; 33(2):179–189.
Brevoort P. Botanical (herbal) medicine in the United States. Pharm News 1996; 3:26–28.
Brevoort P. The U.S. botanical market. An overview. HerbalGram 1995; 36:49–59.
Brinker F. Herb Contraindications and Drug Interactions. 2nd edn. Sandy, OR: Eclectic Medical Publications, 1998.
British Medical Association. Complementary Medicine. New Approaches to Good Practice. Oxford: Oxford University Press, 1993:9–36.
Brown RG. Toxicity of Chinese herbal remedies. Lancet 1992; 340:673.
Buchman DD. Herbal Medicine. New York: Gramercy Publishing Company, 1980.
Bulletin of the World Health Organization. Regulatory situation of herbal medicines. A worldwide review. Geneva, 1998:1–43.
Bulletin of the World Health Organization. Research guidelines for evaluating the safety and efficacy of herbal medicine. Geneva, 1993:1–86.
Bullock RJ, Rohan A, Straatmans J-A. Fatal royal jelly-induced asthma [letter]. Med J Aust 1994; 160:44.
But PP-H. Need the correct identification of herbs in herbal poisoning [letter]. Lancet 1993; 341:637.
Catlin DH, Sekera M, Adelman DC. Erythroderma associated with the ingestion of an herbal product. West J Med 1993; 159:491–493.
Champpault G, Patel JD, Bonnard AM. A double-blind trial of an extract of the plant Serenoo repens in benign prostatic hyperplasia. Br J Clin Pharmacol 1984; 18(3):461–462.
Chan TYK, Chan AYW, Critchley JAJH. Hospital admissions due to adverse reactions to Chinese herbal medicines. J Trop Med Hyg 1992; 95:296–298.
Charles V, Charles SX. The use and efficacy of Adadirachta indica ADR (neem) and Curcuma longa (turmeric) in scabies. Trop Geograph Med 1991 Nov; 43:178–181.

Chaudhury RR. Herbal Medicine for Human Health. World Health Organization (SEARO, No. 20), 1992.
Chen X. Protective and FFA metabolic effect of ginsenosides on myocardial ischemia. J Med Cell Cardiol 1983; 25:121–123.
Chen X, Zhu QY, Li LY, Tang XL. Effect of ginsenosides on cardiac performance and hemodynamics of dogs. Acta Pharmacol Sinica 1982; 3:235–239.
Cheng XJ, Liu YL, Lin GF, Luo XT. Comparison of action of Panax ginsenosides and Panax quinquonosides on anti-warm stress in mice. J Shenyang Coll Pharmacy 1986; 3(3):170–172.
Cheng XJ, Shi XR, Lin B. Effects of ginseng root saponins on brain Ach and serum corticosterone in normobaric hypoxia stressed mice. Presented at the 5th Southeast Asian and Western Pacific Regional Meeting of Pharmacologists, Chinese Pharmacological Association, Beijing, 1988.
Chopra RN, Chopra IC. A review of work on Indian medicinal plants. Indian Council of Medical Research Special Report Series No. 1, 1959:99, 107.
Coeugniet EG, Elek E. Immunomodulation with Viscum album and Echinacea purpurea extracts. Onkologie 1987; 10(3 suppl):27–33.
Corsi S. Report on Trial of Bilberry Anthocyanosides (Tegens-inverni della beffa) in the Medical Treatment of Venous Insufficiency of the Lower Limbs. Casa di Cura S. Chiara, Florence, Italy, 1987.
Cragg GM, Newman DJ, Snader KM. Natural products in drug discovery and development. J Nat Products 1997; 60:52–60.
Cui J, Garle M, Eneroth P, Bjorkhem I. What do commercial ginseng preparations contain? [letter] Lancet 1994; 344:134.
Dahanukar SA, Karandikar SM, Desai M. Efficacy of Piper longum in childhood asthma. Indian Drugs 1984; 21:384–388.
D'Arcy PF. Adverse reactions and interactions with herbal medicines. 2. Drug interactions. Adverse Drug Reactions Toxicol Rev 1993; 12:147–162.
De Smet PAGM. Health risks of herbal remedies. Drug Saf 1995a; 13:81–93.
De Smet PAGM. Should herbal medicine-like products be licensed as medicines [editorial]? BMJ 1995b; 310:1023–1024.
De Smet PAGM, Smeets OSNM. Potential risk of health food products containing yohimbe extracts. BMJ 1994; 309:958.
De Smet PAGM. The role of plant-derived drugs and herbal medicines in healthcare. Drugs 1997; 54:801–840.
De Smet PAGM, Tognoni G. Drugs used in non-orthodox medicine. In: Dukes MNG, ed. Meyler's Side Effects of Drugs. 12th ed. Amsterdam: Elsevier, 1992:1209–1232.
De Smet PAGM. Toxicological outlook on the quality assurance of herbal remedies. In: De Smet PAGM, Keller K, Hansel R, eds. Adverse Effects of Herbal Drugs, I. Heidelberg: Springer-Verlag, 1992:1–72.
Dean A. Herbalist's patient had heart attack. Sydney Morning Herald 1994, 21 July:4.
Diamond JR, Pallone TL. Acute interstitial nephritis following use of tung shueh pills. Am J Kidney Dis 1994; 24:219–221.
Diehm C, Trampisch HJ, Lange S, Schmidt C. Comparison of leg compression stocking and oral horse-chestnut seed extract therapy in patients with chronic venous insufficiency. Lancet 1996; 347:292–294.
Drew A, Myers SP. Safety issues in herbal medicine: implications for the health professions. Med J Aust 1997; 166:538–541.
Duke J. *Lupinus perennis*: hypotensive wildflower under hypertensive wires. Coltsfoot 1993; 14(6):4–5.
Duke JA. Handbook of Biologically Active Phytochemicals and their Activities. Boca Raton, FL: CRC Press, Inc., 1992b.
Duke JA. Ginseng: A Concise Handbook. Algonac, MI: Reference Publications Inc., 1989.
Duke JA. Handbook of Northeastern Indian Medicinal Plants. Lincoln, MA: Quarterman Press, 1986.
Duke JA. Handbook of Nuts. Boca Raton, FL: CRC Press, Inc., 1988.
Duke JA. Handbook of Phytochemical Constituents of GRAS Herbs and Other Economic Plants. Boca Raton, FL: CRC Press, Inc., 1992a.

Duke JA, Ayensu ES. Handbook of Medicinal Herbs. Boca Raton, FL: CRC Press, Inc., 1985.
Duke JA, Martinez RV. Amazonian ethnobotanical dictionary. In: Handbook of Ethnobotanicals (Peru). Boca Raton, FL: CRC Press, Inc., 1994.
Dumont E, Petit E, Tarrade T, Nouvelot A. UV-C irradiation-induced peroxidative degradation of microsomal fatty acids and proteins: protection by an extract of Ginkgo biloba (EGb 761). Free Radic Biol Med 1992; 13(3):197–203.
Editorial. Pharmaceuticals from plants: great potential, few funds. Lancet 1994; 343:1513–1515.
EEC Directive 75/318/EEC as amended.
Eisenberg DM, Kessler RC, Foster C, Norlock FE, Calkins DR, Delbanco TL. Unconventional medicine in the United States. N Engl J Med 1993; 328:246–252.
Ernst E, Rand JI, Stevinson C. Complementary therapies for depression—an overview. Arch Gen Psychiatry 1998; 55:1026–1032.
Eskinazi D, Blumenthal M, Farnsworth N, Riggins CW. Botanical Medicine: Efficacy, Quality Assurance, and Regulation. New York: Mary Ann Liebert, Inc. Publisher, 1999.
Fairfield KM, Eisenberg DM, Davis RB, Libman H, Phillips RS. Patterns of use, expenditures, and perceived efficacy of complementary and alternative therapies in HIV-infected patients. Arch Intern Med 1998; 158:2257–2264.
Fang X, Shen N, Lin B. Beneficial changes in prostacyclin and thromboxane A2 by ginsenosides in myocardial infarction and reperfusion of dogs. Acta Pharmacol Sinica 1986; 7:226–230.
Farnsworth NR. Relative safety of herbal medicines. HerbalGram 1993; 29(special suppl):36A–36H.
Farnsworth NR, Akerele O, Bingel AS, Soejarta DD, Eno Z. Medicinal plants in therapy. Bull World Health Organ 1985; 63(6):965–981.
Farnsworth NR, Morris RW. Higher plants—the sleeping giant of drug development. Am J Pharm Educ 1976; 148:46–52.
Feher J, Lang I, Nekam K, Gergely P, Muzes G. Hepaprotective activity of silymarin therapy in patients with chronic alcoholic liver disease. Orv Hetil 1990; 130:51.
Ferenci P, Dragosics B, Dittrick H, et al. Randomized controlled trial of silymarin treatment in patients with cirrhosis of the liver. J Hepatol 1989; 9(1):105–113.
Fisher P, Ward A. Complementary medicine in Europe. BMJ 1994; 309:107–111.
Flemming K. Therapeutic effect of silymarin on x-irradiated mice. Arzneimittelforschung 1971; 21(9):1373–1375.
Foster S. Ginkgo. Austin, TX: American Botanical Council, 1990.
Foster S, Tyler VE. Tyler's Honest Herbal. A Sensible Guide to the Use of Herbals and Related Remedies. New York: The Howorth Press, 1998.
Gatta L. Controlled Clinical Trial among Patients Designed to Assess the Therapeutic Efficacy and Safety of Tegens 160. Ospedale Filippo del Ponte, Varese, Italy, 1982.
Gertner E, Marshall PS, Dean Filandrinos, et al. Complications resulting from the use of Chinese herbal medications containing undeclared prescription drugs. Arthritis Rheum 1995; 38:614–617.
Gilhooley M. Pharmaceutical drug regulation in China. Food Drug Cosmetic Law J 1989; 44:21–39.
Grasso M, Montesano A, Buonaguidi A, et al. Comparative effects of alfuzosin versus *Serenoa repens* in the treatment of symptomatic benign prostatic hyperplasia. Arch Exp Urol 1995; 48:97–103.
Grünwald J. The European phytomedicines market: figures, trends, analysis. HerbalGram 1995; 34:60–65.
Guerrini M. Report on Clinical Trial of Bilberry Anthocyanosides in the Treatment of Venous Insufficiency of the Lower Limbs. Istituto di Patologia Speciale Medica e Metodologia Clinica, Universit de Siena, Italy, 1987.
Haas H. Brain disorders and vasoactive substances of plant origin. Planta Med 1981; (suppl):257–265.
Han BH, Park MH, Woo LK, Woo WS, Han YN. Studies on antioxidant components of Korean ginseng. Korean Biochem J 1979; 12(1):33.

Handbook of Domestic Medicine and Common Ayurvedic Remedies. Central Council for Research in Indian Medicine and Homeopathy, New Delhi, 1979:91–112.
Harrer G, Schimidt U, Kuhn U, Biller A. Comparison of equivalent between the St. John's wort extract LoHyp-57 and fluoxetine. Arzneimittel-Forschung/Drug Res 1999; 49:289–296.
Hirayama T. Nutrition and cancer—a large-scale cohort study. Prog Clin Biol Res 1986; 206:299–311.
Högel J, Gaus W. Studies on the efficacy of unconventional therapies. Arzneimittel-Forschung/Drug Res 1995; 45:88–92.
Huxtable RJ. The harmful potential of herbal and other plant products. Drug Saf 1990; 5(suppl 1):S126–S136.
Jacob A, Pandey M, Kapoor S, Saroja R. Effect of the Indian gooseberry (amla) on serum cholesterol levels in men aged 35–55 years. Eur J Clin Nutr 1988; 42:939–944.
Johnson ES, Kadam NP, Hylands DM, Hylands PJ. Efficacy of feverfew as prophylactic treatment of migraine. BMJ 1985; 291:569–573.
Josey ES, Tackett RL. St. John's wort: a new alternative for depression? Int J Clin Pharmacol Thera 1999; 37:111–119.
Kang-Yum E, Oransky SH. Chinese patent medicine as a source of mercury poisoning. Vet Hum Toxicol 1992; 34(3):235–238.
Kanowski S, Herrmann WM, Stephan K, Wierich W, Hörr R. Proof of efficacy of the Ginkgo biloba special extract EGb 761 in outpatients suffering from mild to moderate primary degenerative dementia of the Alzheimer type or multi-infarct dementia. Phytomedicine 1997; 4:3–13.
Kao FF. The impact of Chinese medicine on America. Am J Chin Med 1992; 20(1):1–16.
Keller K. Herbal medicinal products in Germany and Europe: experiences with national and European assessment. Drug Inform J 1996; 30:933–948.
Keller K. Legal requirements for the use of phytopharmaceutical drugs in the Federal Republic of Germany. J Ethnopharmacol 1991; 32:225–229.
Kestin M, Miller L, Littlejohn G, Wahlqvist M. The use of unproven remedies for rheumatoid arthritis in Australia. Med J Aust 1985; 143:516–518.
Kew C, Morris C, Aihie A, et al. Arsenic and mercury intoxication due to Indian ethnic remedies. BMJ 1993; 306:506–507.
Kirkland J, Mathews HF, Sullivan III CW, Baldwin K, eds. Herbal and Magic Medicine: Traditional Healing Today. Durham, N.C. and London: Duke University Press, 1992.
Kishore P, Pandey PN, Pandey SN, Dash S. Preliminary trials of certain Ayurvedic drug formulation of Amalpitta. Sachitra Ayurved 1990; 33:40–45.
Kleijnen J. Knipschild P Ginkgo biloba. Lancet 1992; 340:1136–1139.
KPMG Management Consulting. Review of Therapeutic Goods Administration on behalf of the Department of Health and Family Services. Canberra: AGPS, 1997, Jan:126–127.
Kristofferson SS, Atkin PA, Shenfield GM. Uptake of alternative medicine [letter]. Lancet 1996; 347:972.
Kulkarni RR, Patki PS, Jog VP, Gandage SG, Patwardhan B. Treatment of osteoarthritis with a herbomineral formulation: a double-blind, placebo-controlled, cross-over study. J Ethnopharmacol 1991; 33:91–95.
Laakmann G, Schüle C, Baghai T, Kieser M. St. John's wort in mild to moderate depression: the relevance of hyperforin for the clinical efficacy. Pharmacopsychiatry 1998; 31:54–59.
Landis E, Shore E. Yohimbine-induced bronchospasm. Chest 1989; 96:1424.
Lazarowych NJ, Pekos P. Use of fingerprinting and marker compounds for identification and standardization of botanical drugs: strategies for applying pharmaceutical HPLC analysis to herbal products. Drug Inform J 1998; 32:497–512.
Li F, Sun S, Wang J, Wang D. Chromatography of medicinal plants and Chinese traditional medicines. Biomed Chromatograph 1998; 12:78–85.
Li JC, Li YP, Xue LS. Influence of ginseng saponins on the circadian rhythm in brain monoamine neurotransmitters. Presented at the 5th Southeast Asian and Western Pacific Regional Meeting of Pharmacologists, Chinese Pharmacological Association, Beijing, 1988.
Linde K, Raminez G, Mulrow CD, Pauls A, Weidenhammer W, Melchart D. St. John's wort for depression—an overview and meta-analysis of randomized clinical trials. BMJ 1996; 313:253–258.

Liu C, Xiao P. Recent advances on ginseng research in China. J Ethnopharmacol 1992; 36:27–38.
Lu G, Cheng XJ, Yuan WX. Effect of ginseng root saponins on serum corticosterone and neurotransmitters of hypobaric hypoxic mice. IRCS Med Sci Libr Compend 1988; 5(6):259.
MacLennan AH, Wilson DH, Taylor AW. Prevalence and cost of alternative medicine in Australia. Lancet 1996; 347:569–572.
Majno GM. Healing Hand: Man and Wound in the Ancient World. Cambridge, MA: Harvard University Press, 1975.
Masshour NH, Lin GI, Frishman WH. Herbal medicine for the treatment of cardiovascular disease. Arch Intern Med 1998; 158:2225–2234.
Moerman DC. Geraniums for the Iroquois. A Field Guide to American Indian Medicinal Plants. Algonac, MI, Reference Publications, 1982:242.
Morris J, Burke V, Mori TA, Vandogen R, Beilin LJ. Effects of garlic extract on platelet aggregation: a randomized placebo-controlled double-blind study. Clin Exp Pharmacol Physiol 1995; 22:414–417.
Mousavi Y, Adlercreutz H. Genistein is an effective stimulator of sex hormone-binding globulin production in hepatocarcinoma human liver cancer cells and suppresses proliferation of these cells in culture. Steroids 1993; 58:301–304.
Murphy JJ, Heptinstall S, Mitchell JRA. Randomised double-blind placebo-controlled trial of feverfew in migraine prevention. Lancet 1988; 2:189–192.
Nath D, Sethi N, Singh RK, Jain AK. Commonly used Indian abortifacient plants with special reference to their teratologic effects in rats. J Ethnopharmacol 1992; 36:147–154.
Pandey BL, Goel RK, Pathak NKR, Biswas M, Das PK. Effect of *Tectona grandis* linn. (common teak tree) on experimental ulcers and gastric secretion. Indian J Med Res 1982; 76(suppl):89–94.
Pei YQ. A review of pharmacology and clinical use of piperine and its derivatives. Epilepsia 1983; 24:177–181.
Perharic L, Shaw D, Colbridge M, et al. Toxicological problems resulting from exposure to traditional remedies and food supplements. Drug Saf 1994; 11:284–294.
Perry LM. Medicinal Plants of East and Southeast Asia: Attributed Properties and Uses. Cambridge, MA: MIT Press, 1980.
Peterson G, Barnes S. Genistein and biochanin A inhibit the growth of human prostate cancer cells but not epidermal growth factor receptor tyrosine autophosphorylation. Prostate 1993; 22:335–345.
Petrovick PR, Marques LC, Paula IC. New rules for phytopharmaceutical drug registration in Brazil. J Ethnopharmacol 1999; 66:51–55.
Qian BC, Zhang XX, Li B, Xu CY, Deng XY. Effects of ginseng polysaccharides on tumor and immunological function in tumor-bearing mice. Acta Pharmacol Sinica 1987; 8:6–14.
Qu JB, Cao YN, Ma XY. Effects of ginseng leaves and root saponins on animals in acute hypoxia due to negative air pressure. Presented at the fifth Southeast Asian and Western Pacific Regional Meeting of Pharmacologists, Chinese Pharmacological Association, Beijing, 1988.
Quality control methods for medicinal plant materials. Geneva, World Health Organization (unpublished document WHO/PHARM/92. 559/rev. 1; available on request from Division of Drug Management and Policies, World Health Organization, 1211 Geneva 27, Switzerland), 1992.
Raabe A, Raabe M, Ihm P. Therapeutic follow-up using automatic perimetry in chronic cerebroretinal ischemia in elderly patients: prospective double-blind study with graduated dose Ginkgo biloba treatment (EGb 761). Klin Monatsbl Augenheilkd 1991; 199(6):432–438.
Rai GS, Shovlin C, Wesnes KA. A double-blind, placebo-controlled study of *Ginkgo biloba* extract in elderly outpatients with mild to moderate memory impairment. Curr Med Res Opin 1991; 12(6):350–355.
Rawlins MD, Thompson JW. Pathogenesis of adverse drug reactions. In: Davies DM, ed. Textbook of Adverse Drug Reactions. Oxford: Oxford University Press, 1977:44.
Roberts JE, Tyler VE. Tyler's Herbs of Choice. The Therapeutic Use of Phytomedicinals. New York: The Haworth Press, Inc., 1998.

Roesler J, Emmendorffer A, Stienmuller C, Luettig B, Wagner H, Lohmann-Matthes ML. Application of purified polysaccharides from cell cultures of the plant *Echinacea purpurea* to test subjects mediates activation of the phagocyte system. Int J Immunopharmacol 1991; 13(7):931–941.
Sawyer MG, Gannoni AF, Toogood IR, et al. The use of alternative therapies by children with cancer. Med J Aust 1994; 160:320–322.
Schaffler K, Reeh P. Long-term drug administration effects of *Ginkgo biloba* on the performance of healthy subjects exposed to hypoxia. In: Effects of *Ginkgo biloba* Extracts on Organic Cerebral Impairment. London: Eurotext Ltd., 1985.
Schaumburg HH, Berger A. Alopecia and sensory polyneuropathy from thallium in a Chinese herbal medication [letter]. JAMA 1992; 268:3430–3431.
Schulz V, Hänsel R, Tyler VE. Rational Phytotherapy. A Physicians' Guide to Herbal Medicine. 3rd edn. Berlin: Springer-Verlag, 1996.
Shen JJ, Jin YY, Wu YS, Zhou X. Effects of ginseng saponins on 14C-arachidonic acid metabolism in rabbit platelets. Acta Pharm Sinica 1987; 22:166–169.
Shu YZ. Recent natural products based drug development: A pharmaceutical industry perspective. J Nat Products 1998; 61:1053–1071.
Shukla AK. Endocrine Response of Certain Species on Albino Rats (doctoral thesis). Banaras Hindu University, Varanasi, India, 1984.
Snider S. Beware the unknown brew: herbal teas and toxicity. FDA Consumer 1991 May; 25:31–33.
Solecki RS. Shanidar IV, a Neanderthal flower burial of northern Iraq. Science 1975; 190:880.
Spanu F, Tava A, Pacetti L, Piano E. Variability of oestrogenic isoflavone content in a collection of subterranean clover from Sicily. J Genet Breeding 1993; 47:27–34.
Stimpel M, Proksch A, Wagner H, Lohmann-Matthes ML. Macrophage activation and induction of macrophage cytotoxicity by purified polysaccharide fractions from the plant Echinacea purpurea. Infect Immun 1984; 46(3):845–849.
Sumathikutty MA, Rajaraman K, Sankarikutty B, Mathew AG. Chemical composition of pepper grades and products. J Food Sci Technol 1979; 16:249–254.
Swanston-Flatt SK, Day C, Flott PR, et al. Glycemic effects of traditional European plant treatments for diabetes: studies in normal and streptozotocin diabetic mice. Diabetes Res 1989; 10(2):69–73.
Taillandier J. Ginkgo biloba extract in the treatment of cerebral disorders due to aging. In: Funfgeld EW, ed. Rokan (*Ginkgo biloba*): Recent Results in Pharmacology and Clinic. Berlin: Springer-Verlag, 1988.
Takahashi M, Matsuo I, Ohkido M. Contact dermatitis due to honeybee royal jelly. Contact Dermat 1983; 9:452–455.
Tang BY, Adams NR. Oestrogen receptors and metabolic activity in the genital tract after ovariectomy of ewes with permanent infertility caused by exposure to phytoestrogens. J Endocrinol 1981; 89(3):365–370.
Therapeutic Goods Administration. Government response to recommendations arising from the Therapeutic Goods Administration Review. Canberra: AGPS, 1997, April: 37–38.
Tong LS, Chao CY. Effects of ginsengoside Rg1 of *Panax ginseng* on mitosis in human blood lymphocytes in vitro. Am J Chem Med 1980; 8(3):254.
Tsumura A. Kampo, How the Japanese Updated Traditional Herbal Medicine. Tokyo and New York: Japan Publications, Inc., 1991.
Upadhyaya SD, Kansal CM, Pandey NN. Clinical evaluation of Piper longum on patients of bronchial asthma—a preliminary study. Nagarjuna 1982; 25:256–258.
Ura H, Obara T, Okamura K, Namiki M. Growth inhibition of pancreatic cancer cells by flavonoids. Gan To Kagaku Ryoho 1993; 20(13):2083–2085.
van Ypersele de Strihou C, Vanherweghem JL. The tragic paradigm of Chinese herbs nephropathy [editorial]. Nephrol Dial Transplant 1995; 10:157–160.
Vanhaelen M, Vanhaelen-Fastre R, But P, Vanherweghem J-L. Identification of aristolochic acid in Chinese herbs [letter]. Lancet 1994; 343:174.
Vanherweghem J-L, Depierreux M, Tielemans C, et al. Rapidly progressive interstitial renal fibrosis in young women: association with slimming regimen including Chinese herbs. Lancet 1993; 341:387–391.

Varkonyi T. Brain edema in the rat induced by triethyltin sulfate. 10 Effect of silymarin on the electron-microscopy picture. Arzneimittelforschung 1971; 21(1):148–149.
Verma SK, Bordia A. Effect of Commiphora mucul (gum guggul) in patients with hyperlipidemia with special reference to HDL cholesterol. Indian J Med Res 1988; 87:356–360.
Vitiello B. Hypericum perforatum extracts as potential antidepressants. J Pharmacy Pharmacol 1999; 51:513–517.
Vogel VL. American Indian Medicine. Norman, Okla: University of Oklahoma Press, 1970.
Vogler BK, Pittler MH, Ernest E. Fever few as a preventive treatment for migraine: a systematic review. Cephalalgia 1998; 18:704–708.
Wagner H, Geyer B, Kiso Y, Hikino H, Rao GS. Coumestans as the main active principles of the liver drugs *Eclipta alba* and *Wedelia calendulacea*. Planta Med 1986; 52:370–377.
Wall Street Journal. Vital statistics: disputed cost of creating a drug. 1993, 9 Nov.
Wang BX, Cui JC, Lui AJ. The effect of polysaccharides of root of Panax ginseng on the immune function. Acta Pharm Sinica 1980; 17:312–320.
Warshafsky S, Kamer RS, Sivak SL. Effect of garlic on total serum cholesterol: a meta-analysis. Arch Intern Med 1993; 119:599–605.
Weitbrecht WV, Jansen W. Double blind and comparative (Ginkgo biloba versus placebo) therapeutic study in geriatric patients with primary degenerative dementia—a preliminary evaluation. In: Agnoli A et al., eds. Effects of *Ginkgo biloba* Extract on Organic Cerebral Impairment. London: John Libbey Eurotext Ltd., 1985.
WHO Expert Committee on Specifications for Pharmaceutical Preparations. Thirty-second report. Geneva, World Health Organization, 1992:44–52; 75–76 (WHO Technical Report Series, No. 823).
Wilt TJ, Ishani A, Stark G, MacDonald R, Lau J, Mulrow C. Saw palmetto extracts for treatment of benign prostatic hyperplasia: a systematic review. J Am Med Assoc 1998; 280:1604–1609.
Witte S, Anadere I, Walitza E. Improvement of hemorheology with Ginkgo biloba extract: decreasing a cardiovascular risk factor. Fortschr Med 1992; 110(13):247–250.
Woelk H, Burkard G, Grünwald J. Benefits and risks of the Hypericum extract LI 160: drug monitoring study with 3250 patients. Geriat Psychiatry Neurol 1994; 7(suppl 1):S34–S38.
World Health Organization. Guidelines for the Assessment of Herbal Medicines. Programme on Traditional Medicines, Geneva, 1991.
Wu M-S, Hong J-J, Lin J-L, et al. Multiple tubular dysfunction induced by mixed Chinese herbal medicines containing cadmium. Nephrol Dial Transplant 1996; 11: 867–870.
Yabe T, Chat M, Malherbe E, Vidal PP. Effects of Ginkgo biloba extract (EGb 761) on the guinea pig vestibular system. Pharmacol Biochem Behav 1992; 42(4):595–604.
Yanagihara K, Ito A, Toge T, Numoto M. Antiproliferative effects of isoflavones on human cancer cell lines established from the gastrointestinal tract. Cancer Res 1993; 53(23): 5815–5821.
Yang Y, Chen Z, Luo G, Zhang Y. The mechanism of inhibitory effects of panaxadiol saponin on rabbit platelet aggregation. Presented at the 5th Southeast Asian and Western Pacific Regional Meeting of Pharmacologists, Chinese Pharmacological Association, Beijing, 1988.
Youngken HW. A Textbook of Pharmacognosy. New York: McGraw-Hill, 1950.
Zeisel SH. Regulation of "nutraceuticals." Science 1999; 285:1853–1855.
Zhang FL, Chen X. Effects of ginsenosides on sympathetic neurotransmitter release in pithed rats. Acta Pharm Sinica 1987; 8:217–220.
Zhang FL, Meehan AG, Rand MJ. Effects of ginsenosides on noradrenergic transmission, histamine response and calcium influx in rabbit ear isolated artery. Presented at the fifth Southeast Asian and Western Pacific Regional Meeting of Pharmacologists, Chinese Pharmacological Association, Beijing, 1988.
Zhang JT. Progress of research on three kinds of anti-aging drugs. Informa Chin Pharmacol Soc 1989; 6(3–4):4.

Glossary of Terms

Accelerated testing   Studies designed to increase the rate of chemical degradation or physical change of a drug substance or drug product by using exaggerated storage conditions as part of the formal stability studies. Data from these studies, in addition to long-term stability studies, can be used to assess longer term chemical effects at nonaccelerated conditions and to evaluate the effect of short-term excursions outside the label storage conditions, such as might occur during shipping. Results from accelerated testing studies are not always predictive of physical changes.

Acceptance criteria   Numerical limits, ranges, or other suitable measures for acceptance of the results of analytical procedures that the drug substance or drug product or materials at other stages of manufacture should meet.

Action limit   An internal (in-house) value used to assess the consistency of the process at less critical steps.

Active constituent   The chemical constituent in a botanical raw material, drug substance, or drug product that is responsible for the intended pharmacological activity or therapeutic effect.

Active pharmaceutical ingredient (API) (or drug substance)   Any substance or mixture of substances intended to be used in the manufacture of a drug (medicinal) product and that, when used in the production of a drug, becomes an active ingredient of the drug product. Such substances are intended to furnish pharmacological activity or other direct effect in the diagnosis, cure, mitigation, treatment, or prevention of disease or to affect the structure and function of the body.

ADMET   In vitro absorption, distribution, metabolism, elimination, and toxicology.

Batch (or lot)   A specific quantity of material produced in a process or series of processes so that it is expected to be homogeneous within specified limits. In the case of continuous production, a batch may correspond to a defined fraction of the production. The batch size can be defined either by a fixed quantity or by the amount produced in a fixed time interval.

Batch number (or lot number)   A unique combination of numbers, letters, and/or symbols that identifies a batch (or lot) and from which the production and distribution history can be determined.

BDCS   The Biopharmaceutics Drugs Classification System.

Bioburden   The level and type (e.g., objectionable or not) of microorganisms that can be present in raw materials, API starting materials, intermediates or APIs. Bioburden should not be considered contamination unless the levels have been exceeded or defined objectionable organisms have been detected.

Biological activity   The specific ability or capacity of the product to achieve a defined biological effect. Potency is the quantitative measure of the biological activity.

Botanical drug product A botanical product that is intended for use as a drug; a drug product that is prepared from a botanical drug substance. Botanical drug products are available in a variety of dosage forms, such as solutions (e.g., teas), powders, tablets, capsules, elixirs, and topical.

Botanical drug substance A drug substance derived from one or more plants, algae, or macroscopic fungi. It is prepared from botanical raw materials by one or more of the following processes: pulverization, decoction, expression, aqueous extraction, ethanolic extraction, or other similar processes. It may be available in a variety of physical forms, such as powder, paste, concentrated liquid, juice, gum, syrup, or oil. A botanical drug substance can be made from one or more botanical raw materials (see Single-herb and Multi-herb botanical drug substance or product). A botanical drug substance does not include a highly purified or chemically modified substance derived from natural sources.

Botanical ingredient A component of a botanical drug substance or product that originates from a botanical raw material.

Botanical raw material Fresh or processed (e.g., cleaned, frozen, dried, or sliced) part of a single species of plant or a fresh or processed alga or macroscopic fungus.

Bracketing The design of a stability schedule such that only samples on the extremes of certain design factors (e.g., strength, package size) are tested at all time points as in a full design. The design assumes that the stability of any intermediate levels is represented by the stability of the extremes tested. Where a range of strengths is to be tested, bracketing is applicable if the strengths are identical or very closely related in composition (e.g., for a tablet range made with different compression weights of a similar basic granulation, or a capsule range made by filling different plug fill weights of the same basic composition into different size capsule shells). Bracketing can be applied to different container sizes or different fills in the same container closure system.

Caco-2 A model for human drug intestinal permeability.

CADD Computer-assisted drug design, also called computer-assisted molecular design (CAMD).

Calibration The demonstration that a particular instrument or device produces results within specified limits by comparison with results produced by a reference or traceable standard over an appropriate range of measurements.

Chemical development studies Studies conducted to scale-up, optimize, and validate the manufacturing process for a new drug substance.

Chiral Not superimposable with its mirror image, as applied to molecules, conformations, and macroscopic objects, such as crystals. The term has been extended to samples of substances whose molecules are chiral, even if the macroscopic assembly of such molecules is racemic.

Climatic zones The four zones in the world that are distinguished by their characteristic, prevalent annual climatic conditions. This is based on the concept described by W. Grimm (Drugs Made in Germany, 28:196–202, 1985 and 29:39–47, 1986).

Combination product A drug product that contains more than one drug substance.

Commitment batches Production batches of a drug substance or drug product for which the stability studies are initiated or completed postapproval through a commitment made in the registration application.

Confirmatory studies These are undertaken to establish photostability characteristics under standardized conditions. These studies are used to identify precautionary measures needed in manufacturing or formulation and whether light-resistant packaging and/or special labeling is needed to mitigate exposure to light. For the confirmatory studies, the batch(es) should be selected according to batch selection for long-term and accelerated testing, which is described in the parent guideline.

Conjugated product A conjugated product is made up of an active ingredient (e.g., peptide, carbohydrate) bound covalently or noncovalently to a carrier (e.g., protein, peptide, and inorganic mineral) with the objective of improving the efficacy or stability of the product.

Constituents with known therapeutic activity These are chemically defined substances or groups of substances that are generally accepted to contribute substantially to the therapeutic activity of a herbal substance, a herbal preparation, or a herbal medicinal product.

Container closure system The sum of packaging components that together contain and protect the dosage form. This includes primary packaging components and secondary packaging components if the latter are intended to provide additional protection to the drug product. A packaging system is equivalent to a container closure system.

Contaminants Any adventitiously introduced materials (e.g., chemical, biochemical, or microbial species) not intended to be part of the manufacturing process of the drug substance or drug product.

Contamination The undesired introduction of impurities of a chemical or microbiological nature, or of foreign matter, into or onto a raw material, intermediate, or API during production, sampling, packaging, or repackaging, storage, or transport.

Contract manufacturer A manufacturer who performs some aspect of manufacturing on behalf of the original manufacturer.

Critical Describes a process step, process condition, test requirement, or other relevant parameters or items that must be controlled within predetermined criteria to ensure that the API meets its specification.

Cross-contamination Contamination of a material or product with another material or product.

Degradation product A molecule resulting from a change in the drug substance (bulk material) brought about over time. For the purpose of stability testing of the products described in this guideline, such changes could occur as a result of processing or storage (e.g., by deamidation, oxidation, aggregation, proteolysis). For biotechnological/biological products, some degradation products may be active.

Delayed release Release of a drug (or drugs) at a time other than immediately following oral administration.

Design space The design space is the established range of process parameters that has been demonstrated to provide assurance of quality. In some cases, design space can also be applicable to formulation attributes. Working within the design space is not generally considered as a change of the approved ranges for process parameters and formulation attributes. Movement out of the design space is considered to be a change and would normally initiate a regulatory postapproval change process.

Desired product   (1) The protein that has the expected structure, or (2) the protein that is expected from the DNA sequence and anticipated post-translational modification (including glycoforms), and from the intended downstream modification to produce an active biological molecule.

Deviation   Departure from an approved instruction or established standard.

Dosage form   A pharmaceutical product type, for example, tablet, capsule, solution, or cream, which contains a drug ingredient (substance) generally, but not necessarily, in association with excipients.

Drug (medicinal) product   The dosage form in the final immediate packaging intended for marketing (Reference Q1A).

Drug extract ratio (DER)   This means the ratio between the quantity of herbal substance used in the manufacture of a herbal preparation and the quantity of herbal preparation obtained.

Drug product   A finished dosage form, for example, tablet, capsule, solution, and so on (21 CFR 210.3(b)(4)).

Drug substance (bulk material)   The material that is subsequently formulated with excipients to produce the drug product. It can be composed of the desired product, product-related substances, and product- and process-related impurities. It may also contain excipients including other components, such as buffers.

Enantiomeric impurity   A compound with the same molecular formula as the drug substance that differs in the spatial arrangement of atoms within the molecule and is a nonsuperimposable mirror image.

Excipients   Ingredients added intentionally to the drug substance which should not have pharmacological properties in the quantity used.

Expiration date   The date placed on the container label of a drug product designating the time prior to which a batch of the product is expected to remain within the approved shelf life specification, if stored under defined conditions, and after which it must not be used.

Expiry date (or expiration date)   The date placed on the container/labels of an API designating the time during which the API is expected to remain within established shelf life specifications if stored under defined conditions and after which it should not be used.

Extended release   Products that are formulated to make the drug available over an extended period after administration.

Extraneous contaminant   An impurity arising from any source extraneous to the manufacturing process.

Forced degradation   These testing studies are undertaken to degrade the sample deliberately. These studies, which may be undertaken in the development phase normally on the drug substances, are used to evaluate the overall photosensitivity of the material for method development purposes and/or degradation pathway elucidation.

Formal experimental design   A structured, organized method for determining the relationship between factors (X) affecting a process and the output of that process (Y). This is also known as "design of experiments."

Formal stability studies Long-term and accelerated (and intermediate) studies undertaken on primary and/or commitment batches according to a prescribed stability protocol to establish or confirm the retest period of a drug substance or the shelf life of a drug product.

Formulation A formula that lists the components (or ingredients) and composition of the dosage form. The components and composition of a multiherb botanical drug substance should be part of the total formulation.

Genuine (native) herbal preparation This refers to the preparation without excipients.

GPCR G-protein-coupled receptor signaling.

Herbal medicinal product Medicinal product containing, as active ingredients, exclusively plant material and/or preparations. This term is generally applied to a finished product. If it refers to an unfinished product, this should be indicated.

Herbal preparations These are obtained by subjecting herbal substances to treatments, such as extraction, distillation, expression, fractionation, purification, concentration, or fermentation. These include comminuted or powdered herbal substances, tinctures, extracts, essential oils, expressed juices, and processed exudates.

Herbal products Medicinal products containing, exclusively, plant material and/or vegetable drug preparations as active ingredients. In some traditions, materials of inorganic or animal origin can also be present.

Herbal substances All mainly whole, fragmented or cut plants, plant parts, algae, fungi, lichen in an unprocessed, usually dried form but sometimes fresh. Certain exudates that have not been subjected to a specific treatment are also considered to be herbal substances. Herbal substances are precisely defined by the plant part used and the botanical name according to the binomial system (genus, species, variety, and author).

Highly water-soluble drugs Drugs with a dose/solubility volume of less than or equal to 250 mL over a pH range of 1.2–6.8. [Example: Compound A has its lowest solubility at $37 \pm 0.5°C$, 1.0 mg/mL at pH 6.8, and is available in 100, 200, and 400-mg strengths. This drug would be considered a low solubility drug, as its dose/solubility volume is greater than 250 mL (400 mg/1.0 mg/mL = 400 mL).]

HTL Hit-to-lead.

HTS High throughput screening.

Identification threshold A limit above (>) which an impurity should be identified.

Identified impurity An impurity, for which a structural characterization has been achieved.

Immediate (primary) pack This is constituent of the packaging that is in direct contact with the drug substance or drug product, and includes any appropriate label.

Immediate release Allows the drug to dissolve in the gastrointestinal contents, with no intention of delaying or prolonging the dissolution or absorption of the drug. Impurity: (1) Any component of the new drug substance that is not the chemical entity defined as the new drug substance. (2) Any component of the drug product that is not the chemical entity defined as the drug substance or an excipient in the drug product.

Impermeable containers Containers that provide a permanent barrier to the passage of gases or solvents (e.g., sealed aluminum tubes for semisolids, sealed glass ampoules for solutions).

Impurity (1) Any component of the new drug substance that is not the chemical entity defined as the new drug substance. (2) Any component of the drug product that is not the chemical entity defined as the drug substance or excipients in the drug product.

Impurity profile A description of the identified and unidentified impurities present in a new drug substance.

In-house primary reference material An appropriately characterized material prepared by the manufacturer from a representative lot(s) for the purpose of biological assay and physico-chemical testing of subsequent lots, and against which in-house working reference material is calibrated.

In-house working reference material A material prepared similarly to the primary reference material that is established solely to assess and control subsequent lots for the individual attribute in question. It is always calibrated against the in-house primary reference material. Potency: The measure of the biological activity using a suitably quantitative biological assay (also called potency assay or bioassay), based on the attribute of the product, which is linked to the relevant biological properties.

In-process control (or process control) Checks performed during production to monitor and, if appropriate, to adjust the process and/or to ensure that the intermediate or API conforms to its specifications.

In-process tests Tests that may be performed during the manufacture of either the drug substance or drug product, rather than as part of the formal battery of tests that are conducted prior to release.

Intermediate A material produced during steps of the processing of an API that undergoes further molecular change or purification before it becomes an API. Intermediates may or may not be isolated. (Note: this guidance only addresses those intermediates produced after the point that a company has defined as the point at which the production of the API begins.)

Intermediate testing Studies conducted at 30°C/65% relative humidity (RH) and designed to moderately increase the rate of chemical degradation or physical changes for a drug substance or drug product intended to be stored long-term at 25°C.

IVIVC In vitro–in vivo correlation.

LO Lead optimization.

Lifecycle All phases in the life of a product from the initial development through pre- and postapproval until the product's discontinuation.

Ligand An agent with a strong affinity to a metal ion.

Long-term testing Stability studies under the recommended storage condition for the retest period or shelf life proposed (or approved) for labeling.

Manufacture All operations of receipt of materials, production, packaging, repackaging, labeling, re-labeling, quality control, release, storage, and distribution of APIs and related controls.

Markers These are chemically defined constituents of a herbal substance that are of interest for control purposes independent of whether they have any therapeutic activity or not. Markers may serve to calculate the quantity of herbal substance(s) or herbal preparation(s) in the finished product if the marker has been quantitatively determined in the herbal substance(s) or herbal preparation(s) when the starting materials were tested.

Marketing pack   This is the combination of immediate pack and other secondary packaging, such as a carton.

Mass balance   The process of adding together the assay value and levels of degradation products to see how closely these add up to 100% of the initial value, with due consideration of the margin of analytical error.

Material   A general term used to denote raw materials (starting materials, reagents, and solvents), process aids, intermediates, APIs, and packaging and labeling materials.

Matrixing   The design of a stability schedule such that a selected subset of the total number of possible samples for all factor combinations is tested at a specified time point. At a subsequent time point, another subset of samples for all factor combinations is tested. The design assumes that the stability of each subset of samples tested represents the stability of all samples at a given time point. The differences in the samples for the same drug product should be identified as, for example, covering different batches, different strengths, different sizes of the same container closure system, and, possibly in some cases, different container closure systems.

Mean kinetic temperature   A single derived temperature that, if maintained over a defined period of time, affords the same thermal challenge to a drug substance or drug product as would be experienced over a range of both higher and lower temperatures for an equivalent defined period. The mean kinetic temperature is higher than the arithmetic mean temperature and takes into account the Arrhenius equation.

Metabolomics   The emerging field of metabolomics seeks to understand all the small molecules found within cells and tissues.

Modified release   Dosage forms whose drug release characteristics of time course and/or location are chosen to accomplish therapeutic or convenience objectives not offered by conventional dosage forms, such as a solution or an immediate-release dosage form. Modified-release solid oral dosage forms include both delayed- and extended-release drug products.

Mother liquor   The residual liquid that remains after the crystallization or isolation processes. Mother liquor may contain unreacted materials, intermediates, levels of the API, and/or impurities. It can be used for further processing.

Multidrug efflux   Whereby a single transporter is capable of recognizing and transporting multiple drugs, with no apparent common structural similarity.

Multiherb (botanical drug) substance or product   A botanical drug substance or drug product that is derived from more than one botanical raw material, each of which is considered a botanical ingredient. A multiherb botanical drug substance may be prepared by processing together two or more botanical raw materials, or by combining two or more single-herb botanical drug substances that have been individually processed from their corresponding raw materials. In the latter case, the individual single-herb botanical drug substances may be introduced simultaneously or at different stages during the manufacturing process of the dosage form.

NAEs   New active entities.

New drug product A pharmaceutical product type, for example, tablet, capsule, solution, cream, and so on that has not previously been registered in a region or Member State, and that contains a drug ingredient generally, but not necessarily, in association with excipients.

New drug substance The designated therapeutic moiety that has not been previously registered in a region or Member State (also referred to as a new molecular entity or new chemical entity). It can be a complex, simple ester, or salt of a previously approved drug substance.

New molecular entity An active pharmaceutical substance not previously contained in any drug product registered with the national or regional authority concerned. A new salt, ester, or noncovalent bond derivative of an approved drug substance is considered a new molecular entity for the purpose of stability testing under this guidance.

PAMPA Parallel artificial membrane permeability analysis.

PAT Process analytical technologies—a system for designing, analyzing, and controlling manufacturing through timely measurements (i.e., during processing) of critical quality and performance attributes of raw and in-process materials and processes with the goal of assuring final product quality.

Pharmacoproteomics Proteome analysis comprises three sequential steps: sample preparation, protein separation and mapping, and protein characterization.

Pilot-scale batch A batch of a drug substance or drug product manufactured by a procedure fully representative of and simulating that to be applied to a full production scale batch. For solid oral dosage forms, a pilot scale is generally, at a minimum, one-tenth that of a full production scale or 100,000 tablets or capsules, whichever is larger.

Plant material A plant or plant part (e.g., bark, wood, leaves, stems, roots, flowers, fruits, seeds, or parts thereof) as well as exudates thereof.

Plant preparations Comminuted or powdered plant material, extracts, tinctures, fatty or essential oils, resins, gums, balsams, expressed juices, and so on, prepared from plant material, and preparations whose production involves a fractionation, purification or concentration process, but excluding chemically defined isolated constituents. A plant preparation can be regarded as the active ingredient whether or not the constituents having therapeutic activities are known.

Polymorphic forms Different crystalline forms of the same drug substance. These can include solvation or hydration products (also known as pseudo-polymorphs) and amorphous forms.

Polymorphism The occurrence of different crystalline forms of the same drug substance. This may include solvation or hydration products (also known as pseudo-polymorphs) and amorphous forms.

Potential impurity An impurity that theoretically can arise during manufacture or storage. It may or may not actually appear in the new drug substance.

Primary batch A batch of a drug substance or drug product used in a formal stability study, from which stability data are submitted in a registration application for the purpose of establishing a retest period or shelf life, respectively. A primary batch of a drug substance should be at least a pilot-scale batch. For a drug product, two of the three batches should be at least pilot-scale batch, and the third batch can be smaller if it is representative with regard to the critical manufacturing steps. However, a primary batch may be a production batch.

Process aids Materials, excluding solvents, used as an aid in the manufacture of an intermediate or API that do not participate in a chemical or biological reaction (e.g., filter aid, activated carbon).

Process-related impurities Impurities that are derived from the manufacturing process. They may be derived from cell substrates (e.g., host cell proteins, host cell DNA), cell culture (e.g., inducers, antibiotics, or media components), or downstream processing (e.g., processing reagents or column leachables).

Product-related impurities Molecular variants of the desired product (e.g., precursors, certain degradation products arising during manufacture and/or storage) that do not have properties comparable with those of the desired product with respect to activity.

Qualification Action of proving and documenting that equipment or ancillary systems are properly installed, they work correctly, and actually lead to the expected results. Qualification is part of validation, but the individual qualification steps alone do not constitute process validation.

Qualification threshold A limit above (>) which an impurity should be qualified.

Quarantine The status of materials isolated physically or by other effective means pending a decision on their subsequent approval or rejection.

Racemate A composite (solid, liquid, gaseous, or in solution) of equimolar quantities of two enantiomeric species. It is devoid of optical activity.

Rapidly dissolving products An immediate-release solid oral drug product is considered rapidly dissolving when not less than 80% of the label amount of the drug substance dissolves within 15 minutes in each of the following media: (1) pH 1.2, (2) pH 4.0, and (3) pH 6.8.

Raw material A general term used to denote starting materials, reagents, and solvents intended for use in the production of intermediates or APIs.

Reference standard, primary A substance that has been shown by an extensive set of analytical tests to be authentic material that should be of high purity. This standard can be: (1) obtained from an officially recognized source, (2) prepared by independent synthesis, (3) obtained from existing production material of high purity, or (4) prepared by further purification of existing production material.

Reference standard, secondary A substance of established quality and purity, as shown by comparison with a primary reference standard, used as a reference standard for routine laboratory analysis.

Reporting threshold A limit above (>) which an impurity should be reported. Reporting threshold is the same as reporting level in Q2B.

Reprocessing Introducing an intermediate or API, including one that does not conform to standards or specifications, back into the process and repeating a crystallization step or other appropriate chemical or physical manipulation steps (e.g., distillation, filtration, chromatography, and milling), which are part of the established manufacturing process. Continuation of a process step after an in-process control test has shown that the step is incomplete, is considered to be part of the normal process, and is not reprocessing.

Retest date The date after which samples of the drug substance should be examined to ensure that the material is still in compliance with the specification and thus suitable for use in the manufacture of a given drug product.

Retest period The period of time during which the drug substance is expected to remain within its specification and, therefore, can be used in the manufacture of a given drug product, provided that the drug substance has been stored under the defined conditions. After this period, a batch of the drug substance destined for use in the manufacture of a drug product should be retested for compliance with the specification and then used immediately. A batch of drug substance can be retested multiple times and a different portion of the batch used after each retest, as long as it continues to comply with the specification. For most biotechnological/biological substances known to be labile, it is more appropriate to establish a shelf life than a retest period. The same may be true for certain antibiotics.

Reworking Subjecting an intermediate or API that does not conform to standards or specifications to one or more processing steps that are different from the established manufacturing process to obtain acceptable quality intermediate or API (e.g., recrystallizing with a different solvent).

Risk management Systematic application of quality management policies, procedures, and practices to the tasks of assessing, controlling, and communicating risk.

Risk reduction Actions taken to lessen the probability of occurrence of harm and the severity of the corresponding harm.

Semipermeable containers Containers that allow the passage of solvent, usually water, while preventing solute loss. The mechanism for solvent transport occurs by absorption into one container surface, diffusion through the bulk of the container material, and desorption from the other surface. Transport is driven by a partial pressure gradient. Examples of semipermeable containers include plastic bags and semirigid, low-density polyethylene (LDPE) pouches for large volume parenteral (LVPs), and LDPE ampoules, bottles, and vials.

Shelf life (also referred to as expiration dating period) The time period during which a drug product is expected to remain within the approved shelf life specification, provided that it is stored under the conditions defined on the container label.

Specification A list of tests, references to analytical procedures, and appropriate acceptance criteria that are numerical limits, ranges, or other criteria for the test described. It establishes the set of criteria to which a material should conform to be considered acceptable for its intended use. Conformance to specification means that the material, when tested according to the listed analytical procedures, will meet the listed acceptance criteria. See International Conference on Harmonization (ICH) Q6A and Q6B.

Specification, release The combination of physical, chemical, biological, and microbiological tests and acceptance criteria that determine the suitability of a drug product at the time of its release.

Specification, shelf life The combination of physical, chemical, biological, and microbiological tests and acceptance criteria that determine the suitability of a drug substance throughout its retest period, or that a drug product should meet throughout its shelf life.

Specified impurity An identified or unidentified impurity that is selected for inclusion in the herbal preparation (herbal substance) or herbal medicinal product specification and is individually listed and limited in order to assure the quality of the herbal preparation (herbal substance) or herbal medicinal product.

Spectroscopic and/or chromatographic fingerprint   A spectroscopic and/or chromatographic profile of a botanical raw material, drug substance, or drug product that is matched qualitatively and quantitatively against that of a reference sample or standard to ensure the identity and quality of a batch and consistency from batch to batch.

Standardization   This means adjusting the herbal preparation to a defined content of a constituent or a group of substances with known therapeutic activity respectively by adding excipients or by mixing herbal substances or herbal preparations (e.g., standardized extract from the European Pharmacopoeia).

Starting material   A material used in the synthesis of a new drug substance that is incorporated as an element into the structure of an intermediate and/or of the new drug substance. Starting materials are normally commercially available and are of defined chemical and physical properties and structure.

Storage condition tolerances   The acceptable variations in temperature and relative humidity of storage facilities for formal stability studies. The equipment should be capable of controlling the storage condition within the ranges defined in this guidance. The actual temperature and humidity (when controlled) should be monitored during stability storage. Short-term spikes due to opening of doors of the storage facility are accepted as unavoidable. The effect of excursions due to equipment failure should be addressed and reported if judged to affect stability results. Excursions that exceed the defined tolerances for more than 24 hours should be described in the study report and their effect assessed.

Stress testing (drug substance)   Studies undertaken to elucidate the intrinsic stability of the drug substance. Such testing is part of the development strategy and is normally carried out under more severe conditions than those used for accelerated testing.

Supporting data   Data, other than those from formal stability studies that support the analytical procedures, the proposed retest period or shelf life, and the label storage statements. Such data include: (1) stability data on early synthetic route batches of drug substance, small-scale batches of materials, investigational formulations not proposed for marketing, related formulations, and product presented in containers and closures other than those proposed for marketing; (2) information regarding test results on containers; and (3) other scientific rationales.

Taxa   A grouping of organisms given a formal taxonomic name, such as species, genus, family, and so on.

Unidentified impurity   An impurity for which a structural characterization has not been achieved and that is defined solely by qualitative analytical properties (e.g., chromatographic retention time).

Universal test   A test that is considered potentially applicable to all new drug substances, or all new drug products; for example, appearance, identification, assay, and impurity tests.

Unspecified impurity   An impurity that is limited by a general acceptance criterion, but not individually listed with its own specific acceptance criterion, in the new drug substance specification.

Validation   A documented program that provides a high degree of assurance that a specific process, method, or system will consistently produce a result meeting predetermined acceptance criteria.

Validation protocol   A written plan stating how validation will be conducted and defining acceptance criteria. For example, the protocol for a manufacturing process identifies processing equipment, critical process parameters and/or operating ranges, product characteristics, sampling, test data to be collected, number of validation runs, and acceptable test results.

Yield, expected   The quantity of material or the percentage of theoretical yield anticipated at any appropriate phase of production based on previous laboratory, pilot scale, or manufacturing data.

Yield, theoretical   The quantity that would be produced at any appropriate phase of production based upon the quantity of material to be used, in the absence of any loss or error in actual production.

Index

Abbott Laboratories
　calcitriol patent expiration, 7, 34
　chemical diversity, 3
Abciximab
　composition, 337
Absorption
　vs. assay complexity, 147
Absorption and distribution in
　　body, metabolism and
　　elimination from body
　　(ADME), 120, 149
Accelerated testing
　stability, 303
Acceptance criteria
　drug product dissolution, 245
　polymorphs, 208
Acid–base theory, 88–89
Acids
　pKa, 217
Activation energy, 304
Active pharmaceutical ingredients
　　(API), 59, 65, 141, 322
　expiry, 324
Actual reduction to practice
　defined, 36
Adalimumab
　composition, 337
Additives
　solubility, 112
ADME. *See* Absorption and distribution in
　　body, metabolism and elimination
　　from body
ADMET studies, 1
Adsorbate, 252, 255
AFM. *See* Atomic force microscope
Aggregation, 370, 387
　preventive actions, 371
AHF (recombinant)
　composition, 337
Aldesleukin
　composition, 337

Alemtuzumab
　composition, 337
Alteplase
　composition, 337
American Chemical Society technical
　　databases, 43
American ginseng, 391
American Herbal
　　Pharmacopoeia, 19
Amino acid
　composition, 385
　sequence, 385
Amorphous forms
　FDA, 212
Amoxicillin patent expiration
　Smith Kline Beecham, 34
Anakinra
　composition, 337
Analytical methods, 361
　validation, 382–383
Anhedral, 199
Animal model testing,
　　153–154
Antibodies
　patents, 41
Antigen combinations
　patents, 41
Antihemophilic factor
　　(recombinant)
　composition, 337
Antioxidants
　selection, 267
APEX II detector, 200
API. *See* Active pharmaceutical
　　ingredients
Approval
　international rules
　　herbal medicines, 393–396
Approval trends, 7–8
Approved biological products
　composition, 337–348

Approved products transfer
  Center for Biologics Evaluation and Research to Center for Drug Evaluation and Research, 331–332
Arcitumomab
  composition, 338
Arrhenius equation, 303, 304, 305, 307
Assay, 291
  complexity vs. absorption, 147
Atomic force microscope (AFM), 265
Aventis
  combinatorial chemistry, 3

Baseline Natural Health Products Survey, 391
Basiliximab
  composition, 338
BCS. *See* Biopharmaceuticals Classification System
Becaplermin
  composition, 338
BET. *See* Braunauer, Emmet, and Teller (BET) method
Beta-elimination, 366, 367
Beta-hydroxyethylenediaminetriacetic acid, 278
BFDH. *See* Bravais, Friedel, Donnay, and Harker
Biogen
  interferon patent, 35
Bioinformatics, 8, 14, 16–17
Biological activity, 379–380
Biological drugs
  patent expiration, 7
Biological pathways, 11–12
Biological products
  approved composition, 337–348
Biopharmaceutical drug
  characterization, 329–387
  packaging and materials, 348–384
  physiochemical characterization tests, 385–386
  preformulation studies, 333–347
  design, 387
Biopharmaceutical drug classification systems
  release, dissolution, permeation, 156–158
Biopharmaceuticals Classification System (BCS)
  FDA, 141, 157

Board of Patent Appeals and Interferences (BPAI), 54
Botanical drugs, 1, 391
  chemistry, manufacturing, control evidence
  clinical trials, 401
  preformulation studies, 60–61
  starting material, 403–404
Botanical guidelines
  EMEA, 60
BPAI. *See* Board of Patent Appeals and Interferences
Bracketing, 317
Bracket method, 304, 305
Bradford method, 355
Bragg's diffraction, 224
Bragg's equation, 224
Braunauer, Emmet, and Teller (BET) method, 252
Bravais, Friedel, Donnay, and Harker (BFDH)
  model, 203
  morphology, 75
Bravais lattice system, 200
British Herbal Pharmacopoeia, 21
Bronsted-Lowry theory, 88–91
Brownian Motion, 263
Brucker SHellXTL software, 200

Caco-2 cells model, 65, 79–80
Caco-2 drug transport assays, 150–161
CADD. *See* Computer-assisted drug design
Calcineurin, 18
Calcitriol patent expiration
  Abbott Laboratories, 7, 34
Calorimetry
  differential scanning, 58, 77, 211, 254
  modulated differential scanning, 219–220, 271
  solution, 221
Cambridge Crystallographic Data Centre (CCDC), 258
Cambridge Structural Database (CSD), 258
CAMD. *See* Computer-assisted molecular design
Canadian Patent Office, 41
Candidate drugs, 296
  screening, 2
  selection, 4

Capillary zone electrophoresis, 119
Capromab pendetide
  composition, 338
Carbamylation, 365
Carbohydrate structure, 385
Carbon
  oxidation numbers, 302
Cations, 372
CCDC. *See* Cambridge Crystallographic Data Centre
CDER. *See* Center for Drug Evaluation and Research
Cellular imaging
  high-resolution probes, 14
Center for Biologics Evaluation and Research to Center for Drug Evaluation and Research
  approved products transfer, 331–332
Center for Drug Evaluation and Research (CDER), 329
CERIUS program, 203
Characterization methods, 355
Chelating agents, 267–268
Chemical crystallography, 199
Chemical diversity
  Abbott Laboratories, 3
Chemical drug substances
  characterization, 287–325
  good manufacturing practice, 322–325
  impurities, 317–321
  scheme of, 288–316
Chemistry
  patents, 39
Chemistry-manufacturing-control
  botanical drugs
    clinical trials, 401
  NCCAM clinical trials, 401
CheqSol®, 65, 67, 144
Chirality, 290
  decision tree, 290
Chloroquinine
  mode of action, 21
Choriogonadotropin alpha
  composition, 338
Chromatography, 357–358. *See also* High-performance liquid chromatography
  IEX, 358
  thin layer, 1
Claims
  definiteness, 49
  dependent, 49–50

[Claims]
  dominant-subservient, 50
  improvement, 51
  Jepson-type, 51
  Markush alternatives, 51
  Markush group, 51
  means-plus-function clauses, 50
  mixed-class, 52
  narrowing, 49
  negative limitations, 51
  patents, 46–48
  process, 50
  product-by-process, 52
  punctuating, 49
  ranges, 51
  relative and exemplary terminology, 51
  step-plus-function clauses, 50–51
  understanding, 49–50
Classification system
  patent, 38
Cleaning validation, 362
Clinical studies, 5
Clinical trials, 5
Clusters of Orthologous Groups (COG), 17
Coagulation Factor VIIa (rFVIIa)
  composition, 339
Code of Federal Regulations
  Title 21, 330
COG. *See* Clusters of Orthologous Groups
Collateral efficacy, 16
Colorimetry
  tristimulus, 259–260
Combinatorial chemistry, 3
  Aventis, 3
Combinatorial pharmacogenetics, 18
Combined simple forms morphology (CSM), 75
Compaction
  powders, 258–259
Compatibility
  dosage form in preformulation, 275
Compendia specifications
  pH measurement, 94–96
Compound testing, 4
Computational biology, 14
Computer-assisted drug design (CADD), 15
Computer-assisted molecular design (CAMD), 15

Confirmatory studies, 269–270, 315
ConQuest®, 258
Constructive reduction to practice
  defined, 36
Container closure system, 310
Contaminants, 382
  process-related, 386–387
Creutzfeldt-Jakob disease, 348
Crixivan, 206
Cryogranulation, 349
Cryopreservation, 349
Crystal habit, 199, 201, 202
Crystal lattice, 197–198
Crystalline index of refraction, 210
Crystalline particles, 199
Crystallinity
  solid stability, 307
Crystallography
  chemical, 199
Crystal morphology
  solid-state properties, 197–203
Crystal screening
  high-throughput
    solid-state properties, 207–209
Crystal systems, 199
CSD. See Cambridge Structural Database
CSM. See Combined simple forms
    morphology
CV. See Cyclic voltammetry
Cyanobacteria, 22
Cyanovirin-N, 22
Cyclic voltammetry (CV), 267
Cyclophilin, 18
Cyclosporin A
  mode of action, 21
Cysteinyl residue loss
  preventive action, 368
Cysteinyl residues, 367
Cytochrome P4503A4 (CYP3A4), 153

Darbepoetin alpha
  composition, 339
Databases
  electronic
    patents, 38
  technical, 43
DDSC. See Dynamic differential
    scanning calorimetry
Deamidation, 334, 335, 363
  preventive actions, 364
Degradation
  kinetics, 302

Degradation products, 387
  decision tree, 292
Degradation testing
  forced, 269, 315, 316, 375
Delphion technical databases, 43
Denaturation, 369
  preventive action, 369
Denileukin diftitox
  composition, 339
DePEGylation, 387–388
Derwent technical databases, 43
Design qualification (DQ), 326, 361
Detergents, 145
Development phases
  drug discovery trends, 2–5
Dialog technical databases, 43
Differential diagnosis, 218–219
Differential scanning calorimetry
    (DSC), 58, 77, 211, 254
Differential thermal analysis
    (DTA), 58
Different light scattering (DLS), 250
Diffuse reflectance Fourier transform
    (DRIFT), 222
Diffusion layer model of
  dissolution, 142
Dilatometry, 206
Dimyristoylphosphatidylcholine
    (DMPC), 107
Dissociation
  phenylacetic acids, 99
Dissociation, partitioning, solubility,
    87–121, 96–97
  ionization principle, 87–97
  measurement strategies, 113–121
  partitioning, 105–112
  quantitative structure-activity
    relationships, 98–104
Dissolution
  salt, 217–218
  thermodynamic driving force,
    142–143
Dissolution rates, 68
Dissolution testing, 229–230
Distribution coefficient, 106–107
DLS. See Different light scattering
DMPC. See Dimyristoylphosphatidylcholine
DNA sequence databases, 18
DNA sequencing, 8
Dominant-subservient claims, 50
Dornase alpha
  composition, 339

Dosage forms
  preformulation, 241–274
    compatibility, 275
    emulsion formulations, 263–270
    freeze-dried formulations, 271
    pulmonary delivery, 274
    solid dosage form, 241–261
    solution formulations, 262
    suspensions, 272
    topical, 273
  selection criteria, 242
  storage, 349–350
  study factors, 289
Double-sink PAMPA permeability assay, 150
DQ. *See* Design qualification
DRIFT. *See* Diffuse reflectance Fourier transform
Drotrecogin alpha
  composition, 339
Drug approval guidelines, 1
Drug delivery systems, 7
Drug discovery trends, 1–23
  development phases, 2–5
  new pathways, 11–23
  product life cycle, 6–10
Drug efflux, 77–78
Drug molecules
  lipophilicity, 78
Drug Price Competition and Patent Restoration Act of 1984, 37
Drug product dissolution, 242
  acceptance criteria, 243, 244, 245
Drugs
  FDA approved, 8, 9–10
  functional groups, 298
Drug selection
  genomics, 3
Drug substance
  characterization guidelines
    ICH, 58
  hydrate forms, 211
DSC. *See* Differential scanning calorimetry
DTA. *See* Differential thermal analysis
DVS. *See* Dynamic vapor sorption
Dye-binding method, 355
Dynamic differential scanning calorimetry (DDSC), 271
Dynamic scattering, 250
Dynamic vapor sorption (DVS), 58, 227–228, 246
  lactose, 247
  microcrystalline cellulose, 246

EDS. *See* Energy dispersive X-ray spectrometry
EDTA. *See* Ethylenediaminetetraacetic acid
Efalizumab
  composition, 339
Efficacy
  collateral, 16
  regulatory filing procedure, 400–403
Efflux transporters, 159
Elan, 7
Electrode potential
  pH, 95
Electronic databases
  patents, 38
Electron spectroscopy, 72
Electrophoresis, 356–357, 357
  2D, 357
  patterns, 386
Electrostaticity
  powders, 260–261
Elemental analysis, 58
EMEA. *See* European Medicines Evaluation Agency
Emulsion formulations
  dosage form in preformulation, 263–270
Emulsion particle size, 264
Enantiotropes, 203
  metastable phase, 205
  temperature, 204
Energy
  activation, 304
Energy dispersive X-ray spectrometry (EDS), 72
Equipment
  validation, 361
Erythropoietin
  composition, 340
  patent expiration, 7
ESCOP. *See* European Scientific Cooperative on Phytotherapy
Etanercept
  composition, 340
Ethane
  ethene reduction to, 301
Ethene
  reduction to ethane, 301
Ethylenediaminetetraacetic acid (EDTA), 278
Eukaryotic cells
  patents, 39

European Medicines Evaluation
    Agency (EMEA)
  botanical guidelines, 60
  drug approval guidelines, 1
  herbal guidelines, 60
European Patent Convention, 37
European Phytojournal, 21
European Scientific Cooperative on
    Phytotherapy (ESCOP), 19
  phytomedicine monographs, 20–21
Excel for Chemists, 117
Excipients, 293, 336–337
  biological products, 348
  compatibility
    preformulation studies, 77
  control, 406
Expression system characterization, 362
Extinction coefficient, 386

FAST HT technology, 207
FCS. *See* Fluorescence correlation
    spectroscopy
Feldkamp cone-beam algorithm, 259
FID. *See* Flame ionization detector
Filgrastim
  composition, 340
Filing Patents Online, 38
Filter plate method
  solubility testing, 121
Filter probe method, 117–118
Flame ionization detector (FID), 249
Flow
  powder properties, 71
  powders, 258–259
FlowSorb III®, 257
Fluids
  pH, 93
Fluorescence correlation spectroscopy
    (FCS), 73
Follicle-stimulating hormone (FSH)
  composition, 340
Forced degradation testing, 269, 315, 316, 375
Fourier transform
  diffuse reflectance, 222
Fourier transform infrared (FT-IR)
    spectroscopy, 73, 222
Fraunhofer theory of light scattering, 251
Freeman Technology, 72
Freeze-dried formulations
  dosage form in preformulation, 271
FSH. *See* Follicle-stimulating hormone
FT4 Powder Rheometer, 72

GAB. *See* Guaranteed adjustment basis
Galactosidase alpha
  composition, 341
Gas adsorption system
  static volumetric, 255
Gas chromatography (GC), 65
Gas physisorption, 252
Gas pycnometry, 257, 258
GC. *See* Gas chromatography
Genetically modified microorganisms
  patents, 39
Genome sequencing, 8
Genomics, 16–17
  drug selection, 3
Ginkgo biloba, 404
Ginseng
  American, 391
Glaxo-Wellcome
  patent expiration, 7
Glucagon
  composition, 341
Good manufacturing practice
    (GMP), 322–323
  chemical drug substances
    characterization, 322–325
G-protein-coupled receptor (GPCR)
  signaling, 15
Guaranteed adjustment basis (GAB), 54

Hammett constants, 102
Hansch analysis, 103–105
Hatch/Waxman Act, 37, 52
HEDTA. *See* Hydroxyethylenediamine-
    triacetic acid
Hemophilus b Conjugate
  composition, 341
Henderson-Hasselbach equation, 92
Hepatitis B (recombinant) vaccine
  composition, 341
Herbal medicines, 391
  efficacy and safety, 399–400
  EMEA guidelines, 60
  international approval rules,
    393–396
  specifications, 397–398
  stability testing, 406–407
    criteria, 407
  standardization, 398–399
Herbal preparations
  stability testing, 408–409
Herbal substances
  control, 405

HFSH
  composition, 340
High-performance liquid chromatography (HPLC), 58, 107, 117, 118–119, 230–231, 357–358
  IEX, 358
High-resolution probes
  cellular imaging, 14
High-throughput crystal screening
  solid-state properties, 207–209
High-throughput screening (HTS), 1, 13, 22, 120, 151
  solubility, 227
Hit-to-lead (HTL), 1
HIV protease inhibitors
  polymorphs, 206
Hot stage microscopy (HSM), 58, 206, 220
HPLC. See High-performance liquid chromatography
HTL. See Hit-to-lead
HTS. See High-throughput screening
Human Genome Project, 17
Hydrates
  drug substance, 211
  solid-state properties, 210–211
Hydrolysis, 296, 335, 368
  preventive action, 370
  procaine, 299
Hydroxyethylenediaminetriacetic acid (HEDTA), 278

ICH. See International Conference on Harmonization (ICH)
ICH Harmonized Tripartite Guideline on Stability of New Drug Substance, 315
ICH Q3C Impurities, 318
ICP. See Inductively coupled plasma-atomic emission spectroscopy
Identification and qualification
  decision tree, 320
Identification testing, 290
IEX. See Ion-exchange
IEX-HPLC. See Ion-exchange high-performance liquid chromatography
IGC. See Inverse gas chromatography
IL-11. See Interleukin eleven
Imiglucerase
  composition, 342
Immunochemical properties, 380–381

Impurities, 291, 381–382
  chemical drug substances characterization, 317–321
  decision tree, 293, 321
  process-related, 386–387
  profile, 46
  qualification, 320
  specified, 319
  thresholds, 322
IND. See Investigational new drug
Indinavir sulfate (Crixivan), 206
Inductively coupled plasma-atomic emission spectroscopy (ICP), 72
Infliximab
  composition, 342
Information portals, 42
Infrared spectroscopy (IRS), 206, 222
Insoluble drugs, 67
Installation qualification (IQ), 326, 361
Instruments
  microprocessor-driven, 1
  particle size, 255
  porosity, 255
  surface area, 255
Insulin aspart
  composition, 342
Insulin glargine
  composition, 341
Insulin glulisine
  composition, 342
Insulin lispro
  composition, 342
Intellectual property, 33–55
  claims, 49–54
  FDA, 55
  patent application, 44–48
  patenting strategies, 33
  patenting systems, 35
  patent myths, 36–37
  patents defined, 35
  patent search, 38–43
  services, 44
Interferogram, 222
Interferon alpha-2a
  composition, 342
Interferon alphacon-1
  composition, 343
Interferon beta-1a
  composition, 343
Interferon gamma-1b
  composition, 343

Interferon patent, 35
  Biogen, 35
Interleukin eleven (IL-11)
  composition, 343
International approval rules
  herbal medicines, 393–396
International Conference on Harmonization (ICH)
  drug substance characterization guidelines, 58
  recombinant DNA products, 63
International Union for the Protection of New Varieties of Plants (UPOV), 38
Internet search engines, 42
Invention
  defined, 36
Inverse gas chromatography (IGC), 246, 247–248, 249
  principles, 248
Investigational new drug (IND)
  phase III filing, 375
Investor's Digest, 44
In vitro–in vivo correlation (IVIVC), 78–79, 153–155
Ion-exchange (IEX), 357
  chromatography, 358
Ion-exchange high-performance liquid chromatography (IEX-HPLC), 358
Ionization principle
  dissociation, partitioning, solubility, 87–97
Ion pair Log P, 113
IQ. See Installation qualification
IRS. See Infrared spectroscopy
Isoform pattern, 386
Isothermal microcalorimetry, 221
IVIVC. See In vitro–in vivo correlation

Jepson-type claims, 51

Kinetics
  degradation, 302

Labeling, 314
Large molecular weight lipophilic drugs, 69
Large molecule drugs, 1
  preformulation studies, 62–63
Laronidase
  composition, 344
Lattice structure
  scalars, 198

Laws
  patents
    terminology, 36
LC. See Liquid chromatography
LC-MS/MS. See Liquid chromatography-tandem mass spectroscopy
Lead compound outsourcing, 3
Lead optimization (LO), 1
Lepirudin
  composition, 344
Letter patent, 35
Lewis theory, 91–92
Libraries
  Roche, 3
Light scattering
  Fraunhofer theory of, 251
  Mie theory of, 251
  quasi-elastic, 250, 251
Linear free-energy, 99
Lipophilic drugs
  large molecular weight, 69
Lipophilicity, 65
  drug molecules, 78
Liposomal solubilization, 68
Liquid chromatography (LC), 65, 386
Liquid chromatography-tandem mass spectroscopy (LC-MS/MS), 117, 359
LO. See Lead optimization
Log D, 106–107, 118
  membrane, 107
Log P, 114, 115–116
Lyophilization, 271, 350–351

MAbs. See Monoclonal antibodies
Macroscopic operations, 198
Madey versus Duke, 53
Malvern Mastersizer series, 250
Manual titration, 116–117
Markers, 19
Market potential, 2
Marketing strategies, 6
Markush alternatives
  claims, 51
Markush group claims, 51
Mass spectroscopy (MS), 1, 58, 117, 359
Massachusetts Institute of Technology (MIT)
  Website, 44
MDSC. See Modulated differential scanning calorimetry

Means-plus-function clauses
  claims, 50
Measurement strategies
  dissociation, partitioning,
    solubility, 113–121
Melting point
  salt, 215–216
  thermal analysis, 220
Mercury, 258
Metabolomics technology development, 12
Metathesis reactions, 301
Microbial diversity, 8
Microbiology, 292
  decision tree, 296
  patents, 39
Microcalorimetry
  isothermal, 221
Microcrystalline cellulose
  DVS, 246
Microfluidization, 263
Micrometrics, 257
Micronization, 146, 244
Microorganisms
  genetically modified
    patents, 39
Micropatent, 44
Microprocessor-driven instruments, 1
Microscope
  atomic force, 265
Microscopic operations, 198
Microscopy
  preformulation studies, 72–73
  scanning electron, 58
Microthermal analysis, 73
Mie theory of light scattering, 251
Miller indices, 224
Minerals
  control, 405
MIT. See Massachusetts Institute of
    Technology (MIT)
Mixed-class claims, 52
MLR. See Multiple linear
    regression
MLSCN. See Molecular Libraries Screening
    Centers Network
Modulated differential scanning
    calorimetry (MDSC),
    219–220, 271
Moisture
  solid stability, 307
Moisture isotherm
  preformulation studies, 75–76

Molar absorptivity, 386
Molecular biology
  patents, 39
Molecular diversity, 3
Molecular imaging, 12–13
  probes, 13–14
Molecular libraries, 12–13
Molecular Libraries Screening Centers
    Network (MLSCN), 13
Molecular Libraries Screening
    Instrumentation (MLSI), 13
Molecular size
  solubility, 112
Molecular spectroscopy, 73–74
Monoclonal antibodies (MAbs), 62, 334
  patents, 40
Monotropes, 203
  temperature, 204
m-PEGylation
  N-hydroxysuccinimide ester, 355
MS. See Mass spectroscopy
Multidrug efflux, 16
Multidrug resistance studies, 77–78, 145
Multiple linear regression (MLR), 74
MultiScreen Caco-2 assay system,
    151–153
Muromonab-CD3
  composition, 344

NAE. See New active entities
NanoCrystal® Technology, 68, 69
Nanomedicine, 14
Nanonization, 68, 146
Nanoparticles, 146
National Institutes of Health (NIH)
  new drug development, 11
  Roadmap, 12, 14
  technical databases, 43
National Library of Medicine (NLM)
  bibliographies, 43
  technical databases, 43
National Technology Centers for
    Networks and Pathways, 12
Natural products, 19
NCCAM clinical trials
  chemistry-manufacturing-control, 401
Near-infrared (NIR) spectroscopy, 74
Nektar, 7
Nektar Advanced PEGylation, 353
Nerac technical databases, 43
Nesiritide
  composition, 344

New active entities (NAE), 1
New drug development
  description, 290
  most active categories, 11
NIH, 11
New lead compounds
  specifications, 288
New pathways
  drug discovery trends, 11–23
NIH. See National Institutes of
    Health
NIR. See Near-infrared
    spectroscopy
NLM. See National Library of Medicine
NMR. See Nuclear magnetic resonance
Nonlinear Mixed Model, 306
Norvir®, 206
Noyes-Whitney equation, 146, 252
Nuclear magnetic resonance
    (NMR), 58

Operational qualification (OQ), 326, 361
Organic solvents, 145
Osmolality, 269–270
Oxidation, 266–267, 300, 365
  numbers, 301
    carbon, 302
  preventive actions, 365

Packaging material, 348
  biopharmaceutical drug
    characterization, 348–384
Palivizumab
  composition, 344
Parallel artificial membrane
    permeability analysis (PAMPA),
    149–150
Parathyroid hormone
  composition, 344
Partial charged surface area (PSA), 109
Particle size, 291–293
  decision tree, 294
  detection, 250
  distribution, 250–251
  instrumentation, 255
  studies, 242–244
Partitioning
  dissociation, partitioning, solubility,
    105–112
Partitioning solvent, 108–109
PAT. See Process analytical technology

Patent Cafe®, 44
Patents
  antigen combinations, 41
  application
    defined, 36
    intellectual property, 44–48
  chemistry, 39
  claims, 46–48
  classification system, 38
  copies, 44
  defined, 35, 36
  electronic databases, 38
  eukaryotic cells, 39
  expiration
    amoxicillin, 34
    biological drugs, 7
    calcitriol, 7, 34
    erythropoietin, 7
    Glaxo-Wellcome, 7
    sildenafil citrate, 35
    Zantac®, 7
  expiry periods, 6
  FDA, 55–56
  formats, 45
  genetically modified
    microorganisms, 39
  infringement case, 53
  laws
    terminology, 36
  letter, 35
  microbiology, 39
  molecular biology, 39
  monoclonal antibodies, 40
  myths
    intellectual property, 36–37
    prescription medicines, average
      life, 52
  recombinant virus, 41
  search, 44
    intellectual property, 38–43
  strategies
    intellectual property, 33
  structurally modified antibodies, 40
  structurally modified
    immunoglobulin, 40
  systems
    intellectual property, 35
  technical databases, 43
  term adjustment, 52, 54
  titles, 45
  United States laws, 6
  utility statement, 46

Index

PCS. *See* Photon correlation spectroscopy
PD. *See* Polydispersity
PDx-IVIVC, 155
Pegfilgrastim
  composition, 345
Peginterferon alpha-2a
  composition, 345
Peginterferon alpha-2b
  composition, 345
Pegvisomant
  composition, 345
PEGylation, 353
  stabilization, 352
Peptide map, 385
Percent ionization, 97
Performance qualification (PQ), 326, 361
Perkin-Elmer DSC 7, 272
Permeability assays, 148–149
Permissive antagonism, 16
PGDP. *See* Propylene glycol dipelargonate
P-glycoprotein (Pgp), 65, 153
pH
  common fluids, 93
  electrode potential, 95
  measurement
    compendia specifications, 94–96
  meters, 96
  scale, 92–93
Pharmaceutical research-based
    manufacturers (PhRMa), 7
Pharmacoproteomics, 18–19
Phase III filing
  IND, 375
Phase I/II studies
  plant substance, 402
Phase II studies, 5
Phase III studies, 5, 6
  plant substance, 402
Phase IV studies, 6
Phase solubility analysis, 227
Phenotypic inventorying, 8
Phenylacetic acids
  dissociation, 99
Photon correlation spectroscopy
  (PCS), 250, 263
  disadvantages, 264–265
Photostability, 269–270, 308–309
Photostability testing, 315
PhRMa. *See* Pharmaceutical
    research-based manufacturers
Physicochemical characterization
  tests, 385

Physicochemical properties, 291, 379
Phytomedicines, 1, 19
  characterization, 391–418
    efficacy and safety, 399
    regulatory filing procedure,
      400–402
    regulatory status, 392
    stability testing, 406–409
    starting material, 403–405
  monographs
    ESCOP, 20–21
  preformulation studies, 60–61
PION pSOL, 65, 66
pKa, 114, 117
  acids, 217
Plant substance, 400
  phase I/II studies, 402
  phase 3II studies, 402
Plate method
  solubility testing, 120
Pluronics, 145
Polydispersity (PD), 353
Polyethylene glycol
  structure, 352
Polymorphic forms, 292
Polymorphism
  decision tree, 295
  powders, 261–262
Polymorphs
  acceptance criteria, 208
  HIV protease inhibitors, 206
  manufacturing factors, 206
  stability, 204
  thermal analysis, 218
  thermodynamic rules, 205
Polymorph screening
  preformulation studies, 65–66
Porosity, 254–255
  instrumentation, 255
Postmarket surveillance, 6
Post-translational modifications (PTM)
  proteins, 16
Powders
  caking, 261
  color, 259–260
  electrostaticity, 260–261
  flow and compaction, 258–259
  polymorphism, 261–262
  properties, 69–70, 71
  solubility, 262–263
PQ. *See* Performance qualification
Praying mantis model, 223

PRCA. *See* Pure red cell aplasia
Precipitation, 371
  preventive actions against, 373
Preclinical studies, 4
Preclinical trials, 4, 5
Predicting aqueous solubility
  silico tools, 68
Preformulation studies, 4, 57–78
  biopharmaceutical drug
    characterization, 333–347
  botanical drugs, 60–61
  design, 387–388
    biopharmaceutical drug
      characterization, 387
  excipient compatibility, 77
  large molecule drugs, 62–63
  microscopy, 72–73
  moisture isotherm, 75–76
  phytomedicines, 60–61
  pKa, partitioning, solubility, 66–67
  polymorph screening, 65–66
  preservatives, 63
  recombinant DNA products, 63–64
  regulatory requirements, 58–64
  salt screening, 69–70
  solid-state characterization, 69–76
  stability testing, 75–76
  testing criteria, 57
  testing systems, 65–68
  thermal analysis, 73–74
  transport across biological
    membranes, 77–78
PreQuest®, 258
Prescription medicines
  average patent life, 52
Preservatives
  preformulation studies, 63
Preventive action
  aggregation, 371
  cysteinyl residue loss, 368
  deamidation, 364
  denaturation, 369
  hydrolysis, 370
  oxidation, 365
  precipitation, 373
  proteolysis, 363
  racemization, 367
Prior art
  defined, 36
Probes
  high-resolution
    cellular imaging, 14

Procaine
  hydrolysis, 299
Process analytical technology (PAT), 289
  FDA, 59
Process controls, 383
Process-related impurities and
  contaminants, 386–387
Process validation, 360–361
Product-by-process claims, 52
Product life cycle, 6
  drug discovery trends, 6–10
Product-related impurities, 387
Propylene glycol dipelargonate
  (PGDP), 108
PROSITE®, 17
Protein Data Bank, 258
Proteins
  PTM, 16
  stability, 335
Proteolysis, 362
  preventive actions, 363
Proteome analysis, 18
PSA. *See* Partial charged surface area
Pseudopolymorphs, 203
PTM. *See* Post-translational modifications
PubChem®, 13
Pulmonary delivery
  dosage form in preformulation, 274
Pure red cell aplasia (PRCA), 348
Purity, 381
  measurement, 113

Q3A Impurities in New Drug
  Substances, 310
Q6A Specifications, 59, 310
Q6B Specifications, 310
QELS. *See* Quasi-elastic light scattering
Q-Rule, 305
QSAR. *See* Quantitative structure–activity
  relationships
QSurf quick surface area analyzer, 256
Quality management, 324–326
Quantified extracts, 404
Quantitative structure–activity
  relationships (QSAR), 98–99
  database, 105
  dissociation, partitioning, solubility,
    98–104
Quasi-elastic light scattering (QELS),
  250, 251
Quinine
  mode of action, 21

Racemization, 367
  preventive actions, 367
Raman spectroscopy, 309
Rapamycin
  mode of action, 21
RAPID-Pharma®, 290
Rasburicase
  composition, 346
Rational drug design, 17
Rayleigh scattering, 250
Recombinant bacterium
  patents, 41
Recombinant DNA products
  FDA, 63
  ICH, 63
  preformulation, 63–64
Recombinant drugs, 22–23
Recombinant virus
  patents, 41
Reduction to practice
  defined, 36
Reference standards and materials, 382
Registration
  application, 318
  health authorities, 5
Regulatory filing procedure
  efficacy and safety, 400–403
  phytomedicines characterization, 400–402
Regulatory requirements
  preformulation studies, 58–64
Regulatory status
  phytomedicines characterization, 392
Relative humidity (RH), 76, 213, 249
Release, dissolution, permeation, 141–158
  assay systems, 147–155
  biopharmaceutics drug classification systems, 156–158
Release limits vs. shelf-life limits, 383–384
Required reduction bases (RRB), 55
Research
  outsourcing, 2
  planning, 3
  trends, 15
Restsrahlen bands, 223
Reteplase
  composition, 346
Reverse-phase (RP) chromatography, 358
Reverse-phase high-performance liquid chromatography (RP-HPLC), 358
Reversing heat flow component, 220
RH. *See* Relative humidity

Rietveld refinement procedure, 207
Ritonavir (Norvir®), 206
Rituximab
  composition, 346
Robotic systems, 1
Roche
  interferon patent, 35
  libraries, 3
Rofecoxib, 6
RP. *See* Reverse-phase chromatography
RP-HPLC. *See* Reverse-phase high-performance liquid chromatography
RRB. *See* Required reduction bases
Rubotherm system, 228
Rule 5 of Lipinsky, 156

Safety
  herbal medicines, 399–400
  phytomedicines characterization, 399
  regulatory filing procedure, 400–403
Salt
  decision flow diagram, 216
  dissolution, 217–218
  form, 214–215
  melting point, 215–216
  screening
    preformulation studies, 69–70
  selection tree, 70
  solubility, 214
  thermal analysis, 218
Sargramostim
  composition, 346
Sauter mean, 251
Scalable processor architecture (SPARC), 117
Scalars
  lattice structure, 198
Scanning electron microscopy (SEM), 58, 72
Scattering
  dynamic, 250
  light, 250–251
  static, 250
Screening, 4
  high-throughput, 1, 13, 22, 120, 151
    solubility, 227
  high-throughput crystal
    solid-state properties, 207–209
  polymorph
    preformulation studies, 65–66
  salt
    preformulation studies, 69–70

SDS-PAGE, 356
Search engines
  Internet, 42
SEC-HPLC, 359
SEM. *See* Scanning electron microscopy
Sennae folium, 404
Sermorelin acetate
  composition, 346
SGA-100 Symmetrical Gravimetric Analyzer, 228
Shake-flask method, 118
Sildenafil citrate
  patent expiration, 35
Silico tools
  predicting aqueous solubility, 68
Sirius GLpK, 65, 66, 67, 115–116
Skyscan-1172, 258, 259
Small molecule drugs, 1
Small Molecule Repository, 13
Smith Kline Beecham
  amoxicillin patent expiration, 34
Software validation, 361–362
Solid-phase scavenger reagents, 3
Solid stability
  crystallinity, 307
  moisture, 307
Solid-state characterization
  preformulation studies, 69–76
Solid-state properties, 197–230
  amorphous forms, 212
  crystal morphology, 197–203
  high-throughput crystal screening, 207–209
  hydrates, 210–211
  hygroscopicity, 212
  polymorphism, 203–206
  solubility, 213–217
  solvates, 210
  study methods, 218–230
Solid-state stability, 306–307
Solubility, 110–111
  additives, 112
  HTS, 227
  measurement, 113
  method, 117
  modulation, 144–145
  molecular size, 112
  powders, 262–263
  salt, 214
  solid-state properties, 213–217
  temperature, 112

[Solubility]
  testing
    filter plate method, 121
    plate method, 120
Solubility classification
  United States Pharmacopeia, 111
Solubilizers, 68, 145
Solution calorimetry, 221
Solution formulations
  dosage form in preformulation, 262
Solvates
  solid-state properties, 210
Solvents, 318
  organic, 145
  partitioning, 108–109
Somatropin
  composition, 347
Sorptomatic 1990, 255
SPARC. *See* Scalable processor architecture
SPC. *See* Supplementary protection certificate
Specifications, 310, 378–379
  herbal medicines, 397–398
  justification, 384
Spectrophotometer, 260
Spectroscopy, 117, 355, 386
  electron, 72
  fluorescence correlation, 73
  FT-IR, 73, 222
  molecular, 73–74
  NIR, 74
  photon correlation, 250, 263
    disadvantages, 264–265
  Raman, 309
  UV-VIS, 355
Spray drying, 349–350
Stability, 266, 334–335, 362–363
  accelerated testing, 303
  commitment, 312–313
  evaluation, 293
  measurement, 113
  polymorphs, 204
  proteins, 335
  solid-state, 306–307
Stability testing
  herbal medicines, 406–407
    criteria, 407
  herbal preparations, 408–409
  phytomedicines characterization, 406–409
  preformulation studies, 75–76
  regulatory considerations, 309

Index

Stabilization
  PEGylation, 352
Standardization
  herbal medicines, 398–399
Standardized extracts, 404
Static scattering, 250
Static volumetric gas adsorption
    system, 255
Step-plus-function clauses
  claims, 50–51
Storage, 311–312
  dosage forms, 349–350
  freezer, 312
  minimum time period, 314
  refrigerator, 312
  statement, 314
Stress testing, 309–310, 335, 376
Structural biology, 14
Structural characterization, 385
Structurally modified antibodies
  patents, 40
Structurally modified immunoglobulin
  patents, 40
Sulfhydryl groups and disulfide
    bridges, 385
Supplementary protection certificate
  (SPC), 55–56
Surface activity, 269
Surface area, 252
  instrumentation, 255
Suspensions
  dosage form in preformulation, 272

TAM. *See* Thermal activity monitor III
TCD. *See* Thermal conductivity detector
Technical databases
  patents, 43
Temperature
  effects, 77
  enantiotropes, 204
  solubility, 112
Tenecteplase
  composition, 347
Terahertz pulsed imaging (TPI), 309
Terminal amino acid sequence, 385
Testing. *See also* Stability testing
  accelerated
    stability, 303
  animal model, 153–154
  compound, 4
  criteria
    preformulation studies, 57

[Testing]
  forced degradation, 269, 315,
    316, 375
  frequency, 310–311
  identification, 290
  photostability, 315
  physicochemical characterization, 385
  stress, 309–310, 335, 376
  systems
    preformulation studies, 65–68
TGA. *See* Thermogravimetric analysis
Thermal activity monitor (TAM) III, 308
Thermal analysis, 206
  differential, 58
  melting point, 220
  polymorphs, 218
  preformulation studies, 73–74
  salt, 218
Thermal conductivity detector (TCD),
  249, 256
Thermodynamic driving force
  dissolution, 142–143
Thermodynamic rules
  polymorphs, 205
Thermogravimetric analysis (TGA), 211,
  220, 254
Thin layer chromatography (TLC), 1
Tositumomab
  composition, 347
TPI. *See* Terahertz pulsed imaging
Trace metals, 267–268
Trade related aspects of intellectual
    property section (TRIPS), 38
  agreement, 56
Transporter-enzyme
  intestines, 159
Transporters, 156
Trastuzumab
  composition, 347
TRIPS. *See* Trade related aspects of
    intellectual property section
TriStar 3000®, 257
Tristimulus colorimetry, 259–260
True density, 257–258
Truncated forms, 387
2D electrophoresis, 357

Ultraviolet spectrophotometric
    measurements, 117
Ultraviolet-visible (UV-VIS)
    spectroscopy, 355
Undercooling, 350

United States Food and Drug
  Administration (U.S. FDA)
 amorphous forms, 212
 BCS, 141
 Biopharmaceuticals Classification
  System (BCS), 157
 drug approval guidelines, 1
 drugs approved, 8, 9–10
 intellectual property, 55
 patents, 55–56
 recombinant DNA products, 63
United States Patent and Trademark
  Office, 38
United States patent laws, 6
United States Pharmacopeia solubility
  classification, 111
UPOV. *See* International Union for the
  Protection of New Varieties of
  Plants (UPOV)
Uruguay Round Agreements Act (URAA), 54
U.S. FDA. *See* United States Food and
  Drug Administration
U.S. Patent Office Classification 435, 38–39
Utility
 defined, 36
Utility statement
 patents, 46
UV-VIS. *See* Ultraviolet-visible
  spectroscopy

Valerianae radix, 404
Validation
 analytical methods, 382–383

[Validation]
 cleaning, 362
 equipment, 361
 software, 361–362
Vapor sorption, 76–77
Vioxx® (rofecoxib), 6
Vista®, 258
Vitamins
 control, 405
VitriLife®, 352
Voltammetry, 267

Wavelength dispersive X-ray
  spectrometry (WDS), 72
World Intellectual Property
  Organization, 41
World Trade Organization
  regulations, 38

X-ray diffraction (XRD), 73, 74–75
X-ray microtomography, 258
X-ray powder diffraction (XRPD),
  206, 224
 design, 225
 polymorphic forms, 244
 sources of error, 225–227
X-ray spectrometry
 energy dispersive, 72

Zantac®
 patent expiration, 7
Zeta potential, 264, 265